Biodiversity Loss and Conservation in Fragmented Forest Landscapes

The Forests of Montane Mexico and Temperate South America

Biodiversity Loss and Conservation in Fragmented Forest Landscapes

The Forests of Montane Mexico and Temperate South America

Edited by

A.C. Newton

Centre for Conservation Ecology and Environmental Change
School of Conservation Sciences
Bournemouth University
Poole
Dorset
UK

www.cabi.org

CABI is a trading name of CAB International

CABI Head Office
Nosworthy Way
Wallingford
Oxfordshire OX10 8DE
UK

Tel: +44(0)1491 832111
Fax: +44(0)1491 833508
E-mail: cabi@cabi.org
Website: www.cabi.org

CABI North American Office
875 Massachusetts Avenue
7th Floor
Cambridge, MA 02139
USA

Tel: +1 617 395 4056
Fax: +1 617 354 6875
E-mail: cabi-nao@cabi.org

A catalogue record for this book is available from the British Library, London, UK.

Library of Congress Cataloging-in-Publication Data

Biodiversity loss and conservation in fragmented forest landscapes : evidence from tropical montane and south temperate rain forests in Latin America / A.C. Newton, editor.
p. cm.
Includes bibliographical references and index.

ISBN 978-1-84593-261-9 (alk. paper) -- ISBN 978-1-84593-262-6 (ebook)
1. Forest biodiversity--Latin America. 2. Forest biodiversity conservation--Latin America. 3. Rain forests--Latin America. 4. Rain forest conservation--Latin America. I. Newton, Adrian C. II. Title.

QH106.5.B52 2007
577.3'098--dc22

2007006483

ISBN: 978 1 84593 261 9

Typeset by SPi, Pondicherry, India.
Printed and bound in the UK by Biddles Ltd, King's Lynn.

Contents

Contributors

Adison Altamirano, *Instituto de Silvicultura, Universidad Austral de Chile, Casilla 567, Valdivia, Chile*

Claudia Alvarez-Aquino, *Instituto de Genetica Forestal, Universidad Veracruzana, Parque El Haya s/n, Apartado Postal No. 551, CP 91000, Xalapa, Veracruz, Mexico. E-mail: clalvarez@uv.mx*

Juan Carlos Aravena, *Laboratorio de Sistemática y Ecología Vegetal, Facultad de Ciencias, Universidad de Chile, Casilla 653, Santiago; Fundación Senda Darwin, Ancud, Chiloé and Centro de Estudios Avanzados en Ecología y Biodiversidad, Pontificia Universidad Católica de Chile, Departamento de Ecología, Alameda 340, Santiago, Chile*

Juan J. Armesto, *CMEM, Universidad de Chile, Ecología Forestal, Facultad de Ciencias, Casilla 653, Santiago, Chile and Centro de Estudios Avanzados en Ecología y Biodiversidad, Pontificia Universidad Católica de Chile, Departamento de Ecología, Alameda 340, Santiago, Chile*

Sarah Bekessy, *RMIT University, GPO Box 2476V, Melbourne, Victoria, 3001, Australia*

Alejandra Blanco-Macías, *Instituto de Ecología, Universidad Nacional Autónoma de México, Ciudad Universitaria, Coyoacán, Mexico DF 04510, Mexico*

Mayra Caldiz, *Southern Swedish Forest Research Centre, PO Box 49, S 230 53 Alnarp, Sweden*

Angélica Camacho-Cruz, *Biodiversidad: Conservación y Restauración, A.C. (BIOCORES, A.C.) Tapachula 17, El Cerrillo, 29229 San Cristóbal de Las Casas, Chiapas, Mexico; Departamento Interuniversitario de Ecología, Facultad de Biología, Universidad Complutense de Madrid, Madrid 28040, Spain; Departamento Interuniversitario de Ecología, Universidad de Alcalá, 28871, Alcalá de Henares, Spain. E-mail: acamacho@sclc.ecosur.mx*

Luis Cayuela, *Departamento de Ecología, Universidad de Alcalá, Carretera de Barcelona km 33,600, E-28871 Alcalá de Henares, Madrid, Spain. E-mail: luis.cayuela@uah.es*

David A. Coomes, *Department of Plant Sciences, University of Cambridge, Cambridge CB2 3EA, UK*

Marco Cortés, *Laboratorio de Dendrocronología, Universidad Católica de Temuco, Casilla 15-D, Temuco, Chile*

Rafael F. del Castillo, *CIIDIR Oaxaca Instituto Politécnico Nacional, Hornos 1003, Santa Cruz Xoxocotlán, Oaxaca 68130, Mexico*

Cristian Echeverría, *Iniciativa Científica Milenio FORECOS, Instituto de Silvicultura, Universidad Austral de Chile, Casilla 567, Valdivia, Chile.* Present address: *Departamento Manejo de Bosques y Medio Ambient Facultad de Ciencias Forestales, Victoria 631, Barrio Universitario Casilla 160-C, Universidad de Concepción, Concepción, Chile*

Carlos Fragoso, *Departamento de Biología de Suelos, Instituto de Ecología A.C. Km. 2.5 Carretera Antigua a Coatepec, #351, Congregación El Haya 91070 Xalapa, Veracruz, Mexico. E-mail: carlos.fragoso@inecol.edu.mx*

Luis Galindo-Jaimes, *Biodiversidad: Conservación y Restauración, A.C. (BIOCORES, A.C.) Tapachula 17, El Cerrillo, 29229 San Cristóbal de Las Casas, Chiapas, Mexico; Departamento Interuniversitario de Ecología, Facultad de Biología, Universidad Complutense de Madrid, Madrid 28040, Spain; Departamento Interuniversitario de Ecología, Universidad de Alcalá, 28871, Alcalá de Henares, Spain. E-mail: lgalindo@sclc.ecosur.mx*

Duncan Golicher, *Departamento de Ecología y Sistemática Terrestres, Área de Conservación de la Biodiversidad, El Colegio de la Frontera Sur (ECOSUR), Apartado Postal 63, 29200 San Cristóbal de Las Casas, Chiapas, Mexico. E-mail: dgoliche@sclc.ecosur.mx*

Mario González-Espinosa, *Departamento de Ecología y Sistemática Terrestres, Área de Conservación de la Biodiversidad, El Colegio de la Frontera Sur (ECOSUR), Carretera Panamericana y Periférico Sur, 29290 San Cristóbal de Las Casas, Chiapas, Mexico. E-mail: mgonzale@sclc.ecosur.mx*

Marc Gracia, *CREAF-Centre for Ecological Research and Forestry Applications, Autonomous University of Barcelona, Bellaterra 08913, Spain. E-mail: mgracia@creaf.uab.es*

Álvaro G. Gutiérrez, *UFZ Centre for Environmental Research, Department of Ecological Modelling, Permoserstr. 15, 04318 Leipzig, Germany. E-mail: alvaro.gutierrez@ufz.de*

Andreas Huth, *UFZ Centre for Environmental Research, Department of Ecological Modelling, Permoserstr. 15, 04318 Leipzig, Germany*

Antonio Lara, *Iniciativa Científica Milenio FORECOS, Instituto de Silvicultura, Universidad Austral de Chile, Casilla 567, Valdivia, Chile*

Fabiola López-Barrera, *Departamento de Ecología Funcional, Instituto de Ecología, A.C., km 2.5 Carretera Antigua a Coatepec No. 351, Congregación el Haya Xalapa, Veracruz 91070, Mexico*

Robert H. Manson, *Departamento de Ecología Funcional, Instituto de Ecología, A.C., km 2.5 Carretera Antigua a Coatepec No. 351, Xalapa, Veracruz, 91070, Mexico*

Camila Martínez-Araneda, *Royal Botanic Garden Edinburgh, 20 A Inverleith Row, Edinburgh EH3 5LR, UK*

Paula Mathiasen, *Laboratorio Ecotono, Universidad Nacional del Comahue, Quintral 1250, 8400 Bariloche, Argentina*

Lera Miles, *UNEP World Conservation Monitoring Centre, 219 Huntingdon Road, Cambridge CB3 0DL, UK. E-mail: lera.miles@unep-wcmc.org*

Guillermo Montoya-Gómez, *Departamento de Gestión de los Recursos Naturales, Área de Sistemas de Producción Alternativos, El Colegio de la Frontera Sur (ECOSUR), Carretera Panamericana y Periférico Sur, 29290 San Cristóbal de Las Casas, Chiapas, Mexico. E-mail: gmontoya@sclc.ecosur.mx*

Miguel A. Muñiz-Castro, *Departamento de Ecología Funcional, Instituto de Ecología, A.C. (INECOL), Km 2.5 Carretera Antigua a Coatepec No. 351, Congregación El Haya, 63, Xalapa, 91070, Veracruz, Mexico*

Simoneta Negrete-Yankelevich, *Departamento de Biología de Suelos, Instituto de Ecología A.C., Km. 2.5 Carretera Antigua a Coatepec, #351, Congregación El Haya 91070, Xalapa, Veracruz, Mexico. E-mail: simoneta.negrete@inecol.edu.mx*

Adrian C. Newton, *Centre for Conservation Ecology and Environmental Change, School of Conservation Sciences, Bournemouth University, Talbot Campus, Poole, Dorset BH12 5BB, UK. E-mail: anewton@bournemouth.ac.uk*

Mariela C. Núñez-Ávila, *CMEM, Universidad de Chile, Ecología Forestal, Facultad de Ciencias, Casilla 653, Santiago, Chile; Centro de Estudios Avanzados en Ecología y Biodiversidad, Pontificia Universidad Católica de Chile, Departamento de Ecología, Alameda 340, Santiago, Chile; Universidad Austral de Chile, Instituto de Silvicultura, Campus Isla Teja, Casilla 567, Valdivia, Chile*

Rosa A. Pedraza, *Instituto de Genetica Forestal, Universidad Veracruzana, Parque El Haya s/n, Apartado Postal No. 551, CP 91000, Xalapa, Veracruz, Mexico. E-mail: rpedraza@uv.mx*

Andrea C. Premoli, *Laboratorio Ecotono, Universidad Nacional del Comahue, Quintral 1250, 8400 Bariloche, Argentina. E-mail: apremoli@crub.uncoma.edu.ar*

Paula Quiroga, *Laboratorio Ecotono, Universidad Nacional del Comahue, Quintral 1250, 8400 Bariloche, Argentina*

Neptalí Ramírez-Marcial, *Departamento de Ecología y Sistemática Terrestres, Área de Conservación de la Biodiversidad, El Colegio de la Frontera Sur,*

Carretera Panamericana y Periférica Sur, 29290 San Cristóbal de Las Casas, Chiapas, Mexico. E-mail: nramirez@sclc.ecosur.mx

Javier Retana, *CREAF-Centre for Ecological Research and Forestry Applications, Autonomous University of Barcelona, Bellaterra 08913, Spain. E-mail: javier.retana@uab.es*

José M. Rey-Benayas, *Departamento de Ecología, Universidad de Alcalá, Carretera de Barcelona km 33,600, E-28871 Alcalá de Henares, Madrid, Spain. E-mail: josem.rey@uah.es*

Rosa María Román-Cuesta, *Oxford University Centre for the Environment, Dyson Perrins Building, South Parks Road, Oxford OX1 3QY, UK. E-mail: rroman@ouce.ox.ac.uk*

Adriana E. Rovere, *Universidad Nacional del Comahue, Laboratorio Ecotono, Quintral 1250, San Carlos de Bariloche, 8400 Río Negro, Argentina*

Nadja Rüger, *UFZ Centre for Environmental Research, Department of Ecological Modelling, Permoserstr. 15, 04318 Leipzig, Germany. E-mail: nadja.rueger@ufz.de*

Cecilia Smith-Ramírez, *CMEM, Universidad de Chile, Ecología Forestal, Facultad de Ciencias, Casilla 653, Santiago, Chile; Centro de Estudios Avanzados en Ecología y Biodiversidad, Pontificia Universidad Católica de Chile, Departamento de Ecología, Alameda 340, Santiago, Chile*

Cintia Souto, *Laboratorio Ecotono, Universidad Nacional del Comahue, Quintral 1250, 8400 Bariloche, Argentina*

Oscar Thiers, *Instituto de Silvicultura, Universidad Austral de Chile, Casilla 567, Valdivia, Chile*

Sonia Trujillo-Argueta, *CIIDIR Oaxaca Instituto Politécnico Nacional, Hornos 1003, Santa Cruz Xoxocotlán, Oaxaca 68130, Mexico*

Guadalupe Williams-Linera, *Instituto de Ecología, Universidad Nacional Autónoma de México, Ciudad Universitaria, Coyoacán, Mexico DF 04510, Mexico*

Kerrie A. Wilson, *The Ecology Centre, The University of Queensland, Brisbane, 4072, Australia. E-mail: k.wilson2@uq.edu.au*

Carlos Zamorano, *Instituto de Silvicultura, Universidad Austral de Chile, Casilla 567, Valdivia, Chile*

Miguel A. Zavala, *Departamento de Ecología, Edificio de Ciencias, Universidad de Alcalá, Alcalá de Henares, Madrid E-28871, Spain. E-mail: ma.zavala@uah.es*

Preface

This book describes the results of a collaborative programme of research, spanning some 10 years, which examined the processes of forest loss, degradation and fragmentation and their impacts on biodiversity, and methods of forest conservation and restoration. The research was undertaken by a multi-disciplinary team of researchers drawn from Chile, the UK, Germany, Argentina, Mexico and Spain, and was funded primarily by two grants from the European Commission, with additional funding obtained from a range of sources from within the partner countries. In order to provide some context for what follows in the rest of the book, I describe here how this initiative originally came about, and how it subsequently developed. I make no apologies for presenting here what is very much a personal account, as it provides me with an opportunity to thank the many people who have helped make the collaboration such a success.

The research focused on the tropical montane forests of eastern and southern Mexico (in the states of Veracruz, Oaxaca and Chiapas) and the temperate rainforests of the southern cone of South America (Chile and Argentina). Why focus on these areas? Undoubtedly, these forests are of international conservation importance, yet they have been neglected by previous research. Over the past two decades, research on forest conservation ecology has focused primarily on lowland tropical rainforests. While this is understandable, given their exceptional biological diversity, other diverse and distinctive forests have been somewhat overlooked.

As detailed in subsequent chapters, forests in both montane Mexico and southern South America have a number of characteristics in common, as they are mixed (including both coniferous and broadleaved elements) and have a high proportion of endemic species. They also face similar pressures resulting from human activities, such as clearance for agriculture, browsing by livestock, timber logging and fuelwood extraction. These are special forests, home to rich and distinctive communities of plants and animals, with long, complex and fascinating evolutionary histories. They also play a major role in supporting the livelihoods of rural communities living nearby. It is worth recording that they are also simply wonderful places to visit.

These parallels suggest that there may be scientific merit in undertaking a comparative investigation of forests in these areas. Are there similarities, for example, in the pattern of forest loss occurring in different areas, and its relationship to the processes influencing biodiversity? Is it possible to identify principles and general findings that might apply to any forests subjected to intense human disturbance? Only by performing such comparative studies could a general synthesis, or theory, of the impacts of anthropogenic disturbance on forests be developed.

It would be wrong to imply, though, that this vision was fully developed at the outset. The roots of this research lie deeper. My own interest in the temperate rainforests of South America was stimulated by the seminal work of Tom Veblen, which I first encountered as an undergraduate more than 20 years ago. Tom's work has had a lasting impact, particularly in terms of understanding the role of natural disturbance in forest dynamics (Pickett and White, 1985). He has also left a lasting legacy in South America, both through his own PhD students and the students that they themselves have supervised. The work described in this book is part of that legacy. I would also like to acknowledge the role that Dr Edmundo Pisano and Matthew Hickman played in providing my first introduction to the Chilean flora, as a newly graduated student, back in 1985.

The research collaboration that produced this book began in 1995, following the award of two grants from the Darwin Initiative, which was created by the UK Government to help implement the Convention on Biological Diversity. Initial research in Chile focused on conservation of threatened conifers, and involved Drs Juan Armesto and Antonio Lara, two internationally recognized forest ecologists who have both continued to play a leading role throughout the research programme described here. This was paralleled by a separate project in Chiapas, Mexico, hosted by El Colegio de la Frontera Sur (ECOSUR). Dr Mario González-Espinosa, latterly Director of the ECOSUR campus at San Cristóbal de Las Casas, has similarly supported the research by his personal involvement. To each of these friends and colleagues, I owe a debt of gratitude. I also thank Martin Gardner, of the Royal Botanic Garden, Edinburgh, who was instrumental in developing the first project in Chile.

Most of the research described here was undertaken in two international collaborative projects supported by the European Commission (INCO programme). These were the SUCRE project ('Sustainable use, conservation and restoration of native forests in southern Mexico and south-central Chile', 1997–2000, ERBIC18CT970146) and its successor the BIOCORES project ('Biodiversity conservation, restoration and sustainable use in fragmented forest landscapes', 2002–2005, ICA4-CT-2001-10095). These provided an opportunity to link partners involved in the previous Darwin Initiative projects, and to commence parallel research activities in the different countries. In addition, the research partnership was extended to include teams headed by Dra Andrea Premoli (Universidad de Comahue, Bariloche, Argentina), Dra

Guadalupe Williams-Linera (Instituto de Ecología, Xalapa, Veracruz, Mexico) and Dr Rafael del Castillo (Instituto Politecnico Nacional, CIIDIR, Oaxaca, Mexico). The latter partner hosted an additional Darwin Initiative project from 1999 to 2002. To each of these people and their colleagues, together with the European partners that have also made such important contributions to the research, I offer my heartfelt thanks. It has been both an honour and a privilege to work with such a dedicated and knowledgeable group of people. With respect to the European partners, special mention should be given to Dr Lera Miles, of the UNEP-World Conservation Monitoring Centre, for her valuable help in coordinating the BIOCORES project. Thanks also to Gillian Myers for helping prepare this book for publication.

I close with some brief personal reflections of the experience of coordinating an international research collaboration that was afforded by these projects. The process has been challenging, complex, sometimes daunting. But, above all, it has been tremendously enjoyable and rewarding. I take particular satisfaction from the large number of undergraduate and postgraduate students that have been able to participate in the research, many of whom have made outstanding individual contributions, as described in later chapters. Their dedication and enthusiasm have been a continual source of inspiration. Another outcome have been the development in the research capacity of the group as a whole. The idea of collaborating with so many partners concurrently was something new for most of the participants. Yet the opportunity to exchange information and expertise by working together has been welcomed enthusiastically, and the legacy of this will outlive the projects themselves. The lesson is that it takes time to build genuinely collaborative partnerships, and the mutual trust on which they depend, but this is time well spent.

Particular challenges have included the need (as required by the donors) to ensure that the research undertaken is policy relevant, something that was also new to many of the partners. Another was the need to link research with the people living in and using the forests being studied. Over the decade, the Latin American partners have all made great progress in strengthening their links with local communities, conservation NGOs, private landholders and land managers. They have also devoted substantial efforts to publicizing research results in national media, and directly supporting practical conservation action on the ground. This book focuses primarily on presenting research results, and is aimed at a technical audience, and therefore it provides only a partial account of the activities that have been undertaken by those involved.

And what of the forests that inspired this work? The research coincided with exceptionally high rates of deforestation in some of the study areas. This fact only became evident towards the end of the period, as results from remote sensing analysis became available. The rate of loss surprised even some of the researchers that were intimately familiar with the study areas. Even in areas that have not yet been deforested, the forests are being severely affected by human activity. One of the clear results of the research described here is that these forests require a great deal of time to recover from human disturbance. Many forests may have already passed a threshold beyond which recovery is impossible. For anyone who values these forests and their associated biodiversity, the situation is very distressing. The protection of remaining forest areas is an urgent priority. The research presented here indicates just how urgent the need for such action has become.

ADRIAN NEWTON
Dorset, October 2006

Reference

Pickett, S.T.A. and White, P.S. (eds) (1985) *The Ecology of Natural Disturbance and Patch Dynamics*. Academic Press, New York.

Some of the participants in the BIOCORES project, photographed in Cambridge, UK, March 2005 (by Lera Miles). Pictured, left to right: Robert Manson, Juan J. Armesto, Antonio Lara, Mariela Núñez-Ávila, Rafael F. del Castillo, Cintia Souto, Luis Cayuela, José M. Rey-Benayas, Adriana Rovere, Nadja Rüger, Cecilia Smith-Ramírez (front row), Guillermo Montoya-Gómez, Adrian Newton, Fabiola López-Barrera (front row), Claudia Alvarez-Aquino, Guadalupe Williams-Linera (front row), Cristian Echeverría, Mario González-Espinosa.

1 Introduction

A.C. Newton

Fragmented forest landscape in the Highlands of Chiapas, Mexico. Photo: Luis Cayuela

Context

The loss and degradation of natural forests, and the associated loss of biodiversity, are now widely recognized as a global environmental concern. This is reflected by the proliferation of policy initiatives developed over the past 15 years aiming to reduce such losses, and support implementation of 'sustainable forest management'. These include the Convention on Biological Diversity (CBD), the Forest Principles and Chapter 11 of Agenda 21, the United Nations Forum on Forests (UNFF) and the many international processes developing criteria and indicators for sustainable forest management. However, despite this policy activity, progress on the ground towards sustainable forest management has been very limited, and high rates of forest loss and degradation are still occurring in many areas (FAO, 2006). This is considered to be a major contributing factor to the current high rate of biodiversity loss, which has been referred to as the 'global extinction crisis' (Ceballos and Ehrlich, 2002; Thomas *et al.*, 2004).

Increasing concern has led to an increase in research effort, aiming to document the processes of forest loss and degradation, and examining their impacts on forest-dwelling organisms. Substantial progress has been made in quantifying the rates and pattern of forest loss, using developing technologies such as remote sensing and GIS. Progress has also been made in identifying the impacts of forest loss on the ecological processes responsible for species decline and extinction. A substantial research literature now exists relating to issues such as deforestation processes (Lambin *et al.*, 2003), forest fragmentation (Fahrig, 2003) and edge effects (Ries *et al.*, 2004), forest dynamics and succession (Shugart, 1998), landscape ecology (Turner, 2005) and metapopulation theory (Hanski, 1999). Despite this, the impacts of human activity on forest biodiversity remain poorly defined.

Latterly, research attention has focused on lowland tropical rainforests in particular. Given their enormous importance for biodiversity, this is understandable. But are the results obtained applicable to other humid forest types? What of the process of biodiversity loss in tropical montane and temperate rainforests? These forest types have received relatively little attention from researchers, despite their global conservation importance. It is this gap in knowledge that the research described in this book was designed to address.

This book is the result of an international collaborative research effort focusing on the tropical montane forests of Mexico and the temperate rainforests of southern South America. The overall aim of the research was to investigate the impact of human activity on the key processes influencing biodiversity in fragmented forest landscapes, and to use the research results to develop practical tools for evaluating land-use decisions, thereby indicating how sustainable forest management might be achieved in practice.

The research sought to answer the following specific questions:

- To what extent have forest loss and fragmentation occurred in these areas during recent decades?
- What other forms of anthropogenic disturbance have these forests been subjected to?

- If forest loss, fragmentation and degradation have occurred, how have they affected different components of biodiversity?
- Given current trends, how can biodiversity be conserved effectively in forest landscapes subjected to human use?

Study Areas

The research was undertaken within a series of study areas located within highland Mexico and southern South America. Details of these study areas are presented in the individual chapters that follow, but some general information about the locations where the research was undertaken is provided below.

Tropical montane forest

Neotropical montane forests are widely recognized as being of exceptional conservation importance, being a centre of high diversity and endemicity for many different groups of organisms (Rzedowski, 1993; Churchill *et al.*, 1995; Hamilton *et al.*, 1995; Challenger, 1998; Bubb *et al.*, 2004). For example, some 45,000 plant species are thought to be found in neotropical highlands, nearly a fifth of all species known (Churchill *et al.*, 1995). Cloud forests, being those montane forests in the humid tropics that are frequently covered in clouds or mist, have attracted particular conservation concern (Bubb *et al.*, 2004). In Mexico, cloud forests cover less than 1% of the land surface of the country, but are thought to contain about 12% of the country's 30,000 plant species (Rzedowski, 1996). Some 30% of these species are endemic to the country. Mexican montane forests are recognized as a biodiversity 'hotspot' in the global assessment performed by Conservation International (Myers *et al.*, 2000), and as a priority ecoregion by the WWF (Olson *et al.*, 2001).

Research focused on montane forests in three areas of Mexico (Fig. 1.1).

(i) Xalapa, Veracruz

The study area is situated between 1200 and 2000 m of altitude in the eastern Sierra Madre mountains, and is located between 19° 13′ and 19° 41′ N, and 96° 51′ and 97° 01′ W, with an area of 842 km^2. Total annual precipitation in this region varies between 1300 and 2200 mm, while mean annual temperature is between 12 and 18°C. Typically, there are three well-defined seasons, the relatively dry-cool season lasting from October–November to March, the dry-warm season during April and May, and the wet-warm season from June to September–October (Williams-Linera, 1997). Soils are andosols. Some important canopy tree species in the study area are *Quercus xalapensis*, *Liquidambar styraciflua*, *Quercus leiophylla* and *Carpinus caroliniana*. Current land cover is a mosaic of cloud forest, secondary forest, coffee plantations, pastures, agricultural crops and human settlements. Most of the land is privately owned.

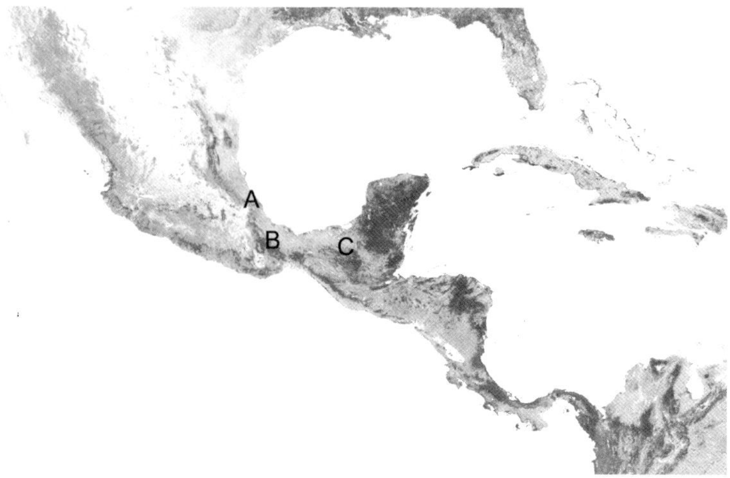

Fig. 1.1. Location of study areas in Mexico. A: Xalapa, Veracruz; B: El Rincón Alto, Sierra Norte, Oaxaca; C: Highlands of Chiapas. The figure illustrates forest cover, produced using MODIS satellite remote sensing data at a spatial resolution of 500 m. The depth of shading on the image relates to the density of tree cover. (Data from Hansen *et al*., 2003.)

(ii) Highlands of Chiapas

The Highlands of central Chiapas are a limestone massif, situated 16° 15′–17° 10′ N, and 91° 45′–92° 50′ W, at altitudes of 1500–2840 m. The climate is temperate subhumid, with a mean annual temperature of 13–17°C and mean annual rainfall typically in the range 1100–1600 m. Soils are a mixture of thin rendzinas, deeper humic acrisols and infertile chromic luvisols. Vegetation includes a number of highly diverse forest formations including seasonal pine and pine–oak forests, montane rainforests (800–2500 m elevation) and evergreen cloud forests (>2500 m; Miranda, 1952; Breedlove, 1981; Ramírez-Marcial *et al*., 2001; González-Espinosa *et al*., 2006). Oaks and pines are usually dominant in the forest canopy, including species such as the oaks *Quercus laurina*, *Quercus rugosa* and *Quercus crassifolia*, and the pines *Pinus oocarpa*, *Pinus pseudostrobus* and *Pinus ayacahuite*. The understorey is typically dominated by a diverse shrub community, including species such as *Myrcia jurguensenii*, *Oreopanax xalapensis*, *Fuchsia* spp. and *Litsea glaucescens*. Most inhabitants belong to Mayan ethnic groups, principally the Tzotzil, Tojolobal and Tzeltal. Land is communally owned. Traditional agriculture involves slash-and-burn (*milpa*) involving the cultivation of maize, beans and squash, producing a landscape mosaic of vegetation at different successional stages.

(iii) El Rincón Alto, Sierra Norte, Oaxaca

The study area is the Sierra Madre de Oaxaca Mountain Range, located in the north of the state of Oaxaca, between the parallels 17° 18′ and 17° 23′ N and the meridians 96° 15′ and 96° 21′ W. The area is part of the El Rincón

Alto region and lies at 1850 ± 150 m altitude, where tropical montane cloud forest is the primary vegetation. Topography is mountainous and the slopes are usually steep (15–64%). The climate is temperate-humid to subhumid, with mean annual temperature ranging between 20 and 22°C and mean precipitation around 1700 mm $year^{-1}$, with a rainy season in summer and a dry season in winter. Soils lie on a bedrock of Mesozoic schist and are classified as entisols, inceptisols and dystrudepts. The land cover is a mosaic of primary forests, secondary forests of different ages after abandonment and agricultural fields. Successional forests are dominated by the conifer *Pinus chiapensis*, which occurs in association with other species such as *Clethra integerrima*, *Gaultheria acuminata*, *L. styraciflua* and *Phyllonoma laticuspis*. Late-successional forests are dominated by broadleaved species such as *Persea americana*, *Quercus* spp., *Rapanea* spp., *Ternstroemia hemsleyi* and *Quetzalia occidentalis*. Human population density is relatively low compared to the other Mexican study areas. The area is inhabited by Zapotecs, a native Mexican ethnic group. Land tenure is communal.

South temperate rainforest

The temperate rainforests of southern Chile and areas in Argentina adjacent to southern Chile are recognized as a biodiversity 'hotspot' in the global assessment performed by Conservation International (Myers *et al.*, 2000), and as a priority ecoregion by the WWF (Dinerstein *et al.*, 1995; Olson *et al.*, 2001). These forests are distributed along the coastal mountain range of Chile and the main Andean range from 38° to 56° S. The forests are home to more than 900 vascular plant species, including 60 tree species, over 90% of which are endemic (Arroyo *et al.*, 1995). Within this area, a number of different forest types may be differentiated, including the Valdivian evergreen forests that extend for 250 km from the Tolten River (40° 50′ S) to south of the Llico River (41° 30′ S) (Smith-Ramírez, 2004). Notable elements of the flora include the long-lived conifers monkey puzzle (*Araucaria araucana*) and alerce (*Fitzroya cupressoides*), with some trees of the latter species living for more than 3620 years (Lara and Villalba, 1993). Over the past 30 years, the Chilean forestry sector has become a driving force in the national economy, with forest exports increasing from approximately US$40 million in 1970 to US$2.2 billion in 2000 (Neira *et al.*, 2002). Other main threats to native forests have been the conversion to pasturelands, human-set fires, highgrading (selective felling), fuelwood cutting and other logging practices (Lara *et al.*, 2000). Most land is privately owned.

While research activities were distributed throughout southern Chile and adjacent locations in Argentina, investigations were particularly focused on the following three areas of Chile (Fig. 1.2).

(i) Los Muermos area, Region X, Chile

The study area corresponds to 503,287 ha located between 41° 30′ S, 73° W and 42° 20′ S, 74° W in southern Chile. The zone is characterized by a rainy

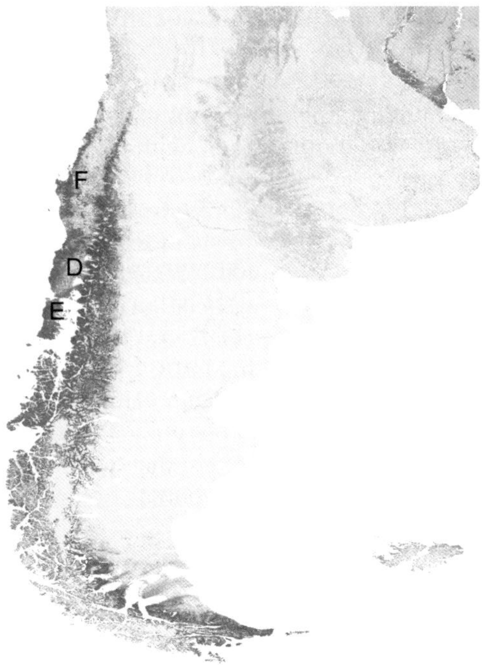

Fig. 1.2. Location of study areas in Chile. D: Los Muermos, Region X; E: Chiloé Island, Region X; F: Regiόn del Maule, Regions VII and VIII. The figure illustrates forest cover, produced using MODIS satellite remote sensing data at a spatial resolution of 500 m. The depth of shading on the image relates to the density of tree cover. (Data from Hansen *et al.*, 2003.)

temperate climate with an oceanic influence and without dry periods, with a mean annual precipitation of around 2000 mm. The landscape is dominated by Valdivian temperate rainforests, surrounded by crops and pasture lands. Many of the remaining forests occur on acidic, shallow, poorly drained soils referred to as *ñadis*, which are classified a gleysols. The forests are characterized by the presence of broadleaved evergreen tree species such as *Drimys winteri* (Winteraceae), *Nothofagus dombeyi* (Nothofagaceae), *Laurelia philipiana* (Monimiaceae), *Amomyrtus luma*, *Amomyrtus meli* (both Myrtaceae) and *Eucryphia cordifolia* (Eucryphiaceae). In some sites, long-lived conifers such as *F. cupressoides* and *Pilgerodendron uvifera* (both Cupressaceae) can also be found. Anthropogenic disturbance has led to early successional stages of the forest being widespread, which are characterized by a high abundance of *D. winteri* and *Nothofagus nitida*. In some degraded sites, shrub species such as *Berberis* spp. (Berberidaceae), *Baccharis* spp. (Asteraceae) and *Gaultheria* spp. (Ericaceae) are abundant.

(ii) Chiloé Island, Region X, Chile

The study area comprises about 400 km^2 in the north-eastern corner of Chiloé Island, approximately 20 km north of the city of Ancud (41° 50′ S, 73°

50′ W). The landscape is characterized by an undulating topography with altitudes ranging from 50 to 100 m. Soils are generally thin (<1 m), originating from Pleistocene moraine fields and glacial outwash plains, often with poor drainage. The prevailing climate is described as wet temperate with a strong oceanic influence. Meteorological records at Senda Darwin Biological Station (45° 53′ S, 73° 40′ W) indicate an annual rainfall of 2090 mm and a mean annual temperature of 12°C. Maximum monthly temperatures (January) are 16°C and minimum monthly temperatures (July–August) are 5°C. Rainfall occurs throughout the year, but 64% of the precipitation is concentrated from April to September. Lowland forests in the area have been logged since the early 1800s, but land clearing became more intense in the second half of the 20th century. The present-day rural landscape is characterized by a mosaic of remnant forest fragments and grazing pastures. The major forms of human impact on forests during the last century have been selective logging of valuable timber trees, widespread use of fire to clear land for pastures and increasing forest fragmentation (Willson and Armesto, 1996).

The forests are dominated by evergreen, broadleaved trees, but include narrow-leaf conifers such as *Saxegothaea conspicua* and *Podocarpus nubigena*. Floristically, many forests belong to the North Patagonian forest type as defined by Veblen *et al.* (1983). *N. nitida*, *D. winteri* and *P. nubigena* are widespread. Some sites are more typical of a Valdivian rainforest with canopy dominants such as *E. cordifolia*, *Laureliopsis philippiana* and *N. nitida*. All forests have an understorey of Myrtaceae trees, often with abundant regeneration of tree seedlings and saplings and abundant cover of the native bamboo *Chusquea quila*, especially in tree-fall gaps.

(iii) Región del Maule, Chile

This study area covers around 578,164 ha in the Coastal Range of the Maule and Bío-Bío (VII and VIII) regions of south-central Chile, at latitudes between 35° and 36° 30′ S and longitudes between approximately 72° and 73° W. The climate is of Mediterranean type, with an average annual rainfall of 700–800 mm concentrated in the winter; the summers are dry from September to April, with high luminosity. The mean annual temperature is 14°C. The two main types of soil are well-developed alfisols, which have evolved from granite substrate, and thinner inceptisols, usually originating from marine sediment rock layers. The natural forest is mainly dominated by secondary forest of *Nothofagus* species (*N. obliqua* and *N. glauca*) (Fagaceae) and sclerophyllous species including *Acacia caven* (Mimosaceae), *Quillaja saponaria* (Rosaceae) and *Maytenus boaria* (Celastraceae). Also, many endangered tree species such as *Nothofagus alessandri*, *Pitavia punctata* (Rutaceae) and *Gomortega keule* (Gomortegaceae) occur in the study area. Anthropogenic disturbance is intense: in particular, the region suffered massive forest clearance in the middle of the 20th century for the cultivation of wheat crops, and non-sustainable extraction of firewood in more recent decades. The area therefore currently is characterized by a highly fragmented forest landscape.

Research Approach

The research was based on the assumption that the conservation of biodiversity depends on the maintenance of key ecological processes, which determine the composition and structure of biological communities (Fig. 1.3). In areas subjected to human use, the main factor determining the scope for biodiversity conservation is the extent to which these processes are influenced by patterns of land use. Areas where deforestation is occurring at a high rate are generally characterized by conversion of forest to agricultural land uses, such as crop cultivation and grazing, often in addition to logging and the use of fire. Clearance of forest for agriculture leads to a decline in forest area and fragmentation of forest habitat. Remnant patches of forest may be further degraded by extraction of forest products, and by alteration of environmental conditions in newly created forest edges.

Although research was undertaken in a number of different study areas, a common investigative approach was adopted, incorporating the following features:

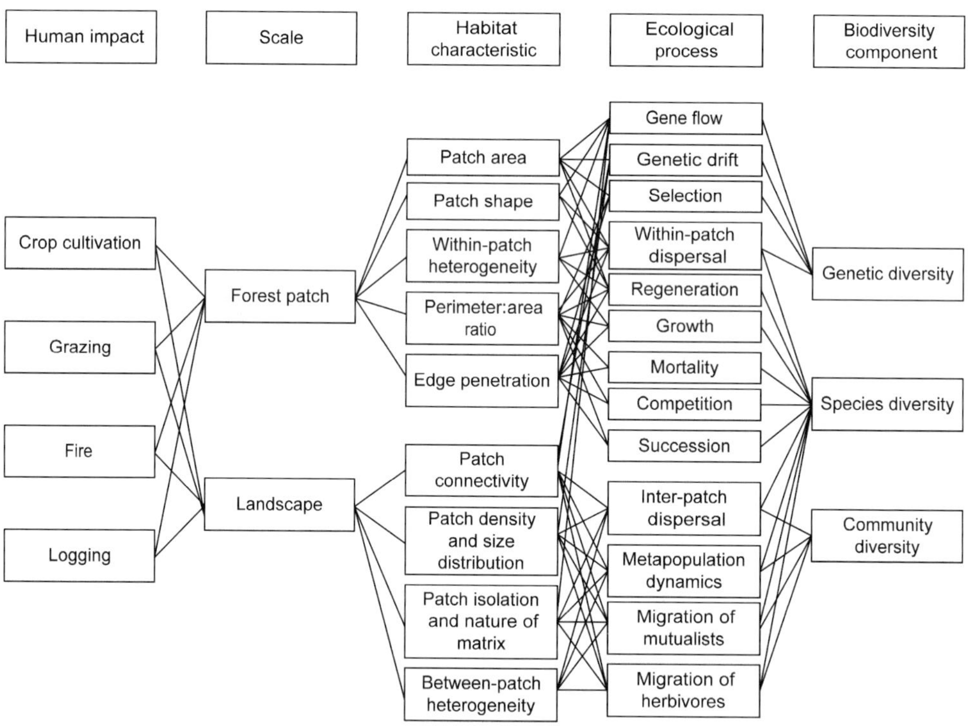

Fig. 1.3. Schematic diagram indicating the hypothesized impacts of different human activities on biodiversity in fragmented forest landscapes. Human activities influence forest habitat characteristics at a range of scales, which in turn affect the key ecological processes influencing the different components of biodiversity.

Landscape approach. Recent progress in landscape ecology and its application to conservation management has highlighted the importance of assessing threats to biodiversity, and their impacts on ecological processes, at the landscape scale. Research therefore examined changes in the spatial characteristics of forest landscapes, and their impacts on processes such as dispersal, pollination, predation, gene flow and succession. Results of the research should therefore support development of management approaches at the landscape scale, such as forest landscape restoration (Mansourian *et al*., 2005).

Integrated assessment of biodiversity. Biodiversity is generally considered to comprise three elements, namely variation at the scale of communities, species and within-species. However, these elements are rarely examined concurrently. The research described here adopted an integrated approach, considering each of these three elements together.

Focus on floristic diversity. Ideally, a comprehensive assessment of biodiversity would involve assessment of a wide range of different species groups, including fungi, insects, mammals, birds, reptiles, etc. This is difficult to achieve in practice, and therefore the focus here is on vascular plants, with particular reference to tree species. Given that forest-dwelling organisms are usually dependent on trees for habitat, the focus on tree species and the floristic communities of which they are a part should provide insights into potential impacts on other species groups. The research did, however, incorporate studies specifically on soil macroinvertebrates, plant–animal interactions with reference to birds, insects and mammals, and the conservation ecology of selected threatened fauna.

Multi-disciplinarity. A range of different techniques and approaches were employed in the research, including molecular markers, GIS, remote sensing, spatial analysis and a variety of different modelling approaches. These were supported by an extensive programme of field-based survey and experimental investigations. By employing a variety of complementary approaches, it was intended that a more complete analysis would be obtained. All of the research activities were collaborative, typically involving inputs from a range of different partners.

Conservation through use. While it is recognized that protected areas represent the most important approach for biodiversity conservation, the research explicitly focused on forest landscapes subjected to human use. If conservation is to be effective, protected areas cannot be viewed in isolation. Methods need to be developed for combining biodiversity conservation with use of forests in surrounding landscapes. While the research did examine the coverage and effectiveness of protected area networks, the main focus was on forest areas that are not protected, but are being actively used to support the livelihoods of local communities.

Policy relevance. The research was designed from the outset to support the development and implementation of policies relating to the conservation, restoration and sustainable use of native forests. This was achieved through the development of tools such as indicators, models and scenarios, to help identify and communicate the practical implications of the research results and to support the process of environmental decision making.

Structure of the Book

The book features two types of chapters: those that integrate information collected across a range of study areas, and those that focus on a single study area. Throughout the research programme, the intention was to perform parallel analyses across a range of study areas using a common set of approaches, with the aim of identifying general patterns and principles. However, some issues of particular importance in a single area were addressed by individual investigations, which also provided an opportunity to develop and test novel analytical approaches.

Chapter 2 describes the pattern and extent of deforestation that has occurred in study areas in both highland Mexico and southern Chile, including an assessment of forest fragmentation, based on analysis of satellite remote sensing imagery. Chapter 3 then considers the impact of these processes of forest loss and fragmentation on floristic diversity. Chapter 4 examines the ecological impact of fragmentation in more depth, profiling field-based research investigations that examined a range of ecological processes, including dispersal, pollination and predation. Particular emphasis is given to the analysis of edge effects. Further analyses of edge effects are presented in Chapter 5, which focuses on the influence of edge effects on reproductive ecology of selected tree species.

One of the key features of this research initiative was to include the genetic component of biodiversity. This is considered in Chapter 6, which presents research performed using a variety of molecular markers to assess patterns of genetic variation in selected tree species. The implications of forest loss and fragmentation for genetic diversity are also considered, with reference to the processes influencing patterns of genetic variation.

In addition to deforestation and forest fragmentation, forests in the study areas are affected by a range of human activities that may affect forest structure and composition, even when forest cover is maintained. The following six chapters consider the impacts of different forms of anthropogenic disturbance on forest communities, using a variety of approaches. Chapter 7 describes a series of successional chronosequences in Oaxaca, Mexico, which have provided unique insights into the processes of forest recovery following localized clearance of tropical montane cloud forest, including both vegetation and soil dynamics. The same chronosequences are considered further in Chapter 8, with respect to the dynamics of the soil macroinvertebrate fauna, and with particular reference to the impacts of logging.

Chapters 9, 10 and 11 present investigations of forest dynamics under human disturbance using three different modelling approaches. In Chapter 9, a forest succession (or 'gap') model is used to explore the impact of different anthropogenic disturbance regimes in Chiapas, whereas Chapter 10 employs transition matrix models to assess vegetation dynamics in the same study area. Chapter 11 describes application of a sophisticated, process-based forest growth model FORMIND to assess forest dynamics in southern Chile and Veracruz, Mexico, with the aim of identifying sustainable approaches to forest management.

Indicators are one of the principal tools used to achieve sustainable forest management in practice, providing a means of monitoring the impacts of management interventions. Chapter 12 tests the effectiveness of some commonly used indicators of forest biodiversity, through a comparative analysis of anthropogenic disturbance gradients in four study areas.

The implications of research results for practical conservation action and sustainable forest management are addressed in all of the individual chapters. However, various response measures are considered in greater depth in Chapters 13–15. The impacts of both natural and human-set fire are considered in Chapter 13, which describes research undertaken in Chiapas, where fires have presented a particularly severe problem in recent years. The implications of current fire regimes for protected areas receive particular attention. Protected areas are further considered in Chapter 14, which describes new approaches to systematic conservation planning, illustrated by research undertaken in southern Chile. Chapter 15 considers an alternative type of management intervention: forest restoration. The chapter summarizes results from a series of experimental investigations examining different restoration approaches used in the study areas.

Another feature of the research described here was the aim of using the results to inform policy development and implementation. Scenarios are increasingly being used as a tool for this purpose. Consequently Chapter 16 describes results of a workshop exercise in which the actual and potential threats to biodiversity within each of the study areas are considered. The potential impact of policy initiatives and management interventions are then explored through alternative scenarios in each of the study areas, informed by the research results described in the preceding chapters. The final chapter presents a brief synthesis of the book, highlighting some of the main conclusions of the research presented.

References

Arroyo, M.K., Lohengrin, C., Peñaloza, A., Riveros, M. and Faggi, A.M. (1995) Relaciones fitogeográficas y patrones regionales de riqueza de especies en la flora del bosque lluvioso templado de Sudamérica. In: Armesto, J.J., Villagrán, C. and Arroyo, M.K. (eds) *Ecología de los Bosques Nativos de Chile*. Editorial Universitaria, Santiago, Chile, pp. 71–92.

Breedlove, D.E. (1981) Introduction to the flora of Chiapas. In: Breedlove, D.E. (ed.) *Flora of Chiapas. Part 1*. California Academy of Sciences, San Francisco, California, pp. 1–35.

Bubb, P., May, I., Miles, L. and Sayer, J. (2004) Cloud Forest Agenda. UNEP-WCMC Biodiversity Series 20. United Nations Environment Programme – World Conservation Monitoring Centre, Cambridge, UK. http://www.unep-wcmc.org/resources/publications/UNEP_WCMC_bio_series/20.htm

Ceballos, G. and Ehrlich, P.R. (2002) Mammal population losses and the extinction crisis. *Science* 296, 904–907.

Challenger, A. (1998) *Utilización y Conservación de los Ecosistemas Terrestres de México. Pasado, Presente y Futuro*. Comisión Nacional para el Conocimiento y Uso de la Biodiversidad, UNAM, Agrupación Sierra Madre, SC México, DF, Mexico.

Churchill, S.P., Balslev, H., Forero, E. and Luteyn, J. (1995) *Biodiversity and Conservation of Neotropical Montane Forests: Proceedings of the Neotropical Montane Forest Biodiversity and Conservation Symposium*. The New York Botanical Garden, 21–26 June 1993. The New York Botanical Garden, New York.

Dinerstein, E., Olson, D., Graham, D., Webster, A., Primm, S., Bookbinder, M. and Ledec, G. (1995) *A Conservation Assessment of the Terrestrial Ecoregions of Latin America and the Caribbean*. World Bank and WWF, Washington, DC.

Fahrig, L. (2003) Effects of habitat fragmentation on biodiversity. *Annual Review of Ecology, Evolution and Systematics* 34, 487–515.

FAO (2006) *The Global Forest Resources Assessment*. FAO, Rome, Italy.

González-Espinosa, M., Ramírez-Marcial, N. and Galindo-Jaimes, L. (2006) Secondary succession in montane pine–oak forests of Chiapas, México. In: Kappelle, M. (ed.) *Ecology and Conservation of Neotropical Oak Forests, Ecological Studies 185*. Springer, Berlin, Germany, pp. 209–221.

Hamilton, L.S., Juvik, J.O. and Scatena, F.N. (1995) *Tropical Montane Cloud Forests, Ecological Studies 110*. Springer, New York.

Hansen, M., DeFries, R., Townshend, J.R., Carroll, M., Dimiceli, C. and Sohlberg, R. (2003) *500m MODIS Vegetation Continuous Fields*. The Global Land Cover Facility, College Park, Maryland.

Hanski, I. (1999) *Metapopulation Ecology*. Oxford University Press, Oxford, UK.

Lambin, E.F., Gesit, H.J. and Lepers, E. (2003) Dynamics of land-use and land-cover change in tropical regions. *Annual Review of Environment and Resources* 28, 205–241.

Lara, A. and Villalba, R. (1993) A 3620-year temperature record from *Fitzroya cupressoides* tree-rings in southern South America. *Science* 260, 1104–1106.

Lara, A., Cortés, M. and Echeverría, C. (2000) Bosques. In: Sunkel, O. (ed.) *Informe País: Estado Actual del Medio Ambiente en Chile*. Centro de Estudios de Políticas Publicas, Universidad de Chile, Santiago, Chile, pp. 131–173.

Mansourian, S., Vallauri, D. and Dudley, N. (2005) *Forest Restoration in Landscapes: Beyond Planting Trees*. Springer, New York.

Miranda, F. (1952) *La Vegetación de Chiapas, Primera Parte*. Ediciones del Gobierno del Estado, Tuxtla Gutiérrez, Chis. Mexico.

Myers, N., Mittermeier, R.A., Mittermeier, C.G., da Fonseca, G.A.B. and Kent, J. (2000) Biodiversity hotspots for conservation priorities. *Nature* 403, 853–858.

Neira, E., Verscheure, H. and Revenga, C. (2002) *Chile's Frontier Forests, Conserving a Global Treasure*. Global Forest Watch, World Resources Institute WRI, Comité Nacional por Defensa de la Fauna y Flora CODEFF, Universidad Austral De Chile UACH, Washington, DC and Valdivia, Chile.

Olson, D.M., Dinerstein, E., Wikramanayake, E.D., Burgess, N.D., Powell, G.V.N., Underwood, E.C., D'Amico, J.A., Itoua, I., Strand, H.E., Morrison, J.C., Loucks, C.J., Allnutt, T.F., Ricketts, T.H., Kura, Y., Lamoreux, J.F., Wettengel, W.W., Hedao, P. and Kassem, K.R. (2001) Terrestrial ecoregions of the world: a new map of life on earth. *BioScience* 51, 933–938.

Ramírez-Marcial, N., González-Espinosa, M. and Williams-Linera, G. (2001) Anthropogenic disturbance and tree diversity in montane rain forests in Chiapas, Mexico. *Forest Ecology and Management* 154, 311–326.

Ries, L., Fletcher, R.J., Battin, J. and Sisk, T.D. (2004) Ecological responses to habitat edges: mechanisms, models, and variability explained. *Annual Review of Ecology, Evolution and Systematics* 35, 491–522.

Rzedowski, J. (1993) Diversity and origins of the phanerogamic flora of Mexico. In: Ramamoorthy, T.P., Bye, R., Lot, A. and Fa, J. (eds) *Biological Diversity of Mexico: Origins and Distribution*. Oxford University Press, New York, pp. 129–144.

Rzedowski, J. (1996) Análisis preliminar de la flora vascular de los bosques mesófilos de montaña de México. *Acta Botánica Mexicana* 35, 25–44.

Shugart, H.H. (1998) *Terrestrial Ecosystems in Changing Environments*. Cambridge University Press, Cambridge, UK.

Smith-Ramírez C. (2004) The Chilean Coastal Range: a vanishing center of biodiversity and endemism in southern temperate rain forests. *Biodiversity and Conservation* 13, 373–393.

Thomas, J.A., Telfer, M.G., Roy, D.B., Preston, C.D., Greenwood, J.J.D., Asher, J., Fox, R., Clarke, R.T. and Lawton, J.H. (2004) Comparative losses of British butterflies, birds, and plants and the global extinction crisis. *Science* 303, 1879–1881.

Turner, M.G. (2005) Landscape ecology: what is the state of the science? *Annual Review of Ecology, Evolution and Systematics* 36, 319–344.

Veblen, T.T., Schlegel, F.M. and Oltremari, J.V. (1983) Temperate broad-leaved evergreen forests of South America. In: Ovington, J.D. (ed.) *Temperate Broad Leaved Evergreen Forests*. Elsevier Science, Amsterdam, The Netherlands, pp. 5–31.

Williams-Linera, G. (1997) Phenology of deciduous and broadleaved-evergreen tree species in a Mexican tropical lower montane forest. *Global Ecology and Biogeography Letters* 6, 115–127.

Willson, M.F. and Armesto, J.J. (1996) The natural history of Chiloé: on Darwin's trail. *Revista Chilena de Historia Natural* 69, 149–161.

2 Spatial and Temporal Patterns of Forest Loss and Fragmentation in Mexico and Chile

C. Echeverría, L. Cayuela, R.H. Manson, D.A. Coomes, A. Lara, J.M. Rey-Benayas and A.C. Newton

Fragmented forest landscape in the Highlands of Chiapas, Mexico. Photo: Luis Cayuela

Summary

The patterns and driving forces of forest loss and fragmentation were assessed in four study areas: two in Mexico (Central Veracruz and the Highlands of Chiapas) and two in Chile (Rio Maule-Cobquecura and Los Muermos-Ancud). For the Highlands of Chiapas, Rio Maule-Cobquecura and Los Muermos-Ancud study areas, three land-cover maps were derived from satellite imagery acquired between 1975–1976 and 1999–2000. For Central Veracruz, two land-cover maps were obtained from the interpretation of aerial photographs and Landsat ETM+ satellite images for 1984 and 2000, respectively. Analysis of these images indicated a reduction in natural forest area of 67% in Rio Maule-Cobquecura, 57% in the Highlands of Chiapas, 26% in Central Veracruz and 23% in Los Muermos-Ancud. These losses are equivalent to annual forest loss rates of 4.4%, 3.4%, 2.0% and 1.1% per year, respectively. Forest fragmentation in the study areas led to a decrease in forest patch size, which was associated with a rapid increase in the density and isolation of forest patches and a decline in area of interior forests and number of large patches. Logistic regression models were used in each study area to identify the factors associated with forest loss. Overall, the probability of an area being cleared of forest was greatest in gently sloping areas and around the margins of forest patches. Additionally, soil fertility appears to be a significant factor associated with deforestation in Central Veracruz. In Maule-Cobquecura and Los Muermos-Ancud the probability of deforestation was higher as size of forest fragments decreased, whereas in the Highlands of Chiapas large fragments were particularly vulnerable to deforestation. Given the current trends of forest loss, we predict that further declines and spatial changes of forest cover will occur in each of the study areas. The patterns observed reveal some of the immediate causes of deforestation in Mexico and Chile such as pasture and crop expansion, forest logging and conversion to plantations of exotic tree species. These changes highlight some weaknesses in the national environmental and economic policies in the countries included in this study.

Introduction

Habitat fragmentation and forest loss have been recognized as major threats to ecosystems worldwide (Iida and Nakashizuka, 1995; Dale and Pearson, 1997; Noss, 2001; Armenteras *et al.*, 2003). These two processes have negative effects on biodiversity by increasing isolation of habitats (Debinski and Holt, 2000), reducing the extent of species habitat and modifying the population dynamics of species (Watson *et al.*, 2004). Fragmentation may also have negative effects on species richness by reducing the probability of successful dispersal and establishment (Gigord *et al.*, 1999) as well as by reducing the capacity of a patch of habitat to sustain a resident population (Iida and Nakashizuka, 1995). For example, fragmentation of Maulino temperate forest in central Chile has affected the abundance of bird richness (Vergara and Simonetti, 2004) and regeneration of shade-tolerant species (Bustamante and Castor, 1998), and has also favoured the invasion of alien species (Bustamante *et al.*, 2003). The ecological consequences of fragmentation can differ, depending on the pattern or spatial configuration imposed on a landscape and how this varies both temporally and spatially (Ite and Adams, 1998; Armenteras *et al.*, 2003). Some studies have shown that the spatial configuration of the landscape and community structure may significantly affect species richness at different scales (Steiner and Köhler, 2003). Other authors emphasize the need to incorporate the spatial configuration and connectivity attributes at a landscape level in order to protect the ecological integrity of species assemblages (Herrmann *et al.*, 2005; Piessens *et al.*, 2005).

The temporal evaluation of forest change based on satellite imagery linked to fragmentation analysis is becoming a valuable set of techniques for assessing the degree of threat to forest ecosystems (Luque, 2000; Franklin, 2001; Imbernon and Branthomme, 2001; Sader *et al.*, 2001; Armenteras *et al.*, 2003). A number of imagery-based studies of deforestation have been conducted in tropical forests (Skole and Tucker, 1993; Turner and Corlett, 1996; Imbernon and Branthomme, 2001; Sader *et al.*, 2001; Steininger *et al.*, 2001; Laurance *et al.*, 2006), including the Amazon (Jorge and García, 1997; Pedlowski *et al.*, 1997; Ranta *et al.*, 1998; Laurance, 1999; Laurance *et al.*, 2000; Sierra, 2000), but few studies of deforestation and fragmentation have been made in temperate forests (Gibson *et al.*, 1988; Staus *et al.*, 2002; Hobbs and Yates, 2003), or in tropical montane forests.

There is a global need to identify the causes of deforestation and forest fragmentation and to understand how these affect the spatial configuration of landscapes over time (Angelsen and Kaimowitz, 1999; Verburg *et al.*, 2002; Bürgi *et al.*, 2004; McConnell *et al.*, 2004; Veldkamp and Verburg, 2004). It is increasingly recognized that simple descriptions of land-cover types are inadequate for conservation planning or resource management, because they do not incorporate information about the patterns of land-use change that can have profound effects on ecological process of interest (Bürgi *et al.*, 2004; Corney *et al.*, 2004). For a more systematic understanding of landscape change, it is necessary to study the driving forces responsible for deforestation, leading to the analysis of processes and not merely patterns (Bürgi *et al.*, 2004). The ability to link a particular driver in the landscape to specific landscape changes is a powerful tool for researchers exploring environmental change (Evans and Moran, 2002). Some researchers add that it is necessary to move beyond the simplistic assessment of the proximate causes of land-use and land-cover change and assess underlying factors such as environment–development policies (Lambin *et al.*, 2001; Silva, 2004).

In this chapter we examine the rates and patterns of deforestation and fragmentation of native forests in each of four study areas in Mexico and Chile. In addition, we analyse the influence of social and environmental driving forces on landscape change in each study area. During the past three decades, expansion of croplands, pasturelands and industrial plantations has resulted in a substantial decline in forest area and in an increase in forest fragmentation. Some research on the ecological consequences of forest fragmentation has previously been undertaken in Chile and Mexico (Willson *et al.*, 1994; Bustamante and Grez, 1995; Donoso *et al.*, 2003; Vergara and Simonetti, 2004; Martínez-Morales, 2005), but few studies have integrated spatial and temporal analyses to assess the pattern and rate of forest loss and fragmentation.

Methods

Study areas

Four study areas were selected from southern Mexico and south-central Chile (Fig. 2.1): (a) the central part of Veracruz in Mexico (Central Veracruz); (b) the

Fig. 2.1. Location of the four study areas in Mexico and Chile.

highlands of the state of Chiapas in Mexico (the Highlands of Chiapas); (c) the Coastal Range in Chile, from Rio Maule (VII region) to Cobquecura municipality in the VIII region (Rio Maule-Cobquecura); and (d) the area situated from Los Muermos to the Chiloé Island in Chile, including the entire municipality of Ancud (Los Muermos-Ancud). These study areas were selected to contrast the effects of different historical patterns of deforestation and different human pressures on the forest ecosystems.

Cloud forests are important to study for a number of reasons. Cloud forest covers less than 1% of the total area of Mexico, yet contains around 2500–3000 plant species, representing about 10–12% of the total number of plant species that occur in Mexico (Rzedowski, 1993; Mitermeier *et al.*, 1997). Moreover, cloud forests have the highest number of mammal species (95) of any type of forest in Mexico and a high rate of endemism in plants (30%) (Ramamoorthy *et al.*, 1993). Despite this high biodiversity, more than 50% of the area of this type of forest has already been converted to other land uses nationally (Challenger, 1998).

The Highlands of Chiapas are also a biologically diverse region, extending over 11,000 km^2, and include 30% of about 9000 vascular plant species of the flora of Chiapas (Breedlove, 1981). Several forest formations are found in the Highlands, including oak, pine–oak, pine and montane cloud forests (Miranda, 1952; Rzedowski, 1978; González-Espinosa *et al.*, 1991). Our study area covers *c.*3550 km^2, with elevation ranging from 600 to 2900 m (mostly above 1500 m). The topography is abrupt, with fairly steep slopes. The climate is cool and humid, with a rainy summer. The region is densely populated by Mayan peasants, who have cleared the forest for shifting cultivation and extracted firewood and other forest resources since pre-Columbian times (Collier, 1975). The main economic activities are traditional agriculture and non-commercial forestry. Slash-and-burn agriculture and long-term exploitation of forests for fuelwood have contributed to the expansion of relatively low diversity pine and mixed pine–oak stands coupled with a reduction in the extent of highly diverse oak and mountain cloud forests (Ramírez-Marcial *et al.*, 2001; Galindo-Jaimes *et al.*, 2002). The cloud forests in Central Veracruz are also highly threatened as a result of deforestation and urban expansion (Williams-Linera *et al.*, 2002). The study area covers *c.*7166 km^2 of the mountainous region (> 800 m) in the centre of the state.

The two Chilean study areas are located in the temperate forest zone, which has been classified as a biodiversity hotspot for conservation (Myers *et al.*, 2000) and has also been included among the most threatened ecoregions in the world in the Global 200 initiative launched by WWF and the World Bank (Dinerstein *et al.*, 1995). The Rio Maule-Cobquecura study area covers *c.*5781 km^2. The natural forest is mainly dominated by secondary forest of *Nothofagus* species (*N. obliqua* and *N. glauca*) (Fagaceae) and sclerophyllous species. At present, approximately 5% of the native forest in the VII region is under the National System of Protected Areas (SNASPE), while the remaining forests lack cohesive protection.

Los Muermos-Ancud covers *c.*5032 km^2 and is characterized by a rainy temperate climate with an oceanic influence and without dry periods (Di Castri and Hajek, 1976). The landscape is dominated by a broadleaved evergreen temperate rainforest within a matrix of agricultural land and pastures. In the middle of the 20th century a significant area of native forests was cut down and burnt as a result of European settlement. Intensive timber exploitation then began in the area, allowing the establishment of areas for grazing and crop cultivation (Donoso and Lara, 1995). In Chiloé Island, the process of deforestation by logging and cultivation occurred mainly in recent decades, its exploitation having been delayed by virtue of its isolation from the mainland.

Generation of spatial data

To analyse the spatial and temporal patterns of land-cover change we used Landsat satellite scenes and aerial photographs, along with the geographic information system (GIS). In Central Veracruz, the analysis of land-cover changes was conducted over a 16-year period using thematic maps generated for the years 1984 and 2000 (Palacio-Prieto *et al.*, 2000). The 1984 coverage was based on digitized topographic maps generated from black and white aerial photographs (INEGI, 1984) and the 2000 map was obtained from the National Forest Inventory (Palacio-Prieto *et al.*, 2000) using Landsat ETM+ (Enhanced Thematic Mapper) satellite images (November 1999 to May 2000).

For the Highlands of Chiapas, Rio Maule-Cobquecura and Los Muermos-Ancud, a set of three satellite images were acquired at different time intervals over the last three decades (Table 2.1). Similar to the methodology described by Fuller (2001), Hansen *et al.* (2001) and Staus *et al.* (2002), the original 79 m MSS (Multispectral Scanner) raster grids were resampled to the resolution of the TM (Thematic Mapper) and ETM+ raster grids (30 m). The resampling enabled the land-cover maps to be produced with consistent resolution, which is essential to develop meaningful comparisons between scenes from different dates. Each image was geometrically, atmospherically and topographically corrected. For the Highlands of Chiapas, the classification of satellite imagery was undertaken applying the Dempster–Shafer theory of evidence (Shafer, 1976), which enabled an increase in the accuracy of classification by the combination of remote sensing data with information derived from expert knowledge (Cayuela *et al.*, 2006a). Classifications of the land-cover types in Chile were conducted using the decision criterion of Maximum Likelihood (Chuvieco, 1996) and set of points of field visits and thematic maps developed by a comprehensive cartographic study of natural vegetation known as Catastro (CONAF *et al.*, 1999).

In Central Veracruz, the 1984 (72 vegetation classes) and 2000 (43 classes) land covers were reclassified using six categories (crop and pasture land, forest, old-field, native forest, urban areas and other) to facilitate comparisons and simplify the evaluation of patterns of land-use change. For this reclassification, disturbed forest (including herbaceous and shrub cover) was grouped into 'old-field', while all other natural vegetation cover types were

Table 2.1. Spatial data used in each study area.

Study area	Type of data	Year
Central Veracruz	Aerial photographs	1984
	Landsat 7 – ETM+	1999–2000
The Highlands of Chiapas	Landsat 1 – MSS	1975
	Landsat 5 – TM	1990
	Landsat 7 – ETM+	2000
Rio Maule-Cobquecura	Landsat 1 – MSS	1975
	Landsat 5 – TM	1990
	Landsat 7 – ETM+	2000
Los Muermos-Ancud	Landsat 1 – MSS	1976
	Landsat 5 – TM	1985
	Landsat 7 – ETM+	1999

MSS, Multispectral Scanner; TM, Thematic Mapper; ETM+, Enhanced Thematic Mapper.

grouped into the 'other' category. In the Highlands of Chiapas, three land-cover classes were defined: (i) native forest, including pine, pine–oak, oak and montane cloud forests; (ii) shade coffee plantations; and (iii) non-forest cover, which corresponded to agriculture fields, pasture lands, recent fallows, cleared areas, bare ground and urban areas (Cayuela *et al.*, 2006b). In the Chilean study areas, the following main categories of land-cover types were obtained: crops and pasture lands, shrublands, arboreous shrublands, native forests, and other land-cover types. In Rio Maule-Cobquecura, native forests corresponded basically to secondary forests, whereas in Los Muermos-Ancud this category included secondary and old-growth forests.

Analyses of forest loss and landscape spatial pattern

The resulting categories of land-cover types were used to analyse forest cover change over time using GIS software. Forest maps for each study year were generated to quantify forest loss and the spatial configuration of native forest fragments. The formula used to determine the annual rate of deforestation was (FAO, 1995):

$$P=\left[\left(\frac{A_2}{A_1}\right)^{1/(t_2-t_1)}-1\right]*100$$

where P is the percentage loss per year, A_1 and A_2 are the forest area at time t_1 and t_2 respectively.

Next, landscape spatial indices were computed using FRAGSTATS (version 3.3) (McGarigal *et al.*, 2002). The following indices were calculated: (i) mean patch size (ha); (ii) patch density (number of patches per 100 ha); (iii) the largest patch index (percentage of area accounted for by the largest patch); (iv) the total edge length (km); (v) total core area (interior area remaining after removing an edge depth of 100 m, in hectares); and (vi) mean proximity index (ratio between the size and proximity of all patches whose edges are within a 1-km search radius of the focal patch). In Central Veracruz, the forest loss analysis was conducted using only undisturbed forest and excluding disturbed forest or old-fields. In the other study areas, forest loss was determined using a unique category of native forest.

Determination of driving forces of deforestation

The question of which environmental and social factors ('drivers') affected the probability of a particular location being deforested was investigated by logical regression analyses. Cover maps from consecutive images (e.g. 1976 and 1985 in Los Muermos-Ancud) were overlain in a GIS, and each pixel of the image was classified as either forested (i.e. forest in both years) or deforested (i.e. forested in the first year and some other cover type in the second year). A random subset of 1000 forested-plus-deforested pixels was then selected from each study area, the pixels being chosen so that the distance between them was at least 1500 m, in order to reduce the degree of spatial autocorrelation within the data. Models were fitted using logistic regression (Crawley, 2005), with a binary response variable ('0' for forested pixels, '1' for deforested pixels), a logit link function and a linear combination of explanatory variables. All the explanatory and response variables used in the analysis were based on pixel sizes of 30 m × 30 m.

Many explanatory variables were available for each site (see below). In the first round of modelling, we tested whether explanatory variables affected deforestation probability by fitting a series of univariate models, and testing the statistical significance of each variable using a χ^2 test. Next, all variables deemed to be significant by this approach ($P<0.05$) were entered into a multivariate analysis, the purpose of which was to test whether the list of significant factors could be reduced because of covariance among variables. A backward selection method was used to test whether the change in deviance associated with dropping terms out of the model was statistically significant (χ^2 test), until a 'minimum adequate model' was produced in which all terms were significant at $P<0.001$.

In Central Veracruz, a total of 15 continuous variables, derived from raster maps (100 m^2 pixels), were included as explanatory variables in the logistic regression analysis. These factors included elevation (m), slope (°), soil fertility (range 1–4 with 0.5 increments), mean annual precipitation (mm), latitude (0.5° increments), distance to roads, rivers and agricultural fields (m), as well as road, river and agriculture density (% pixels) in a 2 km radius, distance to national parks (m), distance to initial forest edge (m), population density (people km^{-2}) and population growth rates (% change). Precipitation, road, river, elevation and slope data were provided by Mexico's National Water Commission (CNA; scale 1:250,000), with the latter two variables based on a digital elevation model generated from topographic maps (1:50,000) with 50 m elevation increments. Soil coverage obtained from the National Commission for the Knowledge and Use of Biodiversity (CONABIO; modified from INIFAP, 1995) was assigned to soil fertility values by a geomorphologist who is familiar with the region (D. Geissert, Instituto de Ecología, Xalapa, Veracruz, Mexico, 2005, personal communication). A map of national parks was also obtained from the CONABIO geographic database (CONANP, 2003). Finally, population data were obtained from the 1995 and 2000 National Censuses (INEGI, 1995, 2000). In the Highlands of Chiapas, elevation (m) and slope (°) were extracted from a 1:50,000 digital elevation model. An index of soil fertility/quality following González-Espinosa *et al.* (2004) was generated based on the interpretation of physical and chemical properties of soil taxa described by FAO–UNESCO maps (Duchaufour, 1987). Road density (m km^{-2}) was calculated using a 1:50,000 digitized road map, giving different weights to paved and unpaved roads. Human population density was obtained by dividing the study area according to the location of human settlements into a meaningful tessellation of Thiessen polygons. Population density was then calculated by dividing total population in each settlement (INEGI, 2000) by the area of its corresponding polygon.

In the Chilean study areas, grid maps for slope (°), elevation (m), distance to roads (km), distance to rivers (km) and distance to urban areas (km) were generated using data set at a scale of 1:50,000 of the National Vegetation Mapping (CONAF *et al.*, 1999). Soil types were acquired at a scale of 1:250,000 (Schlatter *et al.*, 1995).

Road, urban and human population variables were used as surrogate measures of human pressure for all study areas, except for Rio Maule-Cobquecura and Los Muermos-Ancud, where population density is relatively low. Maps of patch size, non-forest density and distance to patch edge were calculated using the forest cover maps derived for each study area. Soil variables were available as categorical variables and the remainder as continuous variables. These data sets are the most comprehensive presently available for the study areas. It is assumed that the contribution of socio-economic and environmental variables to deforestation have operated from 1975–1976 to 1999–2000, and will continue to operate over the next few decades.

Results

Forest loss and land-cover change

Central Veracruz

From 1984 to 2000, the percentage of the landscape represented by crop and pasture land declined from 59.6% to 50.2% (Table 2.2, Fig. 2.2). The total forest area increased from 273,251 ha to 329,908 ha, representing 38.1% and 46% of the landscape, respectively. However, the loss of undisturbed native forest was equivalent to a deforestation rate of 2.0% per year. Conversely, the area of old-fields and disturbed secondary forest increased from 8.7% to 24.2%. Urban areas, included in other categories of land-cover types, showed a substantial increase from 2.3% of the landscape to 3.8%.

The Highlands of Chiapas

Crop and pasture lands covered 18% of the total study area in 1975 (Table 2.2, Fig. 2.3), increasing substantially to 39% in 1990 and to 57% in 2000. An opposite trend was observed for coffee plantations: they declined from 8% in 1975 to 4% in 2000. In 1975, the native forests covered approximately two-thirds of the area of the study landscape. Twenty-five years later, this figure declined to less than one-third. This is equivalent to a total forest loss of 57% between 1975 and 2000, and to a deforestation rate of 3.4% per year for the entire study period. However, there were differences between time intervals. Between 1975 and 1990 the forest loss rate was 1.5% per year, whereas between 1990 and 2000 this rate increased considerably to 6.1% per year.

Rio Maule-Cobquecura

Crop and pasture lands slightly increased during the entire study period, from 18% to 22% (Table 2.2, Fig. 2.4). The shrublands and arboreous shrublands comprised 54% of the landscape in 1975; 25 years later, these land-cover types

Table 2.2. Estimates of area of land-cover types, in hectares and percentage of total classified area, in the four study areas. The land-cover type named as 'other categories' includes urban areas, bare ground and other types of natural vegetation.

Central Veracruz, Mexico

	1984		2000	
Land-cover type	ha	%	ha	%
Crop and pasture land	426,877	59.6	359,631	50.2
Native forest*	273,251	38.1	329,908	46.0
Urban areas and other categories	16,493	2.3	27,081	3.8
Total	716,621	100.0	716,620	100.0

*Includes old-fields (disturbed secondary forests).

The Highlands of Chiapas, Mexico

	1975		1990		2000	
Land-cover type	ha	%	ha	%	ha	%
Crop and pasture land	61,346	17.5	134,579	38.8	193,915	56.8
Coffee plantation	27,689	7.9	19,036	5.5	15,010	4.4
Native forest	231,605	66.0	183,501	52.9	98,339	28.8
Other categories	30,044	8.6	9,811	2.8	34,006	10.0
Total	350,684	100.0	346,927	100.0	341,270	100.0

Rio Maule-Cobquecura, Chile

	1975		1990		2000	
Land-cover type	ha	%	ha	%	ha	%
Crop and pasture land	105,701	18.3	78,482	13.6	124,819	21.6
Shrubland	193,532	33.5	260,607	45.1	104,151	18.0
Arboreous shrubland	112,818	19.5	79,643	13.8	93,261	16.1
Native forest	119,994	20.8	56,133	9.7	39,002	6.7
Exotic species plantation	29,579	5.1	96,777	16.7	211,686	36.6
Other categories	16,541	2.9	6,522	1.1	4,800	0.8
Total	578,164	100.0	578,164	100.0	578,164	100.0

Los Muermos-Ancud, Chile

	1976		1985		1999	
Land-cover type	ha	%	ha	%	ha	%
Crop and pasture lands	46,643	9.3	120,008	23.8	129,008	25.6
Shrubland	101,902	20.2	53,270	10.6	34,642	6.9
Arboreous shrubland	78,349	15.6	66,697	13.3	95,113	18.9
Native forest	266,852	53.0	230,410	45.8	206,736	41.1
Other categories	9,541	1.9	32,902	6.5	37,788	7.5
Total	503,287	100.0	503,287	100.0	503,287	100.0

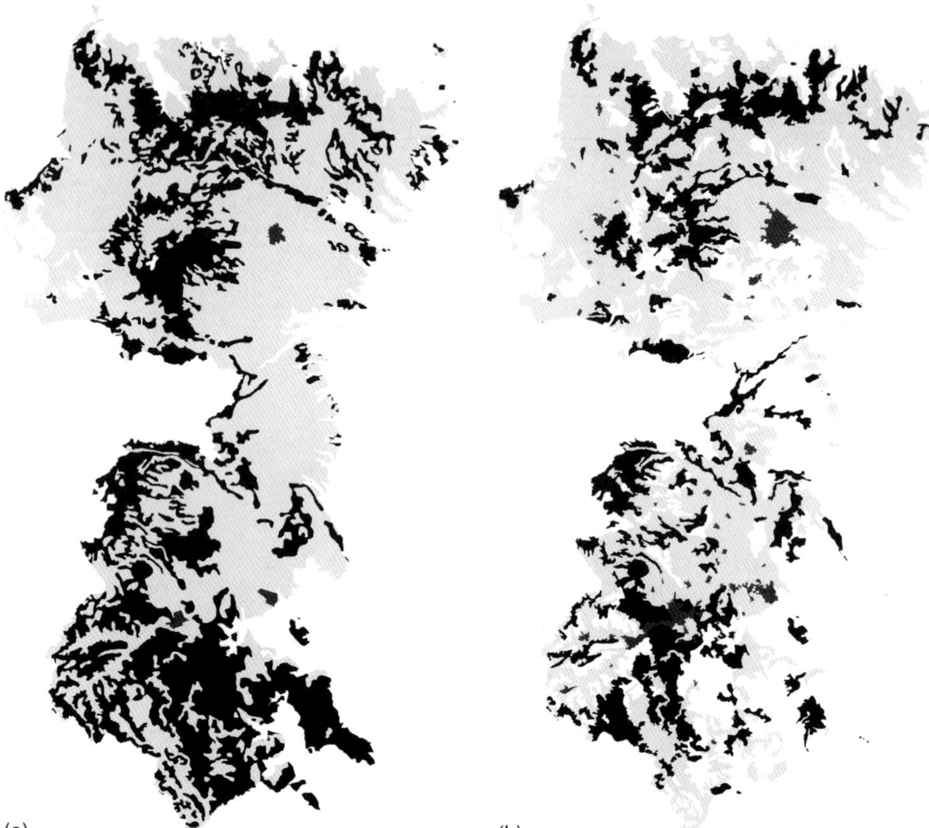

(a) (b)

Fig. 2.2. Major land-cover types in Central Veracruz for the years (a) 1984 and (b) 2000. Light grey, crop and pasture land; black, native forest; medium grey, urban areas.

comprised 34% of the total area. The exotic-species plantations increased from 5% in 1975 to 17% in 1990; by 2000 this land-cover type was the dominant vegetation type on the map, comprising 37% of the land area. During the whole study period, the estimated cover of native forests decreased from 119,994 ha in 1975 to 39,002 ha in 2000. In other words, 67% of the native forest existing in 1975 had disappeared by 2000, which was equivalent to an annual deforestation rate of 4.4% per year. Most of the forest loss was concentrated in the first 15 years of the study period, at a deforestation rate of 5.0% per year. Between 1990 and 2000, the rate decreased slightly to approximately 3.6% per year. Throughout the study period, more than half (53%) of the native forests existing in 1975 had gradually been converted into exotic-species plantations by 2000; another substantial area (40% of native forest in 1975) was transformed into shrublands or arboreous shrublands.

Los Muermos-Ancud

Crop and pasture lands substantially increased during the first time interval, from 9% to 24% (Table 2.2, Fig. 2.5). During the second period of analy-

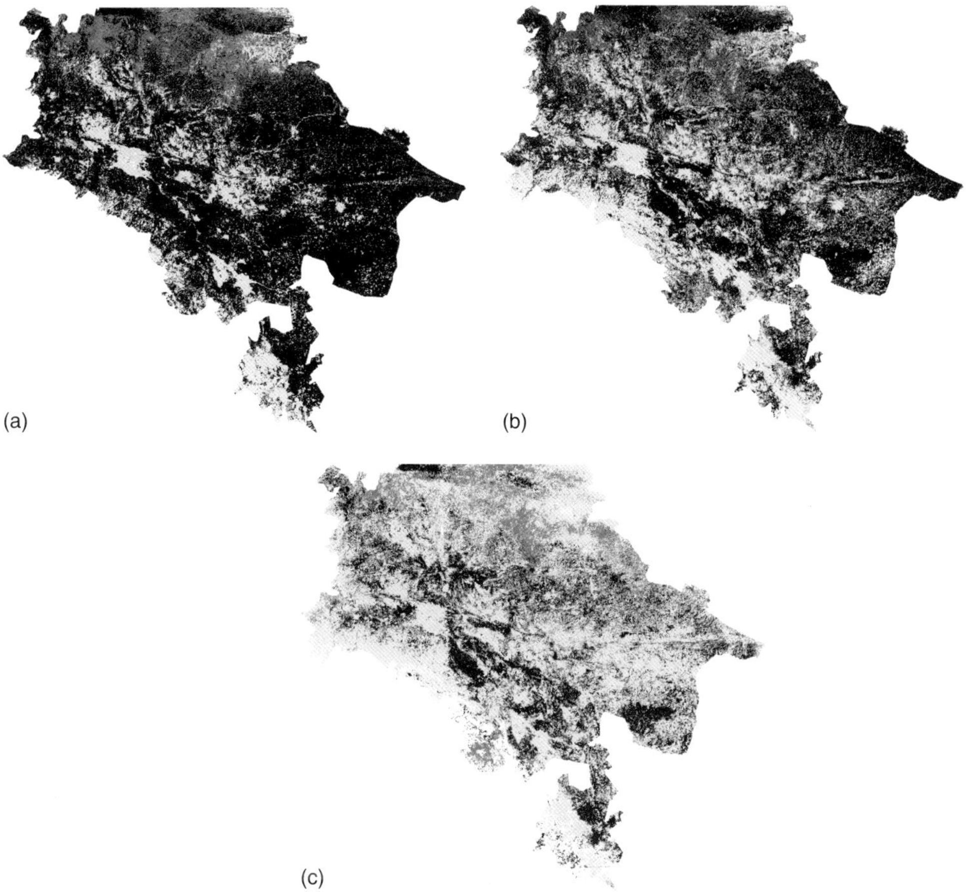

(a) (b) (c)

Fig. 2.3. Major land-cover types in the Highlands of Chiapas for the years (a) 1975, (b) 1990 and (c) 2000. Light grey, crop and pasture; black, native forest; medium grey, coffee plantation.

sis, this cover was relatively stable, representing about 26% of the study area. Between 1976 and 1985, the total area of shrublands and arboreous shrublands decreased from 36% to 24%. This decrease was followed by a subsequent increase to 27% in 1999. The estimated area of native forests decreased from 266,852 ha in 1976 to 206,736 ha in 1999, equivalent to 53% and 41% of the total classified area respectively. This means that approximately 23% of the native forests in 1976 had disappeared by 1999, at an annual forest loss of 1.1% per year. Most of the forest loss was concentrated in the first 9 years of the study period, at a deforestation rate of 1.6% per year. In the second time interval, this rate decreased to 0.8% per year. During the time intervals, 29% of the native forests were replaced by arboreous shrublands and 8% by shrublands. The loss of native forests has been associated with an increasing proportion of arboreous shrublands, and also to an increase in the area of crop and pasture lands.

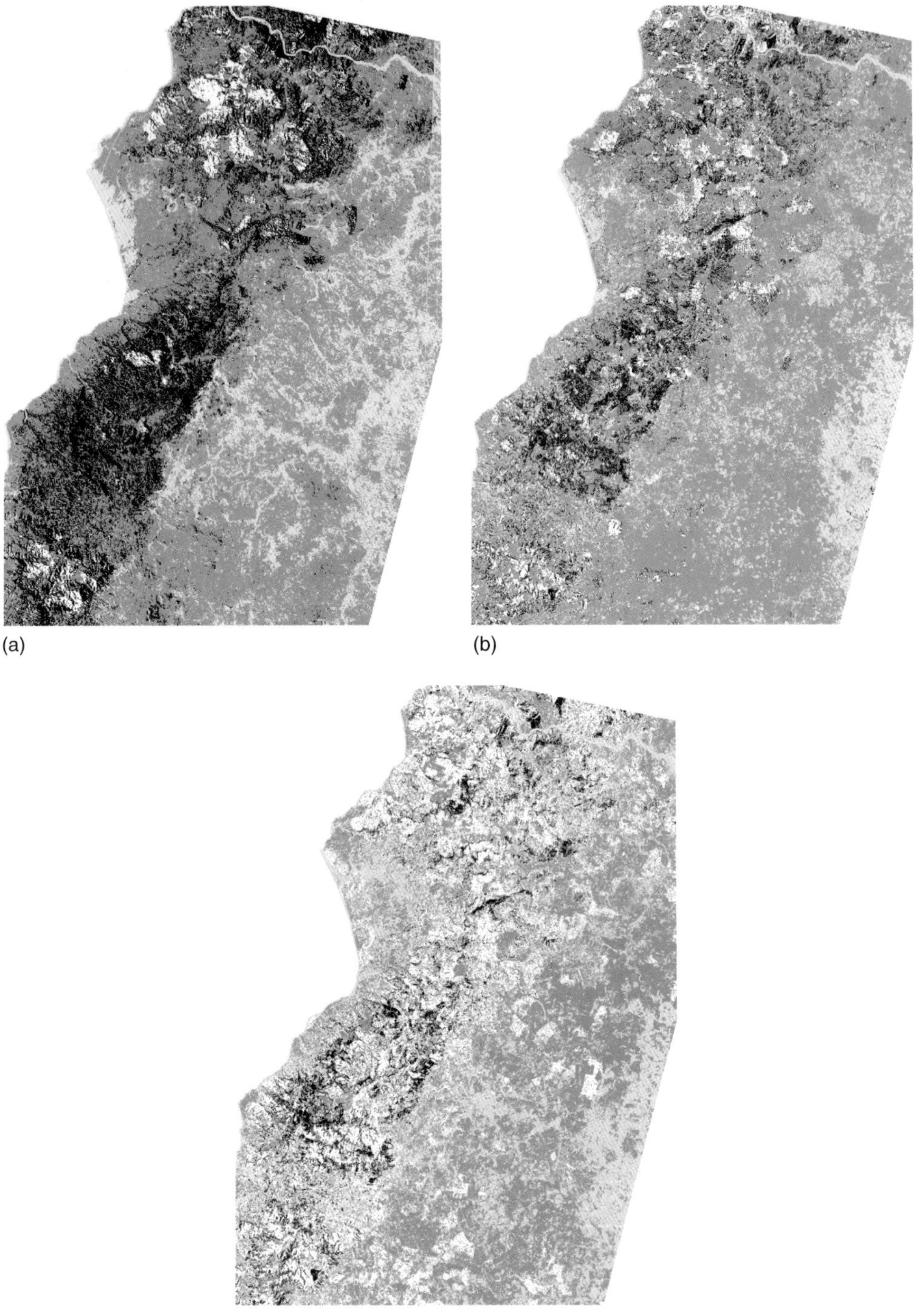

(a) (b) (c)

Fig. 2.4. Major land-cover types in Rio Maule-Cobquecura for the years (a) 1975, (b) 1990 and (c) 2000. Light grey, crop and pasture land; medium grey, shrubland and arboreous shrubland; black, native forest; white, exotic species plantation.

Fig. 2.5. Major land-cover types in Los Muermos-Ancud for the years (a) 1976, (b) 1985 and (c) 1999. Light grey, crop and pasture land; medium grey, shrubland and arboreous shrubland; black, native forest.

Trends in forest fragmentation

The analyses of landscape indices revealed that the loss of native forests of the four study areas has been associated with substantial forest fragmentation. Spatial patterns of forest loss and fragmentation of each study area are reported below.

Central Veracruz, Mexico
Mean size of forest fragments increased slightly from 1176 ha in 1984 to 1291 ha in 2000 (Table 2.3). However, the size of the largest fragment decreased by more than a half during the study period from 75,105 ha to 30,980 ha. Patch density did not show a substantial variation over time. The largest patch index decreased from 5.5% in 1984 to 2.2% in 2000. Similarly, the total edge length and total core area presented a decline during the study period as a result of the fragmentation of remnant forest. These changes in the spatial configuration of the landscape were associated with a decline in mean proximity of more than 60%, as result of the division of forest fragments.

The Highlands of Chiapas, Mexico
During the first time interval, the mean patch size decreased by approximately 58%. Between 1990 and 2000, this index exhibited a greater decline of 67%. Patch density increased overall through time, reaching its maximum value of 3.2 patches per 100 ha in 2000. This pattern was associated with a rapid decline in the largest patch index, from 60.7% in 1975 to 35.1% in 1990, and to 4% in 2000. From 1975 to 1990, the total edge length increased by approximately 100%, as a result of the progressive fragmentation. However, for the second time interval, this index experienced a decline of 23%, owing to the loss of forest fragments. Approximately 70% of the total core area recorded in 1975 had disappeared by 1990. Similarly, a high percentage of decline in total core area (67%) was observed between 1990 and 2000. The process of fragmentation was also accompanied by an increase in fragment isolation, indicated by a rapid decline in the mean proximity over time. Between 1975 and 2000, this index decreased by approximately 98%.

Rio Maule-Cobquecura, Chile
During the first study period, the native forests were mainly affected by severe fragmentation (increasing number of patches) and deforestation (decreasing mean patch size). For the second time interval, deforestation became the dominant process, owing to a decline in both mean patch size and patch density. This trend was associated with a reduction in the size of the largest forest patch, ranging from 7% of the total area in 1975 to 0.2% in 2000. The landscape was also characterized by the presence of more patch edges, which indicates that the shape of native forest patches had become more irregular during the first time interval. However, between 1990 and 2000 the total edge length in the landscape declined as a result of the loss of forest fragments. The native forest fragments showed a substantial decrease in the total amount of core area and in the mean proximity over time. Between 1975 and 1990, the total core area decreased from 21,138 ha to 918 ha, and then to 839 ha in 2000. Similarly, the main change in the mean proximity was recorded in the first time interval. During this period, the neighbourhood of forest patches rapidly became occupied by areas of a different land-cover type, as native forest patches became spatially separated and less contiguous in distribution.

Table 2.3. Changes in landscape pattern indices of the native forests in each study area. Numbers in parentheses correspond to minimum and maximum values.

Central Veracruz, Mexico

Landscape indices	1984	2000
Mean patch size (ha)	1,176 (1–75,105)	1,291 (1–30,980)
Patch density (per 100 ha)	0.013	0.009
Largest patch index (%)	5.5	2.2
Total edge length (km)	5,276	4,222
Total core area (ha)	177,031	131,020
Mean proximity	860 (0.0–37,555)	326 (0.0– 8,951)

The Highlands of Chiapas, Mexico

Landscape indices	1975	1990	2000
Mean patch size (ha)	65.0 (0.5–211,180)	26.9 (0.5–119,516)	8.67 (0.5–13,279)
Patch density (per 100 ha)	1.0	1.9	3.2
Largest patch index (%)	60.7	35.1	4.0
Total edge length (km)	24,781	50,114	38,400
Total core area (ha)	99,422	29,860	9,611
Mean proximity	101,369 (0.02–587,150)	60,017 (0.0 –342,240)	1,405 (0.0 –34,466)

Rio Maule-Cobquecura, Chile

Landscape indices	1975	1990	2000
Mean patch size (ha)	17 (0.5–52,178)	5 (0.5–9,842)	4 (0.5–1,182)
Patch density (per 100 ha)	0.93	1.65	1.36
Largest patch index (%)	6.91	1.30	0.16
Total edge length (km)	20,330	22,337	15,799
Total core area (ha)	21,138	918	839
Mean proximity	5,880 (0.0–145,119.4)	612 (0.0–29,276.4)	73 (0.0–6,031.6)

Los Muermos-Ancud, Chile

Landscape indices	1976	1985	1999
Mean patch size (ha)	47 (0.5–132,971)	24 (0.5–49,767)	19 (0.5–42,785)
Patch density (per 100 ha)	0.36	0.60	0.65
Largest patch index (%)	8.31	3.11	2.67
Total edge length (km)	21,403	30,931	31,072
Total core area (ha)	143,428	89,007	69,900
Mean proximity	19,350 (0.0–369,603.5)	4,380 (0.0–152,583.1)	2,552 (0.0–120,135)

Los Muermos-Ancud, Chile

The mean size of forest patches decreased gradually from 47ha in 1976 to 24ha in 1985 to 19ha in 1999 (Table 2.3). This decline in the patch size was associated with a continuous increase in the patch density over time, reaching its maximum value of 0.65 fragments per 100ha in 1999. This pattern was accompanied by a reduction of the largest forest patch, from 8% in 1976 to 3% of the total area in 1999. This modification of the landscape was also characterized by the presence of more forest patch edges, which increased during the first study period. In the second time interval, the total edge length showed a slight increase. The total core area also showed a gradual decline across the time intervals, by 1999 decreasing by more than 50% of the core area recorded in 1976. The main change in fragment isolation occurred from 1976 to 1985, when mean proximity decreased to almost one-fifth of its initial value. Between 1985 and 1999, the mean proximity presented a further decline.

Drivers of deforestation

Multiple logistic regression models indicated which explanatory variables were significantly related to the probability of deforestation in each study area (Table 2.4). In Central Veracruz the logistic regression model revealed that the probability of an area being cleared of forest for the 1984–2000 interval was highly significant and negatively related to slope and distance to patch edges (Table 2.4). For the same study period, soil fertility was highly positively related to deforested areas. Distance to towns and population density in 1995 were negatively associated with the probability of deforestation, whereas distance to national parks, mean annual rainfall and distance to agricultural areas appeared to be positively related.

In the Highlands of Chiapas the logistic regression model showed that slope and distance to patch edges were significantly associated with deforested areas for the 1975–1990 interval. Conversely, density of non-forest areas and patch size appeared to be positively related to the probability of deforestation. For the 1990–2000 interval, elevation, distance to patch edge and slope were statistically negatively related to deforested areas. Conversely, patch size was positively associated with the probability of deforestation.

In Rio Maule-Cobquecura, the probability of an area being cleared of forest for the 1975–1990 interval was negatively related to distance to patch edge and patch size. For the 1990–2000 interval distance to patch edge and slope were negatively associated with deforested areas. In Los Muermos-Ancud, slope and distance to patch edge were statistically negatively related to the loss of native forests during the first time interval. In the second time interval, distance to patch edge, patch size and slope appeared to be highly negatively associated with the clearance of forested areas. The regression model also revealed that distance to rivers was positively related to the probability of deforestation.

Table 2.4. List of coefficients for variables significantly affecting the probability of deforestation in Mexico and Chile. Results were obtained from multivariate logistic regression modelling. The model is logit (*P*) = *X*, where *P* is the probability of deforestation and *X* is a linear combination of explanatory variables.

Variable	Order	Coefficients	Std Error	χ^2	*P*-value
Central Veracruz, Mexico					
Intercept	1	−0.8771	0.59770	2.1	n.s.
Slope	2	−0.0628	0.00943	44.3	***
Distance to national parks	3	0.0001	0.00001	5.2	*
Mean annual rainfall	5	0.0007	0.00023	9.9	**
Population density in 1995	6	−0.0015	0.00051	8.9	**
Soil fertility	7	0.5664	0.16270	12.1	***
Distance to towns	8	−0.0003	0.00011	7.0	**
Distance to agricultural areas	9	0.0002	0.00009	5.3	*
Distance to patch edge	4	−0.0010	0.00024	17.8	***
Estimated by stepwise logistic regression procedure of SAS. Order of entry into the model is provided. Df = 1.					
The Highlands of Chiapas, Mexico					
Period 1: 1975–1990					
Intercept		$-8.045\ 10^{-1}$	$3.848\ 10^{-1}$	2.091	**
Slope		$-4.577\ 10^{-2}$	$7.770\ 10^{-3}$	5.891	***
Density of non-forest areas		2.112	$4.908\ 10^{-1}$	4.303	***
Distance to patch edge		$-4.869\ 10^{-3}$	$9.277\ 10^{-4}$	5.249	**
Patch size		$3.648\ 10^{-6}$	$1.318\ 10^{-6}$	2.769	***
Period 2: 1990–2000					
Null model					
Intercept		4.034	$5.479\ 10^{-1}$	7.363	***
Elevation		$-1.759\ 10^{-3}$	$2.311\ 10^{-4}$	7.613	***
Distance to patch edge		$-3.087\ 10^{-3}$	$5.740\ 10^{-4}$	5.377	***
Patch size		$4.104\ 10^{-6}$	$1.557\ 10^{-6}$	2.635	***
Slope		$-1.839\ 10^{-2}$	$7.243\ 10^{-3}$	2.540	**
N = 1242 points (1975–1990) and 992 points (1990–2000). Df = 1.					
Rio Maule-Cobquecura, Chile					
Period 1: 1975–1990					
Intercept		1.424	$1.060\ 10^{-1}$	13.436	***
Distance to patch edge		$-4.891\ 10^{-3}$	$1.121\ 10^{-3}$	−4.365	***
Patch size		$-9.009\ 10^{-6}$	$3.186\ 10^{-6}$	−2.827	**
Period 2: 1990–2000					
Null model					
Distance to patch edge		−0.017	0.004	−3.711	***
Slope		−0.032	0.008	−3.912	***
N = 1489 points (1975–1990) and 622 (1990–2000). Df = 1.					
Los Muermos-Ancud, Chile					
Period 1: 1976–1985					
Intercept		0.065	0.113	0.573	n.s.
Slope		−0.049	0.014	−3.371	***
Distance to patch edge		−0.006	0.001	−8.012	***

Continued

Table 2.4. *Continued*

Variable	Order	Coefficients	Std Error	χ^2	*P*-value
Period 2: 1985–1999					
Intercept		$1.102\ 10^{-1}$	$1.088\ 10^{-1}$	1.013	n.s.
Distance to rivers		$2.760\ 10^{-4}$	$1.152\ 10^{-4}$	2.395	*
Distance to patch edge		$-3.341\ 10^{-3}$	$8.134\ 10^{-4}$	−4.108	***
Patch size		$-1.762\ 10^{-5}$	$4.000\ 10^{-6}$	−4.406	***
Slope		$-5.716\ 10^{-2}$	$1.494\ 10^{-2}$	−3.826	***

N = 1000 points in both periods. Df = 1.

* $P < 0.05$; ** $P < 0.01$; *** $P < 0.001$; n.s., not significant.

Discussion

Forest loss and land-cover change in Mexico and Chile

The native forests of the four study areas have undergone relatively high rates of forest loss during the decades analysed, compared to many other forested landscapes in the world (Spies *et al.*, 1994; Zheng *et al.*, 1997; Cushman and Wallin, 2000; Cohen *et al.*, 2002; Staus *et al.*, 2002). These forests have been reduced severely and degraded over time owing to logging for timber and fuelwood, and clearance for cultivation. Of all the study areas, Rio Maule-Cobquecura had the highest rate of deforestation in the last three decades (4.5% per year), followed by the Highlands of Chiapas (3.4% per year). Central Veracruz and Los Muermos-Ancud presented lower rates of 2% and 1.1% per year respectively. By analysing the rate of forest loss by time intervals, in the 1990–2000 interval the Highlands of Chiapas had the highest rate of deforestation (6.2% per year). In the two Chilean study areas, the highest rates of forest loss were recorded during the first time interval. As for the total reduction of native forests, Rio Maule-Cobquecura and the Highlands of Chiapas showed the highest losses of 67% and 57% in the last three decades, respectively. Central Veracruz and Los Muermos-Ancud exhibited lower rates of 26% for the 1984–2000 interval and 23% for the 1976–1999 interval, respectively.

Forest losses recorded in other studies, assuming that they were calculated using the FAO formula, have generally been lower. A forest loss rate of 0.5% per year was estimated for the Klamath-Siskiyou ecoregion, USA, and an overall (cumulative) reduction of forest cover by 10.5% was recorded over the period 1972–1992 (Staus *et al.*, 2002). In western Oregon, deforestation rates by clearcutting between 1972 and 1995 varied from 0.5% to 1.2% per year with almost 20% of the total forest impacted (Cohen *et al.*, 2002). Similarly, in other areas of western Oregon, between 1972 and 1988 the rate of deforestation, primarily by clearcutting, was 1.2% per year of the entire study area including the wilderness area (Spies *et al.*, 1994). A rate of 0.6% per year, slightly lower than that determined for the period 1990–2000 in Los Muermos-Ancud, was found for the 1986–1996 interval in the Napo region of western Amazonia (Sierra, 2000). A rate of 6% per year was determined for lowland deciduous forest in eastern Santa Cruz, Bolivia in the middle

1990s (Steininger *et al.*, 2001), thought to be one of the highest deforestation rates reported anywhere in the world. However, the present study reported a slightly higher rate for the 1990–2000 interval in the Highlands of Chiapas.

All of the study landscapes were affected by substantial changes in the area of the different land-cover types over time. The greatest losses of native forests in the study areas were associated with the conversion to human-induced land-cover types. In the Highlands of Chiapas and Los Muermos-Ancud the loss of native forests has been related to an increase in the area of crop and pasture lands. Conversely, in Central Veracruz there was an expansion of forest cover owing to the abandonment of crops. The loss of native forests was mainly determined by the degradation of pristine forest into old-fields (including disturbed secondary forests with herbaceous and shrub cover), although transformation for crops and cattle ranching also played an important but lesser role. In Rio Maule-Cobquecura, a substantial area of native forests was converted to exotic-species plantations such as *Pinus radiata* and *Eucalyptus* spp.

Spatial patterns of forest loss and fragmentation in Mexico and Chile

Landscape pattern indices provide a useful tool to explore cross-site differences and changes over time. The simultaneous use of class-level and patch-level landscape pattern indices enabled assessment of the spatial configuration of forest cover and its relation to principal land-cover types. It is important to highlight that, owing to the different spatial scale of the data on which these analyses were performed (1:250,000 scale) for Veracruz, the analysis of fragmentation generated some differences compared to the other study areas.

Forest fragmentation has three recognizable components at the landscape level: (i) habitat loss; (ii) reduction of patch size; and (iii) increased isolation of habitats (Bennett, 2003). These three components were shown to occur over the last decades in the four study areas analysed in Mexico and Chile. In particular, the mean size of forest fragments declined consistently over time, except in Central Veracruz, which displayed a similar pattern to the situation recorded in Wisconsin (Pan *et al.*, 1999), where the size of fragments increased due to abandonment of agricultural land. Over the last three decades, the greatest reduction in the mean patch size was recorded in the Highlands of Chiapas, followed by Rio Maule-Cobquecura. However, Rio Maule-Cobquecura reached the smallest size of forest fragments in the last study interval. These results support the statement made by Armenteras *et al.* (2003) that progressive reduction in the size of forest habitats is a key component of ecosystem fragmentation.

Patch density increased gradually in the Highlands of Chiapas and Los Muermos-Ancud, except in Central Veracruz. In Rio Maule-Cobquecura patch density reached its maximum value in 1990 and then decreased by 2000. A similar trend was observed in the total edge length in the Highlands of Chiapas and Rio Maule-Cobquecura, which increased until 1990 and then declined by 2000. This pattern reflects an increase of patch density and edge length in the earliest stages of forest loss and fragmentation and a decline during the later stages of deforestation. Zipperer *et al.* (1990) also observed that the constant action of deforestation led to a decline in patch density in central New York, USA. In Rio Maule-Cobquecura, this process even eliminated forest patches created during the first study period. Similarly to the findings of Ranta *et al.* (1998), the substantial increase of patch density in Rio Maule-Cobquecura was related to the concentration of the forest area in patches less than 100 ha in area.

As recorded elsewhere (Fitzsimmons, 2003), the greatest absolute decline in the largest forest patch size in the Highlands of Chiapas and Los Muermos-Ancud coincided with the time period where the greatest absolute decline in annual forest loss was observed. In Rio Maule-Cobquecura, a slightly higher decline was observed in the time interval that did not present the highest rate of deforestation. The decline of large forest fragments might have a significant effect on the response of some species in the study area. For instance, the size of the largest cloud forest fragments was the most important characteristic influencing the response of bird species in eastern Mexico (Martínez-Morales, 2005). Similarly, higher bird species richness of resident and migrant species occurred in larger forest fragments in Singapore Island (Castelletta *et al.*, 2005).

Interior forest habitat decreased progressively over time in all of the study areas. Also, forest fragments became more isolated as other land-cover types occupied the deforested areas in the study landscapes. Rio Maule-Cobquecura and the Highlands of Chiapas were characterized by substantial reductions in the total core area (96% and 90%, respectively) over the last three decades, while Los Muermos-Ancud and Central Veracruz presented lower reductions (51% and 26%, respectively). Reductions in the mean proximity over the study periods were also higher in Rio Maule-Cobquecura and the Highlands of Chiapas, with 98.7% and 98.6%, respectively. In Los Muermos-Ancud and Central Veracruz, this index declined to a lower percentage of 86.8% and 62%, respectively.

The analysis of spatial patterns of landscape indices needs to be understood as a first step to comprehend ecological processes, and not as an end itself (Li and Wu, 2004). Although these indices allow forest fragmentation to be assessed at the landscape level, it is necessary to explore the relationships between pattern and process. A variety of studies that relate spatial patterns to ecological processes have demonstrated that forest fragmentation may lead to a change in the abundance and richness of some woody (Metzger, 1997, 2000) and bird species (Willson *et al.*, 1994; Cornelius *et al.*, 2000; Drinnan, 2005; Martínez-Morales, 2005; Uezu *et al.*, 2005). Therefore, the loss of forest habitats and the increasing trend of fragmentation over forthcoming decades in the study areas may have negative consequences on the flora and fauna existing in the remnant forests, due to changes in composition of assemblages and changes in ecological processes (Forman and Godron, 1986; Bennett, 2003) (see Chapter 3).

Causal factors of deforestation

The driving factors of deforestation identified by the spatially explicit models are all variables that express geophysical attributes or the 'symptoms' of the underlying causes of forest loss. These factors show how the forest loss has taken place spatially and temporally in the landscapes, but they do not necessarily show the underlying causes. In fact, these factors are the result of cultural and socio-economic processes that have been modifying the landscape over many decades. Bürgi *et al.* (2004) define these factors as 'attractors of change', as they are the primary driving forces likely to induce change at a local scale.

According to the models generated for the different time intervals, deforestation was essentially concentrated in gently sloping areas as a result of the expansion of crop and pasture lands in Mexico and Chile, except for the first time interval in Rio Maule-Cobquecura. In this landscape the process of deforestation was not significantly related to slope owing to the conversion of native forests to exotic plantations in sites of different degrees of slope between 1975 and 1990. Similarly to results obtained for the second time interval in Rio Maule-Cobquecura and for all time intervals in Los Muermos-Ancud and the two study areas in Mexico, Wilson *et al.* (2005) found that slope is a highly significant variable for explaining the probability of deforestation. In particular, these authors found that forested flat areas near towns and roads were highly vulnerable to the conversion of native forest to industrial plantations of exotic species. In contrast to that study, the logistic regression of the present work revealed that distance to towns and distance to roads were not significant in accounting for clearance of forest area, except in Central Veracruz, where the distance to towns was a significant factor influencing deforestation.

Results also revealed that the clearance of forests was concentrated around edges of forest fragments in all of the study areas. A logistic model-based study conducted in Madagascar similarly found that the expansion of agriculture into the remaining natural forest was associated with progressive clearance from forest edges (McConnell *et al.*, 2004).

In most of the study areas small patches became vulnerable to deforestation, owing to the fact that they were mainly concentrated in flat areas, where the process of deforestation was more intense. In particular, the severe fragmentation reported for both time intervals in the Highlands of Chiapas and for the first time interval in Rio Maule-Cobquecura led to an increase in the abundance of smaller patches that were subsequently eliminated by expansion of agriculture and plantations of exotic species respectively. In Los Muermos-Ancud the loss of forests between 1985 and 1999 was concentrated in small fragments located away from rivers or streams. The significance of this driver is related to the fact that clearance of forests is legally prohibited in areas close to rivers. This prohibition was more evident in flat areas where the forest patches were intensely fragmented and left as riparian vegetation. Similarly, the application of logistic regression to analyse the decline of native grassland in Melbourne, Australia revealed that patches close to streams were associated with a low probability of being destroyed (Williams *et al.*, 2005). In the Highlands of Chiapas the loss of native forest was located in lowlands owing to the presence of steep slopes in the highlands.

These cultural and socio-economic factors do not by themselves describe the immediate causes of forest cover change, but are related to various environment policies. In fact, some assessments indicate that neither population growth nor poverty alone constitutes the sole and major underlying cause of land-cover change worldwide (Lambin *et al.*, 2001). Rather, people's responses to economic opportunities, mediated by institutional factors, drive land-cover changes. For instance, population and income variables were found to be significant factors explaining forest area variation between 1970 and 1991 in 67 tropical countries (Uusivuori *et al.*, 2002). However, these results do not explain directly the causes of deforestation, as they need to be linked with international forest policies in order to understand the deforestation at the regional level. At a local level, people's response to institutional support has been documented in Bangladesh, where farmers abandoned extensive shifting cultivation, adapting suitable commercial land uses such as agroforestry, horticulture and forest plantations (Rasul *et al.*, 2004). Conversely, in Chile the context of a strong free market economy, dominated by economically powerful private domestic and international pulp and paper companies, has led to a market-friendly forest policy (Silva, 2004). As a result of this economy, many principles of sustainable development have been violated, causing a series of negative impacts on the environment (Lara and Veblen, 1993; Lara *et al.*, 2000).

Compared to Chile, Mexican industrial timber interests were relatively weak over the interval studied, and the forest peasant sector was much stronger and better organized, with the result that timber interests could not dominate forest communities as they could in Chile (Silva, 2004). The *ejido* land tenure system provided a platform for organizing political and economic activity that was not available in Chile. However, large-scale timber interests gained significant support in Mexico in the 1990s owing to the support of the presidency (Silva, 2004). The increase in the rate of deforestation during the 1990s in the Highlands of Chiapas reflects the effect of changes in social and economical policies in this country over the last decade. For instance, the lack of governance following the Zapatista rebellion in 1994, which allowed rampant illegal clearing for agriculture, livestock ranching and human settlement (González-Espinosa, 2005), did not help forest conservation. The contrasting cases of Chile and Mexico provide significant insight into the conditions needed for an improvement in national forest policies. Although the policy environment and socio-economic circumstances are very different in the two countries, the end result – high rates of deforestation and forest fragmentation – has been the same.

Future trends of deforestation

In the last three decades, substantial changes in the land-cover types and in the spatial configuration of native forests were recorded across the study areas. Expansion of human-induced land-cover types such as crops and pasture lands, forest plantations of exotic species and degraded secondary forests were associated with a considerable loss of native forest in each study area. With progressive forest loss and fragmentation, the native forests presented abrupt changes in their spatial configuration over the whole study period, from a forest habitat formed by complex clusters of large fragments to a sparse distribution of smaller patches.

Based on the current trends of deforestation, and if the primary driving forces of deforestation continue operating, we expect a continuous loss and fragmentation of native forests during forthcoming decades in Mexico and Chile. The forest area of patches corresponding to the smallest size class will tend to decline as a result of the clearance of small fragments that still exist in gently sloping sites. Furthermore, a decline in patch density might be observed in coming decades in Central Veracruz, the Highlands of Chiapas and Los Muermos-Ancud. This decline in the curve of patch density was recorded in the 1990s in Rio Maule-Cobquecura, reflecting the endpoint of the deforestation process: a landscape largely devoid of natural forest.

Conclusions

This study has succeeded in characterizing the major changes in forest configuration that have taken place over the past three decades in Mexico and Chile. The land-cover change analysis demonstrated that the landscapes are becoming increasing dominated by crops and pasture and by forest plantations of exotic species. Results also demonstrated that the patterns of deforestation have had a notable effect on the spatial configuration of the remaining forest fragments. As a result, the study landscapes have become dominated by isolated, small forest fragments. This pattern exposes how native forests are being disturbed spatially, which in turn illustrates the effects of socio-economic drivers of deforestation, such as forest logging and clearance for crops and pasture land. These causes are dependent on underlying social and economic policies, which in reality drive land-cover change. The assessment of these local causal relationships can potentially inform the development of improved land-use policies and management. However, the cases of Mexico and Chile provide evidence of similar patterns of forest loss and fragmentation in four different landscapes affected by human activities, despite contrasting policy environments and socio-economic characteristics.

References

Angelsen, A. and Kaimowitz, D. (1999) Rethinking the causes of deforestation: lessons from economic models. *The World Bank Research Observer* 14, 73–98.

Armenteras, D., Gast, F. and Villareal, H. (2003) Andean forest fragmentation and the representativeness of protected natural areas in the eastern Andes, Colombia. *Biological Conservation* 113, 245–256.

Bennett, A. (2003) *Linkages in the Landscape. The Role of Corridors and Connectivity in Wildlife Conservation*. IUCN, Gland, Switzerland and Cambridge, UK.

Breedlove, D. (1981) *Flora of Chiapas. Part I: Introduction to the Flora of Chiapas*. California Academy of Sciences, San Francisco, California.

Bürgi, M., Hersperger, A.M. and Schneeberger, N. (2004) Driving forces of landscape change – current and new directions. *Landscape Ecology* 19, 857–868.
Bustamante, R. and Castor, C. (1998) The decline of an endangered ecosystem: the ruil (*Nothofagus alessandrii*) forest in Central Chile. *Biodiversity and Conservation* 7, 1607–1626.
Bustamante, R. and Grez, A. (1995) Consecuencias ecológicas de la fragmentación de los bosques nativos. *Ambiente y Desarrollo* 11, 58–63.
Bustamante, R., Serey, I. and Pickett, S.T. (2003) Forest fragmentation, plant regeneration and invasion processes in Central Chile. In: Bradshaw, G. and Marquet, P. (eds) *How Landscapes Change: Human Disturbance and Ecosystem Fragmentation in the Americas*. Springer, Berlin/Heidelberg, Germany, pp. 145–160.
Castelletta, M., Thiollay, J. and Sodhi, N. (2005) The effects of extreme forest fragmentation on the bird community of Singapore Island. *Biological Conservation* 121, 135–155.
Cayuela, L., Golicher, D.J., Salas-Rey, J. and Rey-Benayas, J.M. (2006a) Classification of a complex landscape using Dempster–Shafer theory of evidence. *International Journal of Remote Sensing* 27, 1951–1971.
Cayuela, L., Rey-Benayas, J.M. and Echeverría, C. (2006b) Clearance and fragmentation of tropical montane forests in the highlands of Chiapas, Mexico (1975–2000). *Forest Ecology and Management* 226, 208–218.
Challenger, A. (1998) *Utilización y Conservación de los Ecosistemas Terrestres de México. Pasado, Presente y Futuro*. Comisión Nacional para el Conocimiento y Uso de la Biodiversidad (CONABIO), Instituto de Biología, Universidad Nacional Autónoma de México (UNAM), Mexico City, Mexico.
Chuvieco, E. (1996) *Fundamentos de Teledetección Espacial*, 3rd edn. Ediciones RIALP, S.A., Madrid, Spain.
Cohen, W., Spies, T., Alig, R., Oetter, D., Maiersperger, T. and Fiorella, M. (2002) Characterizing 23 years (1972–1995) of stand replacement disturbance in western Oregon forest with Landsat imagery. *Ecosystems* 5, 122–137.
Collier, G. (1975) *Fields of the Tzotzil: The Ecological Bases of Tradition in Highland Chiapas*. Texas Pan-American Series. University of Texas Press, Austin, Texas.
CONAF, CONAMA, BIRF, Universidad Austral de Chile, Pontificia Universidad Católica de Chile, and Universidad Católica de Temuco (1999) *Catastro y Evaluación de los Recursos Vegetacionales Nativos de Chile. Informe Nacional con Variables Ambientales*. Corporación Nacional Forestal (CONAF), Ministerio de Agricultura, Santiago, Chile.
CONANP (2003) *National Natural Protected Areas* (Escala 1:4 000 000), 4th edn. National Commission of Natural Protected Areas (CONANP), Secretaría de Medio Ambiente y Recursos Naturales (SEMARNAT), México, D.F., Mexico.
Cornelius, C., Cofre, H. and Marquet, P. (2000) Effects of habitat fragmentation on bird species in a relict temperate forest in semiarid Chile. *Conservation Biology* 14, 534–543.
Corney, P.M., Le Due, M.G., Smart, S.M., Kirby, K.J., Bunce, R.G.H. and Marrs, R.H. (2004) The effect of landscape-scale environmental drivers on the vegetation composition of British woodlands. *Biological Conservation* 120, 491–505.
Crawley, M. (2005) *Statistical Computing. An Introduction to Data Analysis Using S-Plus*. Wiley, Chichester, UK.
Cushman, S. and Wallin, D. (2000) Rates and patterns of landscape change in the Central Sikhote-alin Mountains, Russian Far East. *Landscape Ecology* 15, 643–659.
Dale, V.H. and Pearson, S.M. (1997) Quantifying habitat fragmentation due to land use change in Amazonia. In: Laurance, W. and Bierregaard, R. (eds) *Tropical Forest Remnants*. The University of Chicago Press, Chicago, Illinois, pp. 400–414.
Debinski, D. and Holt, R. (2000) A survey and overview of habitat fragmentation experiments. *Conservation Biology* 14, 342–355.

Di Castri, F. and Hajek, E. (1976) *Bioclimatología de Chile*. Ediciones Universidad Católica de Chile, Santiago, Chile.
Dinerstein, E., Olson, D., Graham, D., Webster, A., Primm, S., Bookbinder, M. and Ledec, G. (1995) *A Conservation Assessment of the Terrestrial Ecoregions of Latin America and the Caribbean*. WWF – World Bank, Washington, DC.
Donoso, C. and Lara, A. (1995) Utilización de los bosques nativos en Chile: pasado, presente y futuro. In: Armesto, J.J., Villagrán, C. and Arroyo, M.K. (eds) *Ecología de los Bosques Nativos de Chile*. Editorial Universitaria, Santiago, Chile, pp. 363–387.
Donoso, D., Grez, A. and Simonetti, J. (2003) Effects of forest fragmentation on the granivory of differently sized seeds. *Biological Conservation* 115, 63–70.
Drinnan, I. (2005) The search for fragmentation thresholds in a southern Sydney suburb. *Biological Conservation* 124, 339–349.
Duchaufour, P. (1987) *Manual de Edafología*. Masson, Barcelona, Spain.
Evans, T. and Moran, E. (2002) Spatial integration of social and biophysical factors related to landcover change. *Population and Development Review* 28, 165–186.
FAO (1995) *Forest Resources Assessment 1990. Global Synthesis*. FAO, Rome, Italy.
Fitzsimmons, M. (2003) Effects of deforestation and reforestation on landscape spatial structure in boreal Saskatchewan, Canada. *Forest Ecology and Management* 174, 577–592.
Forman, R.T.T. and Godron, M. (1986) *Landscape Ecology*. Wiley, New York.
Franklin, S. (2001) *Remote Sensing for Sustainable Forest Management*. Lewis, Boca Raton, Florida.
Fuller, D. (2001) Forest fragmentation in Loudoun County, Virginia, USA evaluated with multitemporal Landsat imagery. *Landscape Ecology* 16, 627–642.
Galindo-Jaimes, L., González-Espinosa, M., Quintana-Ascencio, P.F. and García-Barrios, L. (2002) Tree composition and structure in disturbed stands with varying dominance by *Pinus* spp. in the highlands of Chiapas, México. *Plant Ecology* 162, 259–272.
Gibson, D., Collins, S. and Good, R. (1988) Ecosystem fragmentation of oak–pine forest in the New Jersey pinelands. *Forest Ecology and Management* 25, 105–122.
Gigord, L., Picot, F. and Shykoff, J. (1999) Effects of habitat fragmentation on *Dombeya acutangula* (Sterculiaceae), a native tree on La Réunion (Indian Ocean). *Biological Conservation* 88, 43–51.
González-Espinosa, M. (2005) Forest use and conservation implications of the Zapatista rebellion in Chiapas, Mexico. In: Kaimowitz, D. (ed.) *Forests and Conflicts*. European Tropical Forest Research Network, ETFRN News No. 43–44, Wageningen, The Netherlands, pp. 74–76.
González-Espinosa, M., Quintana-Ascencio, P.F., Ramírez-Marcial, N. and Gaytán-Guzmán, P. (1991) Secondary succession in disturbed *Pinus–Quercus* forests of the highlands of Chiapas, México. *Journal of Vegetation Science* 2, 351–360.
González-Espinosa, M.G., Rey-Benayas, J.M., Ramírez-Marcial, N., Huston, M. and Golicher, D. (2004) Tree diversity in the northern Neotropics: regional patterns in highly diverse Chiapas, Mexico. *Ecography* 27, 741–756.
Hansen, M.J., Franklin, S.E., Woudsma, C.G. and Peterson, M. (2001) Caribou habitat mapping and fragmentation analysis using Landsat MSS, TM, and GIS data in the North Columbia Mountains, British Columbia, Canada. *Remote Sensing of Environment* 77, 50–65.
Herrmann, H., Babbitt, K., Baber, M. and Gongalton, R. (2005) Effects of landscape characteristics on amphibian distribution in a forest-dominated landscape. *Biological Conservation* 123, 139–149.
Hobbs, R. and Yates, C. (2003) Impacts of ecosystem fragmentation on plant populations: generalising the idiosyncratic. *Australian Journal of Botany* 51, 471–488.
Iida, S. and Nakashizuka, T. (1995) Forest fragmentation and its effect on species diversity in sub-urban coppice forests in Japan. *Forest Ecology and Management* 73, 197–210.

Imbernon, J. and Branthomme, A. (2001) Characterization of landscape patterns of deforestation in tropical rain forests. *International Journal of Remote Sensing* 22, 1753–1765.
INEGI (1984) *Land Use and Vegetation Cover*. Dirección General de Geografía, Instituto Nacional de Estadística, Geografía y Informática (INEGI), Mexico City, Mexico.
INEGI (1995) *National Census 1995*. Instituto Nacional de Estadística, Geografía y Informática (INEGI), Mexico City, Mexico.
INEGI (2000) *National Census 2000*. Instituto Nacional de Estadística, Geografía y Informática (INEGI), Mexico City, Mexico.
INIFAP (1995) *Mapa Edafológico*. Generated by Instituto Nacional de Investigaciones Forestales y Agropecuarias (INIFAP) for the Comisión Nacional para el Conocimiento y Uso de la Biodiversidad (CONABIO), México DF, Mexico.
Ite, U.E. and Adams, W.M. (1998) Forest conversion, conservation and forestry in Cross River State, Nigeria. *Applied Geography* 18, 301–314.
Jorge, L.A.B. and García, G.J. (1997) A study of habitat fragmentation in Southern Brazil using remote sensing and geographic information systems (GIS). *Forest Ecology and Management* 98, 35–47.
Lambin, E., Turner, B., Geist, H., Agbola, S., Angelsen, A., Bruce, J., Coomes, O., Dirzo, R., Fischer, G., Folke, C., George, P.S., Homewood, K., Imbernon, J., Leemans, R., Li, X., Moran, E., Mortimore, M., Ramakrishnan, P.S., Richards, J.F., Skanes, H., Steffen, W., Stone, G., Svedin, U., Veldkamp, T., Vogel, C. and Xu, J. (2001) The causes of land-cover and land-cover change: moving beyond the myths. *Global Environmental Change* 11, 261–269.
Lara, A. and Veblen, T. (1993) Forest plantations in Chile: a successful model? In: Mather, A. (ed.) *Afforestation. Policies, Planning and Progress*. Belhaven Press, London, UK, pp. 118–139.
Lara, A., Cortés, M. and Echeverría, C. (2000) Bosques. In: Sunkel, O. (ed.) *Informe País: Estado Actual del Medio Ambiente en Chile*. Centro de Estudios de Políticas Publicas, Universidad de Chile, Santiago, Chile, pp. 131–173.
Laurance, W.F. (1999) Reflections on the tropical deforestation crisis. *Biological Conservation* 91, 109–117.
Laurance, W.F., Vasconcelos, H.L. and Lovejoy, T.E. (2000) Forest loss and fragmentation in the Amazon: implications for wildlife conservation. *Oryx* 34, 39–45.
Laurance, W.F., Nascimento, H.E.M., Laurance, S.G., Andrade, A.C., Fearnside, P.M., Ribeiro, J.E.L. and Capretz, R.L. (2006) Rain forest fragmentation and the proliferation of successional trees. *Ecology* 87, 469–482.
Li, H. and Wu, J. (2004) Use and misuse of landscape indices. *Landscape Ecology* 19, 389–399.
Luque, S. (2000) Evaluating temporal changes using Multi-spectral Scanner and Thematic Mapper data on the landscape of a natural reserve: the New Jersey pine barrens, a case study. *International Journal of Remote Sensing* 21, 2589–2611.
Martínez-Morales, M. (2005) Landscape patterns influencing bird assemblages in a fragmented neotropical cloud forest. *Biological Conservation* 121, 117–126.
McConnell, W., Sweeney, S. and Mulley, B. (2004) Physical and social access to land: spatio-temporal patterns of agricultural expansion in Madagascar. *Agriculture, Ecosystems and Environment* 101, 171–184.
McGarigal, K., Cushman, S.A., Neel, M.C. and Ene, E. (2002) Fragstats: spatial pattern analysis program for categorical maps, University of Massachusetts, Landscape Ecology Program. Available at: http://www.umass.edu/landeco/research/fragstats/fragstats.html (accessed 20 January 2003).
Metzger, J.P. (1997) Relationships between landscape patterns structure and tree species diversity in tropical forests of south-east Brazil. *Landscape and Urban Planning* 37, 29–35.
Metzger, J.P. (2000) Tree functional group richness and landscape structure in a Brazilian tropical fragmented landscape. *Ecological Applications* 10, 1147–1161.
Miranda, F. (1952) *La Vegetación de Chiapas*, Primera Parte. Ediciones del Gobierno del Estado, Tuxtla Gutiérrez, Chis. México.

Mitermeier, R.A., Robles-Gil, P. and Mittermeier, C.G. (1997) *Megadiversidad: Los Países Biológicamente más Ricos del Mundo*. Cementos Mexicanos, Mexico City, Mexico.
Myers, N., Mittermeler, R.A., Mittermeler, C.G., da Fonseca, G.A.B. and Kent, J. (2000) Biodiversity hotspots for conservation priorities. *Nature* 403, 853–858.
Noss, R.F. (2001) Forest fragmentation in the southern Rocky Mountains. *Landscape Ecology* 16, 371–372.
Palacio-Prieto, J.L., Bocco, G., Velásquez, A., Mas, J.F., Takaki-Takaki, F., Victoria, A., Luna-González, L., Gómez-Rodríguez, G., López-García, J., Palma, M., Trejo-Vázquez, I., Peralta, A., Prado-Molina, J., Rodríguez-Aguilar, A., Mayorga-Saucedo, R. and González, F. (2000) Technical Note: Current situation of forest resources in Mexico: results of the 2000 National Forest Inventory. *Investigaciones Geográficas, Boletín del Instituto de Geografía, Universidad Nacional Autónoma de México (UNAM)* 43, 183–203.
Pan, D., Domon, G., De Blois, S. and Bouchard, A. (1999) Temporal (1958–1993) and spatial patterns of land use change in Haut-Saint-Laurent (Quebec, Canada) and their relation to landscape physical attributes. *Landscape Ecology* 14, 35–52.
Pedlowski, M., Dale, V.H., Matricardi, E.A.T. and Pereira da Silva Filho, E. (1997) Patterns and impacts of deforestation in Rondonia, Brazil. *Landscape and Urban Planning* 38, 149–157.
Piessens, K., Honnay, O. and Hermy, M. (2005) The role of fragmented area and isolation in the conservation of heathland species. *Biological Conservation* 122, 61–69.
Ramamoorthy, T.P., Bye, R., Lot, A. and Fa, J. (1993) *Biological Diversity of Mexico*. Oxford University Press, New York.
Ramírez-Marcial, N., González-Espinosa, M. and Williams-Linera, G. (2001) Anthropogenic disturbance and tree diversity in montane rain forests in Chiapas, Mexico. *Forest Ecology and Management* 154, 311–326.
Ranta, P., Blom, T., Niemela, J., Joensuu, E. and Siitonen, M. (1998) The fragmented Atlantic rain forest of Brazil: size, shape and distribution of forest fragments. *Biodiversity and Conservation* 7, 385–403.
Rasul, G., Thapa, G. and Zoebisch, M. (2004) Determinants of land-cover changes in the Chittagong hill tracts of Bangladesh. *Applied Geography* 24, 217–240.
Rzedowski, J. (1978) *Vegetación de México*. Editorial Limusa, Mexico, D.F., Mexico.
Rzedowski, J. (1993) Diversity and origins of the phanerogamic flora of Mexico. In: Ramamoorthy, T.P., Bye, R., Lot, A. and Fa, J. (eds) *Biological Diversity of Mexico*. Oxford University Press, New York, pp. 129–144.
Sader, S.A., Hepinstall, D.J.H., Coan, M. and Soza, C. (2001) Forest change monitoring of a remote biosphere reserve. *International Journal of Remote Sensing* 22, 1937–1950.
Schlatter, J., Gerding, V. and Huber, H. (1995) *Sistema de Ordenamiento de la Tierra. Herramienta para la Planificación Forestal Aplicado a la X Región*. Serie Técnica. Facultad de Ciencias Forestales, Universidad Austral de Chile, Valdivia, Chile.
Shafer, G. (1976) *A Mathematical Theory of Evidence*. Princeton University Press, Princeton, New Jersey.
Sierra, R. (2000) Dynamics and patterns of deforestation in the western Amazon: the Napo deforestation front, 1986–1996. *Applied Geography* 20, 1–16.
Silva, E. (2004) The political economy of forest policy in Mexico and Chile. *Singapore Journal of Tropical Geography* 25, 261–280.
Skole, D. and Tucker, C. (1993) Tropical deforestation and habitat fragmentation in the Amazon: satellite data from 1978 to 1988. *Science* 260, 1905–1909.
Spies, T., Ripple, W. and Bradshaw, G. (1994) Dynamics and pattern of a managed coniferous forest landscape in Oregon. *Ecological Applications* 4, 555–568.
Staus, N., Strittholt, J., Dellasala, D. and Robinson, R. (2002) Rate and patterns of forest disturbance in the Klamath-Siskiyou ecoregion, USA between 1972 and 1992. *Landscape Ecology* 17, 455–470.

Steiner, N. and Köhler, W. (2003) Effects of landscape patterns on species richness – a modelling approach. *Agriculture Ecosystems and Environment* 2086, 1–9.
Steininger, M., Tucker, C., Ersts, P., Killeen, T., Villegas, Z. and Hecht, S. (2001) Clearance and fragmentation of tropical deciduous forest in the tierras bajas, Santa Cruz, Bolivia. *Conservation Biology* 15, 856–866.
Turner, I.M. and Corlett, T. (1996) The conversion value of small, isolated fragments of lowland tropical rain forest. *Trends in Ecology and Evolution* 11, 330–333.
Uezu, A., Metzger, J. and Vielliard, J. (2005) Effects of structural and functional connectivity and patch size on the abundance of seven Atlantic Forest bird species. *Biological Conservation* 123, 507–519.
Uusivuori, J., Lehto, E. and Palo, M. (2002) Population, income and ecological conditions as determinants of forest area variation in the tropics. *Global Environmental Change* 12, 313–323.
Veldkamp, A. and Verburg, P.H. (2004) Modelling land use change and environmental impact. *Journal of Environmental Management* 72, 1–3.
Verburg, P.H., Soepboer, W., Veldkamp, A., Limpiada, R., Espaldon, V. and Mastura, S. (2002) Modeling the spatial dynamics of regional land use: the CLUE-S model. *Environmental Management* 30, 391–405.
Vergara, P. and Simonetti, J. (2004) Avian responses to fragmentation of the Maulino in central Chile. *Oryx* 38, 383–388.
Watson, J., Whittaker, R. and Dawson, T. (2004) Habitat structure and proximity to forest edge affect the abundance and distribution of forest-dependent birds in tropical coastal forest of southern Madagascar. *Biological Conservation* 120, 311–327.
Williams, N.G., McDonnell, M.J. and Seager, E. (2005) Factors influencing the loss of an endangered ecosystem in an urbanizing landscape: a case study of native grasslands from Melbourne, Australia. *Landscape and Urban Planning* 71, 35–49.
Williams-Linera, G., Manson, R.H. and Isunza-Vera, E. (2002) La fragmentación del bosque mesófilo de montaña y patrones de uso del suelo en la región oeste de Xalapa, Veracruz, México. *Madera y Bosques* 8, 73–89.
Willson, M., De Santo, T.I., Sabag, C. and Armesto, J.J. (1994) Avian communities of fragmented south-temperate rainforests in Chile. *Conservation Biology* 8, 508–520.
Wilson, K., Newton, A.C., Echeverría, C., Weston, C. and Burgman, M. (2005) A vulnerability analysis of the temperate forests of south central Chile. *Biological Conservation* 122, 9–21.
Zheng, D., Wallin, D. and Hao, Z. (1997) Rates and patterns of landscape change between 1972 and 1988 in the Changbai Mountain area of China and North Korea. *Landscape Ecology* 12, 241–254.
Zipperer, W.C., Burgess, R.L. and Nyland, R.D. (1990) Patterns of deforestation and reforestation in different landscape types in central New York. *Forest Ecology and Management* 36, 103–117.

3 Plant Diversity in Highly Fragmented Forest Landscapes in Mexico and Chile: Implications for Conservation

J.M. Rey-Benayas, L. Cayuela, M. González-Espinosa, C. Echeverría, R.H. Manson, G. Williams-Linera, R.F. del Castillo, N. Ramírez-Marcial, M.A. Muñiz-Castro, A. Blanco-Macías, A. Lara and A.C. Newton

Aerial photograph illustrating clearcuttings and industrial plantations of *Pinus radiata* in the coastal range in south-central Chile. Photo: Cristian Echeverría

Summary

Research addressed a range of issues relating to the regional determinants of species diversity, the effects of fragmentation and human disturbance on tree diversity at different spatial scales, changes of diversity along secondary succession after deforestation, and plausible future scenarios of species decline associated with continued habitat loss across a variety of ecological and socio-economic conditions in Latin America. This analysis was performed using different woody vegetation datasets in combination with various field measurements, remote sensing and GIS data. Regionally, climatic factors emerged as primary predictors of tree diversity. At finer scales, fragmentation and human disturbance better explained patterns of species diversity. These effects were, however, dependent on the time after fragmentation occurred. In the short term, habitat fragmentation was not likely to reduce the overall diversity of a fragment, but could have a slight positive effect on local diversity within fragments. Moreover, we detected a negative effect of human disturbance that far outweighed the effects of fragmentation at this scale. In the long term, however, fragmentation was found to significantly reduce the overall diversity of forest remnants.

Patterns of diversity along chronosequences of abandoned pastures and croplands were consistent across all study areas in Mexico and suggested that vegetation structure and community composition gradually come to mirror those of mature forests. However, species richness strongly depended upon the functional type under consideration.

Using ground-based floristic inventories and forest loss rates derived from satellite imagery, we estimated the percentage of species most likely to disappear, or at least become seriously threatened with extirpation, assuming continued habitat loss until 2025. Alarmingly, the predicted species decline in the Highlands of Chiapas was over 40% using estimated yearly deforestation rates of 4.8%.

Introduction

Identifying the factors driving patterns of species diversity has always fascinated ecologists. Some relevant issues that have emerged since the end of the 19th century are: (i) the description of diversity patterns; (ii) the underlying causes and processes that determine these patterns; (iii) the development of accurate estimators of species richness and diversity indices; and (iv) applied issues related to conservation, restoration and ecosystem management. Studies of diversity are complicated as this variable can be measured in a variety of ways and at different scales. The spatial scale at which biodiversity patterns are measured is directly related to the identification and understanding of underlying causal processes (Allen and Starr, 1982; Cushman and McGarigal, 2004).

The processes that have been suggested as determining patterns of diversity are varied and include phylogenetic, historical, biogeographic and environmental processes, as well as stochasticity (Brown and Lomolino, 1998; Rey-Benayas and Scheiner, 2002). At large spatial scales, factors relating to the flow of energy in the system (e.g. productivity and evapotranspiration) have emerged as primary predictors of species diversity (Wright *et al.*, 1993; Pausas and Austin, 2001; González-Espinosa *et al.*, 2004). At finer-grained scales, however, the type, history and frequency of disturbance, land use, and patch-specific characteristics such as soil type, topography and landscape pattern, as well as processes such as competition and dispersal, may be more relevant for explaining patterns of species diversity (Kerr and Packer, 1997; Lawton *et al.*, 1998; Ricklefs, 2004).

Increasing rates of biodiversity loss and its effects on essential ecosystem services (Heywood, 1995; Costanza *et al.*, 1997; Terborgh, 1999; Tilman, 1999; Bininda-Emonds *et al.*, 2000; Pimm and Raven, 2000; Gaston *et al.*, 2003; Lara *et al.*, 2003) have fuelled increasing concern about biodiversity conservation over the last decade (Ricketts *et al.*, 1999; Cincotta *et al.*, 2000; Myers *et al.*, 2000). Forest loss and fragmentation have been recognized as the main threat to biological diversity worldwide (CBD, 2005). An additional legacy of the extensive removal of native forest is the increased isolation and deterioration of remaining forest habitat owing to edge effects (Forman and Godron, 1986; Reed *et al.*, 1996; Franklin, 2001). Many theoretical and observational studies suggest that habitat fragmentation has a negative impact on the flora and fauna of remnant habitats and it is likely to affect a variety of population and community-level processes (Saunders *et al.*, 1991; Debinski and Holt, 2000). However, the ecological consequences of fragmentation may differ, depending on the peculiarities of particular taxonomic groups or species, the spatial configuration of the relevant landscape, and how it varies both temporally and spatially (Fahrig, 2003). In addition to forest loss and fragmentation, forest resources can be exploited and degraded by forest users to different degrees. These local disturbances alter the ecological processes operating in fragments and may have additive or interactive effects with fragmentation, affecting forest community structure and function (Debinski and Holt, 2000; Laurance and Cochrane, 2001).

This chapter synthesizes research undertaken to understand the mechanisms affecting plant diversity at multiple scales in a variety of highly fragmented forest landscapes in Latin America (Mexico and Chile). We address issues such as the regional determinants of species diversity, the effects of fragmentation and human disturbance on tree diversity at different spatial scales, changes of diversity along secondary succession after deforestation and plausible future scenarios of species decline associated with continuing habitat loss using a variety of case studies under different ecological and socio-economic conditions. We conclude by describing how this knowledge can be applied to the development and implementation of conservation policies.

Study Areas

The highly fragmented forest landscapes we studied were in the tropical mountainous regions of southern Mexico (the Highlands of Chiapas, Central Veracruz and Oaxaca) and temperate zone forests of southern Chile (Los Muermos-Ancud, Fig. 3.1). In Chiapas we also studied regional patterns of diversity for the entire state. Our study areas spanned a variety of biophysical and socio-economic situations that are described in Chapter 1.

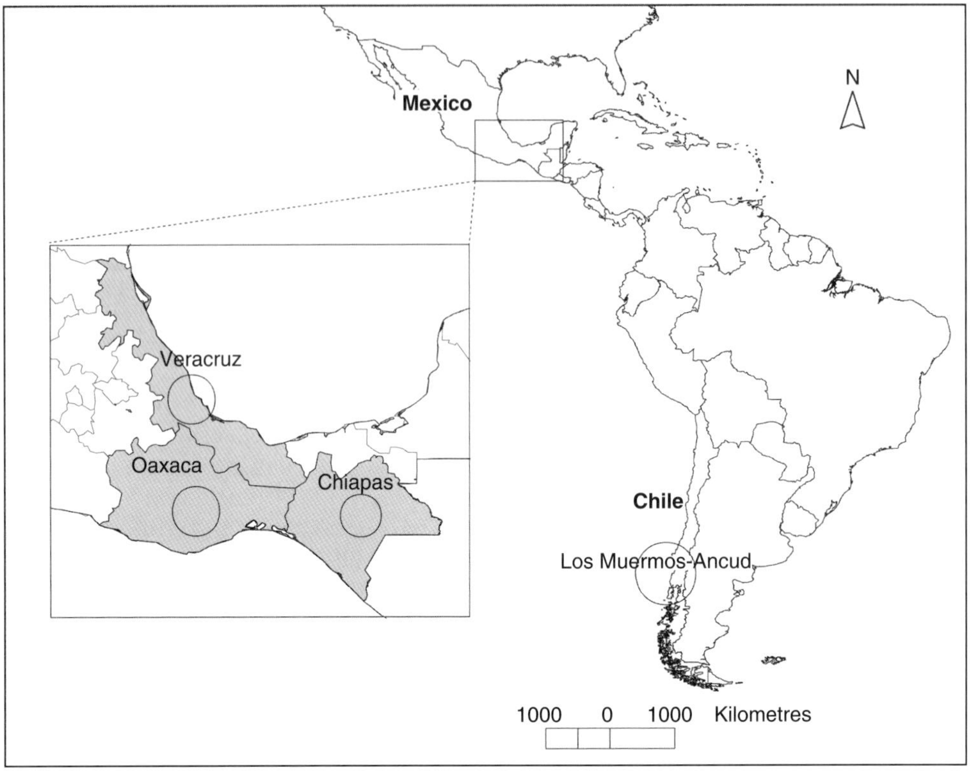

Fig. 3.1. Geographical location of the four study areas in Mexico and Chile.

Case Studies

Regional scale determinants of tree diversity in Chiapas, Mexico

Explaining the distribution of diversity along broad environmental gradients continues to challenge ecologists (Francis and Currie, 2003; Qian and Ricklefs, 2004). Spatial patterns of plant diversity have been related to regional macro-scale processes as well as to local processes. Here we investigate the relationship of tree species diversity and regional-scale environmental factors (e.g. several hundred thousand square kilometres; Mittelbach *et al.*, 2001; Qian and Ricklefs, 2004) in the state of Chiapas, Mexico.

Methods

We compiled a database with information from labels of herbarium sheets of all tree species collected over a period of 135 years (i.e. woody plants with dbh ≥3 cm and height ≥3 m) in Chiapas that are native to the state (González-Espinosa *et al.*, 2004). These data were spatially assigned to grid cells each of 5 minutes latitude × 5 minutes longitude. In each cell we calculated Simpson's index of diversity.

We used several climatic (*precipitation*, *temperature*, *actual evapotranspiration (AET)*, *seasonality*), edaphic (an index of *soil fertility/quality* was calculated based on the interpretation of physical and chemical properties of soil taxa as described in the legend of the map by FAO–UNESCO, 1974) and topographic heterogeneity variables to predict tree diversity (González-Espinosa *et al.*, 2004).

Multiple regression models were fitted to independent explanatory variables to predict tree diversity as tests of hypotheses relating the variation of tree diversity with habitat favourableness, temporal heterogeneity or seasonality and spatial heterogeneity. We also tested whether the shape of the function between tree species diversity and *AET* was curvilinear unimodal with an interior maximum using the Mitchell-Olds and Shaw test (1987), with a null hypothesis of a non-intermediate maximum. We used regression quantiles (Cade *et al.*, 1999) to estimate multiple rates of change (slopes) of tree diversity as a function of subsets of values (upper quantiles, 0.75–0.95) of those variables expected to be affected by known or unknown limiting factors.

Results

A model that accounted for 41.4% of the total variance in tree diversity showed positive effects of *AET* and *seasonality*, whereas *soil fertility/quality* had a negative effect. A curvilinear model described the relationship between tree diversity and *AET* well ($R^2=0.45$), and an intermediate maximum was detected (Fig. 3.2a). The data pattern also suggested an asymptotic relationship, which was confirmed with a two-part regression. Regression quantiles with the upper envelope of the data (0.85–0.90 quantiles) provided better estimates of the effect of *soil fertility/quality*. This analysis indicated that the statistical effect of *AET* is relatively independent of other environmental factors (i.e. it is – or its correlates are – an important limiting factor of tree diversity by itself). However, the effects of *soil fertility/quality* seem to be more dependent on the interactions with other limiting factors, and may be underestimated or obscured by multiple linear models. One such factor is rainfall. Minimum diversity at intermediate *rainfall* values hints at a bimodal model of tree diversity along a rainfall gradient, in opposition to the frequently described positive linear relationship (Fig. 3.2b).

Patch-level effects of forest fragmentation on tree diversity

Habitat fragmentation has serious implications for a variety of population and community processes over a range of temporal and spatial scales. Studies investigating these effects often draw analogies between forest fragments and oceanic islands (Harris, 1984; Laurance and Bierregaard, 1997; Rosenblatt *et al.*, 1999; Ferraz *et al.*, 2003; Hill and Curran, 2003). Their key assumptions are that loss in area, increasing edge effects and reduced connectivity decrease species diversity. We analysed the patch-level effects of fragmentation on tree diversity for the Highlands of Chiapas, Los Muermos-Ancud in southern Chile and Central Veracruz. We used Spearman's rank correlations to test univariate relationships between measures of tree diversity and different spatial

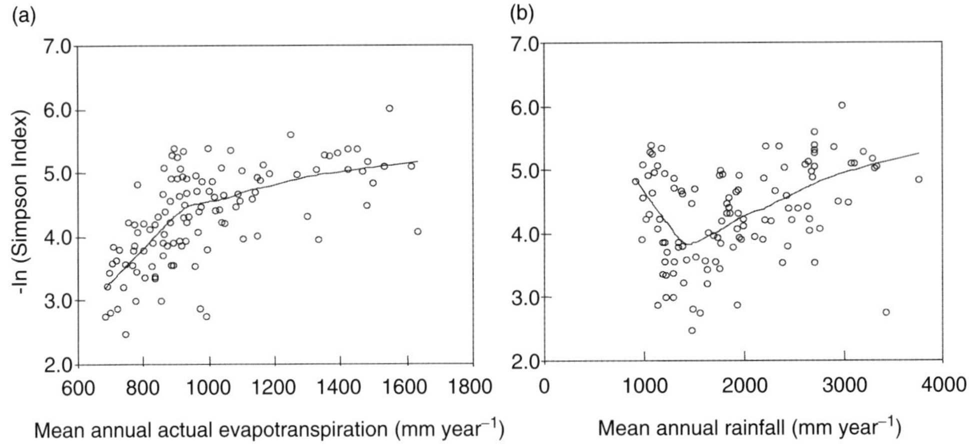

Fig. 3.2. Local weighted curve smoothing (LOWESS) for tree diversity (–ln SI, Simpson's index) as a function of: (a) mean annual actual evapotranspiration (mm $year^{-1}$); and (b) mean annual rainfall (mm $year^{-1}$) (after González-Espinosa *et al.*, 2004).

metrics. These analyses were performed for: (i) all species; (ii) forest interior species that are characteristic of mature forest stands; and (iii) pioneer species that are characteristic of earlier successional stages.

Methods

In the Highlands of Chiapas, floristic inventories were carried out using 204 circular plots of 1000 m^2 each in different forest fragments; 168 of these plots were sampled from January 2003 to May 2004 (Cayuela *et al.*, 2006a), and 36 were sampled in 1998 using the same sampling protocol (Galindo-Jaimes *et al.*, 2002; L. Galindo-Jaimes, unpublished data). The abundance of all tree species with dbh ≥10 cm was recorded. The final database included 230 native tree species. Fragments with fewer than five plots were discarded prior to analysis, leaving 195 plots in 16 forest fragments.

In Los Muermos-Ancud, a total of 51 fragments were randomly sampled throughout the landscape (see Chapter 2). The selection of fragments with an age of at least 23 years with the same spatial attributes in 1976, 1985 and 1999 maximized our chances of recording the ecological impacts associated with fragmentation. Samples were also stratified by soil type. Owing to the different sizes of fragments, the number of sampling plots was weighted by the patch size (one plot in fragments less than 100 ha, two plots in fragments between 100 and 1000 ha, three plots in fragments >1000 ha). Sixty-three 20 m × 25 m plots were established in the core areas of 51 fragments. In fragments larger than 100 ha, these were evenly distributed at a minimum distance of 50 m from each other. Each plot was divided into 20 contiguous 5 m × 5 m subplots and, in each, shrub and tree species were identified and counted to estimate the number of individuals per species.

In Central Veracruz, 21 fragments were sampled: seven abandoned pastures (>12 years old) by means of eight 10 m × 10 m plots, four abandoned coffee plantations using ten 20 m × 20 m plots (10–20 years old), and ten montane cloud forest fragments using ten 10 m × 10 m plots. Fragment size ranged from 1.1 to 54 ha. In each plot, the abundance of all tree species with dbh ≥ 5 cm was recorded. A total of 153 tree species were recorded: 125 native species and 28 non-native species.

In all three study areas, we estimated, when feasible, a variety of diversity measures for each fragment. For α-diversity, total and mean plot species richness were calculated. For β-diversity, we calculated the mean Sørensen's index of dissimilarity between plots within each fragment (Magurran, 1988).

The predictors of diversity used were: (i) *area* (ha); (ii) *core area* (ha remaining after removing a 100 m edge); (iii) *total edge length* (km); and (iv) *proximity index* (ratio between the area and distance of all fragments whose edges are within a 1-km search radius of the focal fragment). Computation of spatial metrics was based on land-cover maps (Chapter 2).

Results

There were hardly any significant correlations between any of the diversity measures and fragment metrics in the Highlands of Chiapas and Central Veracruz (Table 3.1). The exception was the estimated number of tree species in a fragment in the Highlands of Chiapas, which was negatively related to the *proximity index*, indicating that the more isolated a fragment was, the higher the number of tree species it contained. For Los Muermos-Ancud, however, *area, core area, edge length* and *proximity index* were all negatively associated with mean pioneer species richness, and positively associated with forest interior species richness (Table 3.1).

Local effects of human disturbance and fragmentation on tree diversity

In addition to deforestation and fragmentation, forest patches can be degraded by selective logging, ground fires, the impacts of browsing by livestock and overhunting. These local disturbances alter the ecological processes operating in the fragments and may have additive or interactive effects with fragmentation on forest community structure and function (Cochrane *et al.*, 1999; Nepstad *et al.*, 1999; Gascon *et al.*, 2000; Laurance and Cochrane, 2001). In the previous case study we analysed the patch-level effects of fragmentation on tree diversity. Here, we investigated the local effects of fragmentation and habitat disturbance. We focused our study in the Highlands of Chiapas, Mexico. The analysis was hierarchically structured so that the relative effects of climatic gradients on tree diversity could be separated from more subtle human-induced local effects (Cayuela *et al.*, 2006a).

Methods

For the 195 plots described in the previous case study, we calculated Fisher's alpha as a measure of plot diversity. Fisher's alpha is a good estimator of α-diversity because it is independent of the number of individual trees in a sample (Rosenzweig, 1995) and assumes an underlying parametric model for the distribution of species abundances (Fisher *et al.*, 1943).

Table 3.1. Spearman correlation coefficients (*R*) between forest fragment metrics and different tree diversity measures in the Highlands of Chiapas, Los Muermos-Ancud and Central Veracruz. Significance values are also provided (*P*). Total species richness was calculated using Clench accumulation curves (Colwell and Coddington, 1994). In Los Muermos-Ancud in Chile, all species (seedlings, juveniles and adults) sampled in a fragment were used to calculate total species richness. Sørensen's index was not calculated for pioneer and forest interior species because the matrices included too many zero values.

	Diversity measures	Area (ha)		Core area (ha)		Edge length (km)		Proximity index	
		R	*P*	*R*	*P*	*R*	*P*	*R*	*P*
The Highlands of Chiapas (n = 16)									
All species	Total species richness	−0.03	0.930	−0.08	0.784	0.00	0.990	−0.52	0.040
	Mean plot richness	−0.11	0.680	−0.11	0.672	−0.12	0.664	−0.41	0.120
	Sørensen index	0.38	0.146	0.06	0.818	0.42	0.110	0.03	0.917
Pioneer	Total species richness	−0.07	0.788	−0.21	0.433	−0.04	0.895	−0.41	0.117
	Mean plot richness	0.09	0.742	0.05	0.861	0.08	0.792	0.22	0.407
	Sørensen index	0.36	0.171	0.19	0.480	0.38	0.146	−0.07	0.805
Forest interior	Total species richness	0.05	0.865	0.16	0.560	0.00	0.991	−0.31	0.235
	Mean plot richness	−0.09	0.746	−0.11	0.672	−0.08	0.780	−0.39	0.132
	Sørensen index	0.14	0.605	0.11	0.684	0.09	0.738	0.19	0.480
Los Muermos-Ancud (n = 51)									
All species	Total species richness	0.11	0.421	0.08	0.571	0.10	0.467	0.08	0.584
	Mean plot richness	0.07	0.622	0.04	0.767	0.04	0.782	−0.02	0.897
	Sørensen index	−0.27	0.054	−0.25	0.075	−0.26	0.065	−0.25	0.075
Pioneer	Mean plot richness	−0.29	0.037	−0.34	0.014	−0.29	0.039	−0.46	0.001
Forest interior	Mean plot richness	0.44	0.001	0.39	0.004	0.41	0.003	0.44	0.001
Central Veracruz (n = 21)									
All species	Mean plot richness	−0.13	0.587	−0.11	0.627	−0.09	0.678	n.a.	n.a.
Pioneer	Mean plot richness	−0.34	0.127	−0.30	0.187	−0.25	0.280	n.a.	n.a.
Forest interior	Mean plot richness	0.39	0.085	0.33	0.139	0.34	0.126	n.a.	n.a.

We used Non-Metric Multidimensional Scaling (NMDS) to identify major community types in relation to climatic gradients (Cayuela *et al.*, 2006a). We then tested the effects of forest fragmentation and local disturbance on tree species diversity within each of these community types. The effect of fragmentation was measured as *proximity of a plot to the nearest forest edge* (m); this distance was divided by the maximum value in order to produce standardized values ranging between 0 and 1. Surrogates of human disturbance included *canopy closure*, measured as the proportion of forest cover in a 500 m-radius circle centred on each plot (ranges between 0 and 1), and a *degradation index* (DI), ranging between −1 and 1, which was calculated as the relative change in the Normalized Difference Vegetation Index (NDVI) between 1990 TM and 2000 ETM+ Landsat satellite images, respectively. Negative values of this index indicate forest disturbance, e.g. by selective logging of certain species, whereas positive values indicate recent forest recovery.

Effects within fragments were analysed by examining patterns in the deviations from the mean value for Fisher's alpha within each fragment. To do this we used linear mixed-effects models. These models include fixed effects (within-fragments) and additional random-effect terms (between-fragments) that are appropriate for representing clustered and therefore potentially correlated data (Pinheiro and Bates, 2000). In our case, the random variation arose from the grouping of plots within separated fragments.

Results

Using NMDS axes of floristic composition, five major community types were defined in relation to regional climatic gradients (Cayuela *et al.*, 2006a). We hypothesized that within-fragment variability was determined by the local effects of human activity. Thus, we explored in detail the effects of fragmentation and human disturbance within-fragments for those groups for which there was more than one forest fragment. These groups were montane cloud forest, pine–oak–liquidambar forest and pine–oak forest. All three vegetation types show clear differences regarding alpha tree diversity (ANOVA, $F=35.42$, $P<0.001$).

Linear mixed-effects models revealed notable differences in diversity between-fragments (random effect) for montane cloud and pine–oak–liquidambar forests ($P<0.001$). There is also considerable variation in alpha diversity that is not linked to a random effect, but to variables related to fragmentation and local disturbance (Fig. 3.3). *Canopy closure* was, in all cases, highly correlated with the intercept ($r>0.8$), suggesting that this variable might be important in determining differences in diversity between-fragments in addition to within-fragments. For the two remaining forest types (oak and transitional forests), each consisting of one fragment, simple regressions resulted in non-significant relationships between tree diversity and the variables related to fragmentation and local disturbance ($P>0.1$).

An analysis by guilds revealed that effects were more noticeable for forest interior species than for pioneer species (Fig. 3.3). This was particularly relevant in pine–oak forests, where no significant relationships between alpha diversity of pioneer species and any of the variables related to fragmentation and local disturbance were found.

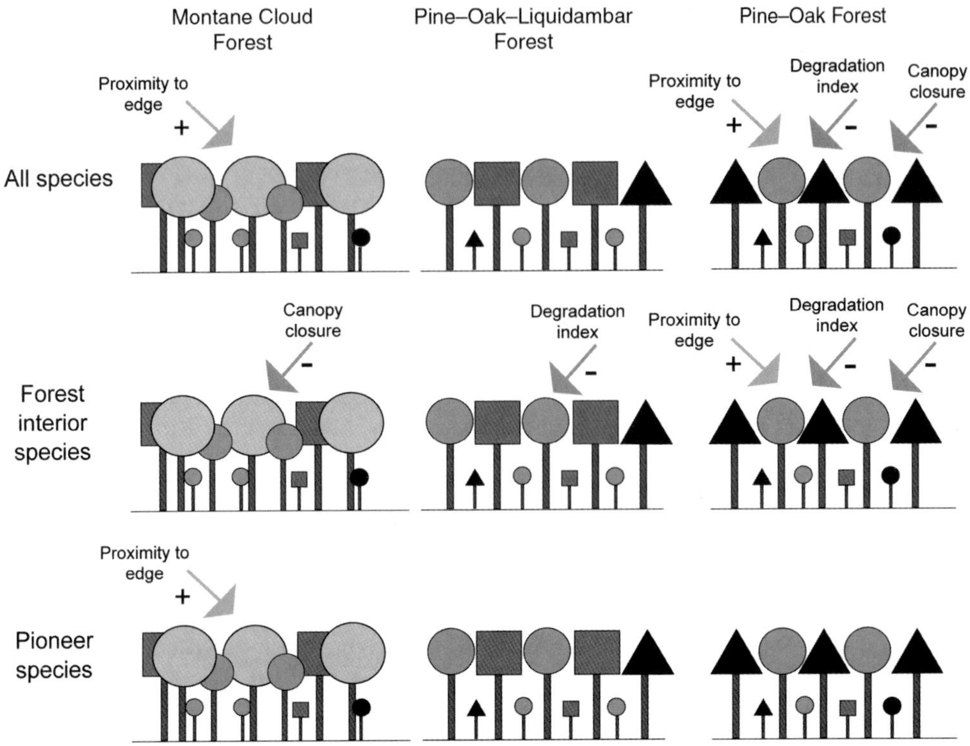

Fig. 3.3. Representation of within-fragment effects (fixed effects) of fragmentation and local disturbance on tree diversity for evergreen cloud forest, pine–oak–liquidambar forest and pine–oak forest considering all tree species (upper), late-successional species (middle) and pioneer species (bottom) (after Cayuela *et al.*, 2006a).

Secondary succession and plant diversity

Montane forests of Central and South America have been subjected for centuries to a wide range of human disturbances. In Mexico, traditional land use drives secondary succession with impacts on forest composition, structure and regeneration through practices such as slash-and-burn agriculture, sparse logging, extraction of saplings and lopping of hardwoods for fuelwood, and sporadic cattle grazing (Ramírez-Marcial *et al.*, 2001; González-Espinosa *et al.*, 2006; Muñiz-Castro *et al.*, 2006). Here, we investigate the importance of secondary succession for plant diversity conservation in the three study areas of Mexico.

Methods

In Central Veracruz, we selected 15 abandoned pastures from 0.25 to 80 years old that were adjacent to a forest fragment (Muñiz-Castro *et al.*, 2006). To assess distance effects, in each old-field 100 m × 10 m parallel bands were located at 0–10 m and at 40–50 m from the forest edge. Four 10 m × 10 m plots were randomly located in each band to sample trees >5 cm dbh. At the centre of each 10 m × 10 m plot, one 4 m × 4 m plot was established to sample woody plants <5 cm dbh, and one 2 m × 2 m plot was sampled for seedlings <1.3 m height. We measured dbh and height and counted the number of individuals per species in all but the 2 m × 2 m plots, where only the number of individuals and basal diameter of tree species were measured. All plants in the plots were identified to species. Also, tree species were classified as forest interior and pioneer following species description in the *Flora de Veracruz* (Sosa and Gómez-Pompa, 1994). We used an analysis of covariance (ANCOVA) to test the effect of distance to forest edge (categorical variable) and age after abandonment (covariate) on plant diversity. A quadratic term of age was also included in the ANCOVA model since a non-linear quadratic relationship was expected between age and some of the response variables.

In Oaxaca, we studied three chronosequences developed in a montane cloud forest area at El Rincón Alto. Here we report an analysis of secondary succession, based on a subset of data of a study described in more detail in Chapter 7. The sampling plots were positioned in forests of approximately 15, 45, 75 and >100 years old, away from their edges. We sampled all plants with ≥3.5 cm dbh in ten 100 m^2 rectangular plots per stand. Plants were identified at species level in most of the cases, classified as lianas and climbing plants, shrubs and understorey trees (with less than 10 m height at adult stage), and canopy trees. We measured dbh and height of each sampled plant identified mostly at the species level. Analyses were based on correlation and ordination techniques.

In Chiapas, we used data on chronosequences following agricultural abandonment, obtained using different methods at 68 sites over a period of 10 years. Information on forest structure, diversity and composition of human-disturbed forests was collected at each site (sources of original data appear in González-Espinosa *et al.*, 2006).

Results

In Veracruz, a total of 164 woody species were recorded in the 15 abandoned pastures sampled: 71 species were trees, 49 shrubs and 44 vines. We recorded 63 tree, 40 shrub and 36 vine species at 0–10 m from the border, and 49 tree, 38 shrub and 29 vine species at the interior of the old-field. The age of the old-field significantly affected tree species richness and diversity along the chronosequence (Fig. 3.4). Richness of trees ≥5 cm dbh increased with time of pasture abandonment with a decrease towards the final stage of the chronosequence (Fig. 3.4a). Richness of saplings and seedlings increased linearly with age (Fig. 3.4b, c). Shrubs and lianas did not display any significant trend along the chronosequence. Richness at the two distances from the forest edge were similar for trees >5 cm dbh, shrubs and lianas. Only tree seedlings displayed higher richness and diversity values close to the edge (Fig. 3.4c). Interestingly, richness of late-successional species was higher close to the edge for trees <5 cm dbh ($F=9.8$, $R^2=0.18$, $P=0.020$, Fig. 3.4d), saplings ($F=15.9$, $R^2=0.33$, $P<0.001$, Fig. 3.4e) and seedlings ($F=14.6$, $R^2=0.27$, $P=0.002$, Fig. 3.4f). Pioneer species richness was similar between the two distances from the forest edge for trees, juveniles and seedlings.

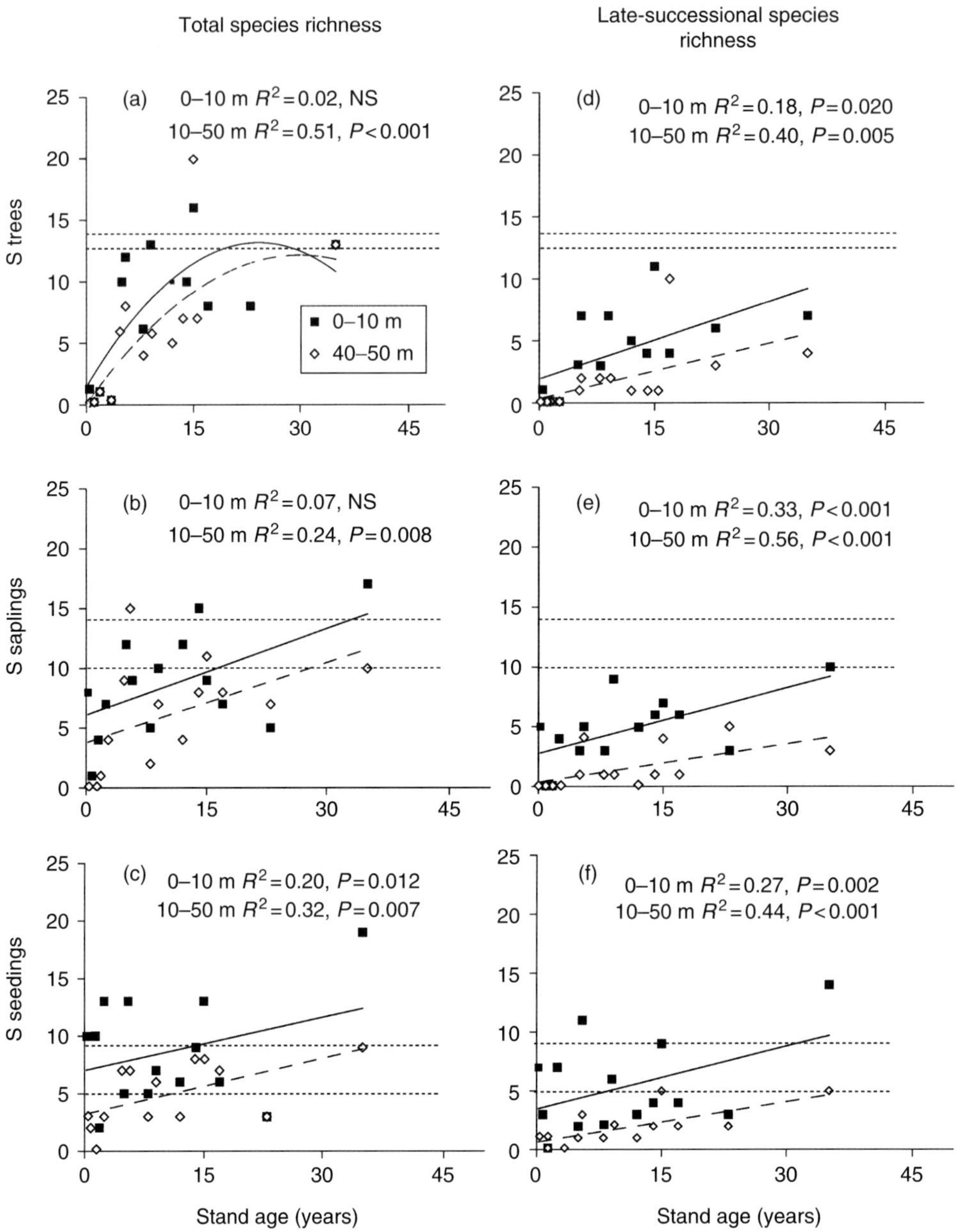

Fig. 3.4. Distance to the forest edge and age effects on tree species richness (S) along a chronosequence of abandoned pastures in Central Veracruz, Mexico. Data from mature forest (two sites) are shown for comparison (dotted lines). (a), (b) and (c) are total tree species; (d), (e) and (f) are late-successional tree species. Distances are 0–10 m (solid lines) and 40–50 m (dashed lines). The lines are derived from the minimal adequate model of ANCOVA; a quadratic term was used in the ANCOVA. When it is significant the relationship is represented by a unimodal curve, when it is non-significant the relationship is linear. Species richness was determined in 400 m^2 for trees, 64 m^2 for saplings and 16 m^2 for seedlings. NS, = not significant ($P > 0.05$) (after Muñiz-Castro *et al.*, 2006).

In Oaxaca, the analysis of changes in composition and structure in the three groups of plants along the three chronosequences identified 209 species of plants distributed in 128 genera and 69 families. Based on a previous structural analysis of the vegetation (Blanco-Macías, 2007), 45-year-old forests had basal area, average height values and a floristic composition similar to those of old-growth forests, but different from those of incipient secondary forests, which were abundant in shrubs and herbs that were absent in older forests. Lianas and climbing plants were the least diverse group. Liana species richness increased with forest age, at least during the first century of forest development (Table 3.2). In the 15-year-old incipient forests, we could not detect any plant in this group with ≥3.5 cm dbh. By contrast, shrubs and short trees peaked in species richness in 15-year-old stands and decreased in later successional stands, presumably as a result of shading and competition of canopy trees. The opposite trend was detected for canopy trees, by far the most diverse group. Species richness was low in 15-year-old stands, but had similar values at 45, 75 and >100 years after abandonment. Pearson correlation analysis in species richness in 0.01 ha sampling plots revealed that canopy tree richness was negatively correlated with that of shrubs and short trees ($r = -0.269$, $P = 0.003$), whereas no significant correlations were detected between canopy trees and liana species richness ($r = 0.023$, $P > 0.05$), or between lianas and shrubs and short trees ($r = 0.024$, $P > 0.05$). Overall these results suggest that secondary succession in tropical montane cloud forest areas involves relatively rapid changes in species richness, particularly during the first 45 years after abandonment. Species richness appears to depend not only on fallow time but also on species composition. In particular, a trade-off exists between the species richness of shrubs and short understorey trees and that of canopy trees, which was detected at both temporal (successional trends) and spatial (within plots) scales. This suggests that negative interactions among

Table 3.2. Species richness (and standard error) in 0.1 ha in four successional stages in three chronosequences for lianas and climbing plants, shrubs and understorey trees, and canopy trees in El Rincón Alto, Sierra Norte Oaxaca, Mexico (after Blanco-Macías, 2007).

Successional stage	Species richness	Range
Early successional forest	0 ± 0.0 for lianas/climbing plants	0–0
(~15 years old)	14 ± 1.0 for understorey trees	12–15
	17 ± 1.7 for canopy trees	15–20
Young successional forest	1 ± 0.0 for lianas/climbing plants	1–1
(~45 years old)	6 ± 1.2 for understorey trees	4–8
	23 ± 3.7 for canopy trees	19–30
Mature successional forest	2 ± 1.0 for lianas/climbing plants	1–3
(~75 years old)	6 ± 1.5 for understorey trees	4–9
	26 ± 0.7 for canopy trees	25–27
Old-growth forest	3 ± 1.5 for lianas/climbing plants	1–4
(≥100 years old)	7 ± 2.3 for understorey trees	3–11
	24 ± 2.4 for canopy trees	21–29

these groups, presumably mediated by competition, regulate the species composition during forest regeneration after disturbance.

The pattern of species richness change along a successional gradient following shifting agriculture in the Highlands of Chiapas is relatively well known (Table 3.3, González-Espinosa *et al.*, 1991, 2006). Under current land-use practices, agricultural land use may now last for many years based on the increased utilization of agrochemicals and pesticides. After abandonment, fallow fields may be replaced by induced grassland communities, depending on sheep and cattle stocking rates. Recruitment of both pine and oak individuals may occur in early open conditions, and they may become dominant in old-growth stages during the same successional series. Yet an almost complete floristic replacement has been recorded between the open and forested seral stages (González-Espinosa *et al.*, 1991). Canopy and understorey tree species may account for 20–30% of the total floristic richness of any given old-growth stand (not including epiphytes), but may be as low as 12–15% in severely disturbed forests (González-Espinosa *et al.*, 1995).

Future scenarios of species decline

Deforestation and habitat loss are widely expected to precipitate an extinction crisis among forest species (Tilman *et al.*, 1994; da Silva and Tabarelli, 2000; Brook *et al.*, 2006; Wright and Muller-Landau, 2006). These extinctions can be inferred by linking deforestation rates with estimates of regional diversity. Here, we use species–area accumulation curves to explore the likely impact

Table 3.3. Species richness (and standard error) in different successional stages in the Highlands of Chiapas, Mexico (after González-Espinosa *et al.*, 2006; N. Ramírez-Marcial, 2006, personal communication).

Successional stage (no. sites)	Mean species richness	Range
Old-field fallow (n = 22)	50 ± 3.6	30–85
Grassland (n = 4)	42 ± 1.7	30–57
Shrubland (n = 17)	76 ± 3.3	51–82
Early successional forest (n = 25)	47 ± 2.3 in the herb layer	30–50*
	30* in the shrub layer	25–35*
	27 ± 1.2 in the tree layer	20–35*
Mid-successional forest	20–30*	15–20[a]
		5–8[b]
Old-growth forest	35–45*	25–45[a,*]
		8–12[b,*]

[a]Understorey tree species only; regional-level richness based on available herbarium vouchers.
[b]Canopy tree species only; regional-level richness based on available herbarium vouchers.
*N. Ramírez-Marcial, El Colegio de la Frontera Sur, San Cristóbal de las Casas, Chiapas, México, 2006, personal observation.

of forest loss on tree species diversity in the Highlands of Chiapas, Central Veracruz, Oaxaca and Los Muermos-Ancud.

Methods

To investigate the potential effects of deforestation and fragmentation on the loss of tree diversity, we constructed species accumulation curves in the four study areas. As seen in Chapter 2, these are currently being subjected to different degrees of accelerated transformation as a result of human activities. Exponential (Fisher *et al.*, 1943) and power function models (Preston, 1962) were fitted to predict the potential loss of tree species linked to current estimated deforestation rates in each of the study areas.

Results

Based upon the forest extent in 2000 and given the estimated annual deforestation rates, we were able to predict the forest extent for 2025 (Table 3.4). Constructed non-asymptotic species–area accumulation curves predict the decline of species richness with the reduction of forest extent according to estimated ongoing deforestation rates (Fig. 3.5). For the 2025 scenario, the predicted decline of tree species ranged between <1% in Oaxaca and 41% in the Highlands of Chiapas (Table 3.4 and Fig. 3.5).

Discussion

The assessment of biodiversity in managed landscapes poses several methodological difficulties since: (i) diversity measures strongly depend on the spatial and temporal scale chosen, and unfortunately the scaling functions applicable to transfer results from one scale to another are not completely satisfactory (Waldhardt, 2003); (ii) it is often impractical to consider all the different ecological, historical and human-related factors that may contribute to patterns of species diversity (Lobo *et al.*, 2001); and (iii) field data are often scarce, particularly in tropical regions, owing to limited accessibility to forests (Stockwell and Peterson, 2003) and limited resources and capacity. However, for different regions and spatial scales we identified some common patterns relating to the environmental determinants of diversity and the effects of deforestation, forest fragmentation and human disturbance.

Regional determinants of plant diversity

There has been considerable research interest in the shape of the relationship and the possible mechanisms underlying the energy–species hypothesis at different spatial scales (Rosenzweig and Abramsky, 1993; Whittaker *et al.*, 2001). At the regional scale, the diversity–evapotranspiration relationship was found here to be significant in all multiple linear models. The proportion of total variance explained by either linear (36%) or quadratic (44%)

Table 3.4. Predicted effects of deforestation on species richness (S) decline in the four target areas. Only tree species were sampled for the Highlands of Chiapas, and all woody species were sampled for the other regions. The power model was not calculated for Central Veracruz owing to the small sample size.

Study region	Number of plots	Plot size (ha)	Forest area in 2000 (ha)	Deforestation rate (% year^{-1})[a]	Predicted forest area in 2025 (ha)	Exponential model			Power model		
						2000 Log S	2025 Log S	Species loss (%)	2000 S	2025 S	Species loss (%)
Highlands of Chiapas	204	0.1	98,340	4.80[b]	19,850	4.01	3.78	41.18	5,144	3,314	35.58
Central Veracruz	21	0.1	69,493[c]	3.04	32,135	4.86	4.66	37.40	–	–	–
Sierra Norte, Oaxaca	89	0.01	4,098	0.13	3,948	2.67	2.66	0.01	249	247	0.01
Los Muermos-Ancud	54	0.05	202,167	0.78	166,223	2.40	2.38	4.20	145	140	3.21

[a]Deforestation rates were estimated for the period 1990–2000 in the Highlands of Chiapas, 1985–2000 in Los Muermos-Ancud, 1984–2000 in Central Veracruz and 1995–2000 in Sierra Norte, Oaxaca.
[b]For the Highlands of Chiapas we used a more conservative corrected deforestation rate (after Cayuela *et al.*, 2006b) than that presented in Chapter 2.
[c]In Central Veracruz only montane cloud forest was considered.

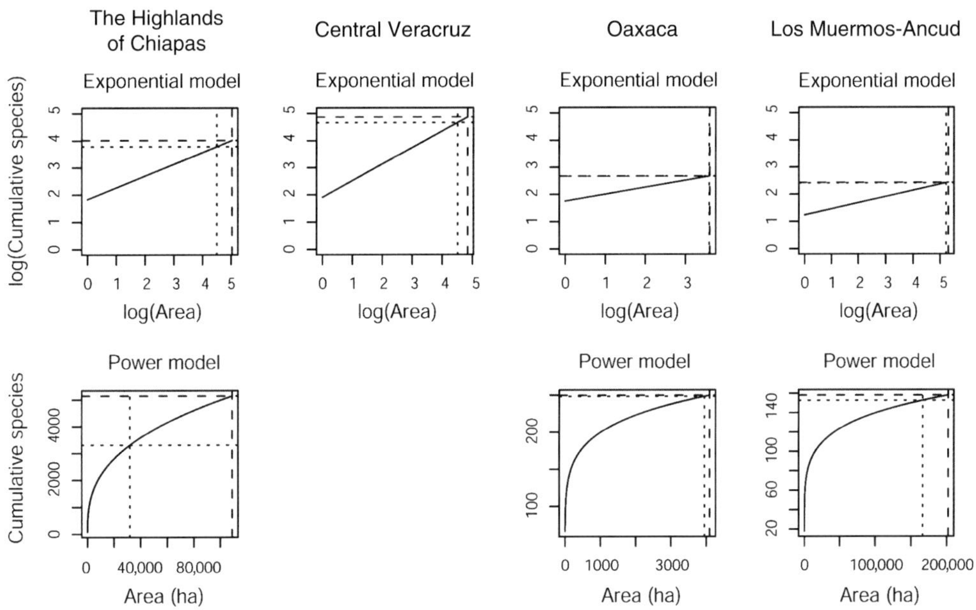

Fig. 3.5. Exponential and power models of species loss associated with deforestation for the Highlands of Chiapas, Central Veracruz, Sierra Norte in Oaxaca and Los Muermos-Ancud. Dashed lines refer to diversity in 2000, whereas dotted lines indicate the estimated loss of diversity in 2025 according to current deforestation rates.

models was higher than a previously reported median of 30% for plant diversity and energy-related factors at similar regional scales (González-Espinosa *et al.*, 2004). A more conservative interpretation of the data pattern (Fig. 3.2a) suggests a pronounced linear increase of tree diversity up to mid-range values of *AET* and a steady but slower increase up to a levelling off at higher values.

Mean annual precipitation showed an interior minimum (Fig. 3.2b). At high precipitation, higher plant diversity may occur owing to the following concatenated events: soil nutrients are depleted through weathering and leaching and growth rates and tree height are reduced; dominance in the canopy is then reduced; and diversity increases as a result of higher canopy richness and increased richness of the shade-tolerant understorey (Austin and Smith, 1989; González-Espinosa *et al.*, 2004).

Soil fertility/quality was found to be negatively related to tree diversity (González-Espinosa *et al.*, 2004). The effects of soil fertility/quality, however, seemed to be more dependent on the interactions with other limiting factors, and may be underestimated or obscured by multiple linear models. Other studies have highlighted different relationships (positive, negative, hump-shaped) between soil characteristics and plant diversity at meso- and landscape scales (Huston, 1980; Clinebell *et al.*, 1995; Clark *et al.*, 1999; Rey-Benayas and Scheiner, 2002). It therefore seems difficult to generalize the response of plant diversity to soil characteristics at the regional scale. We also found positive effects of climate seasonality but not of spatial heterogeneity as measured by elevation range and soil diversity.

Effects of fragmentation and human disturbance on tree diversity

At the patch level, there were significant differences between the study areas. Whereas in the Highlands of Chiapas and Central Veracruz the effects of fragmentation on tree diversity were not directly observable, in Los Muermos-Ancud correlations between mean species richness and fragment metrics were all significant for the forest interior (negative effect) and pioneer species (positive effect, Table 3.1). We attribute these differences to the historical patterns of deforestation. In the Chilean study area, the process of deforestation began during the 1850s and was driven largely by an expansion of agricultural land and Monterrey pine and eucalypt industrial plantations, particularly since the 1970s (Lara *et al.*, 2003). At present, most native forest fragments have been affected by logging for fuelwood and timber (Echeverría *et al.*, 2007). In the Highlands of Chiapas, forest loss was also associated with intensification of traditional agriculture and exploitation of forest resources, particularly since the 1990s (Cayuela *et al.*, 2006b). Because of the slow response of tree populations to recent disturbances, it is likely that the full impact of these changes will not become apparent for some time (Hanski and Ovaskainen, 2002; Helm *et al.*, 2006), thus explaining the current lack of a relationship between diversity measures and fragment metrics.

Lack of detection, however, does not necessarily mean that the effects of fragmentation are not important. Rather it indicates the limitations of statistical and conceptual models. One of these limitations is related to the scale at which species interact with their environment. Cushman and McGarigal (2004), for instance, suggested that bird species interact most strongly with fine-scale habitat, within the range of their immediate perception. This is the scale at which predation, competition and other interspecific interactions occur, and at which the organisms experience their environment (Levey *et al.*, 2005). As a consequence, the fragment scale might not be appropriate for detecting the impacts of fragmentation and local disturbance. Our results in the Highlands of Chiapas support this hypothesis for tree diversity. We found that fragmentation and disturbance act simultaneously on tree diversity at a local scale, yet with opposite effects (Fig. 3.3). Whereas forest edges had a weak but positive effect on tree diversity (a review study by Ries *et al.*, 2004 largely corroborated this response), local disturbance was negatively related to it (Ramírez-Marcial *et al.*, 2001; Galindo-Jaimes *et al.*, 2002). Such a positive response of tree diversity to forest edges can be the result of traditional shifting cultivation, a common practice in many mountainous tropical regions of Central and South America. This creates a matrix dominated by semi-natural vegetation in various states of modification (Kappelle, 2006), which does not create dispersal barriers to most of the species, as opposed to the traditional concept of fragmentation, which implies that high-quality habitat remnants are isolated by a hostile environment to the organisms that thrive in the remnants. Under these circumstances, forest edges do not become hard boundaries between contrasting habitats but allow many species to disperse and flourish (Laurance *et al.*, 1998; Laurance and Cochrane, 2001; López-Barrera and Newton, 2005; López-Barrera *et al.*, 2006). Consequently, tree diversity increases near the forest edges (Fig. 3.3). This increase might occur owing to the increase of the more opportunistic pioneer species near these edges (Laurance *et al.*, 1998; Metzger, 2000; Hill and Curran, 2001; Kupfer *et al.*, 2004). However, we found that the positive effect of forest edges on tree diversity affected both the pioneer and late-successional species. The reason for this might be related to the time lag of tree species colonization (Helm *et al.*, 2006). After a gap is opened in the forest, pioneer species tend to colonize the forest edge. Shade-tolerant, late-successional species have a lower chance of colonizing these sites, but mature trees growing near the forest edge can persist. Consequently these effects are likely to be neglected in the short term, but would be manifest after some decades, as seen in the temperate forests of southern Chile (Table 3.1). It may be significant therefore that fragmentation effects were only detected in the latter study, in which only fragments created at least 23 years ago were assessed.

Secondary succession and tree diversity

Disturbance, particularly deforestation, triggers secondary succession and hence a change in the community type. Consistencies in the patterns of diversity can be found along the chronosequences of abandoned pastures and croplands in all of the study areas. Overall, a longer time since abandonment produced a vegetation structure and community composition more similar to that of mature forests. However, species richness strongly depended upon the functional type under consideration. There are usually negative correlations between species richness of trees and other woody plants such as shrubs and geophytes. We found the largest number of woody species in early successional forests in the Highlands of Chiapas (Table 3.3). Similarly, richness and diversity values of tree species similar to those of the mature forest were achieved in earlier successional stages (*c.*20 years) in Central Veracruz (Fig. 3.4). This pattern could be explained by the shifting balance between late-successional (positive) and pioneer species (negative, and flat for juveniles) along this chronosequence resulting in a peak in species diversity at intermediate successional stages. At a distance closer to the forest border, there are higher richness and diversity of late-successional species (Muñiz-Castro *et al.*, 2006). As the forest matures, late-successional species outcompete pioneer species. Since biomass per unit area is a function of wood density and individual size, the biomass of the secondary forest will progressively become more similar to the biomass of the primary forest as long as the characteristic tall late-successional tree species with dense wood enter into the successional community by replacing light wood pioneer species (Brown and Lugo, 1992; Clark and Clark, 1996). Rapid recovery of species richness in smaller size classes has been reported in other tropical regions (Saldarriaga *et al.*, 1988; Denslow and Guzmán, 2000).

Predicted species loss

Our models predicted the potential impact of deforestation on species diversity (Table 3.4 and Fig. 3.5). Based on ground-based floristic inventories and known forest deforestation rates derived from satellite imagery, and assuming that the drivers of deforestation will not change in the future, we calculated the percentage of species that are destined to disappear – or at least be seriously threatened with extinction – by a future year (2025). We can compare these values with the proportions of species projected to become extinct as a result of global habitat losses. Thomas *et al.* (2004) applied the species–area relationships to changes in global land use based on global rates of habitat loss. Projected conversion of humid tropical forest at an annual rate of 0.43% (Achard *et al.*, 2002) from 1990 to 2050 predicted a value of 6.3% of species destined for extinction, a rate far lower than the rates estimated for the montane tropical forests of Central Veracruz and the Highlands of Chiapas. The amount of diversity decline differed considerably from one region to another. This might be due to differences in: (i) the regional species diversity, which determines the shape of the species–area accumulation curves; and (ii) the annual deforestation rate, which determines the amount of habitat that is lost. These two characteristics vary greatly in the forests in our study regions, and hence may explain the different predictions in Table 3.4 and Fig. 3.5. Alarmingly, the rate of species destined for extinction in the Highlands of Chiapas is above 40% owing to the extraordinarily high recent deforestation rates (see Chapter 2) and the high species diversity in the region. This predicted rate of species extinction is comparable to the rates of plant extinction projected for scenarios of maximum expected climate change in Amazonia (69%, an average of different area methods, Thomas *et al.*, 2004). We note, however, that these models are still quite simplistic, since not all forest types are vanishing at the same rate, nor do they have the same diversity, and in addition the models are not spatially explicit (Cayuela *et al.*, 2006a).

Implications for conservation

Forest fragmentation is most often a direct consequence of deforestation. A study by Fahrig (1997) indicated that the effects of habitat loss on population extinction far outweigh the effects of habitat fragmentation. Forest loss continues to be a major concern in most of our study areas, as indicated in Chapter 2. Therefore an obvious priority for conservation should be preventing further forest loss and fostering habitat preservation and restoration.

Many studies have focused their attention on what spatial pattern a landscape should adopt to enhance connectivity and reduce the adverse effects of fragmentation (Collinge, 1998; Hill and Curran, 2001; Butler *et al.*, 2004; Platt, 2004; Damschen *et al.*, 2006). This is impractical in many mountain regions of Central and South America, where land tenure, poverty and social issues would impede the implementation of effective regional conservation plans.

An important finding of this study is that the patch-level effects of fragmentation will only manifest some decades after this process starts taking place and, possibly, after a certain threshold of habitat loss is surpassed (Fahrig, 2001). It has been suggested that this threshold value is at about 20% of habitat, below which the effects of habitat fragmentation on population persistence may be more evident (Fahrig, 1997, 2001). According to Ewers and Didham (2006), actual empirical measurements of the landscape threshold suggest that a figure such as 20% is far too simplistic. In fact, the threshold for some species is as high as 95%, and varies widely from species to species. Matrix quality can also influence the effects of fragmentation, but unfortunately this variable is rarely included in research investigations (Fahrig, 2001). Ideally, conservation strategies should be aimed at mitigating the external influences on the natural system as much as at preserving the natural system itself (Saunders *et al.*, 1991).

Also of considerable relevance is the management of forest by local people. This study has stressed the negative effects that over-exploitation of forest resources has on tree diversity by triggering shifts in species composition along a successional gradient. However, if enough time is allowed without disturbance, and a source of colonists is available sufficiently nearby, the vegetation structure and community composition should begin to resemble that of mature forests. This provides a potential scenario for natural vegetation recovery.

Acknowledgements

This work was financed by the European Commission BIOCORES Project (INCO Contract ICA4-CT-2001-10095), and received additional support from the FOREST Project (ALPHA Programme II-0411-FA-FCD-FI-FC). We are especially thankful to Duncan J. Golicher, Luis Galindo-Jaimes and Fabiola López-Barrera for their valuable contributions.

References

Achard, F., Eva, H.D., Stibig, H.J., Mayaux, P., Gallego, J., Richards, T. and Malingreau, J.P. (2002) Determination of deforestation rates of the world's humid tropical forests. *Science* 297, 999–1002.

Allen, T.F.H. and Starr, T.B. (1982) *Hierarchy: Perspectives for Ecological Complexity*. University of Chicago Press, Chicago, Illinois.

Austin, M.P. and Smith, T.M. (1989) A new model for the continuum concept. *Vegetatio* 83, 35–47.
Bininda-Emonds, O.R.P., Vazquez, D.P. and Manne, L.L. (2000) The calculus of biodiversity: integrating phylogeny and conservation. *Trends in Ecology and Evolution* 15, 92–94.
Blanco-Macías, A. (2007) Patterns of change in plant guilds during secondary succession in a tropical montane cloud forest area in Oaxaca, Mexico. Master thesis. Universidad Nacional Autónoma de México (UNAM), Mexico City, Mexico.
Brook, B.W., Bradshaw, C.J.A., Koh, L.P. and Sodhi, N.S. (2006) Momentum drives the crash: mass extinction in the tropics. *Biotropica* 38, 302–305.
Brown, J.H. and Lomolino, M.V. (1998) *Biogeography*. Sinauer Associates, Sunderland, Massachusetts.
Brown, S. and Lugo, A.E. (1992) Aboveground biomass estimates for tropical moist forests of the Brazilian Amazon. *Interciencia* 17, 8–18.
Butler, B.J., Swenson, J.J. and Alig, R.J. (2004) Forest fragmentation in the Pacific Northwest: quantification and correlations. *Forest Ecology and Management* 189, 363–373.
Cade, B.S., Terrell, J.W. and Schroeder, R.L. (1999) Estimating effects of limiting factors with regression quantiles. *Ecology* 80, 311–323.
Cayuela, L., Golicher, D.J., Rey-Benayas, J.M., González-Espinosa, M. and Ramírez-Marcial, N. (2006a) Fragmentation, disturbance and tree diversity conservation in tropical montane forests. *Journal of Applied Ecology* 43, 1172–1182.
Cayuela, L., Rey-Benayas, J.M. and Echeverría, C. (2006b) Clearance and fragmentation of tropical montane forests in the Highlands of Chiapas, Mexico (1975–2000). *Forest Ecology and Management* 226, 208–218.
CBD (2005) *Handbook of the Convention on Biological Diversity*, 3rd edn. CBD Secretariat, United Nations Environment Programme, Nairobi, Kenya.
Cincotta, R.P., Wisnewski, J. and Engelman, R. (2000) Human population in the biodiversity hotspots. *Nature* 404, 990–992.
Clark, D.B. and Clark, D.A. (1996) Abundance, growth and mortality of very large trees in Neotropical lowland rain forests. *Forest Ecology and Management* 80, 235–244.
Clark, D.B., Palmer, M.W. and Clark, D.A. (1999) Edaphic factors and the landscape-scale distributions of tropical rain forest trees. *Ecology* 80, 2662–2675.
Clinebell, R.R., Phillips, O., Gentry, A.H., Stark, N. and Zuuring, H. (1995) Prediction of neotropical tree and liana species richness from soil and climatic data. *Biodiversity and Conservation* 4, 56–90.
Cochrane, M.A., Alencar, A., Schulze, M.D., Souza, C.M., Nepstad, D.C., Lefebvre, P. and Davidson, E.A. (1999) Positive feedbacks in the fire dynamics of closed canopy tropical forests. *Science* 284, 1832–1835.
Collinge, S.K. (1998) Spatial arrangement of habitat patches and corridors: clues from ecological field experiment. *Landscape and Urban Planning* 42, 157–168.
Colwell, R.K. and Coddington, J.A. (1994) Estimating terrestrial biodiversity through extrapolation. *Philosophical Transactions of the Royal Society of London B* 345, 101–118.
Costanza, R., Arge, R., de Groot, R., Farber, S., Grasso, M., Hannon, B., Limburg, K., Naeem, S., O'Neill, R.V., Paruelo, J., Raskin, R.G., Sutton, P. and van den Belt, M. (1997) The value of the world's ecosystem services and natural capital. *Nature* 387, 253–260.
Cushman, S.A. and McGarigal, K. (2004) Patterns in the species–environment relationship depend on both scale and choice of response variables. *Oikos* 105, 117–124.
da Silva, J.M.C. and Tabarelli, M. (2000) Tree species impoverishment and the future flora of the Atlantic forest of northeast Brazil. *Nature* 404, 72–74.
Damschen, E.I., Haddad, N.M., Orrock, J.L., Tewksbury, J.J. and Levey, D.J. (2006) Corridors increase plant species richness at large scales. *Science* 313, 1284–1286.
Debinski, D.M. and Holt, R.D. (2000) A survey and overview of habitat fragmentation experiments. *Conservation Biology* 14, 342–355.

Denslow, J.S. and Guzmán, S. (2000) Variation in stand structure, light, and seedling abundance across a tropical moist forest chronosequence. Panama. *Journal of Vegetation Science* 11, 201–212.
Echeverría, C., Lara, A., Newton, A.C., Rey-Benayas, J.M. and Coomes, D. (2007) Impacts of forest fragmentation on species composition and forest structure in the temperate landscape in southern Chile. *Global Ecology and Biogeography* 16, 426–439.
Ewers, R.M. and Didham, R.K. (2006) Confounding factors in the detection of species responses to habitat fragmentation. *Biological Reviews* 81, 117–142.
Fahrig, L. (1997) Relative effects of habitat loss and fragmentation on species extinction. *Journal of Wildlife Management* 61, 603–610.
Fahrig, L. (2001) How much habitat is enough? *Biological Conservation* 100, 65–74.
Fahrig, L. (2003) Effects of habitat fragmentation on biodiversity. *Annual Review of Ecology and Systematics* 34, 487–515.
FAO–UNESCO (1974) *Soil Map of the World, 1:5,000,000. Volume 1: Legend*. United Nations Education, Scientific and Cultural Organization (UNESCO), Paris, France.
Ferraz, G., Russell, G.J., Stouffer, P.C., Bierregaard, R.O., Pimm, S.L. and Lovejoy, T.E. (2003) Rates of species loss from Amazonian forest fragments. *Proceedings of the National Academy of Sciences* 100, 14069–14073.
Fisher, R.A., Corbet, A.S. and Williams, C.B. (1943) The relation between the number of species and the number of individuals in a random sample of an animal population. *Journal of Animal Ecology* 12, 42–58.
Forman, R.T.T. and Godron, M. (1986) *Landscape Ecology*. John Wiley and Sons, New York.
Francis, A.P. and Currie, D.J. (2003) A globally consistent richness–climate relationship for angiosperms. *The American Naturalist* 161, 523–536.
Franklin, S. (2001) *Remote Sensing for Sustainable Forest Management*. Lewis Publishers, Boca Raton, Florida.
Galindo-Jaimes, L., González-Espinosa, M., Quintana-Ascencio, P. and García-Barrios, L. (2002) Tree composition and structure in disturbed stands with varying dominance by *Pinus* spp. in the Highlands of Chiapas, Mexico. *Plant Ecology* 162, 259–272.
Gascon, C., Williamson, G.B. and Fonseca, G.A.B. (2000) Receding edges and vanishing reserves. *Science* 288, 1356–1358.
Gaston, K.J., Blackburn, T.M. and Goldewijk, K.K. (2003) Habitat conversion and global avian biodiversity loss. *Procedures of the Royal Society of London B* 270, 1293–1300.
González-Espinosa, M., Quintana-Ascencio, P.F., Ramírez-Marcial, N. and Gaytán-Guzmán, P. (1991) Secondary succession in disturbed *Pinus–Quercus* forests of the Highlands of Chiapas, Mexico. *Journal of Vegetation Science* 2, 351–360.
González-Espinosa, M., Ochoa-Gaona, S., Ramírez-Marcial, N. and Quintana-Ascencio, P.F. (1995) Current land-use trends and conservation of old-growth forest habitats in the Highlands of Chiapas, Mexico. In: Wilson, M.H. and Sader, S.A. (eds) *Conservation of Neotropical Migratory Birds in Mexico*. Maine Agriculture and Forest Experiment Station, Miscellaneous Publication 727, Orono, Maine, pp. 190–198.
González-Espinosa, M., Rey-Benayas, J.M., Ramírez-Marcial, N., Huston, M.A. and Golicher, D. (2004) Tree diversity in the northern neotropics: regional patterns in highly diverse Chiapas, Mexico. *Ecography* 27, 741–756.
González-Espinosa, M., Ramírez-Marcial, N. and Galindo-Jaimes, J. (2006) Secondary succession in montane pine–oak forests in Chiapas, Mexico. In: Kappelle, M. (ed.) *Ecology and Conservation of Neotropical Montane Oak Forests*, *Ecological Studies 185*. Springer, Berlin, Germany, pp. 209–221.
Hanski, I. and Ovaskainen, O. (2002) Extinction debt at extinction threshold. *Conservation Biology* 16, 666–673.

Harris, L.D. (1984) *The Fragmented Forest: Island Biogeography Theory and the Preservation of Biotic Diversity*. University of Chicago Press, Chicago, Illinois.
Helm, A., Hanski, I. and Pärtel, M. (2006) Slow response of plant species richness to habitat loss and fragmentation. *Ecology Letters* 9, 72–77.
Heywood, V.H. (1995) *Global Biodiversity Assessment*. Cambridge University Press, Cambridge, UK.
Hill, J.L. and Curran, P.J. (2001) Species composition in fragmented forests: conservation implications of changing forest area. *Applied Geography* 21, 157–174.
Hill, J.L and Curran, P.J. (2003) Area, shape and isolation of tropical forest fragments: effects on tree species diversity and implications for conservation. *Journal of Biogeography* 30, 1391–1403.
Huston, M.A. (1980) A general hypothesis of species diversity. *The American Naturalist* 113, 81–101.
Kappelle, M. (2006) *Ecology and Conservation of Neotropical Montane Oak Forests, Ecological Studies No. 185*. Springer, Berlin, Germany.
Kerr, J.T. and Packer, L. (1997) Habitat heterogeneity as a determinant of mammal species richness in high-energy regions. *Nature* 285, 252–254.
Kupfer, J.A., Webbeking, A.L. and Franklin, S.B. (2004) Forest fragmentation affects early successional patterns on shifting cultivation fields near Indian Church, Belize. *Agriculture, Ecosystems and Environment* 103, 509–518.
Lara, A., Soto, D., Armesto, J., Donoso, P., Wernli, C., Nahuelhual, L. and Squeo, F. (2003) *Componentes Científicos Clave para una Política Nacional Sobre Usos, Servicios y Conservación de los Bosques Nativos Chilenos*. Universidad Austral de Chile, Valdivia, Chile.
Laurance, W.F. and Bierregaard, R.O. (1997) *Tropical Forest Remnants: Ecology, Management and Conservation of Fragmented Communities*. University of Chicago Press, Chicago, Illinois.
Laurance, W.F. and Cochrane, M.A. (2001) Synergistic effects in fragmented landscapes. *Conservation Biology* 15, 1488–1489.
Laurance, W.F., Ferreira, L.V., Rankin-de Merona, J.M., Laurance, S.G., Hutchings, R.W. and Lovejoy, T.E. (1998) Effects of forest fragmentation on recruitment patterns in Amazonian tree communities. *Conservation Biology* 12, 460–464.
Lawton, J.H., Bignell, D.E., Bolton, B., Bloemers, G.F., Eggleton, P., Hammond, P.M., Hodda, M., Holt, R.D., Larsen, T.B., Mawdsley, N.A., Stork, N.E., Srivastava, D.S. and Watt, A.D. (1998) Biodiversity inventories, indicator taxa and effects of habitat modification in tropical forest. *Nature* 391, 72–76.
Levey, D.J., Bolker, B.M., Tewksbury, J.J., Sargent, S. and Haddad, N.M. (2005) Effects of landscape corridors on seed dispersal by birds. *Science* 309, 146–148.
Lobo, J.M., Castro, I. and Moreno, J.C. (2001) Spatial and environmental determinants of vascular plant species richness distribution in the Iberian Peninsula and Balearic islands. *Biological Journal of the Linnean Society* 72, 233–253.
López-Barrera, F. and Newton, A. (2005) Edge type on germination of oak tree species in the Highlands of Chiapas, Mexico. *Forest Ecology and Management* 217, 67–79.
López-Barrera, F., Manson, R.H., González-Espinosa, M. and Newton, A.C. (2006) Effects of the type of montane edge on oak seedling establishment along forest-edge-exterior gradients. *Forest Ecology and Management* 225, 234–244.
Magurran, A.E. (1988) *Ecological Diversity and Its Measurement*. Princeton University Press, Princeton, New Jersey.
Metzger, J.P. (2000) Tree functional group richness and landscape structure in a Brazilian tropical fragmented landscape. *Ecological Applications* 10, 1147–1161.
Mitchell-Olds, T. and Shaw, R.G. (1987) Regression analysis of natural selection, statistical inference and biological interpretation. *Evolution* 41, 1149–1161.

Mittelbach, G.G., Steiner, C.F., Scheiner, S.M., Gross, K.L., Reynolds, H.L., Waide, R.B., Willig, M.R., Dodson, S.I. and Gough, L. (2001) What is the observed relationship between species richness and productivity? *Ecology* 82, 2381–2396.
Muñiz-Castro, M.A., Williams-Linera, G. and Rey-Benayas, J.M. (2006) Distance effect from cloud forest fragments on plant community structure in abandoned pastures in Veracruz, Mexico. *Journal of Tropical Ecology* 22, 431–440.
Myers, N., Mittermeler, R.A., Mittermeler, C.G., da Fonseca, G.A.B. and Kent, J. (2000) Biodiversity hotspots for conservation priorities. *Nature* 403, 853–858.
Nepstad, D.C., Verissimo, A., Alencar, A., Nobre, C., Lima, E., Lefebvre, P., Schlesinger, P., Potter, C., Moutinho, P., Mendoza, E., Cochrane, M. and Brooks, V. (1999) Large-scale impoverishment of Amazonian forests by logging and fire. *Nature* 398, 505–508.
Pausas, J.G. and Austin, M.P. (2001) Patterns of plant species richness in relation to different environments: an appraisal. *Journal of Vegetation Science* 12, 153–166.
Pimm, S.L. and Raven, P. (2000) Extinction by numbers. *Nature* 403, 843–844.
Pinheiro, J.C. and Bates, D.M. (2000) *Mixed-Effects Models in S and S-PLUS. Statistics and Computing*. Springer, New York.
Platt, R.V. (2004) Global and local analysis of fragmentation in a mountain region of Colorado. *Agriculture, Ecosystems and Environment* 101, 207–218.
Preston, F.W. (1962) The canonical distribution of commonness and rarity: part I. *Ecology* 43, 185–215.
Qian, H. and Ricklefs, R.E. (2004) Taxon richness and climate in angiosperms: is there a globally consistent relationship that precludes region effects? *The American Naturalist* 163, 773–779.
Ramírez-Marcial, N., González-Espinosa, M. and Williams-Linera, G. (2001) Anthropogenic disturbance and tree diversity in montane rain forests in Chiapas, Mexico. *Forest Ecology and Management* 154, 311–326.
Reed, R., Johnson-Barnard, J. and Baker, W. (1996) Fragmentation of a forested rocky mountain landscape, 1950–1993. *Biological Conservation* 75, 267–277.
Rey-Benayas, J.M. and Scheiner, S.M. (2002) Plant diversity, biogeography and environment in Iberia: patterns and possible causal factors. *Journal of Vegetation Science* 13, 245–258.
Ricketts, T.H., Dinerstein, E., Olson, D.M., Loucks, C.J., Eichbaum, W., DellaSala, D., Kavanagh, K., Hedao, P., Hurley, P.T., Carney, K.M., Abell, R. and Walters, S. (1999) *Terrestrial Ecoregions of North America: A Conservation Assessment*. Island Press, Washington, DC.
Ricklefs, R.E. (2004) A comprehensive framework for global patterns in biodiversity. *Ecology Letters* 7, 1–15.
Ries, L., Fletcher, R.J., Battin, J. and Sisk, T.D. (2004) Ecological responses to habitat edges: mechanisms, models, and variability explained. *Annual Review of Ecology and Systematics* 35, 491–522.
Rosenblatt, D.L., Heske, E.J., Nelson, S.L., Barber, D.M., Miller, M.A. and MacAllister, B. (1999) Forest fragmentation in East-central Illinois: islands or habitat patches for mammals? *The American Midland Naturalist* 141, 115–123.
Rosenzweig, M.L. (1995) *Species Diversity in Space and Time*. Cambridge University Press, Cambridge, UK.
Rosenzweig, M.L. and Abramsky, Z. (1993) How are diversity and productivity related? In: Ricklefs, R.E. and Schluter, D. (eds) *Species Diversity in Ecological Communities: Historical and Geographical Perspectives*. University of Chicago Press, Chicago, Illinois, pp. 52–65.
Saldarriaga, J.G., West, D.C., Tharp, M.L. and Uhl, C. (1988) Long term chronosequence of forest succession in the upper Rio Negro of Colombia and Venezuela. *Journal of Ecology* 76, 938–958.

Saunders, D.A., Hobbs, R.J. and Margules, C.R. (1991) Biological consequences of ecosystem fragmentation: a review. *Conservation Biology* 5, 18–32.
Sosa, V. and Gómez-Pompa, A. (1994) *Flora de Veracruz*. Instituto de Ecología de Xalapa, A.C. Xalapa, Mexico. University of California, Riverside, California.
Stockwell, D. and Peterson, A.T. (2003) Comparison of resolution of methods used in mapping biodiversity patterns from point-occurrence data. *Ecological Indicators* 3, 213–221.
Terborgh, J. (1999) *Requiem for Nature*. Island Press, Washington, DC.
Thomas, C.D., Cameron, A., Green, R.E., Bakkenes, M., Beaumont, L.J., Gollingham, Y.C., Erasmus, B.F.N., de Siqueira, M.F., Grainger, A., Hannah, L., Hughes, L., Huntley, B., van Jaarsveld, A.S., Midgley, G.F., Miles, L., Ortega-Huerta, M.A., Peterson, A.T., Phillips, O.L. and Williams, S.E. (2004) Extinction risk from climate change. *Nature* 427, 145–148.
Tilman, D. (1999) The ecological consequences of changes in biodiversity: a search for general principles. *Ecology* 80, 1455–1474.
Tilman, D., May, R.M., Lehman, C.L. and Nowak, M.A. (1994) Habitat destruction and the extinction debt. *Nature* 371, 65–66.
Waldhardt, R. (2003) Biodiversity and landscape–summary, conclusions and perspectives. *Agriculture, Ecosystems and Environment* 98, 305–309.
Whittaker, R.J., Willis, K.J. and Field, R. (2001) Scale and species richness: toward a general hierarchical theory of species diversity. *Journal of Biogeography* 28, 453–470.
Wright, D.H., Currie, D.J. and Maurer, B.A. (1993) Energy supply and patterns of species richness on local and regional scales. In: Ricklefs, R.E. and Schluter, D. (eds) *Species Diversity in Ecological Communities: Historical and Geographical Perspectives*. Chicago University Press, Chicago, Illinois, pp. 66–74.
Wright, S.J. and Muller-Landau, C. (2006) The future of tropical forest species. *Biotropica* 38, 287–301.

4 Fragmentation and Edge Effects on Plant–Animal Interactions, Ecological Processes and Biodiversity

F. López-Barrera, J.J. Armesto, G. Williams-Linera, C. Smith-Ramírez and R.H. Manson

A fragment of tropical montane forest in central Veracruz, Mexico. Note the high contrast ('hard edge') with the surrounding pasture land. Photo: Adrian Newton

 Biodiversity Loss and Conservation in Fragmented Forest Landscapes: The Forests of Montane Mexico and Temperate South America (ed. A.C. Newton)

Summary

We summarize studies of forest fragmentation and edge effects on a diverse range of ecological processes and abiotic variables in neotropical montane and south temperate rainforests. The main findings from these studies are that: (i) anthropogenic edge effects significantly altered forest regeneration processes occurring over small spatial and temporal scales; (ii) adjacent vegetation type affected not only the probabilities of tree invasion and regeneration, but also the extent of microclimatic edge effects within the forest interior; and (iii) edge structure and function are linked as a habitat for plants and animals and as a front for forest expansion. We found that the main modulators of edge effects were: (i) forest edge to non-forest matrix contrast (hard and soft edges); (ii) edge orientation with respect to biotic or abiotic fluxes; (iii) season (dry or wet) or year of study (temporal variance); and (iv) species-specific responses. Future edge studies should consider the modulators of edge effects for the particular response variable being studied. The consequences of edge effects for the conservation of regional biodiversity and changes in forest structure in fragmented forest landscapes are discussed.

Introduction

Forest fragmentation and its effects

Global forest fragmentation has been documented extensively, with an emphasis on the substantial loss of tropical rainforests in Central Africa and Amazonia (Fearnside, 1996; Justice *et al.*, 2001; Semazzi and Yi, 2001; Zhang *et al.*, 2001). The tropical montane forests of Mexico and Central America and the temperate rainforests of southern South America have been less studied, but are also suffering rapid changes in land use leading to increased forest fragmentation and larger perimeter/forest patch area ratios (see Chapter 2). Such patterns are threatening the conservation of regional biodiversity, especially narrow endemics and forest specialists in each forest type (Chapter 3), as well as the dynamics of biotic interactions in rural landscapes. In general, the main trends associated with anthropogenic forest fragmentation are: (i) increasing habitat loss; (ii) increasing number of forest fragments; (iii) decreasing size of forest fragments; and (iv) increasing isolation of remnant forest habitats (Fahrig, 2003). A higher number of isolated forest patches leads to the creation of more edge habitat or forest–matrix transitions. Loss of forest cover, therefore, changes the habitat mosaic not only by creating new edges, but also by changing the edge contrast, from low to high contrast between the forest and the adjacent degraded or human-dominated habitat (Wiens *et al.*, 1985). These changes are likely to affect wildlife habitat quality, ecological processes and ultimately regional and local biodiversity.

Edges are a transition zone separating two contiguous habitat types that are perceived by some focal organism as being of significantly different quality (Lidicker, 1999). Hence, edge definition and measurement depend upon habitat use by focal species and the spatial scale of the study (Murcia, 1995; Sarlov-Herlin, 2001). Accordingly, the study of habitat edges is subjected to several restrictions (Lidicker, 1999): (i) recognition of habitat edges depends on the human observer; (ii) responses to habitat edges will be species-specific and possibly sex- and age-specific as well; and (iii) assessing the width and length of habitat edges is difficult, as various abiotic and biotic factors, which may influence the focal organism, penetrate to different distances across an edge. Considering these restrictions, it is not surprising that it has been difficult to generate a unifying theory of habitat edges (see reviews by Ries *et al.*, 2004; Harper *et al.*, 2005). However, extensive reviews suggest that the magnitude and distance of edge influences are a direct function of the contrast in structure and composition between adjacent habitats, resulting in different edge types. Therefore edge responses are more predictable where specific focal species, distributed along specific edge types, are defined a priori (Ries *et al.*, 2004).

Different functional edge types have been compared in a small number of experimental studies, the results of which have suggested the following a priori edge type classifications: (i) thinned versus intact (Cadenasso and Pickett, 2000, 2001; Kollmann and Buschor, 2002); (ii) natural versus anthropogenic (Song and Hannon, 1999); (iii) hard versus soft (Fenske-Crawford and Niemi, 1997; López-Barrera and Newton, 2005; López-Barrera *et al.*, 2005); and (iv) border-edge cuts versus uncut edges (Fleming and Giuliano, 1998). The characteristics of the edge itself (thickness, sharpness, etc.) influence not only the movement within or across edges, but also the movement to and from adjacent patches in the landscape (Sarlov-Herlin, 2001). Duelli *et al.* (1990) suggested that permeability to the movement of organisms is an important edge feature and proposed six edge types based on the 'hardness' for the focal organism. However, permeability can also refer to physical influences across edges, such as the effects of atmospheric chemistry and fertilizers derived from activities in the surrounding matrix.

Edge effects and ecological processes

The term 'edge effect' was first used in 1933 by Leopold, a wildlife ecologist, to explain the increased richness of generalist game species at edges between two habitats or 'ecotones' (Sarlov-Herlin, 2001). Later the concept was broadened to include the negative impacts of edges within large and well-preserved forest fragments (Fox *et al.*, 1997; Benitez-Malvido, 1998; Gascon *et al.*, 2000). As applied to tropical countries, studies of reserve design first addressed the issue of edges in planning protected areas (Laurance, 1991). Today the concept comprises a wide range of ecological processes occurring at edges (Murcia, 1995), as mutual influences on physical and biological flows result in changes of species composition and structure (Fagan *et al.*, 1999; Lidicker, 1999; Cadenasso and Pickett, 2000, 2001; Laurance *et al.*, 2001).

Edge effects may be defined by changes in physical or biotic response variables, which occur at the transition between adjoining habitats (Lidicker, 1999). The current use of the edge-effect concept in the literature summarizes a diversity of responses. Edges may have both positive and negative consequences for focal organisms and may produce emergent response properties (Fig. 4.1). Lidicker (1999) differentiated two general edge effects depending on emergent properties: the *matrix* effect and the *ecotone* effect. The matrix effect is an abrupt change in some response variable as the edge is crossed,

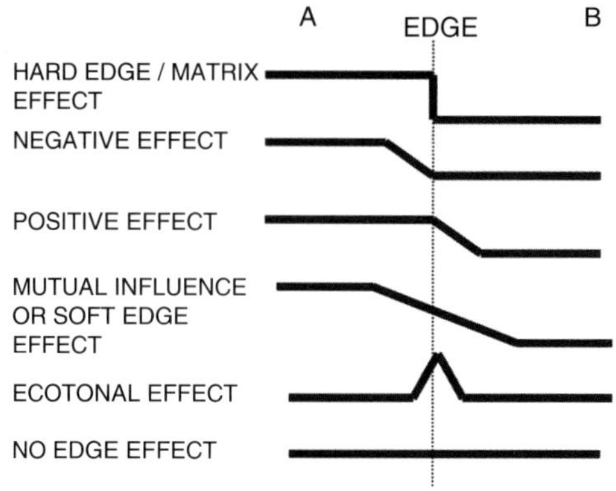

Fig. 4.1. Simple representation of potential edge effects for physical or biotic response variables (thick line). A and B are two juxtaposed habitat types. Modified from Duelli *et al.* (1990).

where the response of organisms at an edge can be explained strictly by the organism's behaviour in the two habitat types (away from the edge). This type of boundary is defined as a 'hard edge' by Duelli *et al.* (1990; Fig. 4.1). The ecotonal effect is characterized by the presence of emergent properties (negative, positive or mutual influences), therefore the response of the organism at the edge cannot be explained solely by its contrasting behaviour in each habitat type (Lidicker and Peterson, 1999). Emergent properties could produce either increasing or decreasing responses near the edge (Fig. 4.1).

Studies of habitat fragmentation have often examined the effects of edges on bird nest predation by mammals and other birds (see review by McCollin, 1998). Edges are believed to be detrimental to some bird species because of reduced reproductive success and increased rates of nest parasitism and predation. However, studies of the effects of distance from edge on the nesting success of birds have produced mixed results. Not all studies have documented edge effects and the general pattern seems to vary according to region, ecosystem, predator assemblage, forest size and type of adjoining habitat (Andren, 1994; Murcia, 1995; Hinsley *et al.*, 1998; Bergin *et al.*, 2000; Brand and George, 2000; Hansson, 2000). It seems that an edge-related increase in nest predation is most common inside small forest patches surrounded by farmland or highly fragmented anthropogenic landscapes, and is rarely present or undetectable in forest mosaics or continuous landscapes (Donovan *et al.*, 1997; McCollin, 1998).

Edge effects on seed and seedling herbivory have been less studied than bird nest predation. There are no comparisons of these effects in forest mosaics or forests surrounded by rural or urban habitats. Small mammals are important seed predators and/or dispersers in many forest landscapes. Habitat use (specialist or generalist) by small mammals will have a great influence on seed–predator interactions in edges (Lidicker, 1999; Manson *et al.*, 1999) by determining the ability of the seed predators and/or dispersers to move between adjacent habitat patches of different quality (Rodriguez *et al.*, 2001). Increased herbivory in edges may also be associated with greater insect activity, as insects may be attracted to greater productivity in open canopy habitats (Chacón and Armesto, 2006).

Edge effects on seed dispersal are associated with changes in vegetation structure (Kollmann and Buschor, 2002); however, most studies of edge effects on tree regeneration do not provide a precise description of edge structure. There is substantial discrepancy among recent studies regarding the existence and intensity of edge effects on seed predation (Kollmann and Buschor, 2002). Some lack of consistency in the results may be attributed to improper design (lack of true replication), differences in edge definition, lack of temporal replication and oversimplification of the perception of edge dynamics (Murcia, 1995) and also to temporal and spatial variability in the occurrence of different seed predators.

Although negative consequences of fragmentation and resulting edge effects have been documented in a large number of studies performed mainly in lowland tropical rainforest (Williams-Linera, 1990a, b; Malcolm, 1994; Fox *et al.*, 1997; Kapos *et al.*, 1997; Laurance, 1997; Turton and Freiburger, 1997; Benitez-Malvido, 1998; Laurance *et al.*, 1998; Didham and Lawton, 1999; Sizer and Tanner, 1999; Gascon *et al.*, 2000; Laurance and Williamson, 2001), research has been sparse in tropical montane and high-latitude temperate rainforests, such as those in Mexico and Chile. Despite their globally recognized conservation importance (Myers *et al.*, 2000), we know little about the impact of fragmentation in these ecosystems that results from the increased land-use change and logging in recent decades (Chapter 2). In this chapter, we present a synthesis of ongoing research to assess the effects of such fragmentation on a variety of ecological processes in these two study regions. Although the studies were not designed to be directly comparable, we consider it important to examine the generality of these findings and their value to forest managers in efforts to conserve regional biodiversity. Finally, we propose several directions for future research in these regions based on the results presented.

General predictions and aims

The following general predictions were tested as part of the research summarized here:

1. Forest edges are habitats with emergent properties different from those of forest interior and tree-fall gaps.
2. Edge-related changes in abiotic conditions (light, soil moisture, etc.) relative to the forest interior or tree-fall gaps may result in measurable differences in forest structure, species composition and ecological interactions along edges.
3. Edges may induce changes in the abundance and distribution of species, which will in turn produce changes in species interactions, such as predation, herbivory, pollination and seed dispersal.

4. Edge orientation modulates the intensity of abiotic effects (such as fog penetration in cloud-dependent forests), thereby affecting resource distribution and consequently forest structure and dynamics.
5. According to landscape ecology theory, as structural similarity between two adjacent habitats increases, the edge created becomes less abrupt and the edge effect less evident. Hence, ecological flows (such as animal movements) across edges may be enhanced by uniformity and reduced by sharpness or abruptness of edges.

Research Approaches

Edges, forests and canopy gaps

Because of the considerable heterogeneity among study systems and research foci, several different experimental and descriptive approaches were employed (Table 4.1). Most of the studies defined a priori the habitat contrasts compared, for instance forest edges versus canopy gaps, edges versus forest interior, trees in forest patches versus isolated trees in pastures, forests versus shrublands, etc. Five of the studies ('experimental habitat') introduced artificial avian nests, seeds or seedlings to evaluate the effect of the habitat on ecological process, and eight studies ('descriptive habitat') estimated attributes of plants and animals (species richness, abundance, etc.) living in contrasting adjacent habitats. Two studies ('descriptive gradient') estimated forest structure and composition along forest-edge–exterior gradients and six studies ('experimental gradient') tested different ecological processes along forest-edge–exterior gradients (using distance from the edge as a factor); most of these studies also tested the effects of edge type.

Varying sized forest patches, isolated trees and riparian corridors

Other studies within this project compared ecological processes such as pollination and seed dispersal, focusing on landscape elements including isolated trees in pastures, remnant forest patches in rural areas (1 ha, small; 8–23 ha medium; > 150 ha large patches; Smith-Ramírez and Armesto, 2003), and the function of riparian vegetation strips in rural landscapes (Box 4.10).

Main Research Findings

Edge effects

Owing to the broad range of methodological approaches, differences among the study regions, and variation in the response variables measured, we summarize the main results in Table 4.2 comparing edge versus interior habitat in forest patches. To integrate the extensive range of response

Table 4.1. Different descriptive and experimental approaches used in the studies of forest edges summarized in this review.

Approach	Response variables or ecological processes measured	Spatial scale (m)	Reference
Experimental Habitat contrasts: Forest edges were defined as habitats and were compared with other habitats such as forest interior, canopy gaps, isolated trees in pastures, shrublands, old-fields	Nest predation risk	5–50	Willson *et al*. (2001)
	Seed and seedling survival	1–50	Guzmán-Guzmán and Williams-Linera (2006)
	Seed predation	1–100	Díaz *et al*. (1999)
	Seedling growth, survival, specific leaf area and foliar damage	<100	Chacón and Armesto (2005, 2006)
	Foliar damage	1–30	Reynoso-Moran and Williams-Linera (Box 4.2)
Descriptive Habitat contrasts: Forest edges were defined as habitats and were compared with other habitats such as forest interior, forest exterior, gaps or riparian corridors	Seed rain density and composition	20–100	Armesto *et al*. (2001)
	Epiphyte cover and diversity	1–100	Salinas and Armesto (Box 4.5)
	Bryophyte diversity	1–300	Larraín and Armesto (Box 4.6)
	Seedling and sapling density	10–50	Gutiérrez *et al*. (Box 4.8)
	Tree composition		
	Tree growth rate	10–50	Gutiérrez *et al*. (Box 4.9)
	Plant species richness and diversity	1–50	Muñiz-Castro *et al*. (2006)
	Avian species richness	1–500	Jaña *et al*. (Box 4.1)
	Seed rain		
	Small mammals; abundance and composition	8–80	Ruán-Tejeda and Manson (Box 4.7)
Descriptive Gradient: Different distances from the edge into the interior	Plant species richness and forest structure	1–60	Guzmán-Guzmán and Williams-Linera (Box 4.4)
		1–60	López-Barrera (Box 4.3)
Experimental Gradient: Different distances from the edge into the interior and into the matrix were compared; edge type distinction made	Germination success	1–50	López-Barrera and Newton (2005)
	Rate of acorn removal	1–60	López-Barrera *et al*. (2005)
	Acorn removal and movement	1–60	López-Barrera *et al*. (2007)
	Distribution of small mammals		
	Seedling establishment	1–60	López-Barrera *et al*. (2006)
	Nesting success	1–50	De Santo *et al*. (2002)
	Fog capture	10–200	del Val *et al*. (2006)
	Seedling and sapling densities		

Table 4.2. Summary of studies measuring ecological processes, patterns and microclimate at the forest edge (zero metres) compared to forests interior (except in the case of ecological flows), showing higher or lower values of the response variable at the edge, or no difference. Modulators are factors that conditioned the observed response.

Response variable	Region	Modulators	Higher	Lower	No difference	Potential effects on species richness	Reference
Dynamic processes							
Seed rain	Chiloé	Only fleshy-fruited species	X			Positive	Armesto *et al.* (2001)
	Chiloé	Landscape type		X		Neutral	Jaña *et al.* (Box 4.1)
Avian nest predation	Chiloé		X			Negative	Willson *et al.* (2001)
	Chiloé			X		Positive	De Santo *et al.* (2002)
Seed germination	Chiapas	Species identity	X			Positive	López-Barrera and Newton (2005)
Seed removal and/or predation	Chiapas	Only non-masting years	X			Uncertain	López-Barrera *et al.* (2005)
	Veracruz			X			Guzmán-Guzmán and Williams-Linera (2006)
	Chiloé	Species identity, season and canopy opening		X		Neutral	Díaz *et al.* (1999)
Seedling herbivory	Chiloé			X		Positive	Chacón and Armesto (2005)
	Chiapas	Edge specific: only in hard edges	X			Negative in hard edges	López-Barrera *et al.* (2006)

	Veracruz	Season specific: only in the soft edge during the dry season	X				Reynoso-Moran and Williams-Linera (Box 4.2)
Seedling performance (survival and growth)	Chiapas		X			Positive	López-Barrera *et al*. (2006)
	Chiloé						
	Veracruz		X				Chacón and Armesto (2006)
			X				Guzmán-Guzmán and Williams-Linera (2006)
			X				Reynoso-Moran and Williams-Linera (Box 4.2)
Tree growth rate		Species-dependent	X			Positive	Gutiérrez *et al*. (Box 4.9)
Ecological flows							
Fog interception and tree regeneration	Fray Jorge	Fog influx direction	X			Positive in windward edges	del Val *et al*. (2006)
Seed movement from edge into the exterior	Chiapas	Edge specific: only in soft edges	X			Positive in soft edges	López-Barrera *et al*. (2007)
Patterns							
Woody species richness/ diversity	Chiapas				X	Uncertain	López-Barrera (Box 4.3)
	Veracruz				X		Guzmán-Guzmán and Williams-Linera (Box 4.4)
					X		Muñiz-Castro *et al*. (2006)
Epiphyte cover and species richness	Chiloé			X		Negative	Salinas and Armesto (Box 4.5)
Bryophyte species richness	Chiloé			X		Negative	Larraín and Armesto (Box 4.6)
Avian species richness	Chiloé				X	Neutral	Jaña *et al*. (Box 4.1)

Continued

Table 4.2. *Continued*

Response variable	Region	Modulators	Higher	Lower	No difference	Potential effects on species richness	Reference
Diversity and abundance of native small mammals	Chiapas	Edge specific: only in soft edges	X			Positive in soft edges	López-Barrera *et al*. (2007)
	Veracruz	Only in fragments with soft edges	X			Positive in soft edges	Ruán-Tejeda and Manson (Box 4.7)
Overall tree seedling and sapling density	Chiloé	Species identity and shade tolerance	X			Positive	Gutiérrez *et al*. (Box 4.8)
	Chiapas				X	Uncertain	López-Barrera (Box 4.3)
	Veracruz				X		Guzmán-Guzmán and Williams-Linera (Box 4.4)
Microclimate							
Soil moisture	Semi-arid Chile	Fog influx direction	X				del Val *et al*. (2006)
	Veracruz			X			Reynoso-Moran and Williams-Linera (Box 4.2)
	Chiapas				X	Uncertain	López-Barrera (Box 4.3)
Light	Chiapas	Edge specific: only in hard edges	X			Uncertain	López-Barrera (Box 4.3)
Air temperature	Veracruz		X				Reynoso-Moran and Williams-Linera (Box 4.2)

variables we developed a conceptual framework to compare the patterns recorded along habitat edges. This framework is based on the description of modulators of edge effects, i.e. the factors that determine the recorded change of the response variable along the edge habitat versus the forest interior. Published work is cited in Table 4.2 and briefly discussed in the text, while an abstract of unpublished (or submitted) work is presented as case studies (included in individual boxes) appended at the end of this chapter.

We reviewed 21 studies that recorded 34 response variables that compared forest edge and forest interior habitats in tropical montane and south temperate rainforests. From these 34 response variables, 19 resulted in positive edge effect (higher values of the response variable in the edge versus forest), nine in a negative edge effect (lower values of the response variable in the edge versus forest) and six recorded no edge effect. Considering all positive and negative edge effects documented, 40% recorded an additional factor (modulator) that determined the occurrence of the edge effect. It is interesting to note that for the 18 response variables that characterized ecological processes (seed rain, seedling herbivory, seed movements across borders, etc.), all recorded higher or lower values of the response variable comparing edges versus forest interior habitats (Fig. 4.2). However, from the 16 studies that characterized ecological patterns (plant and animal species richness, seedling and sapling densities, etc.) in edges and forest interior habitats, six did not detect significant differences between these two habitats (Fig. 4.2). As patterns arise from processes, this difference is difficult to understand, but may reflect the loss of a detectable signal as one moves from one level of an ecological hierarchy to another, much as the pattern observed in trophic cascades (Finke and Denno, 2004). Another possible explanation is that some of the species studied are habitat generalists and have sufficient phenotypic plasticity to use interior and edge habitats, as is the case for most avian species in Chilean forests (Rozzi *et al.*, 1996) and for several tree

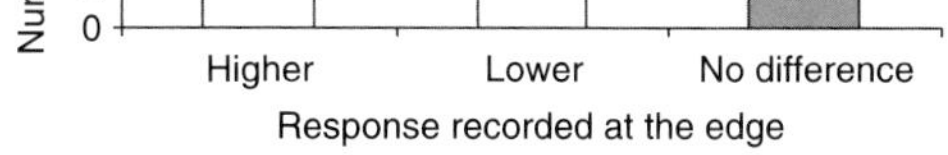

Fig. 4.2. Number of response variables that showed higher, lower or no difference (no edge effect) in the values of a response variable that was compared at the edge versus forest interior. The response variables were divided according to their type; pattern refers to forest attributes such as plant species richness, tree seedling abundance, etc. Process refers to ecological attributes and interactions, such as seed removal by animals, seed dispersal, seedling herbivory, fog capture, etc.

species that are intermediate in light requirements (Figueroa and Lusk, 2001; Aravena *et al.*, 2002).

In terms of edge theory, it has been argued that edge versus matrix contrasts may influence edge effects and therefore edge-related changes in the response variable will be weaker near soft edges (low contrast) than hard edges (high contrast edges; Ries *et al.*, 2004). Differences in edge responses, in the cases reported here, were mainly attributable to different mean vegetation heights and density between adjacent habitats. Studies in Mexico considered this modulator of edge effects in two study areas where hard and soft edges were compared (Table 4.3). In both study areas, studies in both hard and soft edges recorded different emergent properties, but only when ecological processes were studied as a response variable. When static patterns such as forest plant species and tree seedling abundance were studied, no significant differences were recorded in either study region.

The studies that compared ecological processes among forest edges versus forest interior and versus canopy gaps generally found that canopy gaps represented a qualitatively different habitat for forest dynamics compared to the forest interior. Forest edges represented intermediate conditions between these two contrasting environments, but differed from canopy gaps. Changes in vegetation structure generally associated with edge creation increased incident light availability, which in turn promoted plant growth, often to a greater extent than occurs within forest canopy gaps. From all the response variables tested in the different study regions, only tree seedling growth and survival exhibited consistent responses in the three regions, showing enhanced response in edges compared to the forest interior. In all of these experimental studies, nursery-grown tree seedlings were established, to control for age and initial seedling status.

Ecological flows across edges, such as seed movement via biotic vectors and fog penetration, were dependent on edge type. Soft edges were more permeable to the movement of seeds by small mammals compared to hard edges, and windward edges (owing to their orientation) were more important for fog capture by trees, and as a consequence presented higher soil moisture and enhanced tree regeneration.

Discussion

Species-modulated responses

As species differ in their physiological requirements, edge responses may differ between species with respect to processes such as seed germination (López-Barrera and Newton, 2005), seed removal and/or predation by birds and rodents (Díaz *et al.*, 1999), and patterns such as tree seedling and sapling densities relative to the forest interior (Box 4.8). Although the different responses of species may result in changes in tree species composition between edges and the forest interior, these contrasts are not always associated with overall differences in tree species richness or floristic diversity

Table 4.3. Comparisons between study areas in Mexico, in sites with hard (open) and soft edges in each habitat (forest, edge and open) for each of the (*Quercus* spp.) tree regeneration processes and patterns.

	Forest				Edge				Open			
	Hard		Soft		Hard		Soft		Hard		Soft	
	Chiapas	Veracruz	Chiapas	Veracruz	Chiapas	Veracruz	Chiapas	Veracruz	Chiapas	Veracruz	Chiapas	Veracruz
Processes												
Acorn removal	High	High	High	Med	Low	High	High	Med	Low	High	High	High
Seedling performance	Low	Low	Low	Med	Low	Low	Low	Med	Low	Low	High	Med
Patterns												
Forest species richness	High	High	High	High	High	High	High	High	Low	Low	Low	High
Tree seedling abundance	High	High	High	High	High	High	High	High	Low	Low	Low	Med

(Boxes 4.3 and 4.4). However, it is interesting to note that in both temperate and tropical forest types, when seedlings were established experimentally there was an overall positive edge effect on seedling survival and growth. This suggests that the edges should generally exhibit higher densities of tree seedlings and saplings than the forest interior. However, in field studies where local seedling densities were sampled along gradients from edges to forest interior, responses were mixed, and for some species seedling densities were lower in edges than in the forest interior habitat. These findings suggest that, while edges may generally favour seedling growth in controlled experiments, the resulting pattern at the community level is not predictable due to other factors operating at temporal and spatial scales (e.g. disturbance events and herbivory pulses) beyond the period of this study. These factors should be considered in future studies.

Animal species also showed different responses in their habitat preferences, which in turn affected ecological processes such as seed dispersal. Some species preferred edge habitat, while others avoided edges owing to increased predation risks. Consequently the observed species-specific response, such as seed predation rate, was not only attributable to differences in seed mass and chemical quality among plant species, but also to changes in the specific plant–animal interactions along edges. This supports the idea that changes in plant or animal species distribution and abundance along edges may have cascading effects on local assemblage structure (Ries *et al.*, 2004).

Temporal variability

Studies of edge effects should span a range of different timescales (Ries *et al.*, 2004). In this research, intra-annual variability produced different responses in variables relevant to forest dynamics, such as tree seed predation (Díaz *et al.*, 1999) and seedling herbivory (Box 4.2). Inter-annual variability owing to the occurrence of seed masting was also recorded in a study of acorn removal by small mammals (López-Barrera *et al.*, 2005) and in a study of seed rain in Chiloé forests (Armesto *et al.*, 2001). Some temporal changes in response variables along the edges are attributable to temporal differences in resource use and distribution, life cycles, stochastic events, lags in species responses or to relatively gradual changes in the quality of the edges compared to adjacent habitats (Ries *et al.*, 2004). Such differences must be documented by performing studies with longer time frames. Long-term studies of edge effects are an important gap in our current knowledge of this ecological feature in human-modified landscapes, especially considering that edges may change rapidly in structure and composition after creation.

Edge type effects

Differences in edge location and structural differences between edges and adjacent habitat are critical as they influence the strength and quality of edge responses (Ries *et al.*, 2004). Habitat use by different organisms is related to edge-adjacent habitat contrasts and therefore to the edge types, generally characterized as hard (high contrast) and soft (low contrast) edges. In this research we found that hard and soft edges produce different responses in seed removal and seed dispersal across edges (López-Barrera, 2003; Guzmán-Guzmán and Williams-Linera, 2006; López-Barrera *et al.*, 2007), seedling herbivory (López-Barrera *et al.*, 2006 and Box 4.2), diversity and abundance of small mammals (López-Barrera, 2003), and edge microclimate (López-Barrera, 2003). A recent review of the edge literature concluded that the magnitude and distance of penetration of edge influences are a direct function of the contrast in structure and composition between adjacent habitats on either side of the edge (Harper *et al.*, 2005). In this review we provide experimental data supporting the idea that edge structure contrast is a strong modulator of the magnitude of edge effects. For example, we found that edge type affected the permeability of edges to animal movement with soft or regenerating (expanding) edges enhancing small mammal dispersal of seeds from the forest edge into adjacent old-fields, whereas hard or more stable edges tended to concentrate seed dispersal along edges (López-Barrera *et al.*, 2007).

Geographical position and edge orientation can also affect the degree of habitat contrast with respect to the matrix, as orientation relative to prevailing winds can affect the magnitude and distance of physical influences across edges (Harper *et al.*, 2005). This effect was documented by del Val *et al.* (2006), as reflected in greater fog capture by vegetation at the edge facing the incoming wind, in contrast to leeward edges. The difference in fog interception between opposite edges was so great that it affected patterns of tree regeneration and mortality in forest patches located on fog-influenced coastal mountaintops. The presence of fog can also diminish abiotic differences in light, temperature and humidity gradients, and as a consequence biotic responses to such vegetation gradients. This seems to be the case in Veracruz and Chiapas, for example, where soil moisture was similar along the edges compared to forest interiors (Boxes 4.2 and 4.3).

Spatial and temporal scales and region-specific fragmentation effects

To understand the effects of habitat fragmentation it is essential to specify the spatial scale of habitat fragmentation and connectivity, which is related to the habitat area requirements of organisms, home range boundaries and movement patterns of individuals (Andren, 1994). Most of the studies summarized in this chapter were conducted in small areas within forest patches in agricultural or rural mosaics. This may constrain the scaling up of edge effects recorded in these field studies (Manson, 2000). For instance, in many cases only one edge type was represented at an individual study site and different matrix characteristics were not evaluated, making it difficult to test the site or landscape effect. For future experimental designs we recommend that different edge types are examined in the same fragmented landscape. For instance, it may be possible to use large, abandoned clearings within the forest and manipulate part of the vegetation in order to see how vegetation structure at different spatial scales affects response variables at the edge. On the other hand, the effects documented in small forest patches are likely to prevail over entire regions that are being rapidly transformed by increasing deforestation. Effects on regional biodiversity predicted from present and other studies are expected to be large.

Research summarized here suggests that the documentation of simple, descriptive patterns associated with edge creation in fragmented forests should be complemented by testing of mechanistic hypotheses in order to relate vegetation structure to function and understanding interactions between adjacent habitats. This review suggested that edge effects (emergent properties at the edge habitat) were always detected when ecological processes were recorded; however, some studies detected no effect of edges on plant species richness or other descriptive forest attributes. By long-term monitoring of spatial patterns of tree regeneration in edges, we should advance our understanding of the feedbacks from edge dynamics on tree regeneration (Conner and Perkins, 2003).

Another important observation regarding our findings is that their applicability to lowland tropical rainforests, particularly in Amazonia (Laurance and Bierregaard, 1997; Laurance *et al.*, 1998, 2001), may be limited as these landscapes may differ considerably in the degree to which they are subjected to intensive human disturbance (road density, higher population size, etc.). For example, research in Chiapas and Veracruz documented widespread, frequent human disturbance occurring within forest fragments (Ramírez-Marcial *et al.*, 2001; Chapters 3 and 10); precisely the same result was obtained in Region X of Chile (Chapter 3). Such disturbance, including harvesting of timber or firewood and browsing by livestock, makes it more difficult to measure and detect edge effects. Increased disturbance near an edge may increase the edge width and result in edge effects becoming more diffuse. For example, Fox *et al.* (1997) found that Australian rainforest fragments with minor disturbance had an abrupt increase in the abundance of forest interior species with increasing distance from the edge. However, in sites with major disturbance (mainly by cattle) the increase in abundance of core rainforest species was more gradual and the density of pioneer species remained high within the forest as well as at the edge. In effect, the contrast between edge and forest interior is not as pronounced as in areas such as lowland Amazonia, where research has focused on fragmentation of pristine rainforest (Laurance and Bierregaard, 1997).

Fragmentation effects on biodiversity

It has been stated that edges may have negative consequences for biodiversity of forest communities as edges modify forest structure, tree regeneration and mortality, enhancing the loss of forest fragment areas for conservation purposes (Murcia, 1995; Laurance *et al.*, 1998). Nevertheless, the variety of results found in recent years and evidence from the studies summarized in Table 4.2 indicate that edge effects on biodiversity are not straightforward and need to be assessed at a landscape scale. The timing of the study is also important in comparing such effects in old versus new fragments and considering extinction debits and thresholds (Fahrig 2002; Schrott *et al.*, 2005; Helm *et al.*, 2006).

Edge habitats may generate positive or negative conditions for forest biodiversity, and in some ecosystems, such as fog-dependent rainforest fragments on coastal mountaintops of semi-arid Chile or the cloud forests of central Veracruz, edge habitats may represent opportunities for forest regeneration and forest fragment expansion. Fog capture in these patches greatly influences biological diversity, as many species (e.g. ferns, bryophytes) depend on the water dripping from trees. For some tree species in forest edges we have also documented species-rich insect pollinator assemblages, as their flower displays become more attractive than in the forest interior. A similar effect was recorded in edges of riparian vegetation strips (Box 4.1), where frugivorous birds deposit a greater number and diversity of seeds than in the forest interior. In general, in landscapes that may be subjected to frequent large-scale disturbance (such as multiple tree falls, fire or landslides), as may be the case of montane tropical forests and south temperate rainforests, a relatively high proportion of species may be habitat generalists able to exploit the new opportunities. Such species may be able to take advantage of the increasing amount of edge habitats resulting from forest fragmentation.

Future studies of edge effects should examine the changes in ecological processes and structure over long time frames, as edges change because of decline in the size of fragments or successional expansion. Edge effects should be combined with information about landscape variables (fragment size, shape, isolation and habitat contrasts). Attention should be given to the factors that modulate edge effects, in order to conduct a comprehensive analysis of the consequences of fragmentation on forest biodiversity. By comparing different landscapes such as forest mosaics versus intensively managed areas, and by performing cross-scale edge studies, it may be possible to assess the scale-dependency of edge effects (Donovan *et al.*, 1997). Quality of edges, and not just quantity (e.g. width and length), should be considered as a variable when fragmentation patterns are examined. For instance, edge effects may be more severe in forest–prairie mosaics than in relatively homogeneous forests because the edge effect at any point is a function of the nearest edge (Malcolm, 1994). However, if there is longer edge length with lower contrast or abruptness, this may have less negative effects on the forest patches than lower edge length with high contrast or hard edges in a relatively continuous forest. This may also be related to the ability of seeds of forest tree species to reach edges (Armesto *et al.*, 2001) and the ability of seedlings to survive in the new habitats, therefore reducing the contrast between edge and adjacent matrix habitat.

Conclusions

The particular hypotheses addressed in this review and a summary of the main related findings are detailed below.

Forest edges are habitats with emergent properties different from forest interior and tree-fall gaps

Significant differences in response variables occurred between the forest edge and the forest interior or tree-fall gaps, indicating that forest edges are indeed habitats with emergent properties. However, these differences were recorded more frequently when response variables were ecological processes (predation, dispersal, seedling growth, etc.) rather than patterns (seedling abundance, plant diversity, etc.).

Edge-related changes in abiotic conditions (light, soil moisture, etc.) relative to the forest interior or tree-fall gaps may result in measurable differences in forest structure, species composition and ecological interactions along edges

In this review, the abiotic conditions of edges in the three study areas had an effect in the following patterns and processes: seed rain, avian nest predation, seed germination, seed removal and/or predation, seedling herbivory, seedling performance (survival and growth), tree growth rate, fog interception and tree regeneration, seed movement from the edge into the exterior, epiphyte and bryophyte cover and species richness, diversity and abundance of native small mammals and overall tree seedling and sapling density.

In all study areas, tree seedling survival and growth were higher along the edge than in the forest interior (Chacón and Armesto, 2006; Guzmán-Guzmán and Williams-Linera, 2006; López-Barrera *et al.*, 2006; and Box 4.2). However, such experimental differences were not always reflected in higher density and diversity of naturally established tree seedlings along the same edges (Boxes 4.3 and 4.4), highlighting the need for long-term studies that consider all factors that might affect population- and community-level dynamics along such borders.

Edges may induce changes in the abundance and distribution of species, which will in turn produce changes in species interactions, such as predation, herbivory, pollination and seed dispersal

In all the studies where a response variable related to species interactions was compared, an edge effect was recorded (i.e. significant differences were recorded in the response variable between the edge and the forest interior). We need to assess whether these changes are transient or persistent and their overall net impact on ecosystem structure and function along this gradient. Changes in distribution and/or abundance of seed predators and/or dispersers along the forest-edge gradient affected seed predation and dispersal (Díaz *et al.*, 1999; López-Barrera *et al.*, 2005, 2007; Guzmán-Guzmán and Williams-Linera, 2006). However, these edge effects were determined by factors such as species, canopy opening and the occurrence of mast-seeding years. Changes in distribution and/or abundance of insects along the forest-edge gradient affected seedling herbivory (Chacón and Armesto, 2005;

López-Barrera *et al.*, 2006; and Box 4.2), but these differences were determined by factors such as species, edge type and season. Edge effects on patterns of nest predation reflected the different distribution and/or abundance of nest predators (Willson *et al.*, 2001; De Santo *et al.*, 2002). While pollination was not studied along forest edges–interior gradients, results comparing isolated trees in pastures, remnant forest patches in rural areas and riparian vegetation strips in rural landscapes suggest that this ecological interaction is strongly affected by forest fragmentation (Smith-Ramírez and Armesto, 2003; and Box 4.10).

Edge orientation modulates the intensity of abiotic effects (such as fog penetration in cloud-dependent forests), thereby affecting abiotic factors (light, humidity, wind, temperature, etc.), resource distribution and consequently forest structure and dynamics

All forest patches studied in the rainforest patches in the Chilean semi-arid region displayed new tree establishment concentrated in the windward (or fog receiving) edge, where oceanic fog inputs are three times higher than in the opposite edges, providing a suitable microhabitat for seedling and sapling growth. In contrast, tree mortality was largely concentrated in the patch centre and particularly in the edges opposite to the direction of fog input where the oldest trees in the patch are found (del Val *et al.*, 2006). This underlines how edge orientation modulates fog penetration in cloud-dependent forests, thereby affecting resource distribution and consequently forest structure and dynamics.

According to landscape ecology theory, as structural similarity between two adjacent habitats increases, the edge created becomes less abrupt and the edge effect less evident; hence, ecological flows (such as animal movements and seed dispersal) across edges may be enhanced by uniformity and reduced by sharpness or abruptness of edges

This prediction was tested comparing the differences between hard and soft edges. Hard edges presented an abrupt change in micro-environmental conditions at the soil, herb and shrub levels, whereas changes along soft edges were more gradual. These differences, mostly due to changes in microenvironment at the soil level, affected oak regeneration processes (López-Barrera and Newton, 2005; López-Barrera *et al.*, 2005, 2006). Along hard edges the response variables change abruptly and we recorded a diversity of edge effects (negative or matrix effects). In contrast, along soft edges effects were more gradual or not observed at all. A positive relationship was observed between the structural similarity of the matrix and forest vegetation and the permeability of edges to small mammal acorn consumers (López-Barrera *et al.*, 2007).

Forest fragmentation and associated edge effects are widespread in forested landscapes in tropical and temperate South America. The ecological studies presented here were performed at relatively small spatial scales owing to the particular pattern of forest fragmentation in the regions studied (Chapter 3). Remnant forest patches often represent important resources for human populations that have been using these ecosystems for centuries. The loss of forest habitat and biological diversity resulting from land conversion and deforestation threatens the sustainability of productivity and ecosystem services (e.g. water and resource availability and erosion control) provided to local people. However, a general theory about the net consequences of forest fragmentation (including edge creation) on forest biodiversity and ecosystem services remains elusive, as species-specific and site-specific responses are common. Inferred or observed effects of edges on biodiversity in the tropical montane and south temperate areas compared in this study were not always negative. Edge theory is largely derived from data collected in tropical rainforests of South America at larger spatial scales and comparing well-preserved forest fragments with a highly disturbed surrounding matrix. This research highlights the fact that edge effects depend on the specific patterns of forest fragmentation, the spatial scale of the clearings (edges versus canopy gaps), current land-use practices and the relative abundance of regenerating (soft) and static (hard) edges (not always related to the edge age), the additive effects of superimposed edges, the importance of small but frequent canopy gaps occurring along the edges, and the continuous impact of human-related activities within forest fragments, such as collection of firewood and livestock use of edges.

Our review provides evidence supporting the conclusions of Ries *et al.* (2004) and Harper *et al.* (2005) that future edge studies should consider landscape complexity in developing a theoretical, predictive framework for multiple regions. The research summarized here shows that a number of factors modulate the intensity, direction and type of edge effect. These modulators must be taken into account in studies of edge effects on ecological interactions or patterns of diversity. Studies documenting the edge effects of vegetation structure and composition in the same study sites indicate that the responses are modulated by the unique responses of particular species and landscape contexts. This makes it necessary to explicitly incorporate a landscape approach in the study of edges in order to integrate matrix composition and structure and measurable attributes of edges, length, width and connectivity into model building exercises. Particularly important in future studies is to assess the additive effects of humans on edges through the collection of timber and firewood, livestock trampling and grazing, and direct interference with the biotic interactions through hunting and increased predation by domestic animals.

Acknowledgement

This work was financed by the European Commission, BIOCORES Project, INCO Contract ICA4-CT-2001-10095.

Case Studies

Box 4.1. Seed dispersal and avian species richness in riparian vegetation edges (R. Jaña, S. Reid, J. Cuevas and J.J. Armesto)

This study compared avian species diversity and seed rain of fleshy-fruited tree species in riparian and upland forests in a rural landscape of northern Chiloé Island (42° S), southern Chile. Riparian vegetation strips are relevant for the conservation of biodiversity in this and other rural landscapes, as forests are converted to pastures or crops, or become highly degraded by logging and fire. We expected riparian forest habitats to function in a similar way as forest edges, supporting a higher diversity of bird and plant species owing to a more heterogeneous seed rain and colonization from the surrounding matrix. Our study was conducted in two sites (landscape units, LU) in a rural mosaic of pastures and forest patches in northern Chiloé Island. In each LU (*c*.50 ha) we distinguished three habitats: riparian forest strips, edges of upland forest fragments and interior of upland forest fragments. In each habitat and LU we measured: (i) seed rain of fleshy fruits falling in 21 seed traps (30 cm diameter, 1 m above ground) evenly distributed along 500 m transects; (ii) bird species richness and abundance through monthly point censuses along the same transects; and (iii) tree species richness in a 2-m radius circle around each seed trap. Seeds (>3 mm) retrieved bi-weekly from seed traps were counted and identified to species during the entire fruiting period (December–April). All statistical comparisons among habitats were made using two-sample permutation tests based on 10,000 iterations. The two LUs were not exact replicates and differed in some attributes. In one landscape (LU1), seed rain of bird-dispersed propagules was significantly higher by two orders of magnitude in the riparian forest strip than in the edge or interior of upland forests (Fig. 4.1). In the second landscape, however, there was only a marginally significant difference in total seed rain between habitats, and the tendency was the same as in LU1. The deposition of entire fruits (fallen by gravity) did not differ between forest edges and interior, but it was again significantly higher in riparian forests for LU1. Bird-dispersed seed rain was significantly higher than gravity seed rain in the three habitats. No significant differences were detected in avian species richness or abundance (frugivorous species only) among habitats. Consequently differences in seed deposition patterns can be attributed primarily to differences in tree species composition, fruit productivity and forest structure among habitats within landscapes.

Box 4.2. Edge effect and insect folivory on *Quercus xalapensis* seedlings in two cloud forest fragments in central Veracruz, Mexico (J.A. Reynoso-Moran and G. Williams-Linera)

Two tropical cloud forest fragments in Veracruz, Mexico were selected to experimentally contrast insect herbivory on *Quercus xalapensis* seedlings. In each forest edge, four sets of five seedlings were planted along four parallel bands located at the forest interior, forest and field edges, and old-field. The

Continued

Box 4.2. *Continued*

experiments were conducted in the warm-dry (April–May) and warm-wet (July–August) seasons. Herbivore damage was measured at the beginning and end of each season. Weekly environmental variables were measured in the same positions at which seedlings were planted. During the wet season, herbivory was similar on seedlings planted at different distances (*c.*2%), whereas during the dry season, herbivory was higher for seedlings planted at the forest border (3–12%) than for those planted at the old-field border (0.50–0.75%). Herbivory and air temperature were higher, and air relative humidity and soil water content were lower in the abrupt than in the soft forest edge. Herbivory damage was negatively correlated with air and soil humidity. Our results suggest that herbivory level on oak seedlings depends on modulator factors such as edge type and season of the year.

Box 4.3. Edge effects on plant diversity and vegetation structure in a forest mosaic in the Highlands of Chiapas, Mexico (F. López-Barrera)

Forest edges created by scattered-patch clearcutting have become a common landscape feature in neotropical montane forest. A study was carried out in the Highlands of Chiapas, Mexico in order to assess changes in vegetation structure and floristic composition along a gradient from the interior of the forest into adjacent clearings. At six sites an 80 m × 10 m belt transect was established perpendicular to the forest/pasture edge. Plant species presence was recorded and seedling, sapling and tree stem diameter and height were assessed. A single-factor (distance) analysis of variance showed no significant effect of distance from the forest interior towards the edge in plant density and basal area ($P > 0.05$). Richness and floristic composition did not vary with the depth-of-edge influence (chi-square tests, $P > 0.05$). The level of recurrent disturbance may be influencing the response of the vegetation to the edge to interior gradient, and hence the observed results can be explained by the interaction of low but frequent human disturbance and fragmentation. Abandoned grasslands showed the presence of patches of secondary vegetation and recovering forests up to 20 m into the grassland. Expected spatio-temporal changes include the evenness of the forest/grassland edge as time advances. Results suggest that edge effects on vegetation are not measurable with conventional methods in sites where forests are mosaics with small clearings (0.5–2 ha) and widespread, low and frequent human disturbance is occurring.

Box 4.4. Vegetation structure and floristic composition at forest edges in central Veracruz, Mexico (J. Guzmán-Guzmán and G. Williams-Linera)

The fragmentation of tropical montane cloud forest has resulted in patches of forest vegetation connected (or separated) by different land uses. One of the consequences of fragmentation is the increment in forest edges. The objective of this study was to estimate changes in vegetation structure (basal area, density and height) and woody species composition in the forest interior, edge and old-field. In

Continued

Box 4.4. *Continued*

central Veracruz, Mexico, four forest fragments adjacent to old and young abandoned pastures were selected. Basal area, density and height of woody plants ≥5 cm dbh, <5 cm dbh but >2 m height, and <2 m height were determined in plots located in six bands parallel to the border, two in the old-fields and four in the forest edge. Recorded woody species were classified as primary or secondary. Vegetation structure data were analysed using ANOVA and principal component analyses. A total of 158 woody plant species was recorded, and the number of species was similar in the four sites (60–67). Most secondary species were found in edges, 69 species were trees >5 cm dbh, 127 were understorey species ≤5 cm dbh and >2 m height, and 85 were advanced regeneration ≤2 m height. The most abundant tree species were *Carpinus caroliniana*, *Quercus xalapensis*, *Q. germana*, *Q. leiophylla*, *Q. salicifolia*, *Turpinia insignis*, *Liquidambar styraciflua* and *Clethra mexicana*. In addition, *Rapanea myricoides*, *Lippia myriocephala* and *Palicourea padifolia* were recorded in the understorey. Advanced regeneration included *Q. xalapensis*, *Q. leiophylla*, *Hoffmannia excelsa*, *Urera caracasana*, *Ardisia* sp. and *Hampea* sp. Few species were recorded in all study sites and 28 were found both in abandoned pasture and edges such as *Lippia myriocephala*, *Acacia pennatula*, *Cestrum* sp., *Citharexylum* sp., *Leucaena leucocephala*, *Myrica cerifera* and *Cnidoscolus multilobus*. The edge effect was different in each site, but, in general, it was more similar in the sites located at the boundary of forest–older abandoned pasture. The age of the adjacent old-field may be an important factor modulating the depth-of-edge effects.

Box 4.5. Vascular epiphyte diversity associated with *Nothofagus nitida* (Phil.) Krasser (Nothofagaceae) trees in a Chilean temperate rainforest: effects of edges and successional habitat (F. Salinas and J.J. Armesto)

Vascular epiphytes constitute an important functional element of temperate rainforests of southern South America because of their large species richness, high biomass and relevance to ecosystem processes, such as nutrient capture. We assessed vascular epiphyte communities associated with old-growth and mid-successional forests and edges of forest fragments adjacent to pastures in a rural landscape in northern Chiloé Island. To control for differences in host-tree composition and age among habitats, we compared epiphyte assemblages only on adult trees of *Nothofagus nitida* (Phil.) Krasser (Nothofagaceae). Sampling considered only vascular epiphytes (angiosperms and ferns) growing on tree trunks up to 2 m height, because of accessibility limitations. Total epiphyte cover on tree trunks, composed mainly of ferns and vines, was higher in old-growth forest habitat than in mid-successional forest and in edge habitats. However, species richness of epiphytes did not differ among successional forests (old-growth versus mid-successional) or forest edges. Average vascular epiphyte species richness per sample unit was 17, 15 and 12 species for old-growth, mid-successional and forest edge habitats, respectively. A dendrogram of taxonomic affinity of epiphytic floras among trees in the three

Continued

Box 4.5. *Continued*

habitats revealed two statistically different epiphyte assemblages: one formed by old-growth forest species and a second one grouping mid-successional and edge species. Using a discriminant function analysis to determine whether forest habitats differed in their species composition and total epiphyte cover produced two functions. The first canonical function separated forest edge from old-growth forest samples, and the second canonical function separated mid-successional samples from the other two habitats. Most important distinctive taxa were Hymenophyllaceae ferns as different species characterized each habitat type. *Hymenophyllum plicatum*, *Serpyllopsis caespitosa* and *H. dentatum* were most important on *Nothofagus* trees located at forest edges, while *H. plicatum* occurred mainly on trees in mid-successional forest. *H. dicranotrichum* and *Hymenoglossum cruentum* were largely restricted to trees in old-growth forests. Accordingly, the composition of vascular epiphyte communities was strongly affected by forest structure, through microhabitat conditions independent of the host-tree species, and these effects should be considered when managing forests to protect biodiversity.

Box 4.6. The diversity of bryophyte species associated with rural landscapes in northern Chiloé Island (J. Larraín and J.J. Armesto)

We characterized the patterns of species richness of mosses in the rural landscape of northern Chiloé Island (42° 30′ S). We compared moss species assemblages among four common habitats in this human-dominated environment: an abandoned anthropogenic prairie (previously grazed), a secondary shrubland (originated after slash-and-burn), a secondary forest edge (a roughly even age structure, 30–50-year-old trees) and an old-growth forest patch (with uneven age structure and complex vertical layering). The secondary forest habitat occurred along the edge of the old-growth forest and for comparison purposes is considered the forest edge adjacent to an open pasture. In artificial prairies, shrublands and forests, we recorded all the mosses present in the soil within 20 sampling quadrats of 50 cm × 50 cm, located at random distances along a linear transect crossing the habitat patch along its longest axis. In addition, to characterize epiphytic species in shrublands, we randomly selected 20 shrubs (< 1.5 m tall), recording all mosses present in stems and branches of each sample shrub. In second-growth forest edges and in old-growth forests, we sampled the moss species occurring on the bark of the trunks of 20 randomly selected trees (> 10 cm diameter at breast height, dbh), from the base of the main trunk up to a height of 2 m. We were unable to record canopy mosses in either second-growth or old-growth forests because of the height of tree crowns (> 15–20 m) and as a consequence forest epiphyte diversity is underestimated. Old-growth forest habitats accumulated 59% of the total number of species of mosses recorded in all habitats in the study area. From the old-growth moss species, 30.4% were restricted to the old-growth habitat and therefore absent from forest edges and shrublands. The most common and habitat-restricted moss species found in old-growth forests were endemic to South American temperate rainforests. Although some moss taxa are yet to be identified to species, these patterns are unlikely to change significantly. Secondary forest edge species contained a subset of those species found in old-growth habitats.

Box 4.7. Fragmentation effects on small mammal communities in remnants of cloud forest in central Veracruz, Mexico (I. Ruán-Tejeda and R.H. Manson)

Habitat fragmentation is an increasingly prevalent problem affecting both tropical and subtropical forests. Habitat loss, isolation, smaller sized remnants and increased edge often combine to affect microclimate and ecological processes, through which wildlife is also affected. In central Veracruz, tropical montane cloud forests are rapidly being replaced and fragmented by urban expansion and land conversion for economic gain via coffee, cattle ranching, or sugarcane and other crops.

Using a well-known indicator group, this research sought to evaluate the effects of fragmentation on the distribution, abundance and richness of small mammals inhabiting cloud forest remnants in central Veracruz. Three main objectives of this research included: (i) determining the degree to which the species of small mammals in cloud forest fragments are interior forest specialists, generalists or prefer edge habitat; (ii) comparing the richness and diversity of small mammal communities inhabiting cloud forest fragments of different sizes; and (iii) estimating how edge effects for small mammals change as a function of cloud forest fragment shape, disturbance and isolation.

Eight cloud forest remnants in the centre of Veracruz were selected for sampling of small mammals, including four with hard edges delineated by cattle pastures and four with soft edges characterized by old-fields with secondary vegetation at least 1 m high. Two 3-day trapping sessions were conducted at each site using an array of 80 Sherman live-traps distributed in a rectangular 8m × 10m grid with 8 m between traps. The long axis of this grid extended towards the forest interior from each edge under study and was oriented perpendicular to the forest edge in areas where the last trap row did not reach the centre of the fragment. All small mammals captured were identified to species, measured, weighed, sexed and then marked with a small numbered eartag prior to release at the point of capture. In addition, the size, form and degree of isolation of each fragment were calculated using a combination of field measurements and the program ArcView 3.2. A disturbance index was also calculated for all fragments using information about the presence of hunters, trails, trash, cattle, wood extraction and active stewardship of owners.

A total of 694 captures of 325 individuals from nine species (31% of the 29 species registered historically in the region) were registered with a capture success of 18%. These species include *Oryzomys alfaroi*, *Oligoryzomys fulvescens*, *Peromyscus furvus*, *P. aztecus*, *P. leucopus*, *Reithrodontomys fulvescens*, *R. mexicanus*, *Microtus quasiater* and *Cryptotis mexicana*. *O. alfaroi*, *O. fulvescens* and *P. furvus* were captured more frequently in trap rows farther from the forest edge, suggesting a preference for forest interior habitat. In contrast, *P. leucopus* and *R. fulvescens* were captured most frequently in traps located near the forest border, suggesting a preference for edge or open habitats. *M. quasiater* and *C. mexicana* were also captured in traps adjacent to the forest edge, although the small number of individuals captured precluded a statistical test of this pattern. *P. aztecus* and *R. mexicanus* showed no significant changes in abundance or activity with distance from the forest edge and therefore may be habitat generalists. These patterns were consistent irrespective of edge type

Continued

Box 4.7. *Continued*

except for *P. aztecas*, which was captured more frequently in the forest interior and forest edge at sites with soft and hard edges, respectively.

Contrary to expectations, there was an inverse relationship between small mammal species richness and the size of forest fragments in this study. The smallest site (1.9 ha) recorded the highest number of species (7), whereas only three species were captured in the largest forest fragment (18.6 ha). The smallest forest fragment was also the site with the highest diversity of small mammals. Despite a trend of more individuals in smaller fragments, there were no significant differences in the abundance of forest interior species, or those favouring forest edges, across the range of forest fragment sizes included in this study, therefore highlighting the importance of the focal species chosen for study.

Similarly, there was a non-significant trend of increased abundance of both forest interior and edge species in sites with a more circular shape. In contrast, the degree of perturbation appears to be an important explanatory variable, especially for forest interior species. We found a decrease in the abundance of interior species in sites with greater disturbance, while there was no such relationship for forest edge specialists. The degree of forest fragment isolation was not found to be significantly correlated with the abundance or diversity of small mammals in this study.

Our results contrast somewhat with previous studies suggesting that smaller forest fragments have relatively low wildlife conservation value, and highlight the need to consider other factors such as the shape and disturbance of forest fragments simultaneously with measures of forest patch size in predicting the effects of fragmentation on small mammal communities.

Box 4.8. Edge effects on tree recruitment processes in Valdivian and North Patagonian rainforests: anthropogenic edges versus canopy gaps (A.G. Gutiérrez, N.V. Carrasco, D.A. Christie, J.C. Aravena, M. Fuentes and J.J. Armesto)

Regeneration responses of tree species to natural disturbance (canopy openings created by tree falls) and anthropogenic edges of remnant forest fragments in agricultural landscapes were compared in two lowland forests in northern Chiloé Island (42° 30′ S). We sampled the two main types of evergreen rainforest present in southern Chile and hence maximized the regional representation of tree species: North Patagonian, sampled in Senda Darwin Biological Station, and a Valdivian rainforest, sampled in the Guabún Peninsula. The Valdivian rainforest was more diverse than the North Patagonian forest, containing a greater number of epiphytes and vines, as well as canopy trees (Aravena *et al.*, 2002). Owing to differences in habitat conditions between anthropogenic forest edges and canopy gaps (i.e. greater exposure to the matrix along edges), we proposed that the density and composition of tree regeneration should differ between them, presumably leading to changes in forest composition in fragmented forests compared to continuous protected forest.

Continued

Box 4.8. *Continued*

We found interspecific differences in tree regeneration density among tree-fall gaps, forest interior and edge habitats in Valdivian and North Patagonian forests. Canopy gaps concentrated a greater heterogeneity and quantity of microsites suitable for tree regeneration, because of the greater presence of logs and stumps, which are favourable substrates for tree recruitment. The environment under canopy gaps (pooling data from both gap centre and edge) presented higher tree seedling densities and species richness compared to forest interior and anthropogenic edge habitats. We conclude that natural disturbance regimes associated with tree-fall gaps affect tree regeneration differently than anthropogenic forest–prairie edges, primarily because of differences in microsite availability and light environment. Dense stands of juvenile trees along forest edges tend to reduce tree recruitment for many years.

Box 4.9. Growth responses of eight canopy tree species to natural disturbance and anthropogenic edges (A.G. Gutiérrez, M.P. Peña, D.A. Christie and J.J. Armesto)

The effect of anthropogenic forest edges on tree growth responses is important for understanding the impact of habitat fragmentation on forest dynamics. In two old-growth forest patches located in a rural landscape of northern Chiloé Island, Chile, we conducted a comparative analysis of growth responses, measured by tree-ring width increments, of eight canopy tree species to natural (tree-fall gaps) and anthropogenic (edge adjacent to pasture) disturbances. The objective was to determinate the magnitude and duration of changes in tree-ring width following disturbance. The study sites were two large remnant forest fragments in northern Chiloé Island, representing the species composition of Valdivian and North Patagonian rainforests respectively. Five habitats were analysed for comparing tree responses: forest interior under closed canopy, open pastures outside forests, canopy openings within forest patches (tree-fall gaps), forest edges adjacent to open pastures and tree-fall gap edges within forest patches. Light availability was measured in each habitat using a PAR radiation sensor. We quantified the patterns of tree regeneration from the matrix (open pasture) outside the forest match, to the interior of the forests and tree-fall gaps. Regeneration (tree seeds and seedlings) was sampled using 20 1-m^2 plots along 50 m transects. We also collected sections taken at the base or 1.3 m above ground level (dbh), for all those trees <5 cm diameter at breast height (1.3 m). Using standard dendrochronological techniques, we assessed the year of occurrence of disturbance events (tree-fall gaps and edge creation), and the tree-ring growth responses.

The diversity of regenerating trees varied among habitats, with a higher number of recruits and greater species richness in tree-fall gaps (pooling gap centres and edges). These differences were strongly correlated with differential PAR values recorded in each habitat. Shade-tolerant trees regenerated more abundantly in the forest interior, whereas shade-intolerant and early successional

Continued

Box 4.9. *Continued*

trees recruited in edges and open areas. In contrast, tree-fall gaps have greater light heterogeneity, and hence both types of species recruited in these habitats. Time required to reach 1.3 m height differed among species and habitats. For the shade-tolerant conifer *Podocarpus nubigena* (*mañio*), growth was lower in edges and open habitats than inside forests (under the canopy and tree-fall gaps). However, one of the main pioneer species, *Drimys winteri* (*canelo*), did not show differences among habitats in the time required to reach 1.3 m height. These represent two contrasting strategies between late and early successional tree species. Radial increments did not show clear patterns across species. Few species showed marked responses: *Nothofagus nitida* showed notable growth responses to edge creation and *Aextoxicon punctatum* to tree-fall gaps.

Fig. 4.3. Forest edge in Guabun peninsula in Chiloé Island, Chile (Gutiérrez *et al.*, Box 4.9). The forest edge is dominated by *Drimys winteri* and the forest interior by *Eucryphia cordifolia* and *Aextoxicon punctatum* (photo by J.J. Armesto).

We suggest that persistence of each tree species in forest patches may be determined mainly by its ability to survive the seedling stage rather by differences in radial growth among later stages of development.

Box 4.10. Abundance of the rare arboreal marsupial *Dromiciops gliroides* in riparian forest remnants in southern Chile (C. Smith-Ramírez, J.L. Celis-Diez, J. Jiménez and J.J. Armesto)

Riparian vegetation strips found along rivers crossing the predominantly farmed Chilean central valley, separating the relatively more forested coastal and Andean mountains, are important remnants from extensive forest cover that covered the area two centuries ago. We estimated the abundance of the rare arboreal marsupial *Dromiciops gliroides* (only living member of the order Microbiotheridae) along riparian vegetation strips in an intensely managed agricultural landscape of the Chilean Lake District. Riparian vegetation occurred along a major tributary of the large River Bueno, crossing about 49 km through the central valley. We correlated the abundance of this small marsupial, estimated by live trapping over three nights at each site, with the distance from each sampling point to the nearest large tract of continuous forest, found in the foothills of the Andes. In addition, we correlated *Dromiciops* abundance with the width of the riparian vegetation strip as well as with the presence and number of the hemiparasite *Tristerix corymbosus* in the riparian habitat. The fruits of this plant are an important food resource for this arboreal marsupial. Furthermore, we compared the abundance of *Dromiciops gliroides* in the riparian vegetation strip with its abundance in the nearest continuous forest in the Andean foothills and with its abundance in remnant forest fragments in the rural landscape <1 km away from the riparian vegetation strip.

During the 2 years of the study, we captured a total of 32 individuals of the rare *Dromiciops gliroides*. Of these, 22 (70%) were found in the riparian vegetation strip. We found a statistically significant correlation between the width of the riparian vegetation strip and the abundance of *Dromiciops* ($R=0.749$, $P=0.033$). No correlation was found between the local abundance of the hemiparasite *Tristerix corymbosus* (Loranthaceae) in riparian or forest patch habitats and the local abundance of *Dromiciops*. The number of individuals trapped in remnant forest fragments of the central valley and in the continuous Andean forest was similar to the number of individuals trapped in the narrow (around 30 m wide on average) riparian forest strip. We suggest that narrow vegetation strips (minimum width 30 m) along rivers in the intensely farmed and deforested central depression of the Chilean Lake District may be key habitats for the survival of species restricted to forest habitats, such as this endangered arboreal marsupial.

References

Andren, H. (1994) Effects of habitat fragmentation on birds and mammals in landscapes with different proportions of suitable habitat: a review. *Oikos* 71, 355–366.

Aravena, J.C., Carmona, M.R., Pérez, C.A. and Armesto, J.J. (2002) Changes in tree species richness, stand structure and soil properties in a successional chronosequence in northern Chiloé Island, Chile. *Revista Chilena de Historia Natural* 75, 339–360.

Armesto, J.J., Díaz, I., Papic, C. and Willson, M.F. (2001) Seed rain of fleshy and dry propagules in different habitats in the temperate rainforests of Chiloé Island, Chile. *Austral Ecology* 26, 311–320.

Benitez-Malvido, J. (1998) Impact of forest fragmentation on seedling abundance in a tropical rain forest. *Conservation Biology* 12, 380–389.
Bergin, T.M., Best, L.B., Freemark, K.E. and Koehler, K.J. (2000) Effects of landscape structure on nest predation in roadsides of a midwestern agroecosystem: a multiscale analysis. *Landscape Ecology* 15, 131–143.
Brand, L.A. and George, T.L. (2000) Predation risks for nesting birds in fragmented coast redwood forest. *Journal of Wildlife Management* 64, 42–51.
Cadenasso, M.L. and Pickett, S.T.A. (2000) Linking forest edge structure to edge function: mediation of herbivore damage. *Journal of Ecology* 88, 31–44.
Cadenasso, M.L. and Pickett, S.T.A. (2001) Effect of edge structure on the flux of species into forest interiors. *Conservation Biology* 15, 91–97.
Chacón, P. and Armesto, J.J. (2005) Effect of canopy openness on growth, specific leaf area, and survival of tree seedlings in a temperate rainforest of Chiloé Island, Chile. *New Zealand Journal of Botany* 43, 71–81.
Chacón, P. and Armesto, J.J. (2006) Do carbon-based defences reduce foliar damage? Habitat-related effects on tree seedling performance in a temperate rainforest of Chiloé Island, Chile. *Oecologia* 146, 555–565.
Conner, L.M. and Perkins, M.W. (2003) Nest predator use of food plots within a forest matrix: an experiment using artificial nests. *Forest Ecology and Management* 182, 371–380.
De Santo, T.L., Willson, M.F., Sieving, K.E. and Armesto, J.J. (2002) Nesting biology of tapaculos (family Rhinocryptidae) in fragmented south-temperate rainforests of Chile. *Condor* 104, 482–495.
del Val, E., Barbosa, O., Armesto, J.J., Christie, D., Gutiérrez, A.G., Jones, C.G., Marquet, P. and Weathers, K.C. (2006) Rain forest islands in the Chilean semiarid region: fog-dependency, ecosystem persistence and tree regeneration. *Ecosystems* 9, 598–608.
Díaz, I., Papic, C. and Armesto, J.J. (1999) An assessment of postdispersal seed predation in temperate rain forest fragments in Chiloé Island, Chile. *Oikos* 87, 228–238.
Didham, R.K. and Lawton, J.H. (1999) Edge structure determines the magnitude of changes in microclimate and vegetation structure in tropical forest fragments. *Biotropica* 31, 17–30.
Donovan, T.M., Jones, P.W., Annand, E.M. and Thompson, F.R. (1997) Variation in local-scale edge effects: mechanisms and landscape context. *Ecology* 78, 2064–2075.
Duelli, P., Studer, M., Marchand, I. and Jakob, S. (1990) Population movements of arthropods between natural and cultivated areas. *Biological Conservation* 54, 193–207.
Fagan, W.F., Cantrell, R.S. and Cosner, C. (1999) How habitat edges change species interactions. *American Naturalist* 153, 165–182.
Fahrig, L. (2002) Effect of habitat fragmentation on the extinction threshold: a synthesis. *Ecological Applications* 12, 346–353.
Fahrig, L. (2003) Effects of habitat fragmentation on biodiversity. *Annual Review of Ecology, Evolution and Systematics* 34, 487–515.
Fearnside, P.M. (1996) Amazonian deforestation and global warming: carbon stocks in vegetation replacing Brazil's Amazon forest. *Forest Ecology and Management* 80, 21–34.
Fenske-Crawford, T.J. and Niemi, G.J. (1997) Predation of artificial ground nests at two types of edges in a forest-dominated landscape. *Condor* 99, 14–24.
Figueroa, J.A. and Lusk, C.H. (2001) Germination requirements and seedling shade tolerance are not correlated in a Chilean temperate rain forest. *New Phytologist* 152, 483–489.
Finke, D.L. and Denno, R.F. (2004) Predator diversity dampens trophic cascades. *Nature* 429, 407–410.
Fleming, K.K. and Giuliano, W.M. (1998) Effect of border-edge cuts on birds at woodlot edges in southwestern Pennsylvania. *Journal of Wildlife Management* 62, 1430–1437.
Fox, B.J., Taylor, J.E., Fox, M.D. and Williams, C. (1997) Vegetation changes across edges of rainforest remnants. *Biological Conservation* 82, 1–13.

Gascon, C., Williamson, G.B. and da Fonseca, G.A.B. (2000) Receding forest edges and vanishing reserves. *Science* 288, 1356–1358.
Guzmán-Guzmán, J. and Williams-Linera, G. (2006) Edge effect on acorn removal and oak seedling survival in Mexican lower montane forest fragments. *New Forests* 31, 487–495.
Hansson, L. (2000) Edge structures and edge effects on plants and birds in ancient oak–hazel woodlands. *Landscape and Urban Planning* 46, 203–207.
Harper, K.A., MacDonald, E., Burton, P.J., Chen, J., Brosofske, K.D., Saunders, S.C., Euskirchen, E.S., Roberts, D., Jaiteh, M.S. and Esseen, P. (2005) Edge influence on forest structure and composition in fragmented landscapes. *Conservation Biology* 19, 768–782.
Helm, A., Hanski, L. and Patel, M. (2006) Slow response of plant species richness to habitat loss and fragmentation. *Ecology Letters* 9, 72–77.
Hinsley, S.A., Bellamy, P.E., Enoksson, B., Fry, G., Gabrielsen, L., McCollin, D. and Schotman, A. (1998) Geographical and land-use influences on bird species richness in small woods in agricultural landscapes. *Global Ecology and Biogeography Letters* 7, 125–135.
Justice, C., Wilkie, D., Zhang, Q., Brunner, J. and Donoghue, C. (2001) Central African forests, carbon and climate change. *Climate Research* 17, 229–246.
Kapos, V., Wandelli, E., Camargo, J.L. and Ganade, G. (1997) Edge-related changes in environment and plant responses due to forest fragmentation in central Amazonia. In: Laurance, W.F. and Bierregaard, R.O.J. (eds) *Tropical Forest Remnants: Ecology, Management, and Conservation of Fragmented Communities*. University of Chicago Press, Chicago, Illinois, pp. 33–43.
Kollmann, J. and Buschor, M. (2002) Edge effects on seed predation by rodents in deciduous forests of northern Switzerland. *Plant Ecology* 164, 249–261.
Laurance, W.F. (1991) Edge effects in tropical forest fragments: applications of a model for the design of nature-reserves. *Biological Conservation* 57, 205–219.
Laurance, W.F. (1997) Hyper-disturbed parks: edge effects and the ecology of isolated rainforest reserves in tropical Australia. In: Laurance, W.F. and Bierregaard, R.O.J. (eds) *Tropical Forest Remnants: Ecology, Management, and Conservation of Fragmented Communities*. University of Chicago Press, Chicago, Illinois, pp. 71–83.
Laurance, W.F. and Bierregaard, R.O. (1997) *Tropical Forest Remnants: Ecology, Management and Conservation of Fragmented Communities*. University of Chicago Press, Chicago, Illinois.
Laurance, W.F. and Williamson, G.B. (2001) Positive feedbacks among forest fragmentation, drought, and climate change in the Amazon. *Conservation Biology* 15, 1529–1535.
Laurance, W.F., Ferreira, L.V., Rankin-De Merona, J.M., Laurance, S.G., Hutchings, R.W. and Lovejoy, T.E. (1998) Effects of forest fragmentation on recruitment patterns in Amazonian tree communities. *Conservation Biology* 12, 460–464.
Laurance, W.F., Didham, R.K. and Power, M.E. (2001) Ecological boundaries: a search for synthesis. *Trends in Ecology and Evolution* 16, 70–71.
Lidicker, W.Z.J. (1999) Responses of mammals to habitat edges: an overview. *Landscape Ecology* 14, 333–343.
Lidicker, W.Z.J. and Peterson, J.A. (1999) Responses of small mammals to habitat edges. In: Barrett, G.W. and Peles, J.D. (eds) *Landscape Ecology of Small Mammals*. Springer, Berlin, Germany, pp. 211–227.
López-Barrera, F. (2003) Edge effects in a forest mosaic: implications for the oak regeneration in the Highlands of Chiapas, Mexico. PhD thesis. Institute of Atmospheric and Environmental Science, University of Edinburgh, Edinburgh, UK.
López-Barrera, F. and Newton, A.C. (2005) Edge type effect on acorn germination of oak species in the Highlands of Chiapas, Mexico. *Forest Ecology and Management* 217, 67–79.
López-Barrera, F., Newton, A.C. and Manson, R. (2005) Edge effects in a tropical montane forest mosaic: experimental tests of post-dispersal acorn removal. *Ecological Research* 20, 31–40.

López-Barrera, F., Manson, R., González-Espinosa, M. and Newton, A.C. (2006) Effects of the type of montane forest edge on oak seedling establishment along forest-edge–exterior gradients. *Forest Ecology and Management* 225, 234–244.

López-Barrera, F., Manson, R., Newton, A.C. and González-Espinosa, M. (2007) Effects of varying forest edge permeability on seed dispersal in a neotropical montane forest. *Landscape Ecology* 22, 189–203.

Malcolm, J.R. (1994) Edge effects in central Amazonian forest fragments. *Ecology* 75, 2438–2445.

Manson, R.H. (2000) Spatial autocorrelation and the interpretation of patterns of tree seed and seedling predation by rodents in old-fields. *Oikos* 91, 162–174.

Manson, R.H., Ostfeld, R.S. and Canham, C.D. (1999) Responses of a small mammal community to heterogeneity along forest–old-field edges. *Landscape Ecology* 14, 335–367.

McCollin, D. (1998) Forest edges and habitat selection in birds: a functional approach. *Ecography* 21, 247–260.

Muñiz-Castro, M.A., Williams-Linera, G. and Rey-Benayas, J.M. (2006) Distance effect from cloud forest fragments on plant community structure in abandoned pastures in Veracruz, Mexico. *Journal of Tropical Ecology* 22, 431–440.

Murcia, C. (1995) Edge effects in fragmented forests: implications for conservation. *Trends in Ecology and Evolution* 10, 58–62.

Myers, N., Mittermeier, R.A., Mittermeier, C.G., da Fonseca, G.A.B. and Kent, J. (2000) Biodiversity hotspots for conservation priorities. *Nature* 403, 853–858.

Ramírez-Marcial, N., González-Espinosa, M. and Williams-Linera, G. (2001) Anthropogenic disturbance and tree diversity in montane rain forest in Chiapas, Mexico. *Forest Ecology and Management* 154, 311–326.

Ries, L., Fletcher, R.J., Batin, J. and Sisk, T.D. (2004) Ecological responses to habitat edges: mechanisms, models, and variability explained. *Annual Review of Ecology, Evolution and Systematics* 35, 491–522.

Rodriguez, A., Andren, H. and Jansson, G. (2001) Habitat-mediated predation risk and decision making of small birds at forest edges. *Oikos* 95, 383–396.

Rozzi, R., Martínez, D.R., Willson, M.F. and Sabag, C. (1996) Avifauna de los bosques templados de Sudamérica. In: Armesto, J.J., Villagrán, C. and Arroyo, M.T.K. (eds) *Ecología de los Bosques Nativos de Chile*. Editorial Universitaria, Santiago, Chile, pp. 135–152.

Sarlov-Herlin, I. (2001) Approaches to forest edges as dynamic structures and functional concepts. *Landscape Research* 26, 27–43.

Schrott, G.R., With, K.A. and King, A.T.W. (2005) On the importance of landscape history for assessing extinction risk. *Ecological Applications* 15, 493–506.

Semazzi, F.H.M. and Yi, S. (2001) A GCM study of climate change induced by deforestation in Africa. *Climate Research* 17, 169–182.

Sizer, N. and Tanner, E.V.J. (1999) Responses of woody plant seedlings to edge formation in a lowland tropical rainforest, Amazonia. *Biological Conservation* 91, 135–142.

Smith-Ramírez, C. and Armesto, J.J. (2003) Behaviour of nectar-feeding birds visiting *Embothrium coccineum* (Proteaceae) trees on edges of forest fragments in Chiloé Island, Chile. *Austral Ecology* 28, 53–60.

Song, S.J. and Hannon, S.J. (1999) Predation in heterogeneous forests: a comparison at natural and anthropogenic edges. *Ecoscience* 6, 521–530.

Turton, S.M. and Freiburger, H.J. (1997) Edge and aspect effects on the microclimate of a small tropical forest remnant on the Atherton Tableland, Northeastern Australia. In: Laurance, W.F. and Bierregaard, R.O.J. (eds) *Tropical Forest Remnants: Ecology, Management and Conservation of Fragmented Communities*. University of Chicago Press, Chicago, Illinois, pp. 45–54.

Wiens, J.A., Crawford, C.S. and Gosz, J.R. (1985) Boundary dynamics: a conceptual framework for studying landscape ecosystems. *Oikos* 45, 421–427.

Williams-Linera, G. (1990a) Vegetation structure and environmental conditions of forest edges in Panama. *Journal of Ecology* 78, 356–373.
Williams-Linera, G. (1990b) Origin and early development of forest edge vegetation in Panama. *Biotropica* 22, 235–241.
Willson, M.F., Morrison, J.L., Sieving, K.E., De Santo, T.L., Santisteban, L. and Díaz, I. (2001) Patterns of predation risk and survival of bird nests in a Chilean agricultural landscape. *Conservation Biology* 15, 447–456.
Zhang, H., Henderson-Sellers, A. and McGuffie, K. (2001) The compounding effects of tropical deforestation and greenhouse warming on climate. *Climatic Change* 49, 309–338.

5 Habitat Fragmentation and Reproductive Ecology of *Embothrium coccineum*, *Eucryphia cordifolia* and *Aextoxicon punctatum* in Southern Temperate Rainforests

C. Smith-Ramírez, A.E. Rovere, M.C. Núñez-Ávila and J.J. Armesto

Aerial photograph illustrating a matrix of industrial plantations of *Eucalyptus* in zones previously occupied by native forests in the coastal range in south-central Chile. Photo: Cristian Echeverría

Summary

The objective of this work was to study the effect of anthropogenic fragmentation on reproductive variables such as pollination assemblage, frequency of pollinator visits and fructification of three very frequent tree species in southern temperate rainforests. These trees have different pollination and seed dispersal syndromes, and their requirements for pollinators also vary. We expected that fragmentation would negatively affect the reproductive variables studied, diminishing pollinator species richness, frequency of pollinator visits and fructification in isolated trees in relation to large patches, and producing changes in the phenological patterns. At the same time, we assessed the seed rain in isolated trees and reproductive compatibility in one tree species, *Embothrium coccineum*. This study was conducted in southern Chile, specifically in the north of Chiloé Island and the coastal forest of Valdivia, and in northern Chile, specifically Fray Jorge National Park. We found that the responses to richness and frequency of pollinators depend on the time of year (*Eucryphia cordifolia*) and size of the fragment (*Embothrium coccineum*). Fruit production in *Embothrium coccineum* varied in relation to fragment size, being higher in isolated trees and small fragments than in medium and large fragments. The median distance of dispersal of *Embothrium coccineum* seeds was 20 m. Phenology of *Aextoxicon punctatum* was considerably different in the two sites studied and we found no pollinators in spite of the fact that this species produces entomophilous flowers. The reproductive syndromes of *A. punctatum* were also different in the two sites studied. We propose that the different responses of *Embothrium* and *Eucryphia* to fragmentation are related to different pollination syndromes (ornithophilous and entomophilous species respectively). Furthermore the tendencies in seed production in *Embothrium* are positively related to the frequency of pollinator visits. The different phenology and reproductive syndromes found in *Aextoxicon* are related to the long duration since fragmentation of the study sites.

Introduction

Fragmentation of natural habitats is a potential threat to the persistence of animal and plant populations in human-dominated landscapes (Saunders *et al.*, 1991; Andren, 1992; Robinson *et al.*, 1995). Habitat fragmentation changes patch sizes and connectivity among habitat patches in the landscape, in addition to increasing the proportion of edge habitat with respect to patch interior. The increase in the amount of edge habitat may affect species richness in the patches, depending on the contrast between the original habitat and the new anthropogenic matrix. The matrix surrounding the remnant habitat patches may influence the survival of native species within patches because of the introduction of new predators, pathogens and competitors, as well as by restricting the mobility of organisms, pollen and propagules across the landscape.

The restricted patch size, habitat discontinuity and increased edge of fragments may impose strong ecological and genetic effects on plants, both directly and indirectly via animal vectors of pollen and seeds (e.g. Aizen and Feinsinger, 1994a, b). First, plant reproductive output may be affected directly by microclimatic changes induced by fragmentation, such as increased exposure to wind, rain and desiccation (Lovejoy *et al.*, 1986). Plant reproductive outputs may also reflect changes in the visiting rates of pollinators and fruit dispersers, abundances of flower and fruit predators, herbivores, altered seed rain patterns and changes in the amount of habitat available for recruitment. Forest fragmentation and the resulting spatial isolation of tree species alter the activity of pollinators and may have important implications for seed production and mating patterns of the plants that they pollinate. For example, Jennersten (1988) showed that habitat fragmentation resulted in lower flower visitation and reduced seed set in *Dianthus deltoides* when compared to non-fragmented habitats. Similarly, Aizen and Feinsinger (1994a) showed that pollination levels and seed outputs decreased by nearly 20% in plants from continuous forest compared to small fragments in the Chaco region of Argentina. These findings and other studies of tropical plants (Hall *et al.*, 1996; Nason and Hamrick, 1997; Aldrich and Hamrick, 1998) indicate that the reduction of continuous habitat can have negative effects on plant reproductive success.

In contrast, Murcia (1996) reported that pollination biology of tropical plants was not affected by habitat fragmentation, as pollination levels did not differ among forest remnants of different size. Similarly Cascante (1999) reported that forest fragmentation did not affect the rate of pollen deposition in individuals of the tropical dry forest tree *Samaea saman*. However, he found that isolated trees or trees in small forest remnants had fewer pollen tubes growing along their styles and fewer seeds per fruit than trees in continuous forest. On the other hand, Nason and Hamrick (1997) reported that trees of *Spondias mombin* (Anacardiaceae) found in small forest patches suffered a significant reduction in fruit production and seed germination relative to trees in large fragments or continuous forest. They also reported that most of the seeds produced in small patches were sired by trees located in forest stands located >80–1000 m away. Similarly, Aldrich and Hamrick (1998) studied the reproductive success of trees of *Symphonia globulifera* growing in forest remnants and pastures. They reported that most seedlings found in forest remnants were sired by trees in adjacent pastures. Because few trees are typically found in pastures, this situation creates a genetic bottleneck. In addition, they found that the rate of selfing was higher for trees in pastures, further reducing genetic variation.

The reduction of continuous habitat into smaller spatially isolated patches threatens the long-term survival of many plant species (Saunders *et al.*, 1991; Young *et al.*, 1996; Nason *et al.*, 1997). Many studies have demonstrated a lower diversity and/or decreased abundance of various organisms within small habitat fragments compared to larger habitat tracts (e.g. Soule, 1986; Opdam and Schotman, 1987; Matthysen *et al.*, 1995). Forest fragmentation is likely to decrease gene flow, increase endogamy and eventually produce a high genetic differentiation among remnant populations. Few studies have addressed the impacts of changes in the landscape on the reproductive biology and the population genetics of plants (Jennersten, 1988; Templeton *et al.*, 1990; Foré *et al.*, 1992; Aizen and Feinsinger, 1994a, b; Chapter 6). For example, Foré *et al.* (1992) found that after forest fragmentation the genetic difference among adult trees of *Acer saccharum* was greater than among juveniles. They concluded that gene flow was reduced after fragmentation.

This chapter examines the effects of anthropogenic forest fragmentation on the reproduction of forest trees in southern South America. We discuss data on pollination assemblages, frequency of pollinator visits to flowers and fruiting patterns of trees in southern temperate rainforest. We predicted that fragmentation will negatively affect tree reproductive outputs by decreasing pollinator species richness and lowering frequency of pollinator visits to flowers, and also that fruit set will be lower in isolated trees in pastures and small patches than in trees in large forest patches. Additionally, we assessed the effects of fragmentation on tree reproduction by comparing flowering and fruiting phenology of a rainforest tree species (*Aextoxicon punctatum*) between two sites separated by 1200 km. The northern population became segregated from the main species range as a result of climatic change since the late Tertiary period (Chapter 6). We hypothesized that ancient fragmentation between northern (30° S) and southern (40° S) populations in Chile may have produced significant differences in reproductive biology and phenology between the genetically and geographically isolated populations of this rainforest tree.

This research investigated three relatively common tree species from South American rainforests. We report reproductive studies in *Embothrium coccineum* (Proteaceae), *Eucryphia cordifolia* (Eucryphiaceae) and *Aextoxicon punctatum* (the only member of the endemic Aextoxicaceae). These tree species have different pollination syndromes: *Embothrium coccineum* is typically ornithophyllous, *Eucryphia cordifolia* is typically entomophyllous, and both have wind-dispersed seeds, in contrast to *Aextoxicon punctatum*, which is dispersed by birds, but has small flowers with an entomophyllous syndrome.

The species studied

Embothrium coccineum *(Proteaceae)*

Embothrium coccineum J.R. et G. Forster. (Proteaceae), locally known as *notro* (Correa, 1984), is endemic to temperate forests of southern Chile and Argentina. It is most often found in open areas, secondary forests and along forest edges, riparian and waterlogged soils (Fig. 5.1). *Embothrium* has a wide latitudinal range in Chile (35°–55° S) and in Argentina (39°–55° S) (Correa, 1984; Romero *et al.*, 1987), from sea level to 1200 m in elevation. It has bright red tubular hermaphroditic flowers, which remain open for 4 days and are visited mostly by nectar-feeding birds and occasionally by a few insect species. In Chiloé Island, *Embothrium* blooms in spring (September–January) and is primarily pollinated by two bird species: *Elaenia albiceps* (flycatcher) and *Sephanoides sephanoides* (hummingbird) (Smith-Ramírez and Armesto, 2003). Its flowers are protandrous (Humaña and Riveros, 1994), releasing pollen from the anthers before the stigma becomes receptive. The mating system of *Embothrium* was self-incompatible in mountain areas at 41° S (Riveros, 1991) and further south at 50° S (Arroyo and Squeo, 1990), although it may be partly self-compatible in low elevation areas at 39° S (Riveros *et al.*, 1996). The fruit is an almost woody, reddish-brown, oblong follicle (Brion *et al.*, 1988), containing many (average 11) winged seeds.

Eucryphia cordifolia *(Eucryphiaceae)*

Eucryphia cordifolia Cav. (Eucryphiaceae), locally known as *ulmo*, is a tree endemic to temperate forests of southern Chile. It occurs in forest edges and

Fig. 5.1. Flowers of *Embothrium coccineum* (*notro*).

old-growth forests (Fig. 5.2). *Eucryphia* grows between 36° and 43° S, and from sea level to 1200 m in elevation. It is an evergreen tree that reaches up to 30 m in height and 2 m in trunk diameter, and has relatively large symmetric flowers (5–6 cm diameter) that open disc-shaped with many white petals and numerous stamens. It offers nectar and pollen as resources for pollinators (Fig. 5.2). Flowering takes place in the austral summer, starting about the middle of January and lasting until early March. *Eucryphia* has a self-incompatible reproductive system and is highly dependent on pollinators for seed production (Riveros, 1991). Fruits are small woody follicles with many tiny winged seeds that ripen in December and January.

Aextoxicon punctatum *(Aextoxicaceae)*

Aextoxicon punctatum Ruiz and Pav. (Olivillo), locally known as *olivillo*, is an endemic tree species of temperate forests of southern Chile with a small population in Argentina. This species is the only member of the genetically isolated family Aextoxicaceae. *Aextoxicon* has a geographic range (30°–43° S) that exceeds the northern margin of austral temperate rainforests (Chapter 6), extending northwards as a chain of remnant forest patches occurring in

Fig. 5.2. *Eucryphia cordifolia* (*ulmo*).

Fig. 5.3. *Aextoxicon punctatum* (*olivillo*).

coastal gorges of the Mediterranean climate zone in central Chile (32°–39° S) and, farther north, on isolated coastal hilltops in semi-arid Chile (30°–32° S) (Pérez and Villagrán, 1994). It is a dioecious, evergreen tree that reaches up to 25 m in height (Fig. 5.3). The corolla of the female flower has five, almost absent, short petals, a staminode lacking a well-developed anther, and the pistil has a brown-coloured ovary, broader at the base, with a green, bifid style. The male flower has five sepals and five white petals; five stamens with high pollen load in the anthers and a rudimentary ovary in the centre. Flowers have a strong, sweet honey-like smell. The fruit is a fleshy drupe with one seed, which is consumed by frugivorous birds.

Methods

Pollination studies

Embothrium coccineum

We quantified visits to *Embothrium* flowers in edges of forest fragments in four situations: (i) small fragments, about 1 ha in size; (ii) medium size fragments, 20–30 ha; (3) large fragments, >150 ha in size; and (iv) remnant isolated trees in pastures (Table 5.1). We quantified species richness, frequency and identity of flower visitors in 1992, 1993 and 1994. Our observations were concentrated in the period of peak flowering of the species, for 2 weeks in early November in each year. Each tree was observed for a period of 20 minutes (sample unit), with an average of 14 periods by day during the 2–3 week flowering period, from 10 am until 8 pm.

Eucryphia cordifolia

We quantified species richness, frequency and identity of flower visitors over 3 years (2003, 2004 and 2005) for trees in two situations: (i) 14 isolated trees in pastures and (ii) 16 trees along edges of forest fragments. Observations of

Table 5.1. Location and landscape setting of study sites of *Embothrium coccineum* in a rural landscape of northern Chiloé Island, Chile.

Site name	Landscape setting	Fragment area (ha)	Latitude (S)	Longitude (W)
Mandiola	Pasture trees	–	41° 53′	73° 32′
Caipulli	Pasture trees	–	41° 53′	73° 41′
Kesler	Small fragment	1	41° 59′	73° 36′
Seit	Small fragment	1	41° 59′	73° 38′
Grob	Medium fragment	23	41° 55′	73° 41′
Wolf	Medium fragment	18	41° 56′	73° 40′
Koch	Large fragment	>150	41° 55′	73° 39′
Koening	Large fragment	>150	41° 59′	73° 35′

Eucryphia were concentrated during the period of maximum bloom (approximately 1 week in February). Each individual tree flowers for about 3 weeks each year, and the date of the maximum bloom varied by up to 1 month between years of study. The identity of flower visitors to a given plant was recorded during several 20-minute-long observation periods (sample unit). Observations were made with the naked eye from ground level or from a short platform (2 m tall) for trees in pastures and forest edges. Hence, records of visitors were limited to flowering branches located up to 4 m high. Observation periods were uniformly distributed between 10.00 am and 6.00 pm each day. In the first year of study, specimens of all flower visitors were collected for identification, but in other years we only collected new and doubtful specimens. Specimens were identified most often at the family level, and less frequently at the genus or species level, with the help of specialists, and archived in the entomological collection of Senda Darwin biological station.

Aextoxicon punctatum

We recorded flower visitors in the same six female trees where phenology and breeding systems were studied in each locality. Observations were made in Curiñanco (Valdivia, 40° S) and in the northernmost isolated population of Fray Jorge (30° S). In Curiñanco, trees were observed over 2 days because the flowering period was very much shorter than in Fray Jorge. In Fray Jorge observations were made over 14 days, distributed during the years 2004 and 2005, because flowering was scattered over 10 months.

Fruit production

Embothrium coccineum

We quantified fruit production of *Embothrium* trees along edges of forest fragments in four situations: (i) small fragments, around 1 ha; (ii) medium size fragments, 20–30 ha; (iii) large patches, > 150 ha in size; and (iv) isolated remnant trees in pastures (Table 5.1). We quantified percentages of fruiting, i.e.

the proportion of flowers producing fruits, by randomly selecting ten inflorescences in each of five trees during the flowering seasons of 2002 and 2003. Differences among the four situations were assessed by a non-parametric Kruskal–Wallis test because data were not normally distributed.

Aextoxicon punctatum

We quantified fruit production of six *Aextoxicon* trees in Curiñanco and six trees in Fray Jorge. We quantified the proportion of fruits produced by 54 flowers in Curiñanco, with an average of nine flowers per tree, and 699 flowers in Fray Jorge, with an average of 116 flowers per tree. The small number of female trees available limited sample size in Curiñanco.

Breeding systems

Embothrium coccineum

The breeding system of this species was studied between October 2001 and March 2002 at Senda Darwin Biological Station, north-west Chiloé (41° 53′ S, 73° 40′ W, and 50 m elevation). Ten *Embothrium* trees with low branches and numerous flowers were selected. At the beginning of the flowering season, 60 flower buds in each tree (15 by each one of the four treatments) were enclosed in bride tulle bags. Treatments to assess the breeding system were performed in 15 buds per tree and, therefore, on a total of 150 buds. Treatments followed protocols set by Dafni (1992). These were: manual self-pollination (SP), manual cross-pollination (CP), spontaneous or automatic self-pollination (AP) and natural pollination (NP). Treatments differed in their pollen-donor source, except natural pollination (NP) where flower buds were simply labelled and flowers were permanently left exposed to natural pollinators. An index of self-incompatibility (ISI) was calculated (Ruiz and Arroyo, 1978). This index is also known as the self-compatibility index (SCI). It is calculated as the ratio between the percentages of fruits produced from manual self-pollination and those from cross-pollination experiments (%SP/%CP). Species with ratios <0.2 are considered self-incompatible, while higher values indicate that the plant is self-compatible. Breeding success for each treatment was measured by the percentage of fruiting and also by the number of seeds per ripe fruit (Dafni, 1992; Burd, 1994). Differences between treatments were assessed by non-parametric Mann–Whitney and Kruskal–Wallis tests.

Aextoxicon punctatum

The reproductive system of *Aextoxicon punctatum* was evaluated in two of the southern and northern populations by means of four treatments: (i) wind and animal pollinator exclusion; (ii) only animal pollinator exclusion, but exposed to wind; (iii) natural pollination; and (iv) manual pollination. Each treatment was replicated six times in each of six trees in each population (N=72). During the following months, the number of fruits produced by flowers under each treatment was counted.

Seed dispersal

Embothrium coccineum

We evaluated spatial patterns of anemochorous seed dispersal of *Embothrium* in a fragmented forest site in northern Chiloé Island, Chile. Seed traps were placed at regular intervals (2.5 m) around three isolated trees in an open forest area, along transect lines of 25 m following the four cardinal directions.

Phenology

Aextoxicon punctatum

Flowering and fruiting phenology was recorded periodically in six trees in each of two populations, Fray Jorge (30° S, a northernmost isolated fragment) and Curiñanco (41° S, in the Valdivian rainforest region).

Results

Pollination studies

Embothrium coccineum

We found that pollinator visiting rates were negatively correlated with forest patch area and the highest rates were recorded for pasture trees (Smith-Ramírez and Armesto, 2003). This trend was largely due to a decline in the number of visits by the nectar-feeding passerine *Elaenia albiceps*, the main flower visitor, in larger patches. The number of hummingbird visits did not vary with patch size. Lower visitation rates to flowering trees in larger fragments seemed to be a consequence of territorial defence by *Elaenia albiceps* and were unrelated to differences in floral display. No inter-annual differences (1992, 1993 and 1994) in the identity of pollinators and relative percentage of visits were found for trees in forest and isolated trees in pastures (Fig. 5.4).

Eucryphia cordifolia

Species richness of insect pollinators of *E. cordifolia* was identical between isolated trees in pastures and trees in forest patch edges in 2003, but species richness was higher in forests when compared to isolated trees in pastures in 2004 and 2005. In contrast, the total numbers of pollinator visits to flowers was 21% higher in isolated trees in all 3 years. However, when this analysis was made by individual tree, species richness did not differ among trees, or between trees in forest and isolated trees in pastures. The number of visits per tree varied greatly over the study period. In 2003, pollinator visits to isolated trees were higher than visits to forest edge trees, in 2004 no differences were found between these two groups of trees, and in 2005 the number of pollinator visits to forest edge trees was higher than the number of visits to isolated pasture trees. Accordingly, differences in pollinator species richness between forest edge and isolated trees were insignificant, with a weak

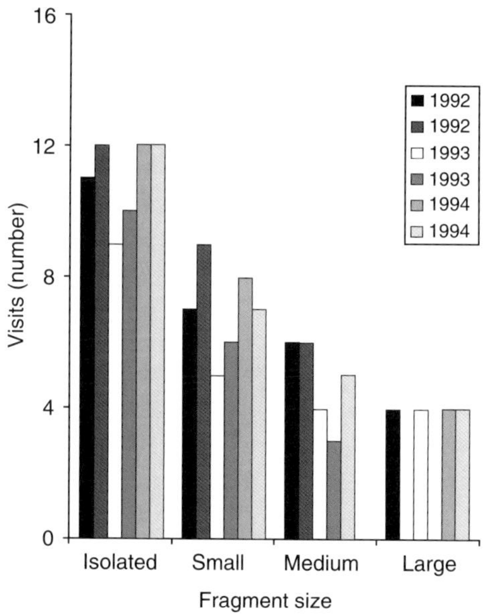

Fig. 5.4. Number of pollinator visits during 3 years of study to *Embothrium coccineum* trees in forest fragments of different size and to isolated trees in pastures (with two replicates by patch size) in a rural landscape of Chiloé Island.

tendency towards higher species richness in forests. The frequency of visits, considering the total and individuals trees, was only slightly higher for isolated trees in pastures than in forests.

The identity of pollinators differed between forest edge and isolated pasture trees. On average, for the 3 years of study, 23–32% of all species were shared between forest and isolated trees. However, when the data for all years are pooled, only 15% of flower visitors were the same in trees in forests and pastures. The frequency of flower visits by pollinator species was low (<5%), except for the main pollinators, *Apis mellifera* (exotic species) and *Bombus dalhbomi* (native species). These species accumulated 9, 47 and 15% of all the visits to flowers each year (*Bombus*) and 43, 4 and 33% of all visits to flowers each year (*Apis*).

Apis mellifera was the main visitor (50% of the visits in 2003 and 2005) to forest and isolated trees, with a lower number of visits to isolated trees (29–31% in 2003 and 2005). In 2004, *Apis* was the main visitor to forest trees, but with a much lower overall abundance (12% forests and 4% isolated trees) in relation to the other 2 years. It was striking that in the same year (2004), the native bumblebee *Bombus* had an increased frequency of visits (47 and 48% of total visits to forest and isolated trees, respectively) relative to 2003 and 2005 (14% of total visits to both forest and isolated trees). Therefore, the main difference in the frequencies of visits to flowers between the main pollinators of *Eucryphia*, *Apis* and *Bombus*, was among years rather than between forest and pasture trees.

Aextoxicon punctatum

Although *Aextoxicon* is thought to have an entomophilous syndrome (Aizen and Escurra, 1998), because of its white-yellow flowers and honey smell, its pollen vectors remain unknown, as we did not record any flower visitors during field observations in Curiñanco or Fray Jorge.

Fruit production

Embothrium coccineum

Results indicated that the fruiting percentages differed among sites in a fragmented landscape (Kruskal–Wallis test, $P<0.0001$). Fruit production decreased from pasture trees and trees in small fragments compared to medium and large fragments. Pasture trees produced more fruits per flower (13%) than small fragments (11%), medium fragments (5%) and large patches (6%) (Fig. 5.5).

Reproductive systems

Embothrium coccineum

Pollination treatments yielded significant differences in fruiting percentages (Kruskal–Wallis test, $H=19$, $P<0.0001$, and subsequent Student–Newman–Keuls test, $P<0.05$). Manual cross-pollination (CP) treatment produced the greatest number of fruits per flower (56%), natural pollination treatment (NP) produced

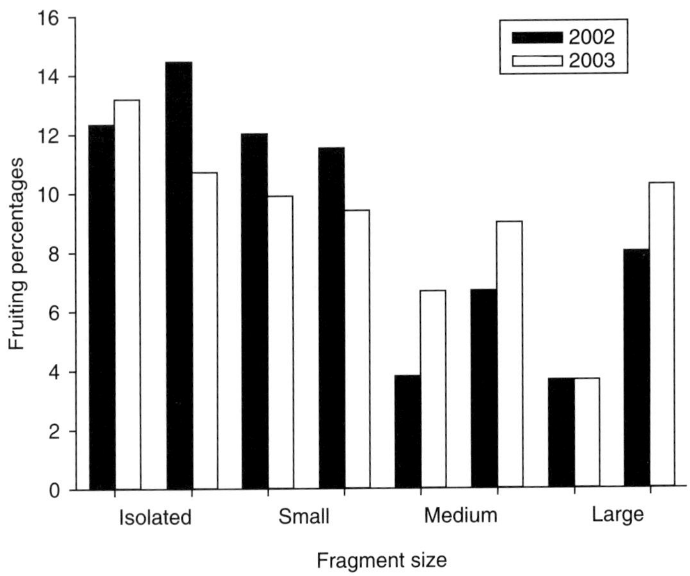

Fig. 5.5. Fruit production (numbers of fruits/flowers) of *Embothrium coccineum* in different study sites in a rural landscape of Chiloé Island. Sites are isolated trees in pastures, small, medium and large size fragments of remnant native forest.

Table 5.2. Results of breeding system assays for *Embothrium coccineum* in a rural landscape of Chiloé Island. Fruit set and seed numbers per fruit are shown for the different pollination treatments. Each pollination treatment was applied to 15 flowers on ten trees. SD = standard deviation, *n* = number of fruits.

Treatments	Total number of fruits	Fruiting percentages per plant ± SD	Number of seeds per fruit ± SD
Manual self-pollination	2	1.33 ± 4.22	14.00 ± 1.00 (*n* = 2)
Automatic pollination	0	0	0
Manual cross-pollination	56	37.33 ± 21.10	12.80 ± 1.83 (*n* = 45)
Natural pollination	26	17.33 ± 11.40	11.90 ± 2.05 (*n* = 23)

intermediate fruit yields (26%) and manual self-pollination (SP) produced a low proportion of fruits (2%) (Table 5.2). In contrast, spontaneous self-pollination (AP) failed to produce fruits (Table 5.2). *Embothrium* breeding success, measured as the number of seeds per fruit, was only assessed for the manual cross-pollination and natural pollination treatments owing to the small quantity of fruits produced by manual self-pollination at this location. The manual cross-pollination treatment yielded a significantly greater number of seeds per fruit and thus displayed higher breeding success than the open cross-pollination (Mann–Whitney test, T=641, P=0.049). The self-incompatibility index was 0.035.

The results confirm that *Embothrium* is self-incompatible and consequently highly dependent on animal pollination agents for sexual reproduction. The absence of fruits in the spontaneous self-pollination treatment shows that *Embothrium* is unable to self-fertilize in the absence of pollinators. The manual self-pollination treatment produced only two fruits, reflecting a high degree of self-incompatibility, and hence the tree was classified as allogamous (Rovere *et al.*, 2006).

Preliminary results on seedling vigour (dry mass) and a population genetics study using isozymes suggest greater vigour and heterozygosity under natural pollination than in cross-pollination treatments (A. Rovere, unpublished data). Perhaps *Embothrium*'s main pollinators (flycatcher and hummingbird) often move distances longer than 500 m between trees, which was the maximum distance used in our cross-pollination experiments. This result may reflect elevated pollen movement and therefore gene flow among *Embothrium* trees in remnants forest, as a result of pollinator activity in this fragmented rural landscape. An alternative explanation is that seedling vigour and heterozygocity are unrelated to current pollen flow.

Aextoxicon punctatum

Results of breeding system tests indicate that, in the northern isolated population of Fray Jorge, only 1% of the fruits were produced by apomixis and/or wind pollination (enclosed treatment), but a greater proportion of fruits per flower (12%) were produced under natural, open pollination (Table 5.3). The southern rainforest population of Curiñanco (Valdivia) displayed a higher proportion of fruits produced by apomixis (15%) and wind pollination (15%). Fruit yields under natural pollination in Curiñanco were lower than in Fray Jorge.

Table 5.3. Reproductive system of the rainforest tree *Aextoxicon puncatatum* (Aextoxicaceae) in its northernmost population of Fray Jorge (30° S) and in the rainforest of Curiñanco (Valdivia, 41° S). Treatments were: –W–P = wind and animal pollinators excluded; +W–P = only animal pollinators excluded, wind allowed; natural = natural, open pollination; and manual = manual pollination.

	Fray Jorge		Valdivia	
Treatment	*n*	%	*n*	%
–W–P	440	1	105	15.4
+W–P	803	1	86	15.2
Natural	699	12	54	1.9
Manual	156	32	72	25

Seed dispersal

Embothrium coccineum

Results showed that the density of winged seeds dispersed declined steeply with increasing distance to the parental tree, thus fitting a leptokurtic seed rain distribution. About 95% of the seeds fell within a 5-m radius around the mother tree, while the longest measured distance of primary dispersal was 20 m (Rovere and Premoli, 2005) (Fig. 5.6). Seed dispersal differed among cardinal points, producing an asymmetrical seed distribution around the parental tree. Seed density was lower and dispersal distances were shorter towards the west, which is the direction of origin of the prevailing winds in the study area.

Flowering and fruiting phenology

Aextoxicon punctatum

According to monitoring of the flowering periods of marked trees over 1 year, the flowering period extended for 4 months in the northern population of Fray Jorge (30° S). In contrast, in the rainforest of Curiñanco (Valdivia), flowering lasted only 2 months (July and August) (Table 5.4). In *Aextoxicon* trees of Fray Jorge, ripe fruits were present with varying abundances among the different months during the entire year (January–December). The fruiting peak was during August. In Curiñanco (Valdivia), ripe fruits were only recorded in 1 month of the year (April) (Table 5.4).

Discussion and Conclusion

The two tree species of southern rainforests, *Embothrium* and *Eucryphia*, which were studied over 3 years, displayed a positive effect of fragmentation on the number of pollinator visits, with more visits to isolated trees in relation to trees in forests. There was some positive effect of fragmentation on the num-

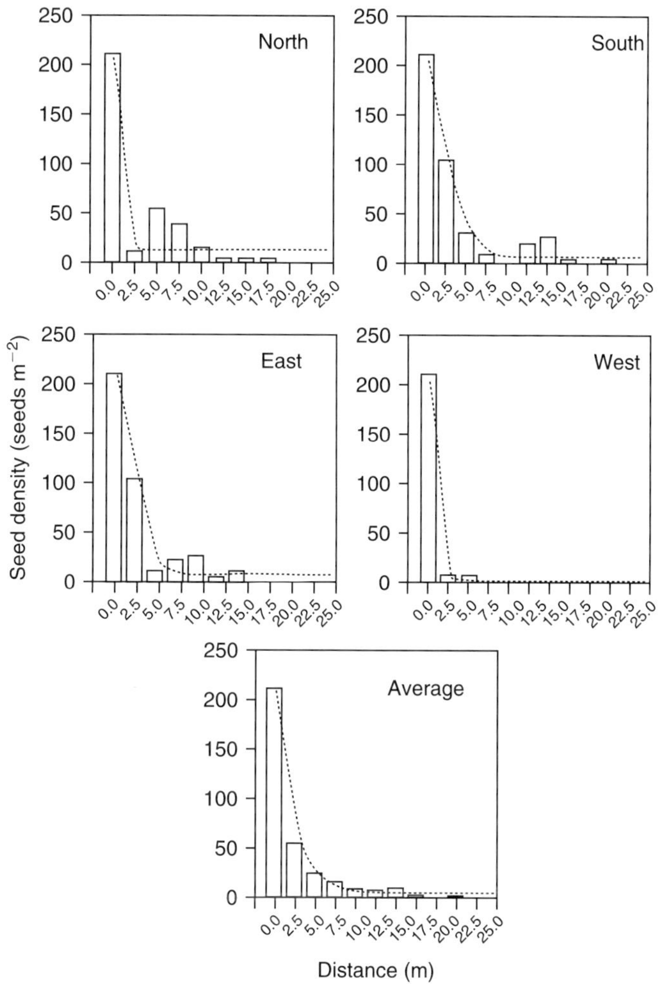

Fig. 5.6. Average densities of *Embothrium coccineum* seeds dispersed to different cardinal directions, as a function of the distance to the parental tree (bars). Fitted curves were calculated for each cardinal direction and for the average of the four directions (dotted line). Regression equations are: North ($y = 14 + 29731\exp(-0.5((x - (-4.9))/1.6)^2)$, $R^2 = 0.92$, F = 26), South ($y = 9 + 229\exp(-0.5((x - (-1.5))/3)^2)$, $R^2 = 0.98$, F = 121), East ($y = 8 + 219\exp(-0.5((x - 0.6)/1.5)^2)$, $R^2 = 0.98$, F = 114), West ($y = 0.6 + 664\exp(-0.5((x - (-2))/1.5)^2)$, $R^2 = 0.99$, F = 3421) and average for all orientations ($y = 7 + 8899\exp(-0.5((x - (-16))/5.7)^2)$, $R^2 = 0.99$, F = 265), where *x* is distance, *y* is seed density.

ber of pollinator species visiting flowers of both *Embothrium* and *Eucryphia* (Table 5.5). These unexpected results are notable because we studied two tree species with very different pollination syndromes. In *Embothrium coccineum* forest fragmentation produces a high genetic differentiation among remnant populations (see Chapter 6; Mathiasen *et al.*, 2007). The differential production of fruits by *Embothrium* in the rural landscape likely reflects a consequence of

Table 5.4. Phenology of coastal populations of *Aextoxicon punctatum* in Fray Jorge and Valdivia, at the northernmost (subtropical) and a southern (temperate) location, covering its latitudinal range in Chile.

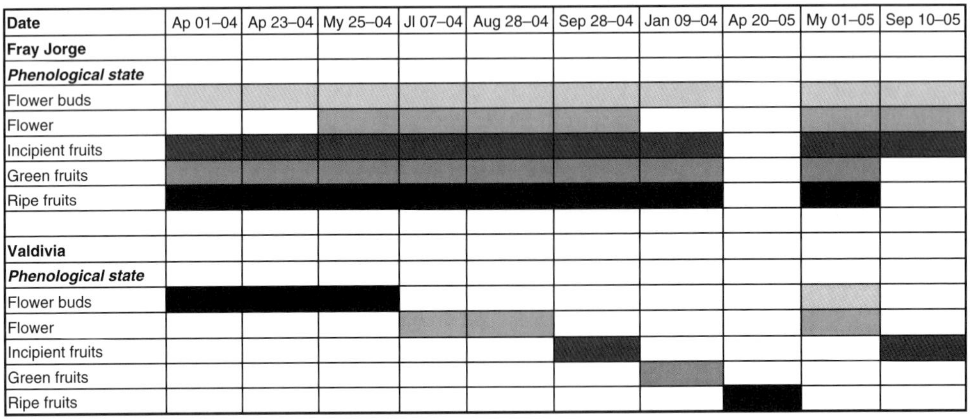

Table 5.5. Effects of forest fragmentation on the identity, species richness and visiting rates of animal pollinators in two common tree species of southern temperate rainforests of South America. Data were obtained for trees found in pastures and different-sized forest fragments in a rural landscape of northern Chiloé Island. Positive effects indicate that fragmentation (accompanied by loss of forest cover) has increased species richness or flower visiting rates by animal pollinators during the period of study.

	Identity	Pollinator species richness	Pollinator species richness per tree	Total number of visits to flowers	Total number of visits per tree
Embothrium coccineum	No effect	Effect depends on year	Positive effect	Positive effect	Positive effect
Eucryphia cordifolia	Effect (different assemblage)	Effect depends on year	No effect	Positive effect	Effect depends on year

higher pollinator visitation rates to isolated trees in pastures and small fragments than to medium- and large-size fragments.

Seed dispersal of *Embothrium*

The restricted seed dispersal and asymmetric seed shadow of *Embothrium*, in the case of isolated trees growing in fields and pastures, may help clarify the population dynamics of this species in fragmented rural landscapes. Other seed dispersal distances known for tree species of southern temperate forests are:

Nothofagus dombeyi, which has dry seeds lacking wings or dispersal structures and dispersed by wind and gravity, between 11 and 14 m; *Austrocedrus chilensis*, with winged seeds dispersed by wind to distances of 16–43 m, depending on site conditions; *Araucaria araucana*, dispersed by birds and ground animals to a maximum dispersal distance of 13 m; *Cryptocarya alba*'s heavy drupes disperse no further than 5 m away from the parental tree; and the wind-disseminated seeds of *Podocarpus saligna* can reach distances of 10 m (Bustamante, 1996).

Owing to the increasing fragmentation of forests in rural areas of Chiloé Island, the maximum distance reached by the winged seeds of *Embothrium* (20 m) is sometimes insufficient for them to arrive at a successional site located away from the fields and pastures, where seedlings could germinate and become established without being eaten by cattle, which dwell in the pastures. However, rare long-distance dispersal events may also occur, but are difficult to detect. *Embothrium* is a successful pioneer tree species invading shrublands in Chiloé Island, but seeding trees must be found not too far from the successional site in order to colonize (J. Armesto and M. Bustamante, 2006, personal communication).

Reproductive systems and phenology of *Aextoxicon punctatum*

Results of preliminary tests show that those populations of *Aextoxicon punctatum* occurring in the opposite extremes of the latitudinal range of the species, separated by 1200 km, exhibit remarkable differences in flowering and fruiting phenologies and reproductive systems. The northernmost populations, found on coastal mountaintops in Fray Jorge, and presumably fragmented from the main range of the species during the entire Quaternary period, present a high genetic differentiation from southern populations (Núñez, 2004) and occur under a markedly different climatic regime. In this northern forest relict population, located at a subtropical latitude (30° S), temperature and humidity conditions remain favourable for phenological activity throughout the year, resulting in extensive flowering and high and prolonged fruit production. In contrast, in Curiñanco (Valdivia, 41° S), under a temperate climatic regime, with cold-windy and very wet winters, the flowering period may be restricted by adverse climatic conditions. It can be argued that these limiting conditions for reproduction at temperate latitudes may favour a trend towards asexual reproductive strategies (e.g. apomixis) in individual trees, as found in this study.

Acknowledgements

Funding and logistic support for these studies were provided by Universidad Nacional del Comahue, Universidad Austral de Chile, CONICET, Senda Darwin Biological Station, BIOCORES project no. ICA4-2000-10029 from the European Community, CMEB Universidad de Chile (P99-103F-ICM) and CASEB, Pontificia Universidad Católica de Chile (Proyecto FONDAP-Fondecyt 1501-0001).

References

Aizen, M. and Escurra, C. (1998) High incidence of plant–animal mutualisms in the woody flora of the temperate forest of southern South America: biogeographical origin and present ecological significance. *Ecología Austral* 8, 217–236.

Aizen, M.A. and Feinsinger, P. (1994a) Forest fragmentation, pollination, and plant reproduction in a Chaco dry forest, Argentina. *Ecology* 75, 330–351.

Aizen, M.A. and Feinsinger, P. (1994b) Habitat fragmentation, native insect pollinators, and feral honeybees in Argentine 'Chaco Serrano'. *Ecological Applications* 4, 378–392.

Aldrich, P.R. and Hamrick, J.L. (1998) Reproductive dominance of pasture trees in a fragmented tropical forest mosaic. *Science* 281, 103–105.

Andren, H. (1992) Corvid density and nest predation in relation to forest fragmentation: a landscape perspective. *Ecology* 73, 794–804.

Arroyo, M.T.K. and Squeo, F. (1990) Relationship between plant breeding systems and pollination. In: Kawano, S. (ed.) *Biological Approaches and Evolutionary Trends in Plants*. Academic Press, London, UK, pp. 205–227.

Brion, C., Puntieri, J., Grigera, D. and Calvelo, S. (1988) *Flora de Puerto Blest*. CRUB, Universidad Nacional del Comahue, Bariloche, Argentina.

Burd, M. (1994) Bateman's principle and plant reproduction: the role of pollen limitation in fruit and seed set. *Botanical Review* 60, 83–139.

Bustamante, R. (1996) Depredación de semillas en bosques templados de Chile. In: Armesto, J.J., Villagrán, C. and Arroyo, M.T.K. (eds) *Ecología de los Bosques Nativos de Chile*. Editorial Universitaria, Santiago de Chile, Chile, pp. 265–278.

Cascante, A.M. (1999) Efecto de la fragmentación del bosque seco sobre el éxito reproductivo de una especie de árbol maderable: *Samanea saman*. Tesis de Magíster en Ciencias. Universidad de Costa Rica, San José, Costa Rica.

Correa, M.N. (1984) *Flora Patagónica, parte IV-a. Dicotiledoneas Dialipetalas*. Colección Científica. INTA, Buenos Aires, Argentina.

Dafni, A. (1992) *Pollination Ecology – A Practical Approach*. IRL, Oxford University Press, New York.

Foré, S.A., Hickey, R.J., Vankat, J.L., Guttman, S.I. and Schaefer, R.L. (1992) Genetic structure after forest fragmentation: a landscape ecology perspective on *Acer saccharum*. *Canadian Journal of Botany* 70, 1659–1668.

Hall, P., Walker, S. and Bawa, K. (1996) Effect of forest fragmentation on genetic diversity and mating system in a tropical tree, *Pithecellobium elegans*. *Conservation Biology* 10, 757–768.

Humaña, A.M. and Riveros, M.C. (1994) Biología de la reproducción en la especie trepadora *Lapageria rosea* R. et P. (Philesiaceae). *Gayana Botánica* 51, 49–55.

Jennersten, O. (1988) Pollination of *Dianthus deltoides* (Caryophyllaceae): effects of habitat fragmentation on visitation and seed set. *Conservation Biology* 2, 359–366.

Lovejoy, T.E., Bierregaard, R.O., Rylands, A.B., Malcolm, J.R., Quintela, C.E., Harper, L.H., Brown, K.S., Powell, A.H., Powell, G.V.N., Schubart, H.O.R. and Hays, M.B. (1986) Edge and other effects of isolation on Amazon forest fragments. In: Soule, M.E. (ed.) *Conservation Biology: The Science of Scarcity and Diversity*. Sinauer Associates, Sunderland, Massachusetts, pp. 257–325.

Mathiasen, P., Rovere, A. and Premoli, A. (2007) Genetic structure and early acting effects of inbreeding in fragmented temperate forests of a self-incompatible tree, *Embothrium coccineum*. *Conservation Biology* 21, 232–240.

Matthysen, E., Lens, L., Van Dongen, S., Verheyen, G.R., Wauters, L., Adriaensen, F. and Dhondt, A.A. (1995) Diverse effects of forest fragmentation on a number of animal species. *Belgian Journal of Zoology* 125, 175–183.

Murcia, C. (1996) Forest fragmentation and the pollination of neotropical plants. In: Schelhas, J. and Greenberg, R. (eds) *Forest Patches in Tropical Landscapes*. Island Press, Washington, DC, pp. 19–36.

Nason, J.D. and Hamrick, J.L. (1997) Reproductive and genetic consequences of forest fragmentation: two case studies of neotropical canopy trees. *Journal of Heredity* 88, 264–276.

Nason, J.D., Aldrich, P.R. and Hamrick, J.L. (1997) Dispersal and the dynamics of genetic structure in fragmented tropical tree populations. In: Laurance, W.F. and Bierregaard, R.O. (eds) *Tropical Forest Remnants: Ecology Management and Conservation of Fragmented Communities*. University of Chicago Press, Chicago, Illinois, pp. 304–320.

Núñez, M. (2004) Diversidad genética de *Aextoxicon punctatum* (Aextoxicaceae) en Chile: implicancias biogeográficas. Tesis de Magíster en Ciencias. Facultad de Ciencias, Universidad de Chile, Santiago, Chile.

Opdam, P. and Schotman, A. (1987) Small woods in rural landscape as habitat islands for woodland birds. *Acta Oecologica* 8, 269–274.

Pérez, C. and Villagrán, C. (1994) Influencias del clima en el cambio florístico, vegetacional y edáfico de lo bosques de 'olivillo' (*Aextoxicon punctatum* R. et Pav.) de la Cordillera de Costa de Chile: implicancias biogeográficas. *Revista Chilena de Historia Natural* 67, 77–90.

Riveros, M. (1991) Aspectos sobre la biología reproductiva en dos comunidades del sur de Chile, 40°S. Tesis de Doctorado en Ciencias. Facultad de Ciencias, Universidad de Chile, Santiago, Chile.

Riveros, M.C., Humaña, A.M. and Arroyo, M.T.K. (1996) Sistemas de reproducción en especies del bosque valdiviano (40° Latitud Sur). *International Journal of Experimental Botany* 58, 167–176.

Robinson, S.K., Thompson, E.R., Donovan, T.M., Whitehead, D.R. and Faaborg, J. (1995) Regional forest fragmentation and the nesting success of migratory birds. *Science* 267, 1987–1990.

Romero, M.M., Riveros, M.C., Cox, C. and Alberdi, A. (1987) Growth dynamics and phenology of *Embothrium coccineum* Forst. at different altitudes. *Revista Brasileira Botánica* 10, 139–145.

Rovere, A. and Premoli, A. (2005) Asimétrica dispersión de semillas de *Embothrium coccineum* (Proteaceae) en el bosque templado de Chiloé, Chile. *Ecología Austral* 15, 1–7.

Rovere, A., Smith-Ramírez, C., Armesto, J. and Premoli, A. (2006) Breeding system of *Embothrium coccineum* J.R. et G. Forster. (Proteaceae) in two populations on different slopes of the Andes. *Revista Chilena de Historia Natural* 79, 225–232.

Ruiz, T. and Arroyo, M.T.K. (1978) Plant reproductive ecology of a secondary deciduous tropical forest in Venezuela. *Biotropica* 10, 221–230.

Saunders, D.A., Hobbs, R.J. and Margules, C.R. (1991) Biological consequences of ecosystem fragmentation: a review. *Conservation Biology* 5, 18–32.

Smith-Ramírez, C. and Armesto, J.J. (2003) Foraging behaviour of bird pollinators on *Embothrium coccineum* (Proteaceae) trees in forest fragments and pastures in southern Chile. *Austral Ecology* 28, 53–60.

Soule, M.E. (1986) *Conservation Biology, the Science of Scarcity and Diversity*. Sinauer Associates, Sunderland, Massachusetts.

Templeton, A.R., Hollocher, H., Lawyer, S. and Johnston, J.S. (1990) The ecological genetics of abnormal abdomen in *Drosophila mercatorum*. In: Barker, J.S.F., Starmer, W.T. and MacIntyre, R.J. (eds) *Ecological and Evolutionary Genetics of Drosophila*. Plenum Press, New York, pp. 17–35.

Young, A.G., Boyle, A.T. and Brown, T. (1996) The population genetics of habitat fragmentation for plants. *Trends in Ecology and Evolution* 11, 413–418.

6 Patterns of Genetic Variation in Tree Species and their Implications for Conservation

A.C. Premoli, R.F. del Castillo, A.C. Newton, S. Bekessy, M. Caldiz, C. Martínez-Araneda, P. Mathiasen, M.C. Núñez-Ávila, P. Quiroga, C. Souto and S. Trujillo-Argueta

A population of the threatened conifer *Fitzroya cupressoides* that has been heavily degraded by timber extraction, and continues to be affected by harvesting of fuelwood and livestock browsing. Photo: Adrian Newton

A population of the threatened conifer *Araucaria araucana* in Argentina, which has been subjected to multiple human impacts, including timber extraction, livestock browsing and fire. The result is a highly degraded population, characterized by small numbers of individuals at low density. Photo: Adrian Newton

Summary

Within-species genetic variability is essential for the maintenance of the evolutionary potential of natural populations. Information from genetic markers can help inform the development of conservation strategies, including those for endemic tree species. This chapter summarizes the results of recent research into the conservation genetics of tree species inhabiting temperate forests of southern Argentina and Chile, and montane forests of northern Argentina and Mexico. Pronounced genetic differences were recorded between populations of all species studied, reflecting their complex biogeographic and evolutionary histories. Species such as *Araucaria*, *Fitzroya* and *Pilgerodendron* appear to have survived in multiple refugia during Pleistocene glaciations, indicated by marked genetic differentiation over small geographical areas. Patterns of variation in the latter two species strongly support the suggestion of local refugia persisting east of the Andes during the last glacial period. Montane species inhabiting subtropical latitudes, such as *Podocarpus parlatorei* and *Pinus chiapensis*, appear to have migrated in elevation during periods of climatic change. In each case, the result is a complex pattern of local population differentiation and adaptation, differing markedly from north temperate tree taxa characterized by large-scale postglacial migrations. Some species, including *Pinus chiapensis*, *Pilgerodendron uviferum* and *Nothofagus pumilio*, displayed evidence of low genetic variation within populations, reflecting the possible occurrence of inbreeding and genetic drift as a result of population isolation. This is despite their possession of life history attributes (such as wind pollination) that imply gene flow over long distances. Results suggest that gene flow may often be restricted, even in wind-pollinated species, in fragmented forest landscapes. However, weak effects of fragmentation were measured in the self-incompatible *Embothrium coccineum*. In this species, higher pollinator activity in fragments reduces selfing, thereby buffering genetic erosion and maintaining adaptive variation. Such results highlight the difficulty of generalizing about the impacts of anthropogenic disturbance on patterns of genetic variation in tree species. These results emphasize the importance of including many populations in conservation strategies and action plans, if the full variation within a species is to be conserved. Particular priorities for conservation include areas of high genetic diversity, which may coincide with putative glacial refugia, such as the coastal mountain range of Chile.

Introduction

In recent years, the protection of genetic diversity within species has become a primary goal of biological conservation, as recognized by international policy initiatives such as the Convention on Biological Diversity (CBD). Information from genetic markers can help inform the development of conservation strategies, including those for endemic tree species (Premoli, 1998; Newton *et al.*, 1999). The theoretical basis of conservation genetics depends on the fact that preservation of genetic variability is essential for the maintenance of the evolutionary potential of natural populations (Frankel and Soulé, 1981). However, some debate has arisen over the relative importance of ecological and genetic factors affecting the survival of species and populations (Lande, 1988; Falk and Holsinger, 1991; Schemske *et al.*, 1994; Hamrick and Godt, 1996). Although the persistence of most species over the short term is believed to depend upon the impact of demographic and environmental threats, genetic factors such as loss of self-incompatible S alleles and inbreeding depression may have important short-term demographic consequences (Young *et al.*, 2000). Distribution patterns of genetic variation therefore need to be considered when planning effective long-term conservation strategies (Mace *et al.*, 1996).

Generalizations on the levels and patterns of genetic diversity in trees have been made in relation to life history traits, which differ among species (Hamrick *et al.*, 1992). For example, factors such as long life-span and high gene flow rates, particularly in species pollinated and seed-dispersed by wind, may result in elevated polymorphism. However, genetic variation may also be affected by the biogeographic history of the species, including the occurrence of migration events or population isolation (Hewitt, 1996; Comps *et al.*, 2001). It is broadly accepted that the mating system and the geographic range of species explain most of the genetic polymorphism found in natural populations. Species that are widespread are expected to maintain higher polymorphism than range-restricted species, which tend to be more affected by genetic drift and isolation, which, in turn, tend to erode genetic diversity (Hamrick *et al.*, 1992). However, Premoli *et al.* (2001) have shown that total geographic range in combination with the degree of population divergence may better predict the patterns of genetic polymorphism in different species than the size of geographic range alone.

Rare and endangered plants usually consist of small, isolated populations that are at high risk of local extinction, particularly under conditions of high forest loss and fragmentation. In particular, loss of continuous habitats and conversion to smaller patches rapidly and severely reduces population size and increases population isolation (Young *et al.*, 1993). As a result, the relatively small, remnant populations are usually reduced in both polymorphism and heterozygosity due to genetic bottlenecks (Barrett and Kohn, 1991). Theoretical studies predict that populations undergoing recent bottlenecks will tend to lose rare alleles owing to random genetic drift more easily than heterozygosity, which may be maintained for several generations (Nei *et al.*, 1975). In addition, small populations may have increased inbreeding, which may result in inbreeding depression for some species (Barrett and Kohn, 1991). These processes working together may lead to fixation of deleterious alleles and to inbreeding depression, affecting individual fitness through reduced viability and fecundity (Young *et al.*, 1993, 1999; Couvet, 2002), which limits the ability of the population to respond to changing selective pressures in the long term (Lesica and Allendorf, 1995; Hamrick and Nason, 1996). However, more empirical information testing these predictions is required, particularly for trees and other long-living plants.

This chapter summarizes the results of recent research into the conservation genetics of tree species inhabiting temperate forests of southern Argentina and Chile and montane forests of northern Argentina and Mexico. Research has ranged from studies at the regional scale along entire species' ranges, analysing levels and distribution patterns of genetic diversity and underlying biogeographic history, studies on the conservation genetics of threatened and rare trees, and the effects of fragmentation on gene flow and genetic diversity and their consequences for adaptive traits. The goal of the research was to assess patterns of variation in selected tree species and to examine the processes influencing these patterns, with the aim of informing the development of conservation strategies in the study regions.

Approaches and Methods for Assessing Genetic Variation

Genetic diversity is one of three components of biodiversity (Chapter 1) and determines the ability of species or populations to respond to environmental change. In theory, the variation in any observable trait detected within the studied individuals can be used for assessing genetic diversity, as virtually any trait has a genetic basis. In practice, however, most of the observable traits are the result of interactions with the environment and, with few exceptions (e.g. chlorophyll deficiency in trees), they are usually determined by many genes. Isolating the genetic component from these traits ideally involves experiments involving transgenerational observations, which are very challenging to perform with tree species, although some insights into patterns of quantitative genetic variation can be obtained using common garden experiments (or their forestry equivalent, progeny and provenance tests). However, division of potential factors determining biological processes into genetic and environmental components has been criticized for being extremely simplistic: genetic and environmental factors are always collinear (Levins and Lewontin, 1985; Peters, 1991). The use of biomolecules as genetic markers has been considered as a useful alternative, based on the assumption of having a simple inheritance basis, and a negligible environment component. While several biomolecules have been used in the past (e.g. terpenes; Glaubitz and Moran, 2000), proteins and DNA are currently favoured for genetic diversity studies in trees. The choice of genetic markers depends on the requirements of the problem to be addressed, and factors such as costs, number of loci required, reproducibility, co-dominance and the degree of polymorphism (Lowe *et al.*, 2004). The research described here was based primarily on the use of protein (isozyme) and DNA markers, although some investigations of quantitative genetic variation have been performed (e.g. Bekessy *et al.*, 2002b; Premoli *et al.*, 2007).

Historical Factors Affecting Biogeographic Patterns of Genetic Diversity

Austral latitudes

The geographic setting of temperate South America consists of pronounced environmental heterogenity. This is due to the presence of the Andean Cordillera, with the Pacific Ocean located to the west and a pronounced rain shadow effect on eastern Andean slopes, and numerous valleys and fjords creating a highly dissected topography. As a consequence, any species inhabiting austral latitudes will encounter highly variable environmental conditions over relatively small geographic distances (Donoso-Zegers, 1987). In addition, historical processes such as climatic change events occurring during the Pleistocene in austral South America have disrupted species ranges, extirpated local populations and changed selective pressures (Premoli *et al.*, 2000a). Therefore spatial heterogeneity and historical processes shape the gene pool of extant populations inhabiting this area.

Studies on distribution patterns of genetic variation in native tree species from Argentina and Chile have been used to test a variety of biogeographic hypotheses. For mid- and high-latitude northern hemisphere species it has been hypothesized that, during glacial cycles, range reductions followed by northward expansion and recolonization by small populations may have diminished genetic variability of populations within newly colonized areas (Critchfield, 1984; Hewitt, 1996). However, the physical setting of austral South America suggests a very different biogeographic history, particularly in relation to Pleistocene climatic changes. In particular, present-day similarities of the southern hemisphere temperate rainforests of South America, Australia and New Zealand suggest less extensive continental ice-sheets, a lower amplitude of Quaternary climate extremes and less persistence of full-glacial environments through interglacials in comparison with the northern hemisphere (Markgraf *et al.*, 1995). Also, pollen records from bogs at mid latitudes from both sides of the Andes indicate rapid forest expansion following the last glaciation (Heusser and Flint, 1977; Heusser, 1981; Markgraf, 1983, 1984, 1991; Villagrán, 1985, 1988, 1991; Villagrán and Armesto, 1993).

Analysis of such pollen records, indicating locally produced tree pollen during full glacial periods in the Chilean lowlands and on Chiloé Island, suggested that the survival of rainforest taxa in glacial refugia was restricted to locations west of the Andes. However, the Quaternary fossil record is not clear for the eastern slopes of the Andes in Argentina at *c.*40–43° S, where many of these same species occur today (Markgraf *et al.*, 1996). In particular, from pollen evidence it is uncertain if rainforest taxa survived the last glaciation in refugia east of the Andes or if they are the result of long-distance migration events from refugia located on the western slopes of the Andes. We tested this hypothesis with a range of endemic tree species, using a variety of molecular markers.

We predicted that populations in refugia will tend to harbour high genetic diversity, whereas recently colonized areas will be genetically depauperate because of genetic erosion occurring through the migratory process. To test this hypothesis, populations were sampled across the entire range of the endemic conifer *Fitzroya cupressoides* (Cupressaceae), which were analysed using isozymes. *Fitzroya* is a monotypic genus that is found in coastal and Andean Chile from 39° 50′ to 43° 30′ S, and on the eastern slopes of the Andes in Argentina occurs as disjunct populations from 41° to 42° 43′ S in remote and humid habitats (Veblen *et al.*, 1995). The results indicated that populations located on the eastern side of the Andes were genetically different (Fig. 6.1) and more variable than those located on the western slopes. Eastern populations had a 13% greater mean number of alleles per locus, a higher total number of alleles and rare alleles (38% and 43% respectively), and 32% higher number of polymorphic loci (0.99 criterion). In addition, seven out of the 25 total rare alleles were restricted to specific populations (i.e. they were unique alleles). Three of these unique alleles were found in the coastal Cordillera (Chile), one on the western slopes of the Andes and the rest in eastern populations.

Based on discriminant analysis of isozyme gene frequency data, we therefore reject the hypothesis of a single refugium in coastal Chile from which *Fitzroya* would have expanded its range after glacial retreat. The presence of unique alleles and the greater degree of genetic variation detected in eastern populations strongly suggest the existence of multiple refugia. In addition, isozyme data indicated that southernmost populations in Argentina have been isolated from western populations for a considerable time, which implies additional locations for glacial refugia east of the Andes. These results were supported by results from RAPD markers, which similarly indicated

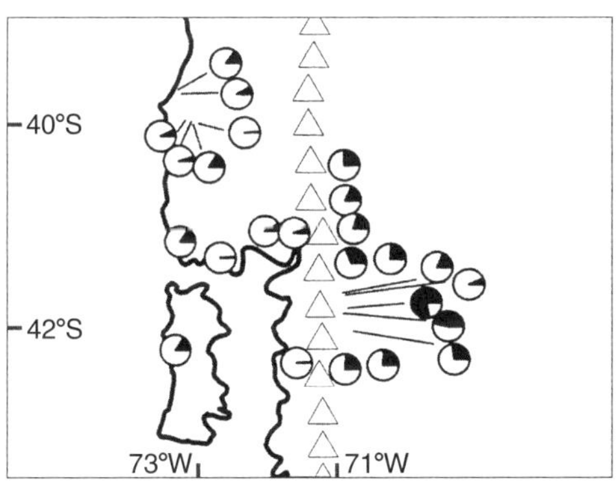

Fig. 6.1. Pattern of genetic variation in *Fitzroya cupressoides* assessed using isozyme markers. Frequency distribution of the Pgi2-1 allele is represented as the filled portion of pie charts in different populations of *Fitzroya cupressoides*. Triangles represent the Andes (Premoli *et al.*, 2000a).

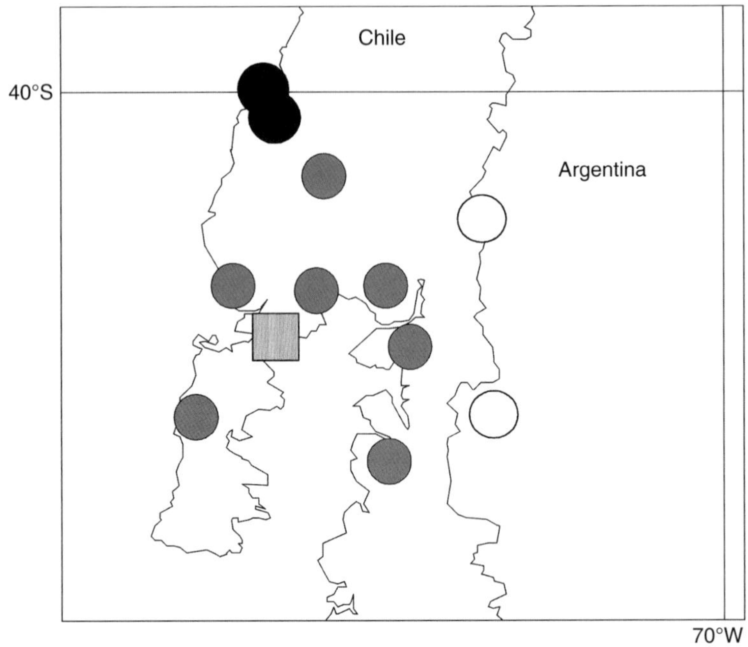

Fig. 6.2. Pattern of genetic variation in *Fitzroya cupressoides* assessed using DNA markers. The different symbols represent groups of populations identified using multivariate statistical techniques (UPGMA) to analyse RAPD marker data. Note the genetic differences detected between populations either side of the Andes (Allnutt *et al.*, 1999).

multiple refugia and genetic differentiation between populations either side of the Andes (Allnutt *et al.*, 1999) (Fig. 6.2). Southern (poleward) location of refugia was an unexpected result and contrasts with the more equatorward locations of tree refugia in the northern hemisphere. Under this northern hemisphere model one would have expected glacial refugia towards the northern limit of taxa in the southern hemisphere. However, a different climatic setting for southern South America, producing minor latitudinal shifts in stormtracks as suggested by Markgraf *et al.* (1995), may have allowed the survival of *Fitzroya* along the Andes during the Pleistocene. This suggests that the continental-scale migration events of the northern hemisphere (Huntley and Webb, 1988) have no analogue at southern latitudes.

Events occurring throughout the Pleistocene may similarly have influenced patterns of genetic variation in another long-lived conifer, *Pilgerodendron uviferum* (D. Don) Florín (Cupressaceae). Although 78% (11/14) of the resolved putative isozyme loci in *Pilgerodendron* were polymorphic (0.95 criterion) in at least one population, approximately half of them were so in only one population (Premoli *et al.*, 2001). Thus, most populations were highly monomorphic, probably reflecting past population bottlenecks and reduced gene flow. Reduced isozyme polymorphism and marked population divergence found in *Pilgerodendron* suggest that historical processes may have played a

Table 6.1. Paired comparisons of mean heterozygosity (He), polymorphism *sensu stricto* (Pss) and among-population divergence (Fst) measured among populations of range-restricted (R) and widespread (W) species at each of four different tree families. Number of analysed populations, loci and polymorphic loci are indicated by Np, Nl and Nlp, respectively. Different letters indicate significant results by t-tests. Values shown are within-population averages (SD).

Family	Species	Range (km)		Np	Nl	Nlp	He	Pss (%)	Fst
Cupressaceae	*Fitzroya cupressoides*	R	201	24	21	11	0.08a (0.03)	33.14a (12.36)	0.08 (0.03)
	Pilgerodendron uviferum	W	1600	20	14	11	0.03b (0.03)	16.79b (12.32)	0.16 (0.06)
Myrtaceae	*Legrandia concinna*	R	163	5	9	6	0.11 (0.04)	35.40a (14.69)	0.41 (0.24)
	Luma apiculata	W	554	6	11	10	0.13 (0.04)	56.06b (16.68)	0.09 (0.02)
Nothofagaceae	*Nothofagus dombeyi*	R	1037	6	8	8	0.20a (0.07)	60.42a (14.61)	0.15 (0.09)
	Nothofagus pumilio	W	2169	20	14	7	0.03b (0.03)	19.29b (9.30)	0.30 (0.24)
Podocarpaceae	*Podocarpus nubigena*	R	728	13	11	10	0.23a (0.06)	66.42a (14.54)	0.22 (0.13)
	Podocarpus parlatorei	W	1203	18	14	8	0.15b (0.03)	41.68b (6.60)	0.11 (0.05)

major role in determining its genetic characteristics (Table 6.1). This finding was again supported by analysis using RAPD markers (Allnutt *et al.*, 2003).

In addition, we tested the hypothesis that the species persisted locally in ice-free areas in temperate South America. It was expected that genetic variation would decrease with latitude, given that ice fields were larger in southern Patagonia and thus refugia were probably located towards the northern distributional limit of the species, as suggested by the fossil record. Isozyme results indicated that southernmost populations tend to be the least genetically variable and were therefore probably more affected by glacial activity than northern ones. This is in agreement with a greater southern extension of ice caps in Patagonia during the LGM (last glacial maximum, about 18,000 C^{14} years BP) (Holling and Schilling, 1981).

The pollen record shows that *Pilgerodendron* might have persisted as small populations during the LGM probably in or near the Chilean coastal range at *c.*42° S latitude (Villagrán *et al.*, 1996). However, fossil information is scarce from both the southern and eastern ranges of the species. Furthermore, *Pilgerodendron* pollen is indistinguishable from *Fitzroya* (Villagrán *et al.*, 1996), leaving open the question of whether *Pilgerodendron* survived in one or several refugia. However, it is noteworthy that a number of different *Pilgerodendron* populations along its current distribution displayed elevated polymorphism and heterozygosity. The retention of such variation in several isolated populations, some of which are genetically distinct and have probably been diverging from the rest for a considerable amount of time, suggests that *Pilgerodendron* probably spread during deglaciation from several surviving populations (Premoli *et al.*, 2002). Under this scenario it is expected that ice caps were not continuous and *Pilgerodendron* populations survived in microclimatologically favourable habitats throughout its current range. RAPD data indicated that southern populations are genetically highly differentiated from those located further north, strongly suggesting that some of the glacial refugia were located in the south (Allnutt *et al.*, 2003). In addition, RAPD analyses identified genetic differences between populations either side of the Andes, closely paralleling results from *Fitzroya* (Allnutt *et al.*, 2003) (Figs 6.3 and 6.4).

Podocarpus nubigena Lindl. is another relatively widespread conifer distributed from 39° 50′ to 50° 23′ S, restricted to high precipitation areas, and generally consisting of small and isolated populations. Eleven isozyme loci analysed in 13 populations sampled throughout the range show high levels of genetic variation (91% *sensu stricto* polymorphism, 3.3 mean number of alleles per locus, 1.5 effective number of alleles, and 0.182 and 0.29 observed and expected heterozygosity, respectively; P. Quiroga, unpublished data). Among-population

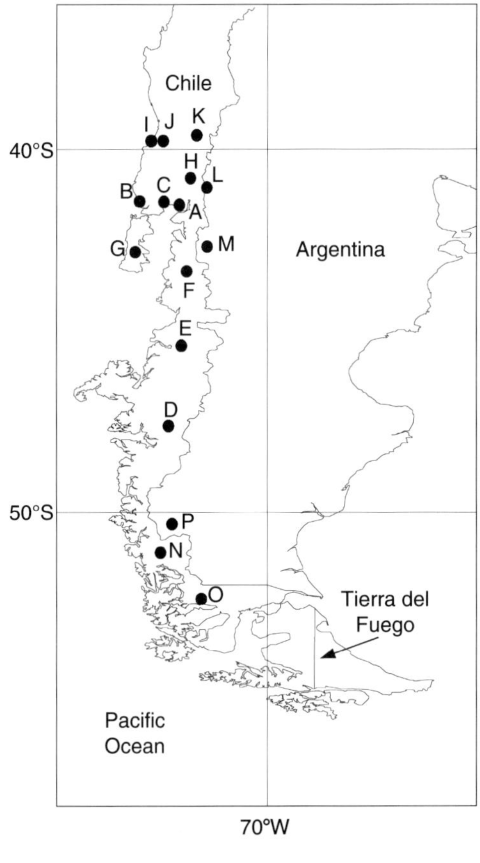

Fig. 6.3. Distribution of populations of *Pilgerodendron uviferum* included in the genetic analysis performed using RAPD markers (see Allnutt *et al.*, 2003).

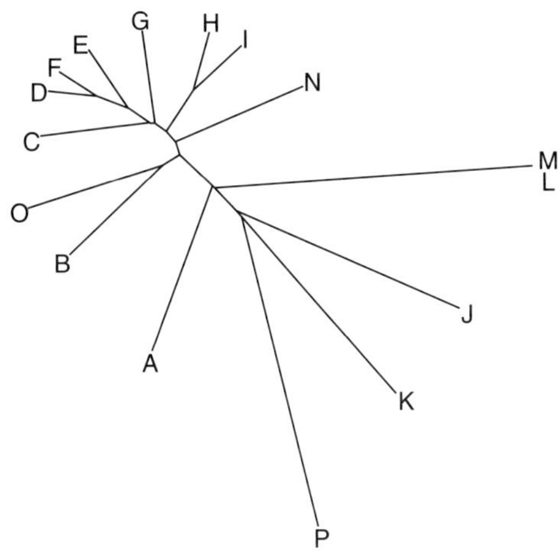

Fig. 6.4. Genetic similarity between populations of *Pilgerodendron uviferum* analysed using RAPD markers (Allnutt *et al.*, 2003). The figure presents results of a UPGMA analysis of pair-wise Phi_{st} values derived from AMOVA of RAPD profiles; those populations that are similar genetically group more closely together on the figure. (For locations of populations, see Fig. 6.3.) Note the similarity of populations L and M to each other, and their difference to the other populations sampled. The genetic differentiation recorded across the Andes directly parallels results obtained with *Fitzroya cupressoides* (Fig. 6.2). Note also the genetic distinctiveness of populations sampled at the extreme of the species' range (J, K, P), suggesting long-term isolation.

divergence was relatively high (Gst=20%) compared to other Podocarpaceae from South America such as *P. parlatorei* (Gst=10%, Quiroga and Premoli, 2007) and *Saxegothaea conspicua* (Gst=11%, P. Quiroga, unpublished data) (Table 6.1). A discriminant analysis based on allelic *P. nubigena* frequencies distinguished among three groups of populations associated with their location relative to the ice limit of the LGM (last glacial maximum). Populations currently located within the area that was covered by the ice have the lowest genetic variation and were clearly separated from the others by the analysis. Most probably, ice-free areas where *P. nubigena* was able to persist during cold periods were located on the western slopes of the Andes (P. Quiroga, unpublished data).

The influence of historical events is also reflected in the gene pool of *Araucaria araucana*. Random amplified polymorphic DNA (RAPD) markers were used to characterize genetic heterogeneity within and among 13 populations of this species from throughout its natural range. Extensive genetic variability was detected and partitioned by analysis of molecular variance, with the majority of variation existing within populations (87.2%), but significant differentiation was also recorded among populations (12.8%) (Table 6.2). Estimates of Shannon's genetic diversity and per cent polymorphism were relatively high for all populations and provide no evidence for a major reduction in genetic diversity from historical events, such as glaciation. Populations

Table 6.2. Comparison of results obtained using dominant DNA markers to assess patterns of genetic variation in woody species native to Mexico and southern South America. P% indicates percentage polymorphic loci, a measure of the mean genetic variation found within populations. (All results obtained with RAPD markers, apart from *obtained with inter-SSR markers.) All species are trees apart from *Berberidopsis*, which is a woody vine.

Species	Sample size	P%	Percentage of variation recorded between populations	Reference
Aextoxicon punctatum	283	63.7	12.1	Núñez-Avila and Armesto (2006)
Araucaria araucana	192	68.4	12.8	Bekessy *et al.* (2002a)
Berberidopsis corallina	44	15.1	54.8	Ehtisham-Ul-Haq *et al.* (2001)
Fagus grandifolia var. *mexicana*	96	39.1	15.6	Rowden *et al.* (2004)
Fitzroya cupressoides	89	72.4	14.4	Allnutt *et al.* (1999)
*Magnolia schiedeana**	64	50.0	25.7	Newton *et al.* (2007)
*Magnolia sharpii**	80	56.0	10.6	Newton *et al.* (2007)
Pilgerodendron uviferum	192	35.7	18.6	Allnutt *et al.* (2003)
Pinus chiapensis	138	24.5	22.6	Newton *et al.* (2002)
Podocarpus salignus	59	47.5	7.0	Allnutt *et al.* (2001)

are currently geographically divided into Chilean Coastal, Chilean Andes and Argentinean regions, but this grouping explained only 1.77% of the total variation. Within Andean populations there was evidence of a trend of genetic distance with increasing latitude, and clustering of populations across the Andes (Fig. 6.5), suggesting postglacial migration routes from multiple refugia (Bekessy *et al.*, 2002a) in common with the other conifers studied.

Araucaria, *Fitzroya* and *Pilgerodendron* are relatively cold-tolerant conifers that may have displayed larger geographical ranges in the past (Villagrán *et al.*, 1996). In contrast, *Aextoxicon punctatum* (Aextoxicaceae) is an endemic tree species of warm temperate rainforests of southern South America. The analysis of the magnitude and geographic distribution of genetic diversity between and within populations of *Aextoxicon punctatum* Ruiz & Pav. (Olivillo) provided clues about the legacy of biogeographic history on the current patterns of genetic variation in a thermophilous tree species. This species exhibits a disjunct distribution along the western margin of southern South America, with a few isolated populations occurring on coastal hilltops of the Chilean Semi-arid Zone (30°–43° S), dependent on coastal fogs. Small populations also persist in some gorges of the central Chilean Mediterranean Coastal Range (32°–39° S), while more continuous forests extend along the southern Temperate Coastal Range (39°–43° S). Some fragmented populations also extend eastward in the south Temperate Central Depression and Andean Foothills (39°–41° S).

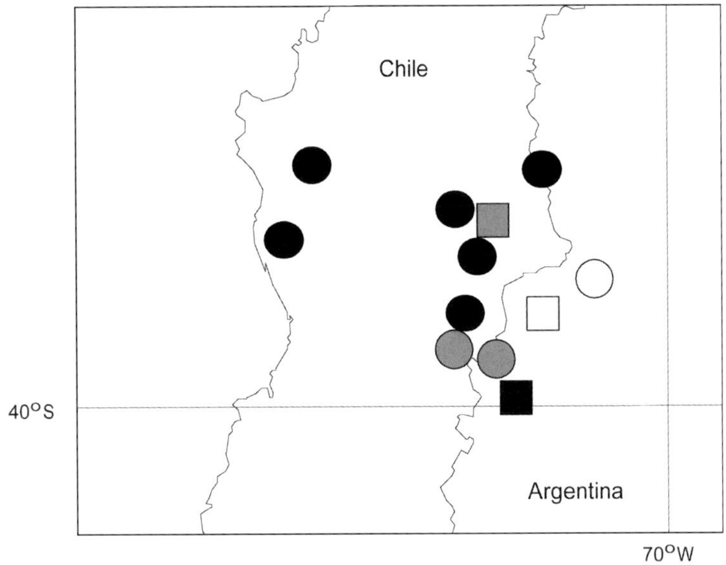

Fig. 6.5. Pattern of genetic variation in *Araucaria araucana*. The different symbols represent groups of populations identified using multivariate statistical techniques (UPGMA) to analyse RAPD marker data. Note the genetic differences detected between populations along the Andes, indicated by the border between Chile and Argentina (Bekessy *et al.*, 2002a).

Random amplified polymorphic DNA (RAPD) markers were used to characterize genetic differences within and among 16 populations of this species throughout its natural range (Table 6.2). The most striking result was the pronounced difference between the isolated populations in coastal hilltops of the Chilean semi-arid zone and the other populations in central and southern Chile (Núñez-Avila and Armesto, 2006). Isolated patches of *olivillo* forest in the Chilean semi-arid zone (Fray Jorge, Talinay and Santa Inés) are thought to represent ancient remnants of preglacial subtropical rainforest that extended along the Chilean coast from subtropical to temperate latitudes (Villagrán *et al.*, 2004a, b). These rainforest islands are currently surrounded by a xerophytic vegetation matrix, but their remarkable floristic affinity with temperate rainforests situated 1000 km to the south suggests that a continuous forest flora may have existed in the past (Villagrán *et al.*, 2004a). The floristic elements of this woody flora have close relatives in the Palaeogene's Mixed Palaeoflora, which was dominated by Australasian-tropical elements that colonized southern South American before the break-up of Gondwana (Hinojosa and Villagrán, 1997). Climatic and tectonic events concentrated during the Plio-Pleistocene transition, such as the onset of west-Antarctic glaciation and the Humboldt Current and the final uplift of the Andes, determined the development of strong aridity in western South America north of 30° S (Hinojosa and Villagrán, 1997), leading to the present isolation of the northern fragments of *Aextoxicon* forests. The high level of genetic divergence between relict populations in the semi-arid zone and the complex of populations of central and southern Chile support this hypothesis (Núñez-Ávila and Armesto, 2006).

At temperate latitudes in southern South America (39–43° S), pollen records show that *Aextoxicon punctatum*, along with other cold-sensitive, Valdivian rainforest species, became restricted to coastal areas during the last glacial maximum (18,000–20,000 years BP, Villagrán, 1991, 2001). Following postglacial climatic warming (between 11,000 and 9500 years BP), palynological evidence indicates that Valdivian tree species expanded south and eastwards from glacial refuges located north of 40° S (Villagrán, 1991, 2001). The fact that the three populations of *Aextoxicon* with the highest genetic diversity (Santa Inés, Los Ruiles and Temuco) are located north of 40° S supports this hypothesis (Núñez-Ávila and Armesto, 2006).

Because the geographic distribution of *Aextoxicon* is currently quite extensive (1200 km) and because it is an obligate outbreeder, a relatively high value of genetic diversity was predicted (Hamrick *et al.*, 1992). However, average genetic diversity for *A. punctatum* populations, using Shannon's index (Spop = 0.36) was lower than RAPD-based Shannon's indices calculated for cold-tolerant conifer tree species from southern South America (Núñez-Ávila and Armesto, 2006). For example, the native Chilean conifers *Fitzroya cupressoides* (Allnutt *et al.*, 1999), *Podocarpus salignus* (Allnutt *et al.*, 2001) and *Araucaria araucana* (Bekessy *et al.*, 2002a) showed higher genetic diversity estimated by RAPDs (Spop = 0.54, 0.64 and 0.65, respectively). We postulate that range contraction, associated with repeated glacial cycles, could have resulted in progressive losses of genetic variability in *Aextoxicon* (Núñez-Ávila and Armesto, 2006) and other Valdivian rainforest species.

RAPD analysis of the thermophilic endemic vine *Berberidopsis corallina* Hook. f. identified an exceptionally high proportion of total genetic variation (54.8%) among populations, and some evidence of reduced genetic variation within populations, suggesting a long history of isolation between remnant populations in the coastal range of Chile (Table 6.2). These results are also consistent with the hypothesis of multiple, restricted refugia for Valdivian rainforest species.

Patterns of genetic diversity for other cold-sensitive woody species of Valdivian rainforests, particularly woody vines, need to be documented to provide a clearer picture of the responses of southern species to glacial and postglacial cycles. However, at least two patterns can be expected from the results discussed in this chapter. Species of cold-tolerant conifers and presumably some species of *Nothofagus* that persisted in glaciated areas and expanded their distribution as Valdivian rainforest species became restricted during the cold glacial cycles (Villagrán *et al.*, 2004b), have maintained relatively high genetic diversity, both within and among populations (Premoli, 1997; Premoli *et al.*, 2000a, 2002). *Pilgerodendron uviferum* presents an exception to this generalization, as reduced within-population variation was recorded using both isozyme and RAPD markers. This may reflect the highly fragmented current pattern of distribution of this species. Cold-sensitive, Valdivian rainforest species that may have suffered repeated range shrinkage during long glacial episodes, followed by short interglacial expansions, may be genetically less diverse and hence more sensitive to present destruction of their habitat.

Montane subtropical areas

Genetic markers were also used to analyse biogeographical hypotheses relating to climatic changes that occurred during the Quaternary in subtropical areas of the southern Andes in South America. Patterns of isozyme diversity were analysed in *Podocarpus parlatorei*, the only conifer inhabiting montane forest of southern Yungas, a cloud forest biome distributed in the subtropics of north-western Argentina, Bolivia and Peru. Populations are restricted to eastern slopes in areas with high precipitation that intercept humid winds from the Atlantic and are surrounded by xerophytic forests. The Yungas consist of different altitudinal vegetation belts, from piedmont forest to rainforest, montane forest and highland pastures. In northern Argentina the Yungas occur across a narrow longitudinal belt of approximately 100 km and are latitudinally discontinuous following N–S orographic patterns.

Results indicate that a marked genetic structure exists in *P. parlatorei* in agreement with isolation-by-distance models. Southern and low elevation populations are the most variable. In addition, southern populations are the most genetically distinct compared to any other group of populations. Genetic diversity declines towards the north and with higher elevations, which may reflect forest migration due to climate change. Northern expansion occurred during relatively cool (i.e. glacial) periods, and range contraction towards the highlands occurred during warming trends (i.e. interglacials; Quiroga and Premoli, 2007). Biogeographic shifts of cold-tolerant conifers, such as *Podocarpus*, growing in the neotropics have been suggested by pollen fossil records. Pollen evidence suggests an altitudinal descent of species including *Podocarpus* between 26,000 and 33,000 years BP. Therefore, cold-tolerant plants that are now montane probably inhabited lowland tropical areas as forest populations throughout a glacial cycle (Colinvaux *et al.*, 2000). Movement of montane species along elevation gradients has been recorded for Amazonian and neotropical plant communities. In particular, evidence suggests that tree species such as *Podocarpus*, *Drimys* and *Alnus* were excluded from lowlands during warming, while persisting in cooler highlands (Colinvaux *et al.*, 2000; Pennington *et al.*, 2000). Hence the genetic characteristics of *P. parlatorei* reflect movements in latitude and elevation, probably associated with changes in climate (Quiroga and Premoli, 2007).

Similarly, the patterns of genetic diversity found in *Pinus chiapensis*, a tropical pine endemic to montane humid areas of southern Mexico and western Guatemala, seem to be explained, in part at least, by the history of this species. *Pinus chiapensis* was found to have higher levels of genetic variation, revealed by both isozymes (del Castillo *et al.*, in prep.) and mtDNA (Newton *et al.*, 2002) (Fig. 6.6), in relatively low altitude areas, such as Chimalapas in the Tehuantepec isthmus. Chimalapas is a likely refuge for this frost-intolerant pine during glacial times owing to its low altitude. Indeed, the entire area has been suggested to have sheltered tropical species during adverse Pleistocenic climatic changes,

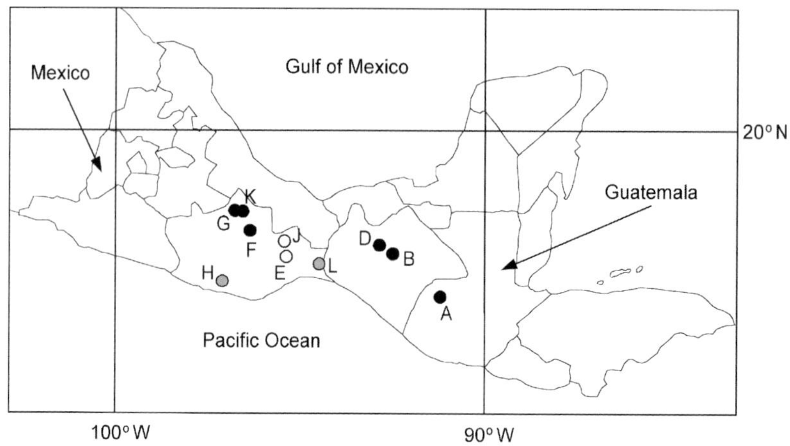

Fig. 6.6. Map illustrating the distribution of two mitotypes of *Pinus chiapensis* detected by RFLP analysis. Populations represented by filled circles were fixed for mitotype A, populations represented by empty circles were fixed for mitotype B. The populations represented by grey circles contained both mitotypes, but in differing proportions: the frequency of mitotype B was 0.125 in population H, and 0.875 in population L. This pattern is interpreted as the result of different migration pathways along mountain chains in southern Mexico during periods of climatic change (Newton *et al.*, 2002).

and is particularly rich in species diversity (Wendt, 1989). The higher genetic diversity in *P. chiapensis* observed in Chimalapas could also be explained by the fact that the Tehuantepec isthmus is a confluence point of the two likely migration paths of this species, which is of putative Holarctic origin, from the north: Sierra Madre del Sur and Sierra Madre de Oaxaca. Indeed, the *CAP2*-slow allele common to the Sierra Madre del Sur studied populations (Yerba Santa and Coatlán), separated by 364 km, appears to be evidence of the Sierra Sur migration path (del Castillo *et al.*, in prep.).

Results obtained with cpDNA PCR-RFLP markers with the threatened Mexican tree *Fagus grandifolia* var. *mexicana* further emphasize the importance of biogeographic history for understanding current patterns of variation in tree species. Three cpDNA haplotypes were identified, two of which were each restricted to an individual population, suggesting that current populations have long been isolated from each other, despite their geographic proximity (Rowden *et al.*, 2004).

Fragmentation Impacts on Gene Flow and Diversity

Austral South America

Temperate forests of austral South America are experiencing increasing rates of deforestation (Armesto *et al.*, 1998; Chapter 2). Loss of primary forests by land conversion, mainly to agricultural uses and plantations of exotic tree species, has affected processes influencing genetic variation in plant and animal species, including the mutualistic interactions influencing gene flow (Smith-Ramírez and Armesto, 2003).

We compared fragmented populations of two monotypic genera endemic to austral forests: (i) small remnant stands of *Fitzroya cupressoides* inhabiting the Central Depression of Chile, and (ii) *Embothrium coccineum* (Proteaceae), occupying different-sized forest fragments and isolated trees immersed in a matrix of agricultural land use in northern Chiloé Island. *Embothrium coccineum* is a widespread fast-growing species that usually occurs as scattered individuals or small populations within the forest. It is considered an early colonizer of open habitats. *E. coccineum* is self-incompatible, and thus dependent upon pollinators (Rovere *et al.*, 2006a, b). Self-incompatibility of *E. coccineum* may favour incoming pollen from other fragments, particularly where few reproductive trees are available for successful pollination, such as in small fragments. This may therefore buffer against the effects of fragmentation.

F. cupressoides presents an extreme case of habitat fragmentation. Small and isolated stands of the species in the Central Depression of southern Chile are remnants of a forest that was once much more extensive (Fraver *et al.*, 1999). Since the 16th century, accessible *F. cupressoides* stands were decimated by timber extraction and by the widespread use of fire for forest conversion into agricultural land. As early as 1850, much of the *F. cupressoides* forest that formerly occurred north of Puerto Montt had been cleared, and by 1890 only a few trees remained in this lowland area (Fonck, 1896; Pérez, 1958; Veblen *et al.*, 1976; Donoso, 1983). As a result, the Andean and coastal populations that were originally connected are currently isolated from each other (Armesto *et al.*, 1995).

The endangered and slow-growing *F. cupressoides* is wind-pollinated, reproduces occasionally by abundant seed production (masting) and depends upon massive disturbances for establishment. *F. cupressoides* is dioecious and, therefore, is an obligate outcrosser. Also, it harbours significant total genetic diversity and reduced within-population inbreeding (Premoli *et al.*, 2000b, 2001), which may be explained in terms of life history and ecological traits, including overall large population size, high fecundity and longevity. The lack of heterozygous deficiency may suggest a general selective advantage of heterozygous individuals (heterosis), or populations with effective sizes large enough that biparental inbreeding is extremely rare, or both. As a result, populations of *F. cupressoides* probably maintain considerable genetic loads given that natural selection has had few opportunities to purge deleterious recessive alleles (Lacy, 1992). Under a habitat-fragmentation scenario, population bottlenecks can be predicted which, in turn, may increase the population's vulnerability to long-term effects of inbreeding. The species may therefore be sensitive to fragmentation.

This was tested by analysis of remnant populations of *F. cupressoides* in the Central Depression. Geographic patterns of isozyme variation of six lowland valley populations maintain elevated within-population isozyme variation in comparison with 30 populations throughout the entire distribution of *F. cupressoides*, and even higher than that of populations located in the Coastal

Table 6.3. Within-population polymorphism (P), genetic diversity (He) and among-population divergence (Fst) of remnant stands of *Fitzroya cupressoides* from the Central Depression, Chile.

Region	P (%)	He	Fst (%)
Central Depression *N* = 6; Np = 8	34.9	0.074	11.2
Coastal Cordillera *N* = 8; Np = 4	28.0	0.063	
Andean Cordillera *N* = 7; Np = 2	28.6	0.067	
Species mean* *N* = 30	33.0	0.075	12.5

N, sampled populations; Np, private, i.e. unique alleles.
*Data from Premoli *et al*. (2000a, b).

Range or the Andes (Table 6.3). In particular, the population of Astillero, located in the Central Depression, has elevated heterozygosity and mean number of alleles compared to other populations in Chile (Premoli *et al.*, 2000b, 2003). Lowland populations were clearly differentiated genetically from those of the Coastal Range or the Andes by the presence of private alleles. These results strongly suggest that ice-free areas existed in lowland valleys during glacial times, which allowed the local survival of cold-temperate woody taxa (Premoli *et al.*, 2003; Villagrán *et al.*, 2004a, b). This, together with the fact that lowland populations appear to be maintaining among-population gene flow rates (Fst = 11%) typical for the species as a whole (Fst = 12%), indicates that genetic characteristics of remnant populations are probably a reflection of pre-fragmentation events.

The genetic distinctiveness of these populations highlights the importance for directing conservation efforts to preserve remnant stands of the Central Depression, which otherwise may result in the loss of genetic diversity of the species as a whole. However, a significant relationship was found between area and log-population size with diversity parameters (Fig. 6.7), which indicate that small relict stands are undergoing population bottlenecks. No such relationship was found for within-population inbreeding, which did not differ significantly from zero, similar to that calculated for other populations of *F. cupressoides*. Studied populations show intense disturbance caused by cattle grazing, logging, fire and/or invasion by *Ulex europaeus*. Despite this, they present abundant regeneration in openings within the evergreen forest dominated by *Drimys winteri* (Winteraceae), *Nothofagus nitida* (Nothofagaceae) and *Eucryphia cordifolia* (Eucryphiaceae), associated with tree stumps, or nearby boggy areas. Adult individuals with diameters at breast height between 20 and 70 cm are scarce and dispersed within the site. Hence, the regeneration process is occurring from a reduced number of remnant trees and thus is suffering from the effects of genetic drift, which tends to erode genetic variation in populations.

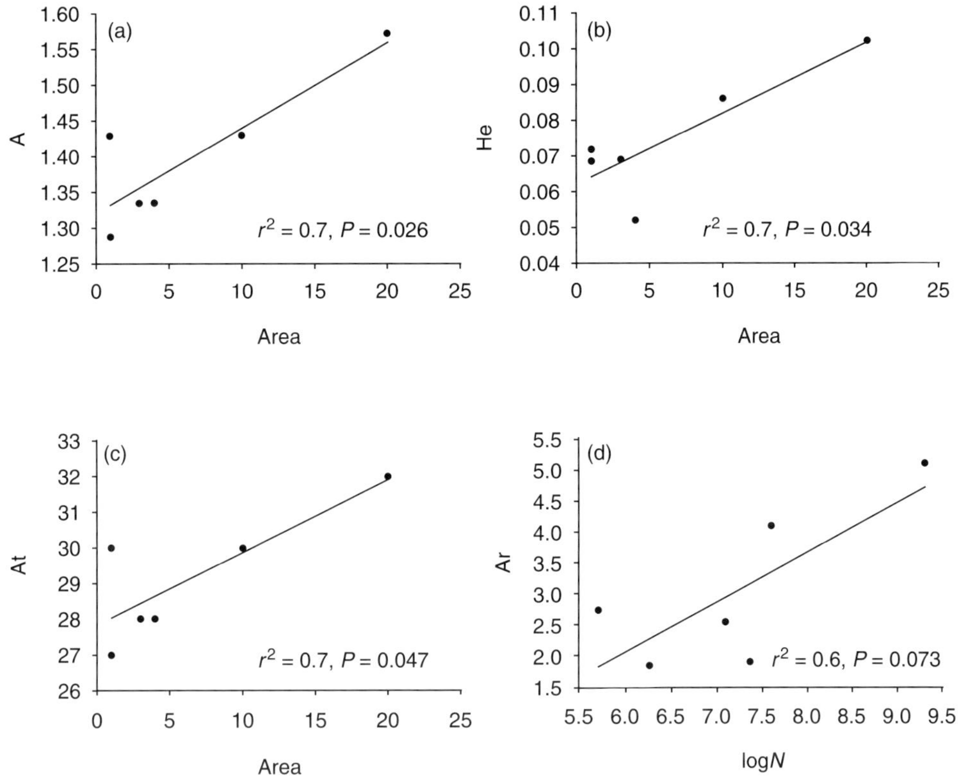

Fig. 6.7. Relationship between levels of within-population genetic diversity parameters and population area (ha) and log-population size of sampled remnant stands of *Fitzroya cupressoides* from the Central Depression in southern Chile. A is the mean number of alleles, He is the expected heterozygosity, At is the total number of alleles and Ar is the number of rare alleles following Premoli *et al.* (2000a).

We studied the effects of reduced population size and increased isolation on population genetic structure and early performance of progeny of *E. coccineum*. Samples were collected from spatially isolated trees and six fragments of differing sizes (small, 1 ha; medium, 20 ha; large, >150 ha). Based on isozyme polymorphisms, we estimated genetic diversity, divergence and inbreeding for adults and greenhouse-grown progeny. We also measured germination, seedling growth and outcrossing rates on progeny arrays. Adult trees and seedlings from the six population fragments differed somewhat in their genetic variation. While adults had more total alleles, seedlings displayed higher polymorphism (Fig. 6.8). Only adults yielded significant within-population inbreeding (Fis) and low, but significant genetic differentiation existed among adult and progeny populations (Fst) (Fig. 6.9). Genetic variation of adults was not correlated with population size, as expected, given that fragmentation occurred relatively recently. The analysis of satellite images from 1976, 1985 and 1999 indicated that studied fragments are 20–25 years old, half of which were

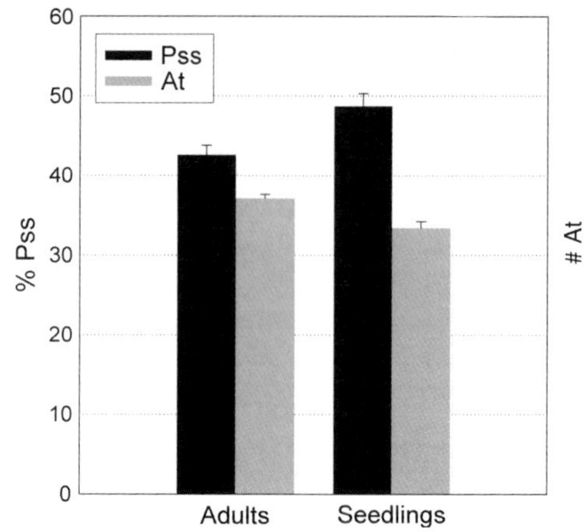

Fig. 6.8. Per cent polymorphic loci (left *Y* axis) and total number of alleles (right *Y* axis) of adults and seedlings of *Embothrium coccineum* sampled from six different-sized fragments in northern Chiloé.

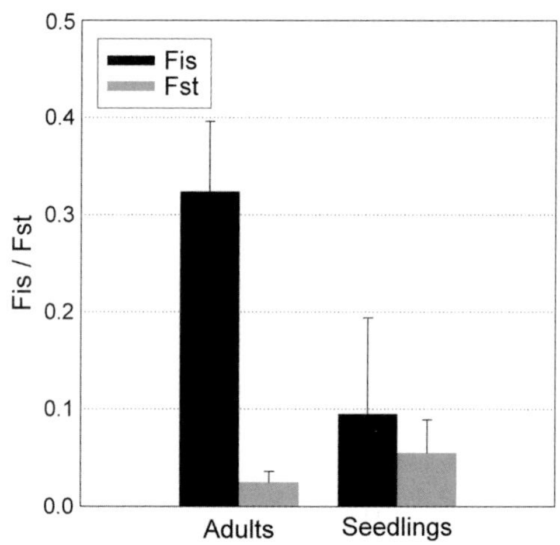

Fig. 6.9. Within-population inbreeding (Fis) and degree of among-population divergence (Fst) of adults and seedlings of *Embothrium coccineum* sampled from six different-sized fragments in northern Chiloé.

produced from one large ancestral population of approximately 130,000 ha (C. Echeverría, Instituto de Silvicultura, Universidad Austral de Chile, Valdivia, Chile, personal communication). Also the oldest and largest *E. coccineum* trees are found in different-sized patches (Mathiasen, 2004). Consequently, in southern Chile not enough time has elapsed to produce noticeable effects on the genetic make-up of adult *E. coccineum* individuals due to fragmentation.

Weak effects of fragmentation, i.e. with population size, were measured on progeny. Seedling growth correlated positively with the effective number of alleles, showing deleterious effects of inbreeding on progeny (Mathiasen, 2004). Seeds from small fragments displayed the highest outcrossing rates and germination success, indicating that higher pollinator activity in such fragments reduces selfing, thereby buffering genetic erosion and maintaining adaptive variation (Mathiasen *et al.*, 2007). Higher reproductive success, assessed as per cent fruit production, was measured in small fragments and isolated trees (11% and 13%, respectively) and compared to that in medium-sized and large fragments (5% and 6%, respectively) within the same study area (A. Rovere, Universidad Nacional del Comahue, San Carlos de Bariloche, Argentina, 2006, personal communication). In comparison with continuous forest and owing to edge effects (see Chapter 4), larger fragments increase the potential area to be colonized by *E. coccineum*, producing large populations that may affect pollinator behaviour. Previous data show that spatially isolated *E. coccineum* individuals and those at forest edges within small fragments are visited more frequently by its main pollinator, the Tyranid *Elaenia albiceps* (Smith-Ramírez and Armesto, 2003). In contrast, pollinators are more territorial in larger fragments, where they tend to defend and feed upon nectar of 3–5 adjacent flowering trees. Nearby *E. coccineum* individuals will tend to be genetically similar, given that primary seed dispersal is highly localized; most seeds are dispersed 20m from the mother tree (Rovere and Premoli, 2005). Therefore, pollinator behaviour in larger fragments favours pollen exchange among closely related individuals, which will increase inbreeding.

Mexico

Within-population genetic variation, assessed with RAPDs and cpDNA PCR-RFLP markers, was positively related to population size in Mexican beech, *Fagus grandifolia* var. *mexicana* (Rowden *et al.*, 2004). As happens with many other species of trees in montane humid forests of Mexico, populations of this species have been severely reduced owing to deforestation. In this case, genetic diversity appears to be affected directly by anthropogenic disturbances, although it is difficult to separate anthropogenic fragmentation effects from natural processes (such as climate change) that may also have resulted in range contraction and the isolation of populations.

The effects of reduced population size were also studied in *Pinus chiapensis*. Population sizes of this conifer were assessed by estimating the number of breeding individuals in different populations, which cover five orders of magnitude, from 1 individual to more than 50,000 trees. Significant and positive correlations between estimates of genetic diversity, estimated at seed stage, and population size were detected after adjusting for differences in sample size (del Castillo *et al.*, in prep.). Nevertheless, some small-sized populations demonstrated high values of genetic diversity. The population of Río Pinal, Chimalapas, with only 36 individuals, is an example. Historical and biogeographic factors, as described above, may account for this result. This population was located in an isolated and preserved tropical rainforest area. This species requires a disturbance for successful establishment (del Castillo, 1996). In this population, the lack of recent disturbance may hinder successful establishment of this species, leaving a population composed only of aged trees that established during the last disturbance (del Castillo *et al.*, in prep.). Similarly, a relative high genetic diversity was detected in the very fragmented and reduced population of Chenalhó, in Chiapas Highlands. In both populations, the reduced population size appears to be recent in terms of numbers of generations after the decline, being in the case of Río Pinal only one generation, and in the case of Chiapas Highlands no more than two, as high deforestation rates in this area were restricted to the last few decades, based on remote sensing analyses (Cayuela *et al.*, 2006). In theory, losses of genetic diversity are expected to occur slowly after population bottlenecks (see Frankham, 2005).

Isolation and reduced population size may dramatically increase the rates of inbreeding. Estimates of mating system in several populations of *P. chiapensis* revealed that this species is predominantly or totally outcrossing (del Castillo *et al.*, in prep.), with multilocus outcrossing values, *tm*, slightly lower than or comparable to those reported for other pines (Delgado *et al.*, 2002). However, in the Río Pinal population, *P. chiapensis* displayed a *tm* value not significantly different from zero. Moreover, correlations of paternity among maternal sibships were very high and not significantly different than one, indicating that maternal sibs are essentially full sibs in this small population. By contrast, in the large-sized populations studied, paternity correlations were always not significantly different from zero. Thus, maternal sibs were in general half sibs, suggesting multiple paternities for a single mother tree. These results provide evidence that significant alterations in breeding systems favouring inbreeding result from population decline.

Small populations of *P. chiapensis* also displayed low rates of seed germinability tested under a common environment (greenhouse). Populations with fewer than 50 breeding individuals had on average germination rates that were 80% lower that those of populations larger than 50 breeding individuals. This result, and the fact that heterozygosity was positively correlated with seed germination, suggests that inbreeding depression is involved. Thus, inbreeding depression may have important demographic implications in *P. chiapensis*. The lack of significant genetic associations between loci detected in this species also suggests the involvement of inbreeding depression. Theoretical studies predict that inbreeding depression decreases the magnitude of linkage disequilibrium (Vargas and del Castillo, 2001).

Geographic Partitioning of Genetic Variation and Among-population Divergence

Spatial heterogeneity along complex gradients

Widespread species show complex genotypic and phenotypic structure reflecting predominant stresses imposed by the environment (Slatkin, 1987). In widely dispersed species, these differences are often the result of natural selection defining ecotypes in relation to moisture and/or temperature gradients (Larsen, 1981; Abrams *et al.*, 1992). Patterns of variation may be adaptive when they either vary in relation to an environmental gradient in different parts of a species' range, and/or vary in a coordinated fashion across the gradient (Jonas and Geber, 1999). Many studies have revealed populations genetically differentiated for distinct characteristics at both small and large spatial scales (Turesson, 1922; Clausen *et al.*, 1948; Hiesey and Milner, 1965; Langlet, 1971; Briggs and Walters, 1984; Abrams, 1988; Abrams and Kubiske, 1990; Linhart and Grant, 1996).

Embothrium coccineum occurs along a widespread latitudinal (*c*.20°) and altitudinal range, from sea level to treeline of the southern Andes. It can be found in contrasting habitats from bogs to dry steppes. Moreover, *E. coccineum* typically occurs most frequently in gaps and open areas, and has been classified as a successful colonizer (Alberdi and Donoso, 2004). Patterns of isozyme variation were used to genetically characterize 34 populations by means of within-population genetic parameters to test the hypothesis that populations occurring in similar environments share genetic traits reflecting adaptation to local habitat conditions. Multivariate discriminant analysis using allelic frequencies grouped populations according to their local environment. The plot of the first two canonical coefficients clearly depicts four distinct groups (Fig. 6.10). These are:

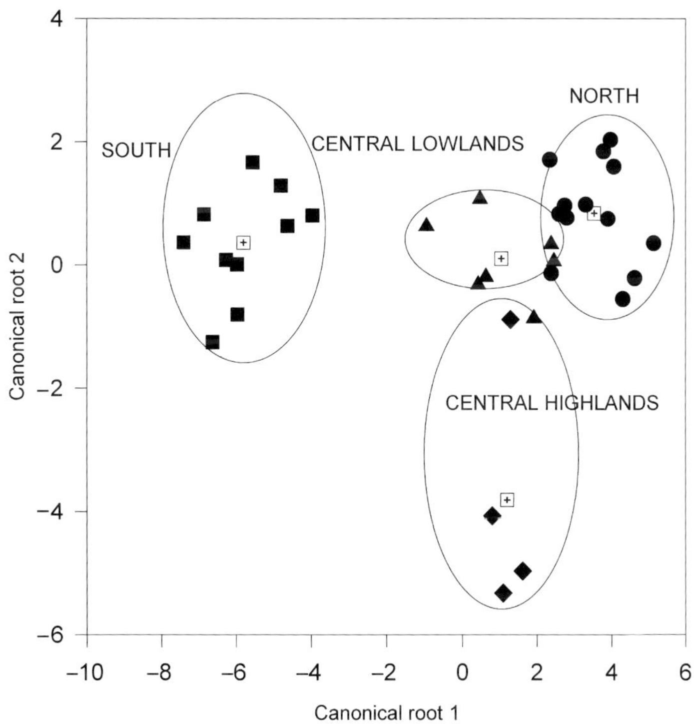

Fig. 6.10. Plot of the first two canonical coefficients generated by multivariate discriminant analysis showing four genetically distinct groups: North, Central Highlands, Central Lowlands and South. The best eight discriminatory variables remaining in the model were allele frequencies at *Mdh3-2*, *Mdh3-3*, *Mnr2-2*, *Pgm1-2*, *Pgm2-3*, altitude, latitude and longitude, all significant at the 0.001 level.

I=North; II=Central Highlands; III=Central Lowlands (West); IV=South (for group description see below). Genetic variation in *E. coccineum* is generally influenced by climate, characterized by colder and wetter conditions in the south and warmer and wetter conditions to the west. An analysis of molecular variance (AMOVA) yielded significant differences between groups ($P<0.03$).

These data, together with morphological information (Souto and Premoli, in prep.) and field observations, allow us to characterize four ecotypes in *E. coccineum* associated with discrete habitat types, commonly found at contrasting locations along latitude, longitude and altitude gradients in Patagonia. Genetic variation of *E. coccineum* therefore reflects a discontinuous geographic structure in response to different environmental conditions. These results suggest that such widespread species maintain levels of adaptation to environmental heterogeneity by specific locally adapted genotypes.

Such adaptation can also be demonstrated in narrow endemic species with a much smaller geographic range. In seedling growth experiments, Bekessy *et al.* (2002b, 2003) demonstrated the occurrence of adaptive variation relating to drought tolerance in the threatened conifer *Araucaria araucana*, including variation in allocation of dry matter to root growth and water-use efficiency. This variation reflects the broad edaphic tolerance of the species, which occurs along a rainfall gradient from coastal Chile (Nahuelbuta) to much drier locations in Argentina, on the eastern side of the Andes. As expected, these adaptive traits were poorly correlated with the selectively neutral variation detected by RAPD markers (Bekessy *et al.*, 2003).

Overall, when comparing a range of species, dominant DNA markers such as RAPDs detected pronounced differentiation between populations in all of the species studied, with an average of around 20% of the variation detected attributable to differences between populations (Table 6.2). Such variation can be attributed to the complex biogeographic histories displayed by these species, as discussed earlier. However, as illustrated by the results from *Araucaria araucana*, where RAPD data failed to detect variation in an adaptive trait (Bekessy *et al.*, 2003), both DNA and isozyme markers are likely to underestimate the extent of variation that occurs within species, particularly those that are distributed over broad environmental gradients.

Widespread versus range-restricted closely related taxa

Large continuous tree populations are expected to maintain greater polymorphism than isolated smaller populations owing to the effects of isolation and drift, which tend to erode genetic variation in the latter (Hamrick *et al.*, 1992). However, it has been suggested that the combination of the total range with the degree of among-population divergence may better predict patterns of genetic polymorphism in different species (Premoli *et al.*, 2001). To test this hypothesis we used distribution patterns of gene diversity, polymorphism and among-population divergence by analysing pairs of range-restricted and widespread closely related taxa belonging to four different tree families endemic to South America. Studied families were Cupressaceae, Myrtaceae, Nothofagaceae and Podocarpaceae.

Corresponding range-restricted and widespread taxa within each family were the monotypic *Fitzroya cupressoides* and *Pilgerodendron uviferum* (Cupressaceae); the rare *Legrandia concinna* (Phil.) Kausel and the common *Luma apiculata* (DC.) Burret (Myrtaceae); *Nothofagus pumilio* (Poepp & Endl.) Krasser, which is restricted to high-elevation and high-latitude forests, and *Nothofagus dombeyi* (Mirb.) Oerst., which is commonly found in lowland temperate habitats (Nothofagaceae); and *Podocarpus* species *P. nubigena* Lindl. and *P. parlatorei* Pilg. (Podocarpaceae) inhabiting wet temperate and montane subtropical latitudes, respectively. Populations of the different species were sampled for genetic analyses by means of isozyme electrophoresis along their latitudinal range (Table 6.1). Total range was calculated from GPS geodetic data specified by latitude, which was converted into geographical distances. We used similar sampling schedules in the field and protocols of genetic analyses for the analysis of a total of 112 populations.

On average 13 loci were resolved with a minimum of 8 and a maximum of 21 per species. Seventy per cent of each were polymorphic at any one population. Except for the Myrtaceae, paired comparisons indicated that the range-restricted species hold higher polymorphism *sensu stricto* (Pss) and gene diversity (He) than their corresponding widespread species (Table 6.1). In contrast, the widespread *L. apiculata* has higher polymorphism than the range-restricted *L. concinna* (Table 6.1). Species with restricted distribution of each studied family show a tendency towards displaying a positive relationship between polymorphism (Pss, adjusted $r^2=0.7$, $P=0.10$; Fig. 6.11a) and increasing total latitudinal range. As a result, range-restricted species fit expectations for increased diversity with distributional range (Hamrick *et al.*, 1992).

As predicted, rare and endangered species such as *Fitzroya cupressoides* and particularly *Legrandia concinna*, which consists of only five known highly isolated populations (Martínez-Araneda, 2004), may suffer the effects of drift that erodes genetic variation. In contrast, the opposite trend was found for widespread species, as those with larger distributions hold the least polymorphism (Pss, adjusted $r^2=0.8$, $P=0.09$; Fig. 6.11b). In particular, *Pilgerodendron uviferum* and *Nothofagus pumilio*, although displaying the widest latitudinal range of all tree species reported here, have significantly lower polymorphism and gene diversity than their range-restricted counterparts, *Fitzroya cupressoides* and *Nothofagus dombeyi*, respectively (Table 6.1).

A plausible explanation for this pattern relies on the fact that the former two widespread species are habitat-restricted. *Pilgerodendron uviferum* is usually found in periglacial environments and permanently boggy terrain under high precipitation regimes. *Nothofagus pumilio* is restricted to high-elevation forests at mid latitudes and it only becomes more common at lower elevations further south under extreme low temperature conditions. Therefore, habitat specialization to patchy environments may result in restrictions for gene flow among populations, which, in combination with historical processes such as range contractions that occurred during glaciation events in Patagonia, probably resulted in population isolation. Therefore, limitations for gene exchange explain the twofold greater genetic divergence among *P. uviferum* (Fst=0.16) and *N. pumilio* (Fst=0.30) populations than their respective range-restricted

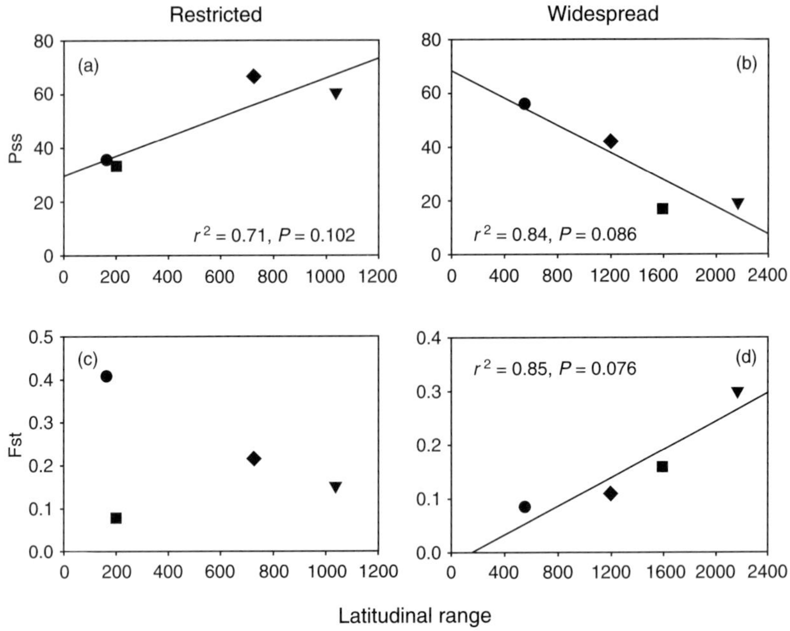

Fig. 6.11. Relationship between latitudinal range (km) and level of within-population polymorphism *sensu stricto* (Pss) and degree of among-population divergence (Fst) of tree species from South America. Pairs of range-restricted and widespread species are Cupressaceae (squares): *Fitzroya cupressoides* and *Pilgerodendron uviferum*; Myrtaceae (circles): *Legrandia concinna* and *Luma apiculata*; Nothofagaceae (diamonds): *Nothofagus dombeyi* and *Nothofagus pumilio*; and Podocarpaceae (triangles): *Podocarpus nubigena* and *Podocarpus parlatorei*.

species *F. cupressoides* (Fst=0.08) and *N. dombeyi* (Fst=0.15), respectively (Table 6.1). This result agrees with previous suggestions that improved predictions on the levels and distribution of genetic polymorphisms may be obtained if the degree of spatial isolation among populations is taken into account (Premoli *et al.*, 2001).

In contrast, other relatively widespread species such as *Luma apiculata* (Fst=0.09) and *Podocarpus parlatorei* (Fst=0.11) have lower among-population genetic divergence than that measured for their range-restricted counterparts *Legrandia concinna* (Fst=0.41) and *Podocarpus nubigena* (Fst=0.22), respectively (Table 6.1). Overall, while the relationship between total range and degree of among-population divergence showed no clear pattern for range-restricted species, widespread species yielded a positive trend (Fig. 6.11). The widespread *Luma apiculata* and *Podocarpus parlatorei* usually consist of large populations. In addition, they have edible fruits and thus animal dispersal may counteract population isolation, contributing to relatively low Fst values (Fig. 6.11). However, the lack of a relationship between Fst and range of narrowly distributed species indicates complex interactions between autoecological traits and historical factors shaping genetic patterns.

Genetic structure at different geographic scales

Significant inbreeding measured in large stands of *Luma apiculata* can be attributed to biparental inbreeding as well as self-fertilization. Estimations of outcrossing rates from progeny arrays of isolated trees from different locations yielded multilocus *t* values that ranged between 0.53 and 0.78 (Caldiz, 1999). This suggests that the mating system of *L. apiculata* fits the mixed mating model, being predominantly an outcrosser. Local seed dispersal and massive establishment of genetically closely related seedlings may result in elevated biparental inbreeding. The formation of family groups has been reported for other tree species at microgeographical scales of hundreds of metres (Linhart *et al.*, 1981; Perry and Knowles, 1991; Schanbel *et al.*, 1991; Shapcott, 1995) or even shorter spatial scales as a response to fine-scale disturbances in *Nothofagus dombeyi* (Premoli and Kitzberger, 2005).

Further evidence of a marked genetic structure in *L. apiculata* comes from a spatial autocorrelation analysis that indicates short-distance clustering of similar genotypes at spatial scales <4 m, which was consistent between different sites within Nahuel Huapi National Park in northern Patagonia (Caldiz, 1999). *L. apiculata* is known to reproduce vegetatively and is insect-pollinated, thus either or both of these two mechanisms may be responsible for producing small-scale genetic structure. Hence large *L. apiculata* stands cannot be considered as genetically homogeneous panmictic populations, and genetic differentiation can occur over short distances. Therefore populations of this species consist of genetically heterogeneous patches resulting from the combined effects of local establishment of related seeds and clonal spread in the species. In addition, isolated populations maintain as much genetic variation as large stands (Caldiz, 1999).

In such species, conservation strategies should seek to conserve isolated and also large stands where natural regeneration processes generate a complex and dynamic genetic structure. Genetic structure at larger geographic scales was found in the rare and range-restricted Myrtaceae *Legrandia concinna*. The species is known only from five isolated populations distributed in the Andes, along a narrow latitudinal range. Multivariate UPGMA cluster analysis and Mantel test of genetic distance against geographic distance suggests geographical structuring among *L. concinna* populations, and clearly separated a northern population (Radal) from southern locations. Also, southern populations showed significant inbreeding (>0.3), while the northern population, Radal, displayed the lowest value (0.15). This northern population consists of a cluster of neighbour populations that contribute genes by pollen or seeds, counteracting the effects of inbreeding. In contrast, the other populations are geographically isolated and inhabit native forest remnants surrounded by plantations that may limit pollination, seed dispersal and therefore gene flow. Historical factors, such as past disjunctions owing to survival in different glacial refugia and/or divergent selective pressures acting in distinct populations, may explain the observed patterns in *Legrandia concinna* (Martínez-Araneda, 2004).

Discussion

Studies of endemic and threatened trees of the temperate and montane forest areas of Argentina, Chile and Mexico provide important insights into patterns of genetic variation, and the processes responsible for these patterns. All of the species studied displayed pronounced genetic differences between populations, reflecting their complex biogeographic and evolutionary histories. For temperate taxa, information strongly suggests that cold-loving species survived in multiple refugia during the last glaciation in Patagonia. Montane species inhabiting subtropical latitudes underwent migrations in elevation, towards higher elevations during climatic warming and lower elevations during cooling. The result is a complex pattern of local population differentiation and adaptation, differing markedly from most northern hemisphere tree taxa that have been studied, which typically underwent large-scale migrations from restricted refugia following the end of the last glacial period (Petit *et al.*, 2003).

The implication of such results is that many populations may need to be included in conservation strategies and action plans, if the full variation within a species is to be conserved. A number of different approaches have been proposed for incorporating such genetic information into conservation planning. *Evolutionarily Significant Units* (ESUs) have been defined as historically isolated populations, which may require separate genetic management (Moritz, 1994, 1995). The concept was initially developed for animals on the basis of differentiation in mitochondrial (mt) DNA, and explicitly on analysis of the spatial distributions of alleles, taking account of their phylogenetic relationships (Moritz, 1994, 1995). Given the low mutation rate of mtDNA in plants, it is uncertain how the concept might usefully be adapted to trees (Newton *et al.*, 1999). The concept of ESUs has also been criticized because molecular marker variation is usually selectively neutral, and therefore it ignores patterns of adaptive variation. As an alternative to ESUs, Crandall *et al.* (2000) suggested that populations be classified according to whether they show recent or historical ecological or genetic exchangeability. This classification is based on whether gene flow is currently occurring between populations, or occurred in the past, and takes into account patterns of adaptive variation. Management recommendations are then based on this assessment. An evaluation of this approach, and the ESU concept overall, is provided by Fraser and Bernatchez (2001). Other approaches that have been proposed include *Management Units* (MU), defined as populations with significant divergence of allele frequencies at nuclear or mitochondrial loci (Moritz, 1994), and *Gene Resource Management Units* (GRMU), which may be defined as areas of land chosen to include a representative sample of the genetic diversity of a species within a particular region, and designated for a particular genetic management objective (Ledig, 1988; Millar and Libby, 1991).

Little attempt has been made to apply such concepts to conservation of forest genetic resources within the study areas. Care should clearly be taken to base conservation decisions on results from more than one type of genetic marker, and on patterns of adaptive variation as well as molecular marker data. However, the information presented here could readily be used to prioritize particular populations for conservation action, such as those that are genetically distinctive, or those of particular high diversity, perhaps reflecting their association with a putative glacial refugium. In this context, the fact that pronounced genetic differentiation was recorded in three different endemic conifer species between populations on either side of the Andes is particularly noteworthy, and highlights the outstanding conservation importance of remaining rainforest areas in southern Argentina. The coastal range of southern Chile also appears to be a centre of genetic diversity for a number of species, although, here too, high differentiation between populations (for example in *Berberidopsis*) indicates that multiple populations should be conserved.

These examples illustrate that within-population levels of genetic variation shared by different species can be used to identify hotspots of diversity within the study regions. For example, in the Andes, *Pilgerodendron* has elevated genetic variation at mid latitudes in Puyehue (40° 45′ S), while *Fitzroya* diversity is concentrated on eastern slopes of the Andes further south (*c*.42° S). The latter coincides with hotspots for *Embothrium* on eastern slopes, which also have elevated diversity in the northern Chilean Andes at 38° S. Most protected areas in Patagonia within both Argentina and Chile are located in the Andes, and therefore these centres are protected. Nevertheless, some populations of these species that hold elevated genetic diversity are located outside protected areas, including *Fitzroya* populations located along the Rio Tigre watershed north of Los Alerces National Park, Argentina (Premoli *et al.*, 2000b) and *Pilgerodendron* populations within Chubut province, Argentina (Rovere *et al.*, 2006b). Elevated polymorphism and heterozygosity exist along the Pacific coast for *Fitzroya* at mid latitudes (41° S) and for *Pilgerodendron* at its northernmost limit (39° 45′–39° 55′ S) and in coastal southern Chile (Puerto Aisen, 45° 23′ S). These populations on the Chilean Pacific coast are in urgent need of conservation action (Smith-Ramírez *et al.*, 2005), and centres of genetic diversity could be used to identify conservation units within this area. In addition, low-elevation populations in the Central Depression of Chile also hold elevated polymorphism for *Pilgerodendron*, *Fitzroya* and *Embothrium* between 41° and 42° S. These lowland populations are surviving remnant stands within a matrix of intense human land use. A network of small remnants of native forests could be effective at preserving diversity, as was demonstrated in fragments of *Embothrium coccineum* (Mathiasen *et al.*, 2007).

Results obtained in South America parallel those obtained for montane species in Mexico, where, again, pronounced genetic differences were recorded between populations separated by relatively small geographic distances. Many of the patterns detected were surprising. For example, in Mexican beech, *Fagus grandifolia* var. *mexicana*, some geographically close populations displayed unique cpDNA haplotypes, whereas others shared a common haplotype with a distant population in the USA, separated by hundreds of kilometres (Rowden *et al.*, 2004). Whereas problems such as homoplasy cannot be ruled out, it is striking that such kinds of patterns are not rare. An autocorrelation analysis using isozymes revealed that very distant populations of *P. chiapensis* are genetically similar, as are very close populations, but populations separated by intermediate distances were negatively genetically correlated (del Castillo *et al.*, in prep.). This pattern is consistent with that observed using mtDNA markers (Newton *et al.*, 2002).

Such surprising results highlight the difficulty of making predictions about patterns of genetic variation in trees. Geographic range has been shown to be a good predictor of the levels of allozyme variation in plants (Karron, 1987; Hamrick and Godt, 1989; Hamrick *et al.*, 1992; Premoli, 1997; Gitzendanner and Soltis, 2000). Geographically restricted species, usually consisting of small, isolated populations, are more susceptible to losses of genetic variation owing to genetic drift and restricted gene flow (Hamrick and Godt, 1989). However, our results show that widespread species occurring in small and disjunct populations may not have elevated polymorphism as predicted (Premoli *et al.*, 2001). The data presented here suggest that size of geographical range in combination with the degree of population continuity may be a better predictor of within-population diversity and divergence than geographic range alone (Premoli *et al.*, 2001). This is because spatial heterogeneity characterizes montane areas, resulting in marked genetic differences among populations. Levels of isozyme variation in *Pinus chiapensis*, for instance, have been shown to be very low and comparable to those found in rare and locally endemic pine species, despite having a range of more than 950 km (del Castillo *et al.*, in prep.). Such low variation is probably the consequence of recent expansion of this species fostered by shifting agricultural practices, which has created open sites in areas previously occupied by montane forest dominated by broadleaved shade-tolerant species. This example illustrates the fact that human activities are influencing patterns of variation within tree species, and such processes need to be incorporated within predictive models.

Genetic effects and demographic consequences of fragmentation are closely linked with life history traits. Responses to habitat loss and increased isolation owing to fragmentation will therefore differ among species. Populations of species such as *Fagus grandifolia* var. *mexicana* and *Magnolia* spp., in common with many other montane tree species in Mexico, have been fragmented and isolated for prolonged periods, as a result of climate change during the Pleistocene. To an extent, these species may be able to tolerate population isolation, and maintain genetic variability despite limited gene flow between populations. This is perhaps most clearly illustrated by *Magnolia sharpii*, which, despite being restricted to only five small populations, still appears to maintain a relatively high degree of variation (Newton *et al.*, 2007). A very restricted geographical distribution is therefore no guarantee that the extent of genetic variation within populations is very limited. On the other hand, the results obtained provide evidence of reduced intraspecific variation within a number of widespread species. Some of these, including *Pinus chiapensis*, *Pilgerodendron uviferum* and *Nothofagus pumilio*, possess life history attributes, such as wind pollination, that suggest gene flow is possible over long distances. This indicates that it can be difficult to predict which species are most vulnerable to genetic impacts of fragmentation from consideration of life history characteristics alone. The pronounced population differentiation recorded in many of the species considered here implies that gene flow may often be restricted, even in wind-pollinated species, in forest landscapes fragmented by human activity.

The research into *Embothrium coccineum* provides detailed insights into how fragmentation can affect genetic variation within a species, by influencing patterns of gene flow. Again, the results were surprising, suggesting that fragmentation can actually increase gene flow. The species vigorously regenerates in cleared areas and therefore population size can increase under fragmentation. Self-incompatibility may favour pollen imported from other fragments, particularly where few reproductive trees are available for successful pollination. The bird *Elaenia albiceps* is the main pollinator of *E. coccineum* in fragmented landscapes of Chiloé Island (Tyrannidae, Passeriformes). It frequently visits spatially isolated individuals and those at forest edges within small fragments (Smith-Ramírez and Armesto, 2003). This feeding behaviour reduces inbreeding, but in large fragments short interplant movements of pollinators results in mating between relatives and thus increased inbreeding (Mathiasen *et al.*, 2007). This provides a clear example of how the ecological and reproductive characteristics of a species may determine the impacts of anthropogenic activity on genetic variability.

Those species that are dependent on mutualistic interactions may be particularly sensitive to fragmentation. In particular, temperate forests of southern Argentina and Chile have elevated incidences of biotic pollination (23%) of woody plants, particularly ornithophily, which is as widespread as in tropical areas (Aizen *et al.*, 2002). However, in contrast to the tropics, ornithophilous plants depend strongly on a restricted number of bird species. For example, approximately 20% of red-flowered woody plants are visited by one hummingbird species, *Sephanoides sephaniodes* (Smith-Ramírez, 1993). Therefore, areas in temperate latitudes with high redundancy of mutualists could be vulnerable to habitat and biodiversity loss due to fragmentation.

Another important result is that populations – even entire species – may be at risk of extinction because of genetic processes. For example, although widespread, *Pilgerodendron uviferum* appears to be suffering from the results of inbreeding and genetic drift, as a result of population isolation. In Mexico, *Pinus chiapensis* provides a clear example of how such genetic factors may be affecting demographic processes such as seed germination, thereby reducing population viability. This species also illustrates how the relationship between genetic diversity and anthropogenic disturbance can be complex. Disturbance is a requirement for successful establishment of this species (del Castillo, 1996; Chapter 7). In the absence of disturbance, populations undergo continuous reduction owing to the lack of open habitats for regeneration. However, overharvesting, habitat destruction and land-use change also lead to population reductions. Both insufficient and excessive disturbance may therefore lead to population size reductions, which are associated with significant declines in genetic diversity and heterozygosity that reduce the probability of population persistence in both the short and long term. Similarly complex impacts of disturbance on genetic variation were observed in *Embothrium coccineum*, which also provided evidence of deleterious effects of inbreeding on seedling growth. Reduction in vital rates such as seed germinability by inbreeding depression is likely to have a major impact on population viability, but has not been investigated to date for most of the species considered here.

The substantial population differentiation recorded in the species studied here suggests that significant genetic variation in many species may already have been lost, as a result of the forest loss and fragmentation that have occurred in the study regions (see Chapters 2 and 3). In the absence of any baseline information, it is difficult to evaluate the losses that have already occurred. However, the exceptionally high deforestation rates reported for the Chilean coastal range and the highlands of Chiapas (Chapter 2) have occurred in areas that may also be considered as 'hotspots' of genetic diversity. *Pinus chiapensis,* for example, now only exists as very small, isolated and degraded populations in Chiapas, despite the species having been named after this part of Mexico. What is clear is that, if current forest losses continue, the losses of species that are projected (Chapter 3) will be accompanied by substantial losses of the genetic variation within the species that survive.

The study of the degraded remnant populations of *Fitzroya cupressoides* in the Central Depression of Chile offers one ray of hope. The species has undergone very severe (and ongoing) population reduction and degradation within this area, to the extent that it was thought to be locally extinct. The small remnant populations discovered by this research were found to house a surprising amount of variation, offering the potential for ecological restoration using locally sourced germplasm (Chapter 15). Such resilience of a species under chronic, intense human pressure over more than a century illustrates that well-targeted conservation action can achieve a great deal, even in situations that appear at first glance to be beyond hope.

References

Abrams, M.D. (1988) Genetic variation in leaf morphology and plant tissue water relations during drought in *Cercis canadensis* L. *Forest Science* 34, 200–207.

Abrams, M.D. and Kubiske, M.E. (1990) Photosynthesis and water relations during drought in *Acer rubrum* L. Genotypes from contrasting sites in central Pennsylvania. *Functional Ecology* 4, 727–733.

Abrams, M.D., Kloeppel, B.D. and Kubiske, M.E. (1992) Ecophysiological and morphological responses to shade and drought in two contrasting genotypes of *Prunus serotina. Tree Physiology* 10, 343–355.

Aizen, M.A., Vazquez, D.P. and Smith-Ramírez, C. (2002) Historia natural y conservación de los mutualismos planta–animal del bosque templado de Sudamérica austral. *Revista Chilena de Historia Natural* 75, 79–97.

Alberdi, M. and Donoso, C. (2004) Variación en *Embothrium coccineum* J.R. et G. Forster (notro o ciruelillo). In: Donoso, C., Premoli, A., Gallo, L. and Ipinza, R. (eds) *Variación Intraespecífica en las Especies Arbóreas de los Bosques Templados de Chile y Argentina*. Editorial Universitaria, Santiago de Chile, Chile.

Allnutt, T.R., Newton, A.C., Lara, A., Premoli, A., Armesto, J.J., Vergara, R. and Gardner, M. (1999) Genetic variation in *Fitzroya cupressoides* (alerce), a threatened South American conifer. *Molecular Ecology* 8, 975–987.

Allnutt, T.R., Courtis, J.R., Gardner, M. and Newton, A.C. (2001) Genetic variation and wild Chilean and cultivated British populations of *Podocarpus salignus* D. Don (Podocarpaceae). *Edinburgh Journal of Botany* 58, 459–473.

Allnutt, T.R., Newton, A.C., Premoli, A.C. and Lara, A. (2003) Genetic variation in the threatened South American conifer *Pilgerodendron uviferum* (Cupressaceae), detected using RAPD markers. *Biological Conservation* 114, 245–253.

Armesto, J.J., Villagrán, C., Aravena, J.C., Pérez, C., Smith-Ramirez, C. and Hedin, L. (1995) Conifer forests of the Chilean Coastal Range. In: Enright, N.J. and Hill, R.S. (eds) *Ecology of the Southern Conifers*. Melbourne University Press, Carlton, Victoria, Australia, pp. 156–170.

Armesto, J.J., Rozzi, R., Smith-Ramírez, C. and Arroyo, M.T.K. (1998) Conservation targets in South American temperate forests. *Science* 282, 1271–1272.

Barrett, S.C.H. and Kohn, J.R. (1991) Genetic and evolutionary consequences of small population size in plants: implications for conservation. In: Falk, D.A. and Holsinger, K.E. (eds) *Genetics and Conservation of Rare Plants*. Oxford University Press, New York, pp. 3–30.

Bekessy, S.A., Allnutt, T.R., Premoli, A.C., Lara, A., Ennos, R.A., Burgman, M.A., Cortes, M. and Newton, A.C. (2002a) Genetic variation in the vulnerable and endemic Monkey Puzzle tree, detected using RAPDs. *Heredity* 88, 243–249.

Bekessy, S.A., Sleep, D., Stott, A., Menuccini, M., Thomas, P., Ennos, R.A., Burgman, M.A., Gardner, M.F. and Newton, A.C. (2002b) Adaptation of Monkey Puzzle to arid environments reflected by regional differences in stable carbon isotope ratio and allocation to root biomass. *Forest Genetics* 9, 63–70.

Bekessy, S.A., Ennos, R.A., Ades, P.K., Burgman, M.A. and Newton, A.C. (2003) Neutral DNA markers fail to detect divergence in an ecologically important trait. *Biological Conservation* 110, 267–275.

Briggs, D. and Walters, S.M. (1984) *Plant Variation and Evolution*. Cambridge University Press, Cambridge, UK.

Caldiz, M.S. (1999) Estructura genética del arrayán, *Luma apiculata* (DC.) Burret (Myrtaceae), una especie endémica del noroeste patagónico: Su relación con el sistema reproductivo y modo de regeneración. Licenciatura thesis. Centro Regional Universitario Bariloche, Universidad Nacional del Comahue, Bariloche, Argentina.

Cayuela, L., Golicher, D.J. and Rey-Benayas, J.M. (2006) The extent, fragmentation and distribution of vanishing montane cloud forest in the highlands of Chiapas, Mexico. *Biotropica* 38, 544–554.

Clausen, J., Keck, D.D. and Hiesey, W.M. (1948) *Experimental Studies on the Nature of Species. III. Environmental Responses to Climatic Races of Achilea*. Carnegie Institute of Washington publication 581, Washington, DC.

Colinvaux, P.A., De Olivera, P.E. and Bush, M.B. (2000) Amazonian and neotropical plant communities on glacial time-scale: the failure of the aridity and refuge hypotheses. *Quaternary Science Reviews* 19, 141–169.

Comps, B., Gömöry, D., Letouzey, J., Thiébaut, B. and Petit, R.J. (2001) Diverging trends between heterozygosity and allelic richness during postglacial colonization in the European beech. *Genetics* 157, 389–397.

Couvet, D. (2002) Deleterious effects of restricted gene flow in fragmented populations. *Conservation Biology* 16, 369–376.

Crandall, K.A., Bininda-Edmonds, O.R.P., Mace, G.M. and Wayne, R.K. (2000) Considering evolutionary processes in conservation biology: an alternative to 'evolutionary significant units'. *Trends in Ecology and Evolution* 15, 290–295.

Critchfield, W.B. (1984) Impact of the Pleistocene on the genetic structure of North American conifers. In: Lanned, R.M. (ed.) *Proceedings of the Eighth North American Forest Biology Workshop*. Utah State University, Logan, Utah, pp. 70–118.

del Castillo, R.F. (1996) Aspectos autoecológicos de *Pinus chiapensis*. In: Garduño, L.L., Chavarria, G.V., Magdaleno, P.L. and Pérez, I.M. (eds) *Memorias del 2do. Coloquio Regional de Investigación, Ciencias Exactas y Naturales; Toluca, Estado de México*. Universidad Autónoma del Estado de México, DF, Mexico, pp. 63–68.

del Castillo, R.F., Trujillo, S. and Newton, A.C. (in prep.) Patterns of genetic diversity and mating systems in a tropical pine with contrasting population sizes.

Delgado, P., Cuenca, A., Escalante, A.E., Molina-Freaner, F. and Piñero, D. (2002) Comparative genetic structure in pines: evolutionary and conservation consequences. *Revista Chilena de Historia Natural* 75, 27–37.

Donoso, C. (1983) Modificaciones del paisaje forestal chileno a lo largo de la historia. In: *Simposio Desarrollo y Perspectivas de las Disciplinas Forestales de la Universidad Austral de Chile*. Facultad de Ciencias Forestales, Universidad Austral de Chile, Valdivia, Chile, pp. 365–438.

Donoso-Zegers, C. (1987) Variación natural en especies de *Nothofagus* en Chile. *Bosque* 8, 85–97.

Ehtisham-Ul-Haq, M., Allnutt, T.R., Armesto, J.J., Smith, C., Gardner, M. and Newton, A.C. (2001) Patterns of genetic variation in the threatened Chilean vine *Berberidopsis corallina* Hook. f. sampled *in* and *ex situ*, detected using RAPD markers. *Annals of Botany* 87, 813–821.

Falk, D.A. and Holsinger, K.E. (1991) *Genetics and Conservation of Rare Plants*. Oxford University Press, Oxford, UK.

Fonck, F. (1896) *Viajes de Fray Francisco de Menéndez a la Cordillera*. Niemeyer, Valparaíso, Chile.

Frankel, O.H. and Soulé, M.E. (1981) *Conservation and Evolution*. Cambridge University Press, Cambridge, UK.

Frankham, R. (2005) Genetics and extinction. *Biological Conservation* 126, 131–140.

Fraser, D.J. and Bernatchez, L. (2001) Adaptive evolutionary conservation: towards a unified concept for defining conservation units. *Molecular Ecology* 10, 2741–2752.

Fraver, S., Gonzalez, M.E., Silla, F. and Lara, A. (1999) Composition and structure of remnant *Fitzroya cupressoides* forests of southern Chile's Central Depression. *Journal of the Torrey Botanical Society* 126, 49–57.

Gitzendanner, M.A. and Soltis, P.S. (2000) Patterns of genetic variation in rare and widespread plant congeners. *American Journal of Botany* 87, 783–792.

Glaubitz, J.C. and Moran, G.F. (2000) Genetic tools: the use of biochemical and molecular markers. In: Young, A.G., Boshier, D. and Boyle, T. (eds) *Forest Conservation Genetics, Principles and Practice*. CSIRO/CABI, Collingwood, Australia, pp. 39–59.

Hamrick, J.L. and Godt, M.J.W. (1989) Allozyme diversity in plant species. In: Brown, A.H.D., Clegg, M.T., Kahler, A.L. and Weir, B.S. (eds) *Plant Population Genetics, Breeding, and Genetic Resources*. Sinauer Associates, Sunderland, Massachusetts, pp. 43–63.

Hamrick, J.L. and Godt, M.J.W. (1996) Conservation genetics of endemic plant species. In: Avise, J.C. and Hamrick, J.L. (eds) *Conservation Genetics, Case Histories from Nature*. Chapman and Hall, New York, pp. 281–304.

Hamrick, J.L. and Nason, J.D. (1996) Consequences of dispersal in plants. In: Rhodes, O.E.J., Chesser, R.K. and Smith, M.H. (eds) *Population Dynamics in Ecological Space and Time*. University of Chicago Press, Chicago, Illinois, pp. 203–236.

Hamrick, J.L., Godt, M.J.W. and Sherman-Broyles, S.L. (1992) Factors influencing levels of genetic diversity in woody plant species. *New Forest* 6, 95–124.

Heusser, C.J. (1981) Palynology of the last interglacial–glacial cycle in mid-latitude of southern Chile. *Quaternary Research* 16, 293–321.
Heusser, C.J. and Flint, R.F. (1977) Quaternary glaciations and environments of northern Isla de Chiloé, Chile. *Geology* 5, 305–308.
Hewitt, G.M. (1996) Some genetic consequences of ice ages, and their role in divergence and speciation. *Journal of the Linnean Society* 58, 247–276.
Hiesey, W.M. and Milner, H.W. (1965) Physiology of ecological races and species. *Annual Review of Plant Physiology* 16, 203–216.
Hinojosa, L.F. and Villagrán, C. (1997) Historia de los bosques del sur de Sudamerica I: antecedentes paleobotánicos, geológicos y climáticos del Terciario del cono sur de América. *Revista Chilena de Historia Natural* 70, 225–239.
Holling, J.T. and Schilling, D.H. (1981) Late Wisconsin–Weichselian mountain glaciers and small ice caps. In: Denton, G. and Hughes, T.J. (eds) *The Last Great Ice Sheets*. Wiley, New York, pp. 179–206.
Huntley, B. and Webb, T. (1988) *Vegetation History*. Kluwer Academic, Dordrecht, The Netherlands.
Jonas, C.S. and Geber, M.A. (1999) Variation among populations of *Clarkia unguiculata* (Onagraceae) along altitudinal and latitudinal gradients. *American Journal of Botany* 86, 333–343.
Karron, J.D. (1987) A comparison of levels of genetic polymorphism and self-compatibility in geographically restricted and widespread plant congeners. *Evolutionary Ecology* 1, 47–58.
Lacy, R.C. (1992) The effects of inbreeding of isolated populations: are minimum viable population sizes predictable? In: Fiedler, P.L. and Jain, S.K. (eds) *Conservation Biology: The Theory and Practice of Nature Conservation Preservation and Management*. Chapman and Hall, New York, pp. 277–296.
Lande, R. (1988) Genetics and demography in biological conservation. *Science* 241, 1455–1460.
Langlet, O. (1971) Two hundred years of genecology. *Taxon* 20, 653–722.
Larsen, J.B. (1981) Geographic variation in winter drought resistance of Douglas-fir (*Pseudotsuga menziesii* (Mirb.) Franco). *Silvae Genetica* 30, 109–114.
Ledig, F.T. (1988) The conservation of diversity in forest trees: why and how should genes be conserved? *BioScience* 38, 471–479.
Lesica, P. and Allendorf, F.W. (1995) When are peripheral populations valuable for conservation? *Conservation Biology* 9, 753–760.
Levins, R. and Lewontin, R.C. (1985) *The Dialectical Biologist*. Harvard University Press, Cambridge, Massachusetts.
Linhart, Y. and Grant, M.C. (1996) Evolutionary significance of local genetic differentiation in plants. *Annual Review of Ecology and Systematics* 27, 237–277.
Linhart, Y.B., Mitton, J.B., Sturgeon, K.B. and Davis, M.L. (1981) Genetic variation in space and time in population of ponderosa pine. *Heredity* 46, 407–426.
Lowe, T.K., Harris, S. and Ashton, P. (2004) *Ecological Genetics: Design, Analysis, and Application*. Blackwell, Malden, Massachusetts.
Mace, G.M., Smith, T.B., Bruford, M.W. and Wayne, R.K. (1996) An overview of the issues. In: Smith, T.B. and Wayne, R.K. (eds) *Molecular Genetic Approaches in Conservation*. Oxford University Press, New York, pp. 3–24.
Markgraf, V. (1983) Late and postglacial vegetational and palaeoclimatic changes in subantarctic, temperate, and arid environments in Argentina. *Palynology* 7, 43–70.
Markgraf, V. (1984) Late Pleistocene and Holocene vegetation history of temperate Argentina: Lago Morenito, Bariloche. *Restschrigt Welten. Dissertaciones Botanicae* 72, 235–254.

Markgraf, V. (1991) Late Pleistocene environmental and climatic evolution in southern South America. *Bamberger Geographische Schriften* 11, 271–281.

Markgraf, V., McGlone, M. and Hope, G. (1995) Neogene paleoenvironmental and paleoclimatic change in southern temperate ecosystems – a southern perspective. *Trends in Ecology and Evolution* 10, 143–147.

Markgraf, V., Romero, E. and Villagrán, C. (1996) History and paleoecology of South American *Nothofagus* forests. In: Veblen, T.T., Hill, R.S. and Read, J. (eds) *The Ecology and Biogeography of Nothofagus Forests*. Yale University Press, New Haven, Connecticut, pp. 354–386.

Martínez-Araneda, C. (2004) Análisis de variabilidad genética en *Legrandia concinna* (Phil.) Kausel a través de su distribución latitudinal. Tesis. Universidad Austral de Chile, Valdivia, Chile.

Martínez-Araneda, C., Premoli, A.C., Echeverría, C., Thomas, P.I. and Hechenleitner, P. (in prep.) Marked among-population divergence and reduced genetic diversity of the rare Myrtaceae *Legrandia concinna* from Central Chile.

Mathiasen, P. (2004) Efectos de la fragmentación del bosque templado sobre la demografía y estructura genética de *Embothrium coccineum* Forst. (Proteaceae) en el Sur de Chile. BSc thesis. Universidad Nacional del Comahue, Bariloche, Argentina.

Mathiasen, P., Rovere, A.E. and Premoli, A.C. (2007) Genetic structure and early effects of inbreeding in fragmented temperate forests of a self-incompatible tree, *Embothrium coccineum*. *Conservation Biology* 21, 232–240.

Millar, C.I. and Libby, W.J. (1991) Strategies for conserving clinal, ecotypic and disjunct population diversity in widespread species. In: Falk, D.A. and Holsinger, K.E. (eds) *Genetics and Conservation of Rare Plants*. Oxford University Press, New York, pp. 149–170.

Moritz, C. (1994) Applications of mitochondrial DNA analysis in conservation: a critical review. *Molecular Ecology* 3, 401–411.

Moritz, C. (1995) Uses of molecular phylogenies for conservation. *Philosophical Transactions of the Royal Society of London, Series B* 349, 113–118.

Nei, M., Maruyama, T. and Chakraborty, R. (1975) The bottleneck effect and genetic variability in populations. *Evolution* 29, 1–10.

Newton, A.C., Allnutt, T., Gillies, A.C.M., Lowe, A. and Ennos, R.A. (1999) Molecular phylogeography, intraspecific variation and the conservation of tree species. *Trends in Ecology and Evolution* 14, 140–145.

Newton, A.C., Allnutt, T.R., Dvorak, W., del Castillo, R. and Ennos, R. (2002) Patterns of genetic variation in *Pinus chiapensis*, a threatened Mexican pine, detected by RAPD and mitochondrial DNA RFLP markers. *Heredity* 89, 191–198.

Newton, A.C., Gow, J., Robertson, A., Williams-Linera, G., Ramírez-Marcial, N., González-Espinosa, M., Allnutt, T.R. and Ennos R. (2007) Patterns of genetic variation in two threatened endemic Mexican trees, *Magnolia sharpii* and *Magnolia schiedeana*. *Oryx* (in press).

Núñez-Ávila, M. and Armesto, J. (2006) Relict islands of the temperate rainforest tree *Aextoxicon punctatum* (Aextoxicaceae) in semi-arid Chile: genetic diversity and biogeographic history. *Australian Journal of Botany* 54, 733–743.

Pennington, R.T., Prado, D.E. and Pendry, C.A. (2000) Neotropical seasonally dry forest and Quaternary vegetation changes. *Journal of Biogeography* 27, 261–273.

Pérez, R.V. (1958) *Recuerdos del Pasado (1814–1860)*. Carlos de Vidts, Santiago, Chile.

Perry, D.J. and Knowles, P. (1991) Spatial genetic structure within three sugar maple (*Acer saccharum* Marsh.) stands. *Heredity* 66, 137–142.

Peters, R.H. (1991) *A Critique for Ecology*. Cambridge University Press, Cambridge, UK.

Petit, R.J., Aguinagalde, I., de Beaulieu, J.L., Bittkau, C., Brewer, S., Cheddadi, R., Ennos, R.A., Fineschi, S., Grivet, D., Lascoux, M., Mohanty, A., Muller-Starck, G., Demesure-Musch, B., Palme, A., Martin, J.P., Rendell, S. and Vendramin, G.G. (2003) Glacial refugia: hotspots but not melting pots of genetic diversity. *Science* 300, 1563–1565.

Premoli, A.C. (1997) Genetic variation in two widespread and a geographically restricted species of *Nothofagus*. *Journal of Biogeography* 24, 883–892.

Premoli, A.C. (1998) The use of genetic markers to conserve endangered species and to design protected areas of more widespread species. In: International Foundation for Science (ed.) *Proceedings of an International Workshop: Recent Advances in Biotechnology for Tree Conservation and Management*. Universidade Federal de Santa Catarina, Florianópolis, Santa Catarina, Brazil, pp. 157–171.

Premoli, A.C. and Kitzberger, T. (2005) Regeneration mode affects spatial genetic structure of *Nothofagus dombeyi* forests. *Molecular Ecology* 14, 2319–2329.

Premoli, A.C., Kitzberger, T. and Veblen, T.T. (2000a) Isozyme variation and recent biogeographical history of the long-lived conifer *Fitzroya cupressoides*. *Journal of Biogeography* 27, 251–260.

Premoli, A.C., Kitzberger, T. and Veblen, T.T. (2000b) Conservation genetics of the endangered conifer *Fitzroya cupressoides* in Chile and Argentina. *Conservation Genetics* 1, 57–66.

Premoli, A.C., Souto, C.P., Allnutt, T.R. and Newton, A.C. (2001) Effects of population disjunction on isozyme variation in the widespread *Pilgerodendron uviferum*. *Heredity* 87, 337–343.

Premoli, A.C., Souto, C.P., Rovere, A.E., Allnut, T.R. and Newton, A.C. (2002) Patterns of isozyme variation as indicators of biogeographic history in *Pilgerodendron uviferum* (D. Don) Florín. *Diversity and Distributions* 8, 57–66.

Premoli, A.C., Vergara, R., Souto, C.P., Lara, A. and Newton, A.C. (2003) Lowland valleys shelter ancient *Fitzroya cupressoides* in the Central Depression of southern Chile. *Journal of the Royal Society of New Zealand* 33, 623–631.

Premoli, A.C., Raffaele, E. and Mathiasen, P. (2007) Morphological and phenological differences in *Nothofagus pumilio* from contrasting elevations. *Austral Ecology* 32, 515–523.

Quiroga, M.P. and Premoli, A.C. (2007) Genetic patterns in *Podocarpus parlatorei* reveal long term persistence of cold tolerant elements in southern Yungas. *Journal of Biogeography* 34, 447–455.

Rovere, A.E. and Premoli, A.C. (2005) Dispersión asimétrica de semillas de *Embothrium coccineum* (Proteaceae) en el bosque templado de Chiloé, Chile. *Ecología Austral* 15, 1–7.

Rovere, A.E., Smith-Ramírez, C., Armesto, J.J. and Premoli, A.C. (2006a) Breeding system of *Embothrium coccineum* J.R. et G. Forster. (Proteaceae) in two populations on different slopes of the Andes. *Revista Chilena de Historia Natural* 79, 225–232.

Rovere, A.E., Souto, C.P. and Premoli, A.C. (2006b) Poblaciones de ciprés de las guaitecas (*Pilgerodendron uviferum* (Don) Florín) fuera de áreas protegidas evidencian elevada variabilidad genética. *Memoria de la XXII Reunión Argentina de Ecología*. Córdoba, 22–25 agosto 2006. Asociación Argentina de Ecología, Buenos Aires, Argentina.

Rowden, A., Robertson, G.P., Allnutt, T.R., Heredia, S., Williams-Linera, G. and Newton, A.C. (2004) Conservation genetics of Mexican beech, *Fagus grandifolia* var *mexicana*. *Conservation Genetics* 5, 475–484.

Schanbel, A., Laushman, R.H. and Hamrick, J.L. (1991) Comparative genetic structure of two co-occurring tree species, *Maclura pomifera* (Moreceae) and *Gleditsia triacanthos* (Leguminoseae). *Heredity* 67, 357–364.

Schemske, D.W., Husband, B.C., Ruckelhaus, M.H., Goowillie, C., Parker, I.M. and Sishop, J.G. (1994) Evaluating approaches to the conservation of rare and endangered plants. *Ecology* 75, 584–606.

Shapcott, A. (1995) The spatial genetic structure in natural populations of the Australian temperate rainforest tree *Atherosperma moschatum* Labill. (Monimiaceae). *Heredity* 74, 28–38.

Slatkin, M. (1987) Gene flow and the geographic structure of natural populations. *Science* 236, 787–792.

Smith-Ramírez, C. (1993) Los picaflores y su recurso floral en el bosque templado de la isla de Chiloé, Chile. *Revista Chilena de Historia Natural* 66, 65–73.

Smith-Ramírez, C. and Armesto, J.J. (2003) Foraging behaviour of bird pollinators on *Embothrium coccineum* (Proteaceae) trees in forest fragments and pastures in southern Chile. *Austral Ecology* 28, 53–60.

Smith-Ramírez, C., Armesto, J.J. and Valdovinos, C. (2005) *Historia, Biodiversidad y Ecología de los Bosques Costeros de Chile*. Editorial Universitaria, Santiago, Chile.

Souto, C.P. and Premoli, A.C. (in prep.) Geographic variation in leaf traits of the widespread Proteaceae *Embothrium coccineum* reflect habitat heterogeneity in Patagonia.

Turesson, G. (1922) The genotypical response of the plant species to the habitat. *Hereditas* 3, 211–350.

Vargas, J.A. and del Castillo, R.F. (2001) Genetic associations under mixed mating systems: the Bennett–Binet effect. *IMA Journal of Mathematics Applied in Medicine and Biology* 18, 327–341.

Veblen, T.T., Delmastro, R.J. and Schlatter, J.E. (1976) The conservation of *Fitzroya cupressoides* and its environment in southern Chile. *Environmental Conservation* 3, 291–301.

Veblen, T.T., Burns, B.R., Kitzberger, T., Lara, A. and Villalba, R. (1995) The ecology of the conifers of southern South America. In: Enright, N.J. and Hill, R.S. (eds) *Ecology of the Southern Conifers*. Melbourne University Press, Carlton, Victoria, Australia, pp. 120–155.

Villagrán, C. (1985) Análisis palinológico de los cambios vegetacionales durante el Tardiglacial y Postglacial en Chiloé, Chile. *Revista Chilena de Historia Natural* 58, 57–69.

Villagrán, C. (1988) Late Quaternary vegetation of southern Isla Grande de Chiloé, Chile. *Quaternary Research* 29, 294–306.

Villagrán, C. (1991) Historia de los bosques templados del sur de Chile durante el tardiglacial y Postglacial. *Revista Chilena de Historia Natural* 64, 447–460.

Villagrán, C. (2001) Un modelo de la Historia de la vegetación de la Cordillera de La Costa de Chile central-sur: La hipótesis glacial de Darwin. *Revista Chilena de Historia Natural* 74, 793–803.

Villagrán, C. and Armesto, J.J. (1993) Full and late glacial paleoenvironmental scenarios for the west coast of southern South America. In: Mooney, H.A., Fuentes, E.R. and Kronberg, B.I. (eds) *Earth System Responses to Global Change. Contrasts Between North and South America*. Academic Press, New York, pp. 195–207.

Villagrán, C., Moreno, P. and Villa, R. (1996) Antecedentes palinológicos acerca de la historia cuaternaria de los bosques chilenos. In: Armesto, J.J., Villagrán, C. and Arroyo, M.K. (eds) *Ecología de los Bosques Nativos de Chile*. Editorial Universitaria, Santiago, Chile, pp. 51–70.

Villagrán, C., Armesto, J.J., Hinojosa, L.F., Cuvertino, J., Pérez, C. and Medina, C. (2004a) El enigmático origen del bosque relicto de Fray Jorge. In: Squeo, F.A., Gutiérrez, J.R. and Hernández, I.R. (eds) *Historia Natural del Parque Nacional Bosque Fray Jorge*. Ediciones Universidad de La Serena, La Serena, Chile, pp. 3–42.

Villagrán, C., León, A. and Roig, F.A. (2004b) Paleodistribución del alerce y ciprés de las Guaitecas durante períodos interestadiales de la Glaciación Llanquihue: Provincias de Llanquihue y Chiloé, Región de Los Lagos, Chile. *Revista Geológica de Chile* 31, 133–151.

Wendt, T. (1989) Las selvas de Uxpanapa, Veracruz-Oaxaca, México: evidencia de refugios florísticos cenozoicos. *Anales del Instituto de Biología serie Botánica* 58, 29–54.

Young, A.G., Merriam, H.G. and Warwick, S.I. (1993) The effects of forest fragmentation on genetic variation in *Acer saccharum* Marsh. (sugar maple) populations. *Heredity* 71, 277–289.

Young, A.G., Brown, A.H.D. and Zich, F.A. (1999) Genetic structure of fragmented populations of the endangered daisy *Rutidosis leptorrhynchoides*. *Conservation Biology* 13, 256–265.

Young, A.G., Brown, A.H.D., Murray, B.G., Thrall, P.H. and Millar, C.H. (2000) Genetic erosion, restricted mating and reduced viability in fragmented populations of the endangered grassland herb *Rutidopsis leptorrhynchoides*. In: Young, A.G. and Clarke, G.M. (eds) *Genetics, Demography and Viability of Fragmented Populations*. Cambridge University Press, Cambridge, UK, pp. 335–359.

7 Secondary Succession under a Slash-and-burn Regime in a Tropical Montane Cloud Forest: Soil and Vegetation Characteristics

R.F. del Castillo and A. Blanco-Macías

Successional forest stand dominated by *Pinus chiapensis;* Sierra Norte, Oaxaca, Mexico. Photo: Adrian Newton

 Biodiversity Loss and Conservation in Fragmented Forest Landscapes: The Forests of Montane Mexico and Temperate South America (ed. A.C. Newton)

Summary

Tropical montane cloud forest (TMCF) areas in southern Mexico are commonly used for growing maize and companion crops under slash-and-burn agriculture. As a result, the landscape is being transformed into a mosaic of crop fields, secondary forest and primary forests. Despite being a widespread process, very little is known about forest regeneration in TMCF areas. This chapter describes secondary succession in the TMCF of El Rincón Alto, Oaxaca, Mexico, with particular reference to soil and vegetation characteristics, taking advantage of successional chronosequences spanning a century of forest development. Cultivation practices cause severe losses of soil and carbon; the original vegetation and soil organic horizons vanish. Such effects are reversed, in part, during earlier stages of secondary succession. Soil layers, including those with organic horizons, begin to accumulate. Epiphytes, low-stature plants and shrubs begin to colonize very early during succession. Species richness of shrubs, geophytes and low-stature plants peak within the first 15 years after abandonment. The largest recorded decline in concentration of soil cations, and the highest annual rates of soil carbon sequestration also characterize this period. A pine-dominated community characterizes the first 10–75 years. Early successional species do not prosper under this forest canopy and an emergent stratum of broadleaf trees eventually replaces the former pine forest. Colonization of trees, lianas and climbing plants exceeds local extinction rates during the first century of forest development. The dominant groups of terrestrial and epiphytic plants are trees and liverworts, respectively, during all studied stages of forest development. Self-thinning of the first colonizing trees took place between 45 and 75 years after abandonment and coincided with a second increase in abundance of shrubs and other low-stature plants, but tree basal area did not decline significantly. Species of terrestrial plants typical of early stages are rare or absent in late successional stages. Thus, disturbances generated by autogenic processes such as natural tree fall are different from allogenic disturbances such as fires and landslides. Soil becomes increasingly acidic and infertile during forest development, a problem that is aggravated by the presence of soluble aluminium. Such acidity fosters mineral hydrolysis, releasing cations to the soil. Soil N/P ratios steadily decrease during forest development. We conclude that environmental changes derived from slash-and-burn processes increase landscape and species diversity under the long fallow regimes observed.

Introduction

Disruption of the original forests of the world by anthropogenic disturbances is becoming increasingly common and widespread. Of these, deforestation, which refers to the complete elimination of the original vegetation cover, is one of the most frequent sources of disturbance in forests. If the disturbance that destroyed the original vegetation ceases, the ecosystem may undergo a series of changes involving the colonization of a diverse suite of organisms, which, in turn, are replaced by others. This process is referred to as secondary succession. Short-term microclimatic and soil changes are inherent features of secondary succession and may critically affect the outcome of succession (Huston and Smith, 1987; Glenn-Lewin and van der Maarel, 1992).

One of the ecosystems most affected by deforestation is tropical montane cloud forest (TMCF) (Aldrich and Hostettler, 2000; Chapter 2). A persistent, frequent or seasonal cloud cover at vegetation level characterizes this ecosystem. Compared to lowland tropical rainforests, TMCFs usually occur at higher elevations, the stature of trees is lower and epiphytes are more common (Hamilton *et al.*, 1995). TMCFs are among the most endangered ecosystems in the world owing to land clearing mostly for agriculture or cattle raising (Churchill *et al.*, 1995; Webster, 1995; Rzedowski, 1996; Bruijnzeel and Hamilton, 2000).

Understanding the underlying process of secondary succession in TMCF areas is critical for several reasons. First, owing to the high rates of deforestation and other types of forest disturbance, most of the forest remnants are becoming secondary. Second, in many tropical areas of the world, including TMCF areas, land is managed under the slash-and-burn rotation system, in which the location of the cultivated area is changed in a regular sequence, thus generating fallow periods (Manshard, 1974). Such periods enable the forest to regenerate to an extent dependent on a range of factors, not all of which have been completely characterized. Secondary succession is therefore a key process in landscapes subjected to shifting cultivation. The search for improved methods of land management requires a deeper understanding of successional processes following forest clearance. Third, ecological restoration is becoming increasingly important given the urgent need to rehabilitate areas of degraded forestlands for conservation purposes and for improved provision of ecosystem services. Secondary succession is the natural and cheapest way to regenerate natural ecosystems. Developing forest regeneration practices requires an understanding of how forests regenerate naturally after anthropogenic disturbance. Restoration and conservation in TMCF areas are particularly important because these ecosystems play a key role in environmental services such as carbon sequestering and water provision (Doumenge *et al.*, 1995), and their disruption may aggravate environmental problems such as global warming (Aldrich and Hostettler, 2000). Moreover, TMCFs are among the richest forests in terms of biodiversity and are characterized by high numbers of endemic species (Rzedowski, 1991; Webster, 1995). Deforestation therefore causes significant losses of biodiversity in TMCF areas. Finally, one of the most important topics of research in ecology is elucidating the factors that drive the dynamics of communities and thus influence species diversity and abundance. The study of secondary succession can be a useful tool for understanding such factors as the entire process involves species replacement. It is still little known to what extent the species assemblage in a community is driven by the characteristics of the species, relating to their niches or functional roles, or by factors such as chance and random dispersal (Hubbell, 2001).

Ideally, the study of secondary succession would involve following an ecosystem over long periods of time after the disturbance that destroyed the original vegetation. However, because of the long timescales involved, this approach is impractical in most cases. For this reason, modelling approaches are widely used to explore successional processes in forests (see Chapters 9–11). Alternatively, the chronosequence approach or space-by-time substitution can be used to study succession, in which neighbouring stands with different ages after disturbance are compared under the assumption that the observed between-stand differences can be attributed to age after abandonment (see Glenn-Lewin and van der Maarel, 1992; Foster and Tilman, 2000). Of course, time is not the only factor that may change the characteristics of such stands. Local environmental variation may have a profound influence on plant community structure and composition (e.g. del Castillo, 1999). It is therefore necessary to estimate the magnitude of the effect of time after disturbance relative to other local factors that may also influence the characteristics of forest stands. This can be achieved by studying several chronosequences composed of similar stages in the same area. In this way, conventional analysis of variance techniques can be used to assess the relative importance of within-age to between-age changes in the properties analysed (e.g. Bautista-Cruz and del Castillo, 2005). Despite the advantages of this approach, relatively few studies of multiple successional chronosequences have been undertaken in forest ecosystems, and we are aware of no such prior investigation performed within TMCFs.

This chapter compiles results from a recent study of secondary succession performed in El Rincón Alto, Oaxaca, Mexico, involving assessments of soil characteristics, terrestrial vegetation and epiphytes, in an area originally occupied by TMCFs, used for maize cropping, and later abandoned. This study was based on analyses of three chronosequences sharing similar ages after disturbance, located within the same area. Such replicates allow the relative effect of time after abandonment on the ecosystem changes studied to be evaluated. This information is used to describe general patterns of change in soil and vegetation during secondary succession and their ecological and conservational implications.

The Study Site

The study area is located in the Sierra Madre de Oaxaca Mountain Range, and is part of El Rincón Alto region in Oaxaca state, southern Mexico, at 1850 ± 150 m altitude, where TMCF is the primary vegetation. Topography is usually steep (15–64%). The climate is temperate-humid to subhumid (Comisión Nacional para el Conocimiento y uso de la Biodiversidad, 1998). The average annual temperature ranges between 20 and 22°C. The average annual precipitation at the nearest meteorological station (*c.*16 km from the study site) is 1719 mm/year, with a rainy season in summer and a dry season in winter (Instituto Nacional de Estadística, Geografía e Informática, 1999). The soil lies on a bedrock of schist from the Mesozoic era (Consejo de Recursos Minerales, 1996).

Secondary Succession and the Slash-and-burn System of Management

Understanding the process of secondary succession in humid montane areas of Mexico can only be fully accomplished by considering the systems of land management in these areas. Maize cropping (*milpa*) in association with companion crops, such as squash and beans, is the main system of land management, used by all ethnic regions of Sierra Madre de Oaxaca, including the Zapotec to which El Rincón Alto belongs (Boege, 1988). Maize is grown under the slash-and-burn method of cultivation. First, an area of the forest is cleared by tree felling, drying of the plant material and burning of the material. Maize is cultivated for several years, after which time cultivation moves to another piece of land. In the previously cultivated land, a fallow period commences, allowing the development of secondary vegetation. The length of the fallow period varies widely from one region to another, and tends to decline as human population size increases. At the study site, however, at least three social factors have allowed the recovery of secondary forest for periods of time sometimes longer than a century, by reducing anthropogenic pressures on the land in certain portions of the Sierra. First, low population densities and emigration rates are common in these areas (Instituto Nacional de Estadística, Geografía e Informática, 2003). Second, entire towns were displaced, reducing the pressure on the land adjacent to the former locations of the towns (López-Chávez, 1953). Finally, shifts in the economic activities of the landholders have resulted in former maizefields at higher altitudes being permanently or semi-permanently abandoned for 60 years or more. In particular, the introduction of coffee plantations at lower altitudes in the Sierra reduced pressure in areas above 1500 m, many of which were used for growing maize. Moreover, certain municipalities, such as that of San Juan Juquila Vijanos, have agreed to leave untouched portions of their forest as a natural reserve, which is secured to avoid furtive exploitation. As a result, the landscape is a mosaic of maizefields, secondary forests of different ages after abandonment, and primary forests.

Three chronosequences were selected for study: Tanetze, Juquila and Yotao. Each chronosequence consisted of a series of stands of different ages after abandonment with the same climate and parent material, and similar topography. The approximate ages were 0, ~15, ~45, ~75 and >100 years after abandonment (Fig. 7.1). These age estimates were based on: (i) the estimated age of the shade-intolerant pioneer tree *Pinus chiapensis*, obtained from ring counts using increment borers; (ii) the floristic composition and vegetation structure of the stands, in particular the abundance and size of tree species typical of primary TMCFs (see below); and (iii) the opinion of the local people regarding the age rank of the stands. The age of *P. chiapensis* gives only an approximate estimate of the time after abandonment, as establishment of this species usually does not take place immediately after abandonment. These three procedures gave the same rank category to each of the studied stands (for details, see Bautista-Cruz and del Castillo, 2005).

Changes in flora

During the maize (*Zea mays*) cropping phase, bracken ferns (*Pteridium* spp.) are the most common weeds. Herbaceous or shrubby weeds of the Asteraceae, Melastomataceae, Phytolacaceae, Poaceae, Rubiaceae and Smilacaceae families are also present. Cultivation at the study site is typically short-lived, lasting 3–5 years. Grasses, shrubs and forbs prosper during the first few years after abandonment, but are eventually shaded out by short-lived light-demanding pioneer species, which dominate the first stages of succession.

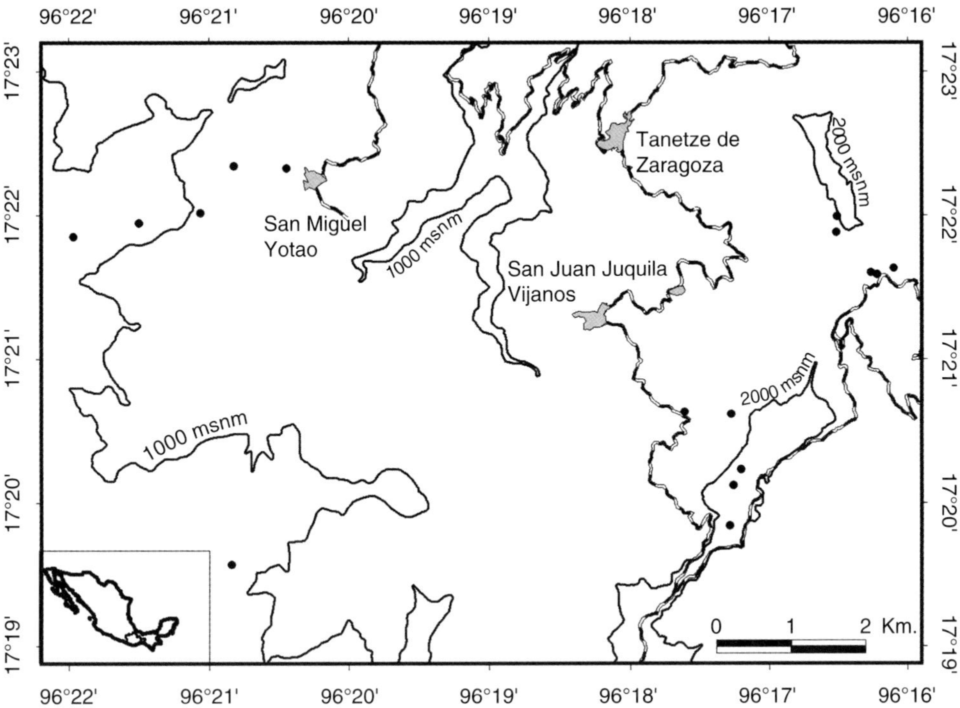

Fig. 7.1. Map of El Rincón and the study sites used for successional studies (from Cordova and del Castillo, 2001).

Ferns such as *Pteridium* spp., *Gleichenia palmata*, *Gleichenia bancroftii* and *Odontosoria schlechtendalii*, and shrubs such as *Tibouchina scabriuscula* are also common in recently disturbed places. Forest regeneration is accomplished not only by incoming propagules but also by plants with subterranean perennial tissues that have persisted during the cultivation phase. Bracken ferns are one of the best examples, as they persist as rhizomes even when their aerial parts are removed during weeding activities. This fern is also the first to be noticed in recently burned forests (R.F. del Castillo, CIIDIR Oaxaca Instituto Politécnico Nacional, Santa Cruz Xoxocotlán, Oaxaca, Mexico, personal observation).

The first forest to appear, approximately 10–15 years after abandonment, is dominated by *P. chiapensis*. Other important species are: *Clethra integerrima*, *Gaultheria acuminata*, *Liquidambar styraciflua*, the sweetgum and *Phyllonoma laticuspis*. Some young localized stands are dominated by sweetgum. The proximity of the trunks of this species leads to the suggestion that many of the plants of sweetgum observed in such stands are clonal. At ~45 years *Bejaria mexicana*, *Clethra kenoyeri*, *P. laticuspis* and *Vaccinium leucanthum* are the most common species. After this stage, self-thinning takes place, and pine trees eventually are replaced by hardwoods. Thus, the previous species, together with *Persea americana*, *Quercus* spp., *Rapanea* spp. and *Ternstroemia hemsleyi*, are common in forests ~75 years old. Old-growth forest 100 years old or older also contains *Beilschmiedia ovalis*, *Freziera* sp., *Osmanthus americana* and *Quetzalia occidentalis* as common species (Blanco-Macías, 2001). In primary forest, *Billia hyppocastanea*, *Oreopanax flaccidus*, *Podocarpus matudae*, *Quercus* sp., *Quercus corrugata*, *Quercus salicifolia*, *Symplocos coccinea*, *Ternstroemia oocarpa* and the species observed in old-growth forest stands are among the most conspicuous species (Cordova and del Castillo, 2001) (Table 7.1). The original vegetation is an upper TMCF *sensu* Webster (1995). With the exception of a few shrubs at early successional stages, the family Leguminosae was virtually absent in the TMCF areas studied. This is one of the most distinctive floristic differences with lowland tropical forest, where legumes are usually one of the richest families in terms of species numbers (Gentry, 1988).

Table 7.1. Common species of successional chronosequences in El Rincón Alto region in Oaxaca.

(a) Species typical of secondary forest, which are absent or rare in old-growth forest, but are among the most conspicuous and abundant in secondary forests:
Brunellia mexicana Standley
Kohleria deppeana (Schldl. and Cham.) Fritsch
Pinus chiapensis (Mart.) Andresen
Podachaenium pachyphyllum (Klatt) Jansen, Naudin
Tibouchina scabriuscula (Schltdl.) Cogn.
Rubus spp.

(b) Species typical of old-growth forest:
Bejaria laevis Benth.
Beilschmiedia ovalis (Blake) C. K. Allen
Begonia hydrocotylifolia Otto and Hook.
Billia hippocastanum Peyr.
Chamaedorea liebmannii Marttens, M.
Dendropanax populifolius (Marchal) A.C. Smith
Drimys granadensis A.C. Smith
Greigia sp.
Maianthemum paniculatum (Mart. and Gal.) La Frankie
Marattia weinmanniifolia Liebm.
Ocotea helicterifolia (Meissn.) Hemsley
Oreopanax xalapensis (Kunth) Decne. and Planchon
Osmanthus americana (L.) Benth. and Hook.
Parathesis tenuis Standley
Passiflora cooki Killip
Persea liebmannii Mez
Psychotria galeottiana (Mart.) Taylor and Lorence
Quetzalia occidentalis (Loes.) Lundell
Quercus leiophylla A. DC.
Quercus salicifolia Née
Styrax argenteus var. *ramirezii* (Greenm.) Gousolin
Symplocos pycnantha Hemsley
Ticodendron incognitum Gómez-Laurito and Gómez-P.
Weinmannia pinnata L.

Changes in terrestrial vegetation

Whereas many studies of succession in forests have focused on trees, there are reasons to consider all groups of terrestrial vascular plants as they have different functional roles and may influence each other during succession. Even in other better-studied systems such as temperate forests, the responses of other groups of plants are not very well understood (e.g. Gilliam *et al.*, 1995). In terms of their functioning roles in the ecosystem, some plant species may be effectively redundant (Wilson, 1999). However, this is not true for plant life forms, which may constitute natural associations that can be examined separately. If community dynamics are shaped by the functional role of species, then each life form should follow different patterns of change during succession that can be predicted in terms of their ecology. Otherwise, such changes should be random and independent of their life form. Classification of the plants on the basis of their functional role is not trivial, as many traits can potentially be considered (e.g. Díaz *et al.*, 1999). Following a system similar to that of Raunkiaer, but including growth habits as well, Blanco-Macías (2007) classified the vascular plants of successional forests at El Rincón as geophytes, shrubs, trees, lianas (including climbing plants) and 'low plants' (including herbs and other low-stature plants, such as trailing plants). Such a classification, though not ideal, has several advantages, as it considers the vertical structure of the community, in particular their relationship with light (see Whittaker, 1975). Height appears to be of chief importance in competition for light and for influencing succession (Huston and Smith, 1987). Plant height is also positively correlated with other important plant attributes, such as seed size, which influence dispersal (Leishman *et al.*, 1995). Furthermore, this classification is easy to apply without an extensive knowledge of the biology of the species. Finally, several simple hypotheses can be constructed regarding the general trends of change in life form abundance with succession. By virtue of their low stature and short life cycles, low plants, in the first place, and geophytes in second should be more abundant at earlier stages of succession before taller plants shade out the forest floor, limiting the availability of space and light. Shrubs should follow, before trees reach their maximum height. Colonization by trees should be slower by virtue of their longer life cycles and their overall larger seed sizes, which might limit dispersal (see Leishman *et al.*, 1995). Therefore, trees should be more abundant and diverse at later successional stages. Similarly, lianas and climbing plants require support from other plants, and therefore should prevail at intermediate or late stages of succession.

Changes in vegetation during succession at El Rincon were studied by means of ten (nine in two sites) 2 m × 50 m plots established for trees in Yotao, Tarantulas and Juquila chronosequences, which were used for sampling plants >3.5 cm diameter at breast height (dbh). Within each of these plots, two subplots of 2 m × 2 m were randomly established in which all freestanding plants were identified, mostly to the species level, measured and mapped (Blanco-Macías, 2007).

Results indicate that the highest rate of increase in tree density takes place during the first 15 years of forest development. This rate decreases over the next 30 years in such a way that tree density peaks in stands of *c.*45 years of age. After that time, self-thinning reduces tree density to approximately 30%, reaching a local minimum at *c.*75 years. A small increase in tree density is detected subsequently (Fig. 7.2). Despite the marked decline in density after 45 years, no equivalent decline was observed in basal area. Tree basal area, an estimate of tree biomass, increased during the first century of forest development, with the highest increase during the first 45 years (Fig. 7.2) (Blanco-Macías, 2001). Thus, these results are consistent with the logistic biomass accretion model in which biomass increases towards a maximum, at least regarding trees, which are by far the dominant group of plants in TMCF areas, and contrast with other models that predict a dramatic decrease in biomass associated with the thinning phase (Peet, 1992). A study performed on *P. chiapensis*, the dominant species at earlier stages of succession in the study area, shows evidence that the highest rate of mortality takes place among the slowest growing trees (del Castillo, 1996). Thus, the loss of biomass by mortality of trees during self-thinning appears not to be great and trees that survive self-thinning, which are probably the largest plants, are likely to have achieved higher growth rates following release from competition. Indeed, some evidence of release of resources is indicated by an increase in concentration of nutrient cations detected in soil after self-thinning in some of the chronosequences (Bautista-Cruz and del Castillo, 2005).

As predicted, the density of geophytes, shrubs and low plants increased at higher rates than that of trees during earlier stages of succession, showing

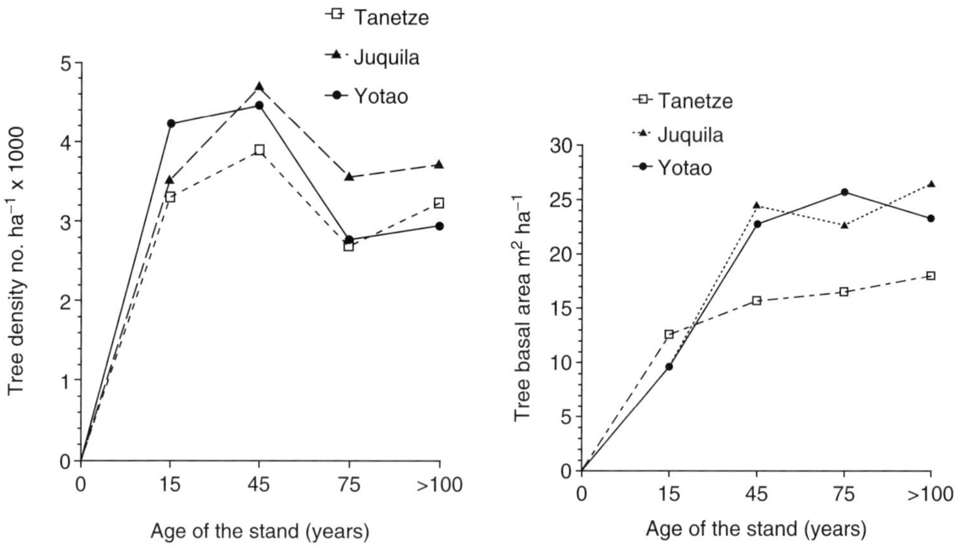

Fig. 7.2. Relationship between tree density (left) and tree basal area (right) with the approximate age of the stand after agricultural use in Tanetze, Juquila and Yotao chronosequences in El Rincón, Sierra Norte, Oaxaca (after Blanco Macías, 2001).

peaks in density during the first 15 years after abandonment, and a decline following this age reaching a minimum between 45 and 75 years (Blanco-Macías, 2007). Such a decline coincides with the period at which trees achieve their maximum density and basal area (Fig. 7.2). Thus it is likely that trees inhibited the development of low-stature plants, probably by diminishing light availability, and reducing soil nutrients and water availability (see below).

From the first 15–45 years of stand development, *P. chiapensis* dominates, and the entire vegetation appears like a typical pine forest. However, shade-tolerant plants such as *B. ovalis*, *Freziera* sp., *Magnolia dealbata*, *O. americana*, *P. americana*, *Quercus* spp. and *Rapanea* spp. establish on the shaded floor of early successional forest and start growing, to create a stratum intermediate in height between that of pine trees that create the forest canopy and that of shrubs and low-stature plants. Pine trees therefore appear to facilitate the establishment of shade-tolerant plants, but at the same time inhibit their own establishment and that of low-stature plants, *P. chiapensis* itself being shade-intolerant as a young plant (del Castillo, 1996). After a century of forest development, only a few large old pine trees remain, and the forest is dominated by broadleaf species of trees. Pine forests are therefore secondary in mid-elevation tropical montane and humid areas. The same conclusion was reached by González-Espinosa and colleagues in Chiapas Highlands (González-Espinosa *et al.*, 1991; Ramírez-Marcial *et al.*, 2001).

A greenhouse study revealed that *P. chiapensis* is a slow-growing plant compared to angiosperm trees such as *Brunellia mexicana*, a species typical of early successional stages, and species such as *T. hemsleyi* and *Ilex pringle* of late successional stages (Hernández Pérez, 2001). Therefore these results do not support the predictions that plants at early stages of succession are relatively fast growing (Huston and Smith, 1987). Perhaps the success of *P. chiapensis* at early successional stages can be explained by its high seed input (Pérez-Ríos and del Castillo, in preparation), and the ability of the seeds to establish successfully on recently created open areas (R.F. del Castillo, in preparation).

Gaps in the forest canopy are created by natural tree falls. However, compared to lowland tropical forest (Martínez-Ramos, 1985), we do not have evidence that gaps opened in secondary forest in TMCF areas can be colonized by pioneer trees typical of secondary forest. As mentioned above, such species are virtually absent in old-growth and primary forests. Gaps are rapidly covered by grasses and ferns, and the shrubs and low-stature plants that have already established expand their canopies, rapidly shading such gaps at ground level. Earlier colonist plants such as *P. chiapensis* rapidly become locally extinct due to their inability to establish successfully in the forest, even in natural gaps. An ongoing study of *P. chiapensis* demography has shown that, although seedlings of this species can establish during the first years after the gap was created, eventually all seedlings die, presumably by shading and competition with faster-growing grasses and other low-stature plants (R.F. del Castillo, in preparation). Without severe disturbance, for instance fires or landslides, early successional species are virtually absent in primary and old-growth forest. This kind of pattern is not compatible with models of forest dynamics such as the shifting-mosaic steady state (see Urban and Shugart, 1992). This model envisioned the forest as a mosaic in which individual patches of the forest are in different stages of succession at any given time by virtue of gap formation. In the TMCFs studied, gap colonization appears to be a different process to the succession that follows the anthropogenic disturbance that destroyed the original vegetation. In other words, no pine forest patches are located within old-growth forest by virtue of tree-fall gap formation. Autogenic processes, such as natural tree falls, appear to generate different patterns of regeneration than allogenic process such as fire or landslides in TMCF areas.

As seen in Fig. 7.3, the patterns of change in species richness during succession obtained from Colwell *et al.*'s (2004) rarefaction method show that each plant group follows different trends of change during the first century of forest development. Trees were, by far, the group with the highest species richness at all successional stages analysed. As predicted, tree species richness increased steadily during the first century of forest regeneration. By contrast, low-stature plants such as geophytes, shrubs and low plants exhibited a peak in species richness during the first 15 years after abandonment, followed by a later decline, displaying a small increase coincident with self-thinning, following a similar trend to that observed for stem density (Blanco-Macías, 2007). As expected, lianas and climbing plants followed a similar pattern to that of trees, which provide support for them. Overall, these results suggest that the rates of colonization of trees, lianas and climbing plants exceed the rates of local extinction during the first century of forest development, whereas colo-

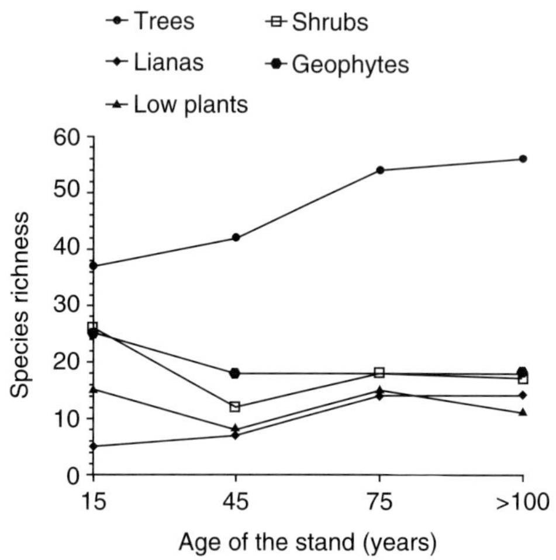

Fig. 7.3. Changes in species richness, obtained from Colwell *et al.*'s (2004) rarefaction method, during secondary succession in TMCFs. Different growth forms are illustrated, namely trees, shrubs, lianas, geophytes and low plants, during the first century of forest development in el Rincón Alto, Sierra Madre de Oaxaca, southern Mexico (from Blanco-Macías, 2007).

nization rates of geophytes, shrubs and low plants exceed local extinction rates mostly during the first 15 years of forest development.

Changes in abundance of epiphytes

The changes in cover of six groups of epiphytes, micro-lichens, macro-lichens, liverworts, mosses, vascular plants and microscopic epiphytes (i.e. unidentified microscopic plants including protonema, green algae and cyanobacteria) were studied in the three chronosequences described above and in an adjacent primary forest. The epiphyte cover area was estimated in four 100 cm^2 grids at four height levels, 0–10, 50–60, 100–110 and 150–160 cm from the base of the trunk of the host plants, in 24 tree trunks in each stage and chronosequence (Cordova and del Castillo, 2001). The relationship of the age of the stand with total epiphyte cover followed a sigmoid pattern, with the highest rates of increase between 15 and 45 years after abandonment (Fig. 7.4). The patterns of colonization during succession were different for all groups studied. The group to colonize the trunks most rapidly were leafy liverworts (order Jungermanniales), which were the dominant epiphytes at all the seral stages studied. Mosses displayed the second highest rates of colonization and were also second in terms of cover at all stages. Vascular plants were the slowest group to colonize tree trunks. Of these, ferns colonized first, and orchids and bromeliads were the latest to colonize. In contrast, the absolute cover of micro- and macro-lichens was not significantly affected by the age of

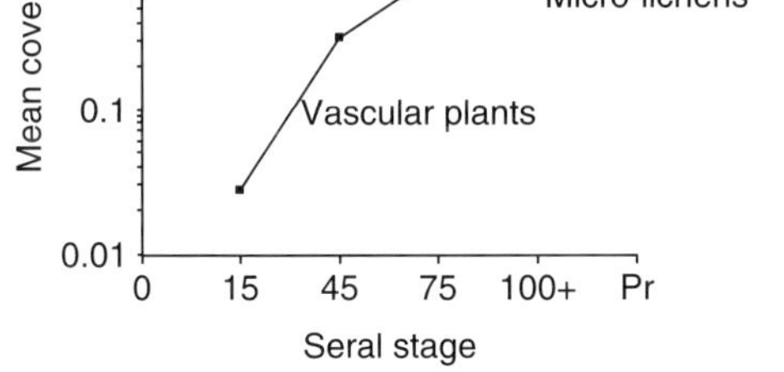

Fig. 7.4. Relationship between the absolute cover of total epiphytes, liverworts, mosses, macro-lichens, micro-lichens, vascular plants and microscopic epiphytes in four 1 dm^2 plots per tree, examined on the lower portion of the host tree with the approximate age of the stand after agricultural use in El Rincón, Sierra Norte, Oaxaca, Mexico. Average cover was examined in 72 trees in each seral stage of three chronosequences and 24 trees from an adjacent primary forest (Pr) (modified from Córdova and del Castillo, 2001).

the stand. However, the relative cover of micro-lichens on tree trunks decreased with stand age, following an opposite trend to vascular plants. Epiphyte cover was negatively related to the diameter of the host trunk in all stands except in the primary forest, suggesting that colonization is not limited by space during the earlier stages of succession, probably because the increase in diameter of the tree trunks takes place at higher rates than the capability of epiphytes to colonize them. The colonization of trunks by epiphytes may take 100 years or more to achieve the values observed in primary forest (Cordova and del Castillo, 2001).

Soil processes during secondary succession

Slash-and-burn imposes severe changes to the soil system in TMCF areas of southern Mexico, affecting soil genesis and mineralization (Bautista-Cruz *et al.*, 2005). Indeed, during the cultivation stage, the O, A and B horizons of the soil are lost by erosion. Soil losses are probably fostered by heavy rains and steep slopes, typical of TMCF areas, in conjunction with the lack of any practice of soil retention and the sparse vegetation cover of maizefields. As a result, soil rejuvenates and the pedogenic processes appear to reinitiate, forming entisols in the maizefields. After the cultivation period, secondary succession promotes soil evolution, in particular the rapid formation of a B horizon and the development of an O horizon. Therefore, all the studied profiles of soils from forest stands were classified as inceptisols (Bautista-Cruz *et al.*, 2005).

The most dramatic changes in soil properties usually took place during the first 15 years of abandonment after agricultural use. These include the highest drop in the soil concentrations of exchangeable K, Mg and Ca (Fig. 7.5).

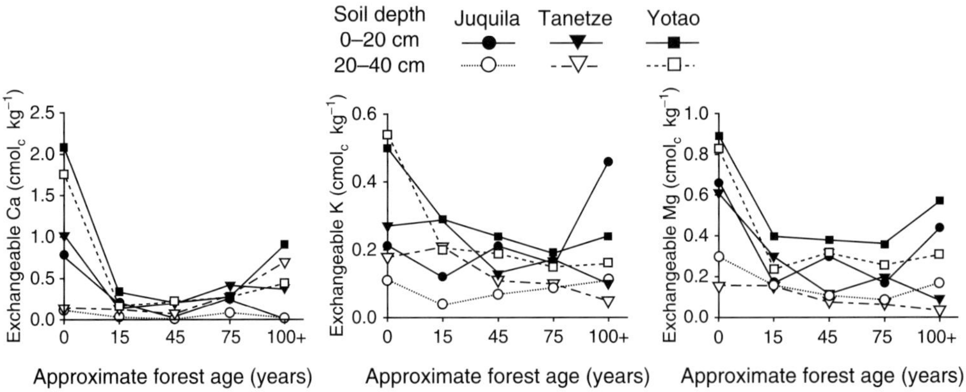

Fig. 7.5. Patterns of response to the age of the stand (mean values) for exchangeable calcium, potassium and magnesium at 0–20 and 20–40 cm soil depths in three chronosequences in a tropical montane cloud forest area in El Rincón, Oaxaca, Mexico (from Bautista-Cruz and del Castillo, 2005). (Reproduced with permission from Soil Science Society of America, 677 S. Segoe Rd, Madison, WI 53711, USA.)

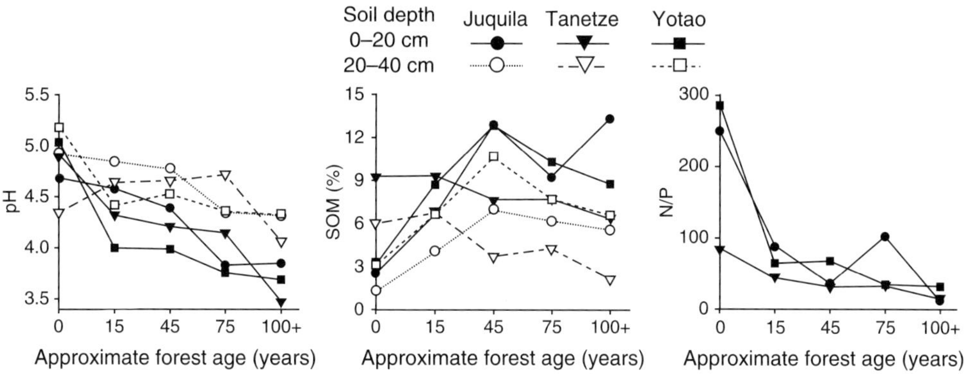

Fig. 7.6. Patterns of response to the age of the stand (mean values) for pH, soil organic matter and N/P ratio, in three chronosequences in a tropical montane cloud forest area in El Rincón, Oaxaca, Mexico (from Bautista-Cruz and del Castillo, 2005). (Reproduced with permission from Soil Science Society of America, 677 S. Segoe Rd, Madison, WI 53711, USA.)

This result can be explained in part by the rapid growth of the vegetation in that time interval, and the increase in soil acidity. Soil organic matter (SOM) accumulated at higher rates than it decomposed (Fig. 7.6). Stands of 45 years or younger had only undecomposed and partially decomposed SOM. By contrast, stands of 75 years or older had undecomposed, partially decomposed and highly decomposed SOM. In the old-growth forests, SOM was strongly humified, and appears to have been relocated to give rise to a Bh horizon (Bautista-Cruz *et al.*, 2005).

Forest soils in TMCF areas appear to be important reservoirs of carbon. Significant amounts of carbon are expected to be released to the atmosphere during forest clearing and subsequent cultivation in TMCF areas, since after that phase virtually all of the original vegetation has disappeared together with the organic layers of the soil. In general, when soil is brought under cultivation, most of the organic matter is oxidized to CO_2 (Schlesinger, 1997). Secondary succession in TMCF areas reverses part of the effects responsible for soil organic carbon losses that occurred when the land was converted to agricultural fields. The highest rates of soil carbon sequestration per year took place during the first 15 years after abandonment in the three studied chronosequences (429 gC m^{-2} $year^{-1}$ at 0–20 cm; and 168 gC m^{-2} $year^{-1}$ at 20–40 cm soil depth; such rates decrease afterwards and may vary from one site to another (Bautista-Cruz and del Castillo, 2005)). The rates of carbon accumulation in soil detected during the first 15 years after abandonment exceed the long-term mean rate observed in forest establishment after agricultural use (33.8 gC m^{-2} $year^{-1}$) (Post and Kwon, 2000), and are comparable to those reported for young soils in tropical volcanic islands (Schlesinger *et al.*, 1998). Other indications of the retention of carbon during secondary succession are the thickness of the litter layer, which increases steadily during the first century of forest development, reaching 10–30 cm in old-growth

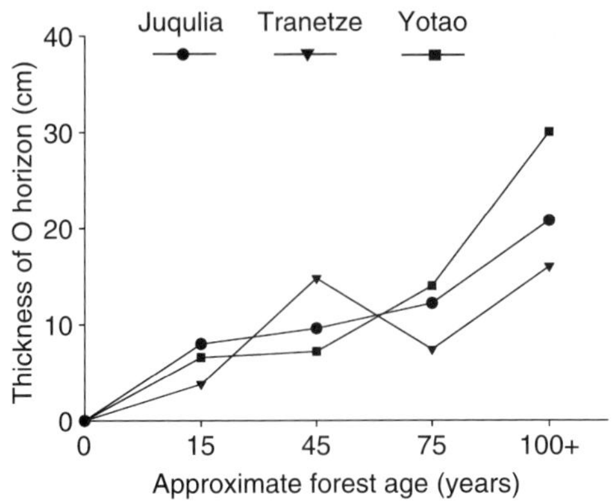

Fig. 7.7. Patterns of response to the age of the stand (mean values) for the thickness of the O horizon in three chronosequences in a tropical montane cloud forest area in El Rincón, Oaxaca, Mexico (from Bautista-Cruz and del Castillo, 2005). The *x* axis indicates approximate forest age in years. (Reproduced with permission from Soil Science Society of America, 677 S. Segoe Rd, Madison, WI 55711, USA.)

forest (Bautista-Cruz and del Castillo, 2005) (Fig. 7.7), and the increase in tree basal area described before (Fig. 7.2). These results are evidence of the importance of secondary forests in TMCF areas for providing environmental services, in particular carbon sequestration.

Soil pH decreases significantly as forest ages (Bautista-Cruz and del Castillo, 2005) (Fig. 7.6). Indeed, soils of old-growth forests are very acidic, with soil pHs (1:2 soil:water) as low as 3.2. Acidification, in turn, appears to affect many ecosystem processes. Primary and secondary minerals are hydrolysed. In particular, muscovite, the dominant mineral of the coarse fraction of the soil, and chlorite, from the fine fraction, decreased with the age of the stand (Bautista-Cruz *et al.*, 2005). In turn, the release of potassium and other nutrient cations from such a process may help to replace part of the ions immobilized by plants and SOM, or lost by leaching. Nevertheless, the availability of nutrient cations in soil decreases as the forests age (Bautista-Cruz and del Castillo, 2005).

There is little cation exchange capability to buffer the soil solution, which together with the continuous supply of acidic litter by the vegetation makes the soil progressively acidic. Other processes that need to be studied may also contribute to enhancing soil acidity. Nitrification may play a key role in the H+ budget, driving soil pH to very low levels and reducing soil fertility (Robertson, 1989). Moreover, processes such as denitrification and nitrification have been shown to change dramatically during secondary succession in other tropical ecosystems (Robertson and Tiedje, 1988). As a consequence of low pH, exchangeable aluminium was high in soils of TMCF areas (Bautista-Cruz and del Castillo, 2005). These results contrast with those found by other studies on secondary succession, indicating that succession improves soil conditions (see Peet, 1992). Plants, particularly those of late successional stands in TMCF areas, are expected to be adapted to conditions of low soil fertility and to cope with problems of aluminium toxicity. Soil fertility appears to be one of the limiting factors in tropical ecosystems. For instance, above-ground net primary productivity in trees has been shown to be positively correlated to soil fertility in a Peruvian tropical rainforest (Cook *et al.*, 2005). In the TMCF area studied, the stands with the lowest basal area for a given forest age were those with the lowest levels of soil fertility (Bautista-Cruz and del Castillo, 2005). Species richness, on the other hand, generally increases with soil fertility in tropical plant communities (Gentry, 1988). In contrast, in TMCF areas studied, species richness increased in older stands, which tend to be less fertile. Care should be taken when trying to make generalizations about soil fertility and species richness.

Phosphorus levels were very low in all stands, something typical of many forests soils (Waring and Schlesinger, 1985). However, the N/P ratio increases significantly with the age of the stand (Fig. 7.6). The opposite trend was expected, as P tends to become largely bound to SOM or secondary minerals, and N fixation is expected to increase during the course of succession, owing to colonization of nitrogen-fixing organisms (see Walker and Syers, 1976; Huston and Smith, 1987; Aerts and Chapin, 2000). The absence of species of Leguminosae (notable for their symbiotic relationship with nitrogen-fixing bacteria) in most of the TMCF areas studied perhaps helps to explain this pattern. The steady increase in N/P during succession in TMCF areas of southern Mexico points to the need for further studies on the dynamics of N and P, and highlights the need for caution when making generalizations about nutrient dynamic trends in secondary succession in forest areas (Bautista-Cruz and del Castillo, 2005).

Discussion

The levels of disturbance inflicted by the slash-and-burn method of cultivation in TMCF areas of southern Mexico can contribute to enhancing biodiversity at least at landscape and species levels. However, this type of land use results in dramatic environmental changes. The soil environment, for instance, differs greatly in crop fields by displaying lower acidity than old-growth forests, a higher exchangeable nutrient cation content and by being poorly developed. Forest soils, by contrast, are well-developed, particularly at later successional stages, and have a very low content of exchangeable nutrient cations, high soluble aluminium content and low N/P ratios. On the other hand, an open area of agricultural fields lacking the shading of trees and shrubs contrasts sharply with the dense stands of trees in 45-year-old forests, and with forests older than 75 years after self-thinning. Therefore it is not surprising that certain species prosper primarily at earlier successional stages, whereas others do so at late successional stages.

Conservation of biodiversity is an urgent environmental priority (e.g. Lubchenco *et al.*, 1991). Therefore, an understanding of the processes influencing diversity is of critical importance. Connell (1978) and Huston (1979) in their classic papers hypothesized that intermediate levels of disturbance may enhance biodiversity by generating a state of non-equilibrium, where competitive exclusion is prevented. This study suggests that one of the sources of disturbance that can enhance diversity is slash-and-burn agricultural practices. In agreement with the Connell–Huston hypothesis, the relationship between slash-and-burn disturbance and diversity may be non-linear, with diversity displaying maximum values at intermediate intensities of disturbance. A landscape composed exclusively of old-growth forest, that is in the absence of slash-and-burn practices or other sources of disturbance, does not sustain most of the common species of young secondary forests. But, at the other extreme, a landscape with high deforestation rates such as that observed in the Chiapas Highlands (Chapter 2) would result in many species being threatened with extinction, including even some typical of earlier successional stages. Both extremes are likely to generate lower environmental variation than a landscape composed of a mixture of forests of different ages and croplands. Thus, a maximum diversity is likely to be achieved in situations in which slash-and-burn is neither very frequent nor very uncommon in both time and space. Spatial heterogeneity is one of the factors that may permit the coexistence of a high number of species (Tilman, 1982). A moderate slash-and-burn practice is a source of disturbance that can prevent a reduction of diversity by generating such spatial heterogeneity in TMCF areas.

Succession seems to depend also on the proximity of source pools of colonists (Cook *et al.*, 2005). The landscape studied consists of a series of adjacent forest fragments and croplands. Therefore the close proximity of sources of old-growth forest in all chronosequences studied may help to explain the relatively rapid replacement of pine forest by hardwoods and the excess of tree colonization over tree local extinction during the first century of forest development. The role of distance to source pools in the outcome of succession has been studied in other humid forest areas of Latin America (see Chapter 2), suggesting that extension and proximity of disturbed lands to seed sources are important for forest regeneration. Indeed, many species typical of old-growth forest have large seed sizes and appear to have restricted seed dispersal capabilities, in contrast to early successional species (Pérez-Ríos and del Castillo, in preparation).

Changes of diversity and abundance in life forms during the course of secondary succession in TMCF areas appear not to be driven by chance but to follow predictable patterns related to environmental changes. Trees, lianas, climbing plants, epiphytic liverworts and vascular epiphytes prosper better at later successional stages, whereas low-stature plants, such as herbs and geophytes, are more abundant and diverse at earlier stages of forest development. These results highlight the importance of the functional role of species in the successional process, and does not support the hypothesis that all plant species or growth forms have the same probability of succeeding at any successional stage. These results clearly point to the need to analyse each functional group separately. The classification of plants in terms of the position of their reproductive organs and growth habits has enough discriminatory power to allow consistent patterns to be identified during secondary succession in TMCF areas.

Secondary forests in TMCF areas appear to be important both as reservoirs of species diversity and as suppliers of ecosystem functions. As mentioned above, secondary forests harbour the highest species richness in some groups of plants. High species diversity in early successional stands has been detected in other forest ecosystems such as temperate forests (e.g. Peet, 1978) and in secondary succession areas of former TMCFs (e.g. Romero-Romero *et al.*, 2000). In the absence of disturbance, many species typical of secondary forest will go extinct. Disturbances might also enhance genetic diversity by allowing shifts in selection regimes (Namkoong and Koshy, 2000).

However, slash-and-burn causes important changes to the soil. Fire, which is used to clear forest areas prior to crop cultivation, plays a major role in making available soil nutrients for plants by releasing nutrients from biomass and SOM, and reducing the levels of aluminium toxicity by increasing soil pH. In the long run, slash-and-burn prepares the land for the colonization of pines, since pine forests, being a transient successional stage, readily colonize recently abandoned crop fields. Adjacent old-growth forest, on the other hand, may be important for supplying a source of colonists to the developing pine forest, thereby facilitating the transition from pine forest to broadleaf forest. It remains to be explored what are the regeneration capabilities of a primary TMCF in the absence of nearby secondary forest, or what would be the outcome of early forests dominated by pine in the absence of nearby primary or old-growth forests as seed suppliers. This point is relevant, as most species of early secondary stages appear not to persist in old-growth forest or established secondary forest; nor do late successional species appear to be good colonizers as they establish only under a forest canopy. Thus, the close conjunction of agricultural activities, as a major source of disturbance, together with adjacent secondary forests and old-growth forests appears to contribute to the maintenance of biodiversity at the landscape level.

Old-growth forests are important suppliers of environmental services, such as operating as a carbon sink, both in terms of tree biomass and soil organic matter. The soil acidity they generate appears to be important for rock weathering, helping to replenish the soil with some of the nutrients immobilized by plant biomass or SOM, or lost by leaching or run-off. Moreover, old-growth forests harbour the highest abundance of epiphytes, and probably other groups of organisms, such as large mammals. The role of old-growth forest in trapping cloud water is expected to be higher than that of young forests, which in general give the impression of being drier habitats. However, the microclimatic changes associated with succession in TMCF have not yet been studied in detail.

The rate of successional change in TMCF areas cleared for maize cropping and later abandoned generally declines with time. The first 15 years of forest re-growth shows the greatest change in several of the characteristics analysed. These include the highest rates of epiphyte colonization on lower tree trunks, the highest rates of colonization of herbs, shrubs and geophytes, the highest decrease in soil contents of exchangeable nutrient cations such as K, Mg and Ca. Also, the rate of C sequestration in soil peaks within this time period. The rapid growth of virtually all plant forms during early stages of succession may explain such results. The decline in rates of successional change with age has been detected in other studies (Foster and Tilman, 2000, and references therein), and suggests that opportunities for colonization are greater than local extinction probabilities for most groups of plants at early succession stages. The highest rates of colonization involve high demands for soil nutrients, thus explaining the highest decline in nutrients detected. Such a decline coupled with the shading at floor level may rapidly shift the balance of colonization and local extinction in favour of the latter for low-stature plants after 15 years of forest development, once trees overtop other life forms. In contrast, tree species richness steadily increases during the first hundred years of forest development, suggesting that colonization opportunities continue to be available for trees, perhaps facilitated by the proximity of sources of colonists, as explained above. Trees also have, on average, deeper roots than shrubs and other low-stature plants, thus permitting greater volumes of soil to be explored for nutrient uptake. Trees also create suitable habitats for epiphyte and liana colonization, a clear example of facilitation *sensu* Connell and Slatyer (1977). Indeed, liana success appears to be controlled by the availability of large trees in other tropical ecosystems (Phillips *et al.*, 2005). Similar trends have been observed in lowland tropical forest, in which a decrease in herbaceous vines was accompanied by increases in shrubs and trees, and epiphytes underwent a dramatic surge in abundance (Guariguata and Ostertag, 2001). Lianas influence the outcome of succession by altering differentially the survival rate of trees in other tropical forests (Pérez-Salicrup, 2001), a relationship that needs to be explored in TMCF areas.

Self-thinning at *c.*45–75 years after abandonment appears to be an important landmark for certain processes that show a shift in trend during this stage. For instance, the density of geophytes, low plants, herbs and shrubs decreased from 15 to 45 years, but increased after 45 years. After the end of self-thinning, that is *c.*75 years after abandonment, nutrient cation concentrations in the soil show a slight increase. As suggested before, such changes may reflect changes in the availability of resources.

Conclusions

Moderate slash-and-burn practices in former TMCF areas may create habitat heterogeneity at the landscape scale, which, in turn, can enhance plant species richness.

The cultivation phase of slash-and-burn agriculture imposes severe losses of soil and carbon. Soil rejuvenates, and organic horizons are lost. Soil pH and cation exchange capacity rises. The original vegetation practically disappears.

Secondary succession reverses, in part at least, the effects of the cultivation phase. Soil carbon is sequestered at high rates during the early stages, and soil layers accumulate. Vegetation starts to regenerate first as shrubs and forbs and other low-stature plants, then as a pine-dominated forest, and, finally, as a hardwood forest. Epiphytes start to colonize the trees very early during succession, but the process may take more than a century to reach the values of cover observed in primary forests.

Old-growth forests do not contain some of those species common in early successional stages. The process of gap formation in such forest is not comparable to forest regeneration following the cultivation phase of the slash-and-burn. Therefore, the persistence of species typical of early secondary forest appears to depend entirely on periodic allogenic disturbances. These species, however, are key to the process of secondary succession as they are the first to colonize abandoned areas.

In contrast to other successional processes in humid forest areas, soil becomes progressively acidic, infertile and toxic as a result of high soluble aluminium concentrations, and the N/P ratio steadily decreases during the first century of forest development. Legumes are virtually absent in all successional stages. The increase in soil acidity allows the weathering of the parent material, hydrolysing primary and secondary minerals and releasing base cations to the soil. Slash-and-burn reverses such trends: soil pH and cation exchange capacity rises and consequently soluble aluminium levels drop.

Periodic disturbances such as those of moderate slash-and-burn create open areas for secondary forest to develop, allowing the persistence of pioneer species in the landscape. Forest development, however, depends on source pools of old-growth forest, as species typical of early successional stages cannot succeed under their own canopy. Thus, periodic moderate disturbances produce a landscape with a mixture of crop fields, early secondary forest and old-growth forest, and appear to maintain stability, resilience and species richness at the landscape scale.

Acknowledgements

This research project was funded by grants from the Darwin Initiative for the Survival of Species (United Kingdom), the European Community INCO-DEV programme (BIOCORES project contract no. ICA4-CT 2001-10095), by CONACyT, Sistema de Investigación Benito Juárez and Instituto Politécnico Nacional (CEGPI and COFAA) and ALFA-FOREST Contract II-0411-FA-FCD-FI-FC. We acknowledge the help of Raul Rivera for field and cartographic work and Salvador Acosta for fieldwork and plant identification. Many students assisted during fieldwork; their help is greatly appreciated. RFDC wishes to thank Angelica Bautista-Cruz for helpful discussions on soil changes related with succession. The authors acknowledge the valuable comments and editorial work of A.C. Newton.

References

Aerts, R. and Chapin, F.S. (2000) The mineral nutrition in wild plants revisited: a re-evaluation of processes and patterns. In: Fitter, A.H. and Raffaelli, D.G. (eds) *Advances in Ecological Research* 30. Academic Press, New York, pp. 1–67.

Aldrich, M. and Hostettler, S. (2000) *Tropical Montane Cloud Forest, Time for Action*. UNEP-World Conservation Monitoring Centre, Cambridge, UK.
Bautista-Cruz, M.A. and del Castillo, R.F. (2005) Soil changes during secondary succession in a tropical montane cloud forest area. *Soil Science Society of America Journal* 69, 906–914.
Bautista-Cruz, M.A., Gutiérrez, C., del Castillo, R.F. and Etchevers, J.D. (2005) Cronosecuencia de un suelo y su clasificación en un área originalmente ocupada por bosque mesófilo de montada. *Terra Latinoamericana* 23, 147–158.
Blanco-Macías, M. (2001) Análisis sucesional del bosque mesófilo en El Rincón, Sierra Norte de Oaxaca. BSc thesis. Universidad Nacional Autónoma de México, Iztacala, Estado de México, Mexico.
Blanco-Macías, A. (2007) Patterns of change in plant guilds during secondary succession in a tropical montane cloud forest area in Oaxaca, Mexico. MSc thesis. Universidad Nacional Autónoma de México, Iztacala, Estado de México, Mexico.
Boege, E. (1988) *Las Mazatecos Ante la Nación. Contradicciones de la Identidad Étnica en el México Actual*. Siglo Veintiuno, México DF, Mexico.
Bruijnzeel, L.A. and Hamilton, L.S. (2000) *Decision Time for Cloud Forest*. UNESCO Humid Tropic Programme (IHPO), series 13. UNESCO, Paris, France.
Churchill, S.P., Griffin, A. and Lewis, M. (1995) Moss diversity of tropical Andes. In: Churchill, S.P., Balslev, H., Forero, E. and Luteyn, J.L. (eds) *Biodiversity and Conservation of Neotropical Montane Forest*. The New York Botanical Garden, New York, pp. 335–346.
Colwell, R.K., Mao, C.X. and Chang, J. (2004) Interpolating, extrapolating, and comparing incidence-based species accumulation curves. *Ecology* 85, 2717–2727.
Connell, J.H. (1978) Diversity in tropical rainforest and coral reefs. *Science* 199, 1302–1310.
Connell, J.H. and Slatyer, R.O. (1977) Mechanisms of succession in natural communities and their role in community stability and organization. *American Naturalist* 111, 1119–1144.
Consejo de Recursos Minerales (1996) *Monografía Geológico Minera del Estado de Oaxaca*. F. Castillo Nieto and E. Rodríguez-Luna (eds). Secretaría de Comercio y Fomento Industrial, Pachuca, Mexico.
Cook, L.M., Yao, J., Foster, B.L., Holt, R.D. and Patrick, L.B. (2005) Secondary succession in an experimentally fragmented landscape community patterns across space and time. *Ecology* 86, 1267–1279.
Cordova, J. and del Castillo, R.F. (2001) Changes in epiphyte cover in three chronosequences in a tropical montane cloud forest in Mexico. In: Gottsberger, G. and Liede, S. (eds) *Life Forms and Dynamics in Tropical Forests. Dissertations Botanicae 346*. J. Cramer in der Gebrüder Borntraeger Verlagsbuchhandlung, Berlin–Stuttgart, Germany, pp. 79–94.
del Castillo, R.F. (1996) Aspectos autoecológicos de *Pinus chiapensis*. In: Garduño, L.L., Chavarria, G.V., Magdaleno, P.L. and Pérez, I.M. (eds) *Memorias del 2do. Coloquio Regional de Investigación, Ciencias Exactas y Naturales*. Universidad Autónoma del Estado de México, Toluca, Estado de México, Mexico, pp. 63–68.
del Castillo, R.F. (1999) Composición y estructura de nopalera bajo situaciones contrastantes de exposición de ladera y herbivoría. *Boletín de la Sociedad Botánica de México* 65, 5–22.
Díaz, S., Cabido, M. and Casanoves, F. (1999) Functional implications of trait environment linkages in plant communities. In: Weiher, E. and Keddy, P. (eds) *Ecological Assembly Rules, Perspectives, Advances, Retreats*. Cambridge University Press, Cambridge, UK, pp. 338–362.
Doumenge, C., Gilmour, D., Ruíz-Pérez, M. and Blockhus, J. (1995) Tropical montane cloud forests: conservation status and management issues. In: Hamilton, L.S., Juvick, J.O. and Scatena, F.N. (eds) *Tropical Montane Cloud Forests*. Springer, New York, pp. 24–37.
Foster, B.L. and Tilman, D. (2000) Dynamic and static views of succession: testing the descriptive power of the chronosequence approach. *Plant Ecology* 146, 1–10.

Gentry, A.H. (1988) Changes in plant community diversity and floristic composition on environmental and geographical gradients. *Annals of the Missouri Botanical Garden* 75, 1–34.
Gilliam, F.S., Turrill, N.L. and Adams, M.B. (1995) Herbaceous-layer and overstory species in clear-cut and mature central Appalachian hardwood forest. *Ecological Applications* 5, 947–955.
Glenn-Lewin, D.C. and van der Maarel, E. (1992) Patterns and process of vegetation dynamics. In: Glenn-Lewin, D.C., Peet, R.K. and Veblen, T.T. (eds) *Plant Succession. Theory and Prediction*. Chapman and Hall, London, UK, pp. 11–59.
González-Espinosa, M., Quintana-Ascencio, P.F., Ramírez-Marcial, N. and Gaytán-Guzmán, P. (1991) Secondary succession in disturbed *Pinus–Quercus* forests in the highlands of Chiapas, México. *Journal of Vegetation Science* 2, 351–360.
Guariguata, M.R. and Ostertag, R. (2001) Neotropical secondary forest succession: changes in structural and functional characteristics. *Forest Ecology and Management* 148, 185–206.
Hamilton, L.S., Juvick, J.O. and Scatena, F.N. (1995) *Tropical Montane Cloud Forests*. Springer, New York.
Hernández Pérez, V. (2001) Influencia del suelo en el crecimiento de cuatro especies arbóreas a lo largo de un gradiente sucesional del bosque mesófilo de montaña, Sierra Norte, Oaxaca. BSc thesis. Universidad Nacional Autónoma de México, Iztacala, Mexico.
Hubbell, S.P. (2001) *The Unified Neutral Theory of Biodiversity and Biogeography*. Princeton University Press, Princeton, New Jersey.
Huston, M. (1979) A general hypothesis of species diversity. *American Naturalist* 113, 81–101.
Huston, M. and Smith, T. (1987) Plant succession: life history and competition. *American Naturalist* 130, 168–198.
Instituto Nacional de Estadística, Geografía e Informática (1999) *Anuario Estadístico del Estado de Oaxaca*. Instituto Nacional de Geografía e Informática, Aguascalientes, Mexico.
Instituto Nacional de Estadística, Geografía e Informática (2003) *Anuario Estadístico, edición 2003*. Instituto Nacional de Geografía e Informática, Aguascalientes, Mexico.
Leishman, M.R., Westoby, M. and Jurado, E. (1995) Correlates of seed size variation: a comparison among temperate floras. *Journal of Ecology* 83, 517–530.
López-Chávez, D. (1953) *Titulo de Propiedad y Demás Documentos Relacionados a los Bienes Comunales de San Juan Juquila Vijanos*. Secretaria de Gobernación, Archivo General de la Nación, México, DF, Mexico.
Lubchenco, J., Olson, A.M., Brubaker, L.B., Carpenter, S.R., Holland, M.M., Hubbell, S.P., Levin, S.A., MacMahon, J.A., Matson, P.A., Melillo, J.M., Mooney, H.A., Peterson, C.H., Pulliam, H.R., Real, L.A., Regal, P.J. and Risser, P.G. (1991) The sustainable biosphere initiative: an ecological research agenda. *Ecology* 72, 371–412.
Manshard, W. (1974) *Tropical Agriculture*. Longman, London, UK.
Martínez-Ramos, M. (1985) Claros, ciclos vitales de los árboles tropicales y regeneración natural de las seklvas altas perennifolias. In: Gómez-Pompa, A.D.A.S. (ed.) *Investigaciones sobre Regeneración de Selvas Altas en Veracruz, México*. Editorial Alhambra, Mexicana, Mexico, pp. 191–239.
Namkoong, G. and Koshy, M.A.S. (2000) Selection. In: Young, A.G., Boshier, D. and Boyle, T. (eds) *Forest Conservation Genetics, Theory and Practice*. CSIRO Publishing/CAB International, Wallingford, UK, pp. 101–111.
Peet, R.K. (1978) Forest vegetation of the Colorado front range: patterns of species diversity. *Vegetatio* 37, 65–78.
Peet, R.K. (1992) Community structure and ecosystem function. In: Glenn-Lewin, D.C., Peet, R.K. and Veblen, T.T. (eds) *Plant Succession. Theory and Prediction*. Chapman and Hall, London, UK, pp. 103–151.

Pérez-Salicrup, D. (2001) Effect of liana cutting on tree regeneration in a liana forest in Amazonian Bolivia. *Ecology* 82, 389–396.

Phillips, O.L., Vásquez-Martínez, R., Monteagudo-Mendoza, A., Baker, T.R. and Nuñez-Vargas, P. (2005) Large lianas as hyperdynamics elements of the tropical forest canopy. *Ecology* 86, 1250–1258.

Post, W.M. and Kwon, K.C. (2000) Soil carbon sequestration and land-use change: processes and potential. *Global Change Biology* 6, 317–327.

Ramírez-Marcial, N., González-Espinosa, M. and Williams-Linera, G. (2001) Anthropogenic disturbance and tree diversity in montane rain forest in Chiapas, Mexico. *Forest Ecology and Management* 154, 314–326.

Robertson, G.P. (1989) Nitrification and denitrification in humid tropical ecosystem: potential controls on nitrogen retention. In: Proctor, J. (ed.) *Nitrification and Denitrification in Humid Tropical Ecosystems: Potential Controls on Nitrogen Retention*. Blackwell Scientific, Oxford, UK, pp. 55–69.

Robertson, G.P. and Tiedje, J.M. (1988) Deforestation alters denitrification in a lowland tropical rain forest. *Nature* 336, 756–759.

Romero-Romero, C., Castillo, S., Meave, J. and van der Wal, H. (2000) Análisis florístico de la vegetación secundaria derivada de la selva húmeda de montaña de Santa Cruz Tepetotutla (Oaxaca), Mexico. *Boletín de la Sociedad Botánica de México* 67, 89–106.

Rzedowski, J. (1991) El endemismo en la flora fanerogámica mexicana: una apreciación preliminar. *Acta Botánica Mexicana* 15, 47–64.

Rzedowski, J. (1996) Análisis preliminar de la flora vascular de los bosques mesófilos de montaña de México. *Acta Botánica Mexicana* 35, 25–44.

Schlesinger, W.H. (1997) *Biogeochemistry: An Analysis of Global Change*. Academic Press, Amsterdam, The Netherlands.

Schlesinger, W.H., Bruijnzeel, L.A., Bush, M.B., Klein, E.M., Mace, K.A., Raikes, J.A. and Whittaker, R.J. (1998) The biogeochemistry of phosphorus after the first century of forest development on Rakata Island, Krakatau, Indonesia. *Biogeochemistry* 40, 37–55.

Tilman, D. (1982) *Resource Competition and Community Structure*. Princeton University Press, Princeton, New Jersey.

Urban, D.L. and Shugart, H.H. (1992) Individual-based models of forest succession. In: Glenn-Lewin, D.C., Peet, R.K. and Veblen, T.T. (eds) *Plant Succession. Theory and Prediction*. Chapman and Hall, London, UK, pp. 249–292.

Walker, T.W. and Syers, J.K. (1976) The fate of phosphorus during pedogenesis. *Geoderma* 15, 1–19.

Waring, R.H. and Schlesinger, W.H. (1985) *Forest Ecosystems Concepts and Management*. Academic Press Inc., Orlando, Florida.

Webster, G.L. (1995) The panorama of neotropical cloud forests. In: Churchill, S.P., Balslev, H., Forero, E. and Luteyn, J.L. (eds) *Biodiversity and Conservation of Neotropical Montane Forest*. The New York Botanical Garden, New York, pp. 53–77.

Whittaker, R.H. (1975) *Communities and Ecosystems*. Macmillan, New York.

Wilson, J.B. (1999) Assembly rules in plant communities. In: Weiher, E. and Keddy, P. (eds) *Ecological Assembly Rules, Perspectives, Advances, Retreats*. Cambridge University Press, Cambridge, UK, pp. 130–164.

8 The Impact of Logging and Secondary Succession on the Below-ground System of a Cloud Forest in Mexico

S. Negrete-Yankelevich, C. Fragoso and A.C. Newton

Use of fire to clear montane cloud forest in the Sierra Norte, Oaxaca, Mexico. Photo: Adrian Newton

 Biodiversity Loss and Conservation in Fragmented Forest Landscapes: The Forests of Montane Mexico and Temperate South America (ed. A.C. Newton)

Summary

Although the process of secondary succession in tropical montane cloud forest (TMCF) has been studied in some detail, very little is known about the consequences of changes in vegetation composition on nutrient budgets and biodiversity below ground. In this chapter we present a conceptual model of how changes in the organic matter provided by the tree community in different stages of secondary succession might mediate the linkage between above-ground and below-ground biodiversities. This model is based on evidence from a study on the litter composition, topsoil nutrient concentration and soil macroinvertebrate fauna in two pristine and two recently logged sites, as well as three successional chronosequences (15–100-year-old forests) of TMCF in Oaxaca, Mexico. Results suggest that low intensity selective logging compromises the close interdependence between the composition, spatial structure and function of above-ground and below-ground biodiversities. The homogenization of the organic matter input to the soil that follows the colonization of pines in disturbed forests may threaten the conservation of the spatial structuring of above- and below-ground communities and possibly the conservation of their high biodiversity. In the absence of further information, conservation measures should pay particular attention to the native species of worm *Ramiellona willsoni* that seem not to be able to re-establish in secondary forest and to the release of P to the soil that results from continuous disturbance. The persistence of late-successional tree species in these forests may partly be associated with their ability to compete in the noticeably phosphorus-poor soils of mature forests.

Introduction

Although the environmental and vegetation shifts associated with post-logging secondary succession in tropical montane cloud forest (TMCF) have been studied in some detail (González-Espinosa *et al.*, 1991; Quintana-Ascencio and González-Espinosa, 1993; Romero-Nájera, 2000; Blanco-Macías, 2001; Galindo-Jaimes *et al.*, 2002), little is known about the consequences of these long-term changes on nutrient budgets and biodiversity below ground.

It is widely recognized that forest disturbance can affect, in the short and long terms, nutrient cycling (Nilsson *et al.*, 1995; Finér *et al.*, 2003) and soil biota (Davies *et al.*, 1999; Lavelle, 2000; Brown *et al.*, 2001; Pietikäinen *et al.*, 2003). The impact of logging activities may be particularly important in forests growing on poor soils such as TMCF. Low photosynthetic capacity, caused by limited solar radiation (given the continuous fog cover), is thought to account for the slow growth, low productivity, poor soils, slow nutrient cycling and low decomposition rates in these mountain forests (Vitousek, 1984; Bruijnzeel and Proctor, 1995; Tanner *et al.*, 1998). Mexican TMCF is increasingly being transformed by human activity, particularly through logging for firewood extraction, which is a moderate type of disturbance that has become a continuous pressure in some areas, directed towards *Quercus* spp. trees (Ramírez-Marcial *et al.*, 2001). As a consequence of tree harvesting and associated canopy opening, the ability of the cloud forest to retain its cloud cover may be reduced and nutrient cycling may become less efficient. Logging produces an above-ground and below-ground flush of nutrient-rich organic matter from residues (Olsson *et al.*, 1996a, b; Finér *et al.*, 2003) and a less diverse pioneer community of plants, dominated by light-demanding species such as pines (González-Espinosa *et al.*, 1991; Walker *et al.*, 1996). These pioneers can be expected to produce, under richer soils, more abundant and more nutrient-rich litter that will increase the rate of decomposition and maintain high nutrient availability. The new soil and environmental conditions triggered by logging disturbance may underlie the capacity of pioneer species to out-compete over prolonged periods the relatively diverse, slow-growing and shade-tolerant community of tropical trees (González-Espinosa *et al.*, 1991).

Below-ground food-web responses to logging are poorly understood (Bengtsson *et al.*, 1997; Wardle *et al.*, 1998). However, the rise in availability of nutrient-rich organic matter may decrease the proportion of fungal-based over bacterial-based food-webs (Wardle, 1992; Siira-Pietikäinen *et al.*, 2001) and produce major changes in macroinvertebrate community composition. Furthermore, after disturbance and in early succession, the diversity of tree species diminishes relative to older forests (Ramírez-Marcial *et al.*, 2001), leading to a decline in the diversity of resources available to the soil system. This, together with the greater abundance of resources, might be expected to lead to a more uniform and less diverse soil macroinvertebrate fauna.

If each tree species generates a particular soil environment under its canopy, a less diverse tree community in early succession may develop more spatially homogeneous soil properties and biotic communities. As the forest recovers through undisturbed succession and the number of dominant species in the canopy increases, not only should significant changes in means of soil properties follow, but also spatial aggregation is expected to increase. The development of high biodiversity, both above and below ground, may depend on the maintenance of a complex structuring of organic matter and nutrient resources.

This chapter presents the findings of a study on the litter composition, topsoil nutrient concentration and soil macroinvertebrate fauna in two pristine and two recently logged sites, as well as three successional chronosequences (15–100-year-old forests) of TMCF in Oaxaca, Mexico (Negrete-Yankelevich, 2004; Negrete-Yankelevich *et al.*, 2006, 2007). Our aim is to integrate the evidence for a linkage between above-ground and below-ground biodiversities after logging disturbance and through secondary succession. Based on our findings, we present a conceptual model of how changes in the organic matter provided by the tree community in different stages of secondary succession might mediate this linkage.

Methods

This section presents a summary of the methods and data treatments described in detail in previous publications (Negrete-Yankelevich, 2004; Negrete-Yankelevich *et al.*, 2006, 2007).

Study sites

The research was carried out in the area of El Rincón (Villa Alta District), in the Sierra Norte of Oaxaca, Mexico (Chapter 1). The study sites were selected within the precise areas where the age of the forest had been established by del Castillo *et al.* (Chapter 6) in the municipalities of Juquila and Tanetze. Two series of plots were selected within Juquila (named here as Juquila and Tarbis chronosequences) and one series in Tanetze (named here as Tanetze chronosequence). Each of these chronosequences was formed by four sites of different successional stage: approximately 15, 45, 75 and 100 years of age. Additionally two pristine forests (Pris and Pris II) and two plots that were logged 2 months before sampling (Tar 0 and Tar 00) were examined. All of the extraction of *Quercus* in El Rincón was performed with a hand-held chainsaw. Fallen trees damaged at least a few other canopy trees during their descent and produced a considerable canopy gap. After the firewood was extracted, the cleared patches were abandoned or used for low intensity and no-input maize cultivation for 3 to 5 years and then abandoned completely (Bautista-Cruz and del Castillo, 2005). The disturbance that was recorded in the recently logged sites in this study represented a mean 18% reduction in canopy cover compared to pristine sites and no cultivation had been performed (Negrete-Yankelevich *et al.*, 2007).

Field and laboratory methods

Sampling was conducted in 30 m × 30 m grids (with 49 intersections every 5 m) established in each successional stage of all chronosequences and in recently logged and pristine sites. First an intensive survey was carried out in the Juquila chronosequence between 11 July and 17 August 2000. In this period all 49 intersections of each of the four grids (15-, 45-, 75- and 100-year-old forests) were sampled. The following year, between 25 June and 3 December 2001, all of the successional stages of Tanetze and Tarbis, recently logged sites (Tar0 and Tar00) and the pristine sites (Pris and Pris II) were sampled. On this occasion, only seven random vertices in each grid were selected for sampling. One monolith was extracted with a box corer in each intersection. The monolith consisted of 25 cm × 25 cm × 5 cm depth of soil, plus all the litter above it. The litter and soil sample was hand sorted *in situ* for macroinvertebrates (defining macroinvertebrates as all invertebrate animals larger than 3 mm in any of its dimensions) and stored in black plastic bags. In the laboratory the litter and soil samples were transferred to paper bags and dried in an oven at 80°C until they reached constant mass. The litter of all the monoliths extracted from the Juquila chronosequence was sorted into six components: *Pinus* needles, *Quercus* leaves, Lauraceae leaves (including the three genera present in all chronosequences: *Persea*, *Ocotea* and *Beilschmedia*), woody and reproductive material, leaves from other genera and unrecognizable leaf material. The dry mass of these components was recorded separately.

All macroinvertebrates collected were counted and classified into Class, Order and groups of immature stages. Additionally all earthworms were identified to species following Fragoso and Reynolds (1997). Although Enchytraeidae and Collembola are currently considered mesofauna, and hand sorting is not the most appropriate method to sample these taxa, we have included them in the analysis because they were a particularly conspicuous component of the faunal community and they have been rarely studied in the montane tropics (Römbke, 2003).

Recently fallen leaves of *Pinus chiapensis*, *Oreopanax xalapensis*, *Beilschmedia ovalis* and *Quercus* spp. were collected from the forest floor at Juquila, dried to constant mass and analysed for nutrient content. For these analyses three replicates per leaf species were analysed. Each replicate consisted of a 30 g sample of dry leaves randomly drawn from the pool of leaves. Both recently fallen leaves and soil samples from Juquila were analysed for the concentration of total C, total P, total N, Ca^{2+} and Mg^{2+}. For leaf litter, acid detergent lignin was also extracted (Negrete-Yankelevich *et al.*, 2007).

Statistical analysis

In order to preserve a balanced design, for comparisons including all chronosequences (Juquila, Tanetze and Tarbis) and sites (Tar0, Tar00, Pris and Pris II), only seven randomly selected samples were considered for each grid in the Juquila chronosequence. For the detailed analysis performed exclusively in the Juquila chronosequence (analysis of litter components, soil nutrient content and spatial distributions), all 49 samples were considered. Two estimates of compositional diversity were calculated for the macroinvertebrate community: the number of elements (macroinvertebrate taxa) and the Shannon–Weiner index (Magurran, 1996).

For taxa diversity indices, the individual abundances of dominant macroinvertebrate taxa (those with a minimum mean abundance of 6 individuals m^{-2} in the soil or litter) and total macroinvertebrate abundances, ANOVAs were used to determine significant differences among forest types (primary, recently logged and secondary) and among secondary forest stages (15-, 45-, 75- and 100-year-old forests). Differences among the Juquila successional stages in the mass per sample of litter components and the chemical composition of soils and leaf litter samples were tested with one-way multivariate analyses of variance (MANOVA). When the MANOVA turned out to be significant, corresponding one-way ANOVAs for each individual variable were then performed followed by post-hoc Tukey's Honest Significant Difference tests (HSD) (Negrete-Yankelevich *et al.*, 2007).

Spatial analysis

A geostatistical analysis (Rossi *et al.*, 1992) of the litter components and soil nutrient concentrations was performed for each grid in the Juquila chronosequence. Omnidirectional variograms were drawn with a minimum pair distance of 5 m and a maximum of 25 m (roughly 60% of the maximum distance available from the data). A variogram model was fitted when there was an initial phase in which semivariance increased (autocorrelated phase). To fit a model to the variograms, the weighted least square method recommended by

Webster and Oliver (1990) was used. A nugget effect ($\gamma(|h|)=C_0$) was combined with one of three functions: Exponential ($\gamma(|h|)=C\bullet[1-\exp(-|h|)/a]$), Linear ($\gamma(|h|)=C\bullet|h|$) or Spherical ($\gamma(|h|)=C\bullet[(3h/2a) - (1/2\bullet(h/a)^3]$ for $h\leq a$ and $\gamma(|h|)=C$ for $h>a$). ($\gamma(|h|)$ is the semivariance at any given distance lag (h). In Exponential and Spherical variogram models the sill (C_0+C) represents the maximum semivariance reached. The range estimates the distance at which maximum semivariance is attained. For Spherical models (a) is the range and for exponential models that approach the sill asymptotically the range was considered to be where semivariance reaches 95% of the sill. Point Kriging was used to draw contour maps of the variables over the experimental grids (Rossi *et al.*, 1992). Original values were retained on the grid intersections (Negrete-Yankelevich *et al.*, 2006).

The most abundant macroinvertebrate taxa in the Juquila chronosequence were analysed with a Spatial Analysis by Distance Indices (SADIE) procedure (Perry, 1998). Macroinvertebrate taxa that reached a mean abundance greater than 1.5 organisms per monolith in any successional stage (in either litter or soil) were defined as dominant. The degree of non-randomness in the distribution of dominant macroinvertebrate taxa in each successional stage was quantified by the indices of Distance to Regularity (I_a) and Distance to Crowding (J_a) using 1950 randomizations (as suggested by Perry, 1998). Tests of significance were performed for each index using the usual two-tailed test and a value of $\alpha<0.05$. The levels of significance of I_a and J_a were adjusted for multiple comparisons with the step-up false discovery rate method (Benjamini and Hochberg, 1995). The values of Clustering Indices (υ_i and υ_j) were mapped and contoured to determine position and size of patches in the distribution of some taxa. The limits of single patches or gaps were considered to be the contours where $\upsilon_i=1.5$ or $\upsilon_j=-1.5$, respectively. In these areas aggregation or gapping is at least 50% greater than expected at random (Negrete-Yankelevich *et al.*, 2006).

Results and Discussion

Changes in leaf litter composition after disturbance

In general TMCFs occur on nutrient-poor soils and under low solar radiation conditions; the dominant tree species are slow-growing and highly efficient in nutrient use (Tanner *et al.*, 1998). The efficient use of resources implies production of nutrient-poor litter (Vitousek, 1984) that is shed infrequently (Hobbie, 1992), and therefore promotes further scarcity of nutrients in the soil. Figure 8.1 summarizes how logging disturbance and secondary succession may disrupt this positive feed-back cycle. Initially, logging provides a flush of nutrient-rich litter in the form of logging debris. This litter decomposes relatively rapidly and frees nutrients for the colonization of pioneer species (such as pines) that are less efficient in nutrient use and produce more litter. In Mexican TMCFs young stands become dominated by light-tolerant and less diverse communities (Quintana-Ascencio and González-Espinosa,

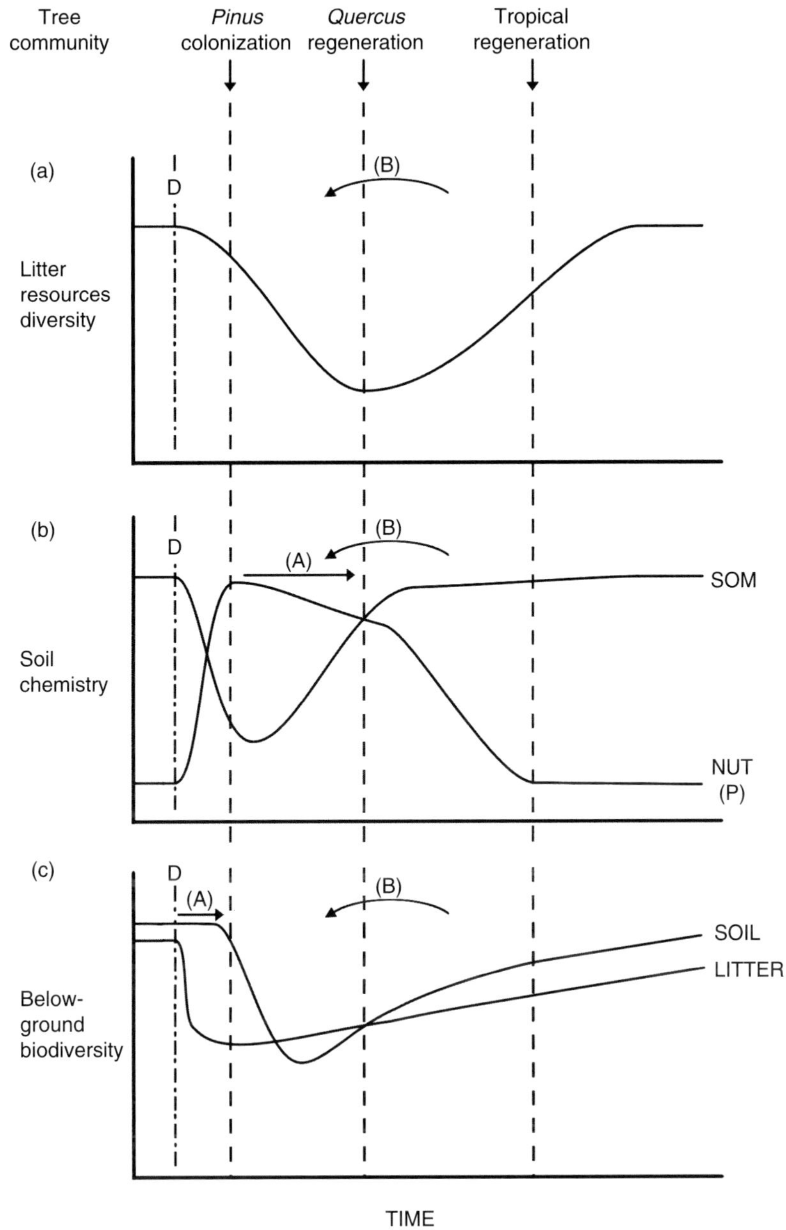

Fig. 8.1. Conceptual model of the effect that selective logging and secondary succession has on (a) litter resources diversity, (b) soil chemistry (soil organic matter (SOM) and nutrients, particularly phosphorus (NUT (P)) and (c) below-ground biodiversity. The first major disturbance to the pristine forest is marked by D. Arrows labelled with (A) represent processes that have long-lasting or delayed effects from disturbance. Arrows labelled with (B) represent the effects of repeated selective logging of oak trees.

1993; Blanco-Macías, 2001; Galindo-Jaimes *et al.*, 2002). In the Juquila chronosequence evidence of greater litter input by dominant trees in younger forests was found when the different litter components were compared between successional stages (Fig. 8.2a and b). Of particular influence was the presence of pines in young forests. Even if they only constituted *c.*16% of the basal area in the 15-year-old forest, pine needles formed the highest proportion (53.8%) of the leaves in this plot. Similarly, pine trees covered *c.*9% of the basal area in the 45-year-old forest and their needles were the second-most abundant component in the litter (39.4%). In contrast, in the 100-year-old forest the canopy appears to have recovered a more diverse tree community with limited litter production. In this forest other genera constituted only 40.4% of the leaves in the litter even if their basal area reached 70% of the total. These results sug-

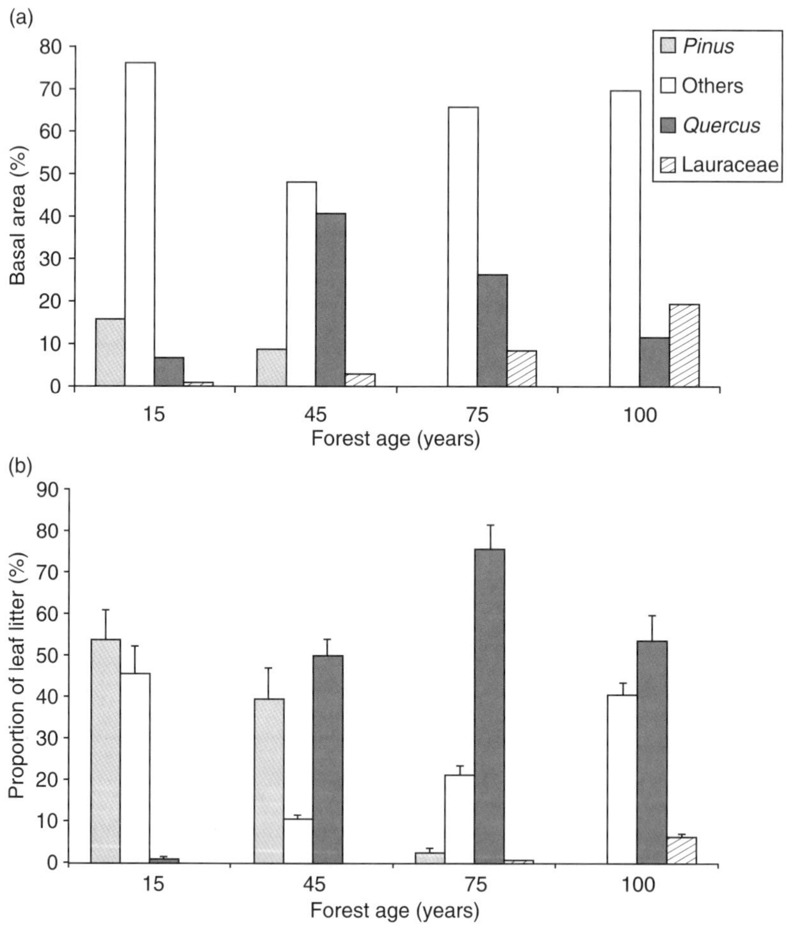

Fig. 8.2. Per cent contribution of tree genera to (a) the total basal area of different successional stages of the Juquila chronosequence compared to their contribution to (b) the mass of leaf litter. Other genera groups 18, 20, 18 and 20% in the 15-, 45-, 75- and 100-year-old forests.

gest that in early succession the litter becomes less diverse not only because the canopy is dominated by fewer species, but also because more productive genera such as *Pinus* are present in the community and their leaves account for an important proportion of the litter mass (Fig. 8.1a).

Consequences of disturbance for the nutrient concentration in the soil

There is no information available for the nutrient status of pristine forests in Oaxaca. However, if these forests are characterized by low nutrient availability, as are other mature TMCF around the world (Vitousek, 1984; Bruijnzeel and Veneklaas, 1998), evidence from the Juquila chronosequence suggests that nutrients progressively become sequestered in undecomposed organic matter and vegetation through undisturbed succession. The increase in total carbon and nitrogen in the topsoil with forest age indicates an accumulation of organic matter (Table 8.1, Fig. 8.1b). Because the decomposition rate in this choronsequence has been found to be generally low, with no differences among successional stages (Negrete-Yankelevich, 2004), the accumulation of litter from productive species in early succession is likely to be the origin of high concentrations of semi-decomposed organic matter in late successional stages. This increase is also reflected in the greater thickness of the O horizon through succession reported by Bautista-Cruz and del Castillo (2005) in these forests. Of particular relevance seems to be the cycling of P, which is particularly scarce in the Oaxacan forest soils (Negrete-Yankelevich *et al.*, 2007). This nutrient was noticeably scarce in the litter of late successional species and more abundant in pine (Table 8.1). This study did not measure the changes in available forms of this element, but it is probable that logging residues

Table 8.1. Mean macronutrient concentration in four leaf litter species and the soil of the Juquila chronosequence. Leaf litter species have been aligned with the successional stage(s) where their litter was most abundant (see Fig. 8.2a). For the litter also lignin content is presented. Only variables with significant differences are presented. Different letters denote significant differences by Tukey's HSD paired comparisons.

	15-year-old	45-year-old	75-year-old	100-year-old	
Leaf litter	*P. chiapensis*		*Q. laurina*	*O. xalapensis*	*B. ovalis*
C	56.76^{b}		56.80^{b}	55.32^{c}	56.15^{a}
P	2.43^{a}		0.95^{c}	1.83^{b}	$1.50b^{c}$
Mg	5.25^{c}		14.18^{b}	35.06^{b}	20.40^{a}
Lignin	39.94^{b}		42.96^{b}	32.68^{c}	51.01^{a}
Soil	**15-year-old**	**45-year-old**	**75-year-old**	**100-year-old**	
C	38.20^{b}	51.42^{a}	54.95^{a}	54.44^{a}	
N	1.21^{b}	1.39^{ab}	1.59^{ab}	1.78^{a}	
P	0.47^{ab}	0.52^{a}	0.27^{b}	0.42^{ab}	
Ca	8.44^{a}	5.98^{b}	2.74^{b}	4.00^{b}	
Mg	3.63^{a}	3.87^{a}	1.67^{b}	4.26^{a}	

release P to the soil, pine litter sustains the availability of this nutrient in early succession and it returns to be trapped in vegetation through mid and late succession (Fig. 8.1b).

The effect that the nutrient content in litter has on the concentration of nutrients in the soil may be long lasting (arrow (A) in Fig. 8.1b). A relatively high concentration of cations was found in the soil of the 15-year-old forest in Juquila (Table 8.1). This could be a result of the increase in high-quality organic matter in harvesting residues added to the decrease in nutrient uptake due to removal of canopy trees 15 years earlier. Similarly, the relatively high phosphorus and low magnesium concentration in pine needles shed during early succession may be responsible for the accumulation of P found in the 45-year-old forest soil and the scarcity of Mg found in the 75-year-old forest soil (Table 8.1).

Evidence from the Juquila chronosequence suggests that TMCF soils are able to recover their efficient nutrient cycling through undisturbed secondary succession. However, the sustained dominance of pine through succession has been associated with a continuous moderate disturbance in Mexican secondary forests through the extraction of wood for firewood, and its presence threatens the recolonization by broadleaved species (Challenger, 1998; Ramírez-Marcial *et al.*, 2001; Galindo-Jaimes *et al.*, 2002). Therefore the continuous presence of pine may also threaten the full recovery of the efficient nutrient cycling (arrows (B) in Fig. 8.1a and b). In a forest in Chiapas, for example, the degree of dominance of pine over oak in the canopy of secondary forests was found to be negatively correlated with the content of organic carbon, cation exchange capacity, total nitrogen content and acidity in the soil (Romero-Nájera, 2000; Galindo-Jaimes *et al.*, 2002), all indicators of the accumulation of organic matter and slow nutrient cycling that are characteristic of mature TMCF. The role that pine plays in the cycling of P could be particularly important because the persistence of late successional tropical trees in these forests may not only depend on the availability of light but may also be associated with their ability to compete in noticeably phosphorus-poor soils. Continuous disturbance through selective logging of oak may promote the release of this element to the soil and facilitate pine persistence (arrows (B) in Fig. 8.1a and b).

Consequences of disturbance for soil biodiversity: the macroinvertebrate example

In the short term, logging activities (particularly if followed by cultivation) can disturb soil and litter faunal communities by physically altering their habitat, increasing the availability of nutrient-rich resources and also by changing microenvironmental conditions due to increased radiation after canopy opening. In the long term (tens of years), the effects of logging often include a delayed response of the soil system to initial disturbance (Bengtsson *et al.*, 1997; Zaitsev *et al.*, 2002) or an indirect consequence of successional changes in the vegetation community composition after abandonment (Switzer and Shelton, 1979; Gross *et al.*, 1995; Fig. 8.1c).

In El Rincón the diversity of higher macroinvertebrate taxa was clearly sensitive to the impact of logging (Fig. 8.3a and b). However, the most conspicuous response to disturbance appears to be delayed in the soil community (but not in the litter). The number of taxa and diversity index in the litter, but not in the soil macroinvertebrate community, were lower in recently logged compared to pristine sites. This was also the case for the individual mean abundances of Chilopoda and Coleoptera larvae in the litter, but no taxon in the soil (Fig. 8.3c and d). The difference may be a reflection of a stronger impact of logging on the more exposed litter community and/or of a vertical migration as a response to the initial environmental perturbation (Zaitsev *et al.*, 2002; Bezkorovainaya and Yashikhin, 2003). An exception to this trend was the Collembola in the litter, which responded quickly to disturbance with a substantial increase in abundance (Fig. 8.3c).

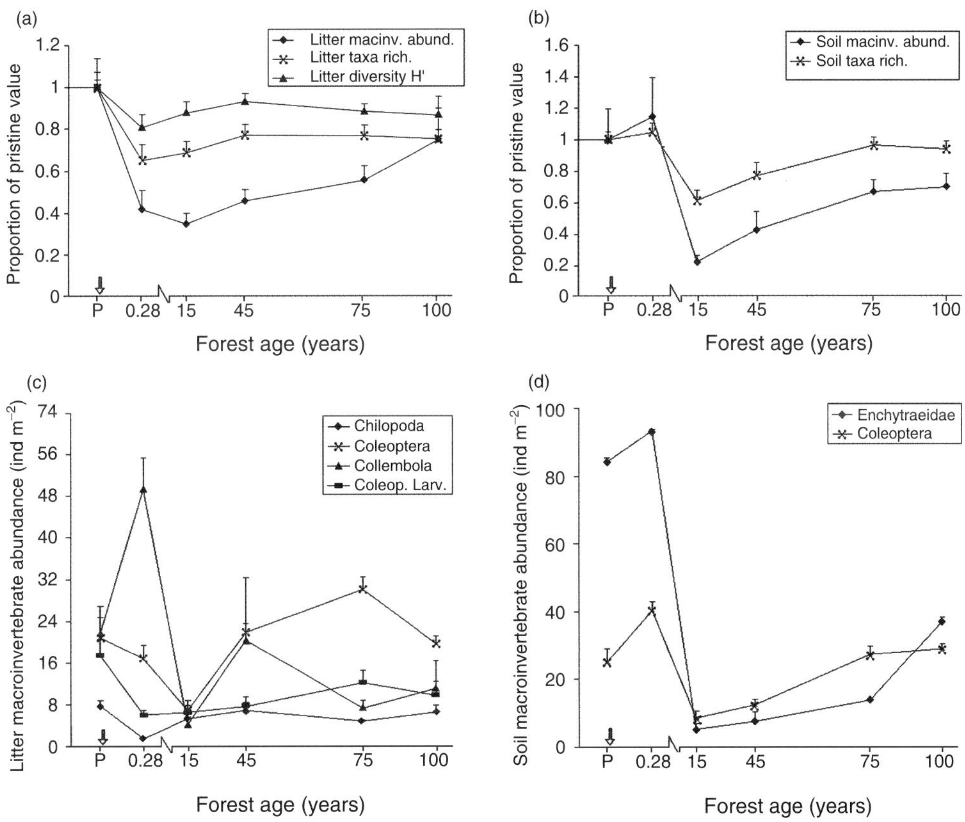

Fig. 8.3. Mean abundances, diversity and richness (± SE) of pristine (n = 2), recently logged (n = 2) and secondary successional forests (n = 3). Values are presented as proportions of mean pristine values in (a) and (b); mean population densities are given in (c) and (d). Only those variables that showed significant differences among forest types or secondary successional stages are presented. P represents the values for pristine forest and the arrow indicates the time of disturbance when forest age equals zero (modified from Negrete-Yankelevich *et al.*, 2007).

In contrast to the response by Collembola, the earthworms were exclusive to the pristine forests (except for six individuals found in secondary forests). Out of the 14 soil samples extracted from the pristine forests, only four did not contain earthworms. Collembola and earthworms could in combination be a valuable target for more detailed conservation research. In particular, the absence of the native worm *Ramiellona willsoni* (the only earthworm species found) from secondary succession could be a result of its low recolonization ability or a product of an extreme sensitivity to differences in environmental or resource conditions between secondary and primary forest.

In early succession, the decline in the diversity of resources available to the soil system, together with the greater abundance of resources, might be expected to lead to a less diverse soil fauna that then recovers gradually as the diversity of resources increases through succession (Fig. 8.1a and c). In the chronosequences of El Rincón, this seems to be the pattern particularly for taxa richness. In the 15-year-old forests, the number of taxa was *c*.60% lower in the litter and *c*.50% lower in the soil compared to pristine sites (Fig. 8.3a and b). Even if the soil macroinvertebrate community appears to become more similar to the pristine community during succession, several diversity and abundance variables in the litter and soil did not recover to the level of pristine forests after 100 years. This was the case for richness and total abundance of litter taxa (both *c*.25% lower in the 100-year-old forests), abundance of litter Coleoptera larvae (44% lower) and abundance of soil Enchytraeidae (56% lower). The total abundance and taxa richness of soil macroinvertebrates increased with succession, but only taxa richness reached the level of pristine sites in the 100-year-old forests. In contrast, Shannon's diversity and the abundance of Chilopoda in the litter had already recovered close to pristine levels in the 45-year-old forests (Fig. 8.3d).

Therefore, even though the tree community composition above ground has been reported to become very similar to pristine conditions after 100 years of succession (Cordova and del Castillo, 2001), the same timescale does not appear to be enough for the macroinvertebrate community to recover its original composition, even in terms of higher taxa. This indicates that for below-ground biodiversity (and possibly for the ecosystem functions performed by the soil system) the consequences of forest disturbance may last longer than for the above-ground community. Further research is required to understand in detail the consequences of disturbance for the soil macroinvertebrate community. However, when conservation decisions concerning TMCFs are made, special attention should be paid to the integrity of the soil community.

Changes in the spatial structuring of the soil system through succession

If trees generate a zone of influence in the soil around their trunks, the decrease in number of tree genera following logging disturbance can be expected to diminish the spatial heterogeneity in the soil at the plot scale.

This should be true for the topsoil in particular, because, in terms of nutrient availability and cycling, the surface soil seems to be where most plant–soil biochemical interactions occur in forests (Hendrick and Pregitzer, 1996). The homogenization in the soil can also be expected to reverse as the tree community becomes more diverse through succession (Fig. 8.1a). These predictions were true in the Juquila chronosequence for the degree of aggregation in litter resources and macroinvertebrate taxa, but not for the majority of nutrients in the soil.

In terms of litter, the more tree species that coexist, the more likely it would be to find patches with distinct proportions of different litter components. This is because, while one tree species may produce more abundant leaf litter, its neighbouring species may lose more twigs or heavier fruits. Consistent with this hypothesis, in Juquila there were more litter components with a patchy distribution in mid and late succession (four or five variables) than in the 15-year-old forest where only *Pinus* needles and other genera displayed spatial structures (Table 8.2). Even if only the litter components that were present in all successional stages (total litter mass, unidentifiable material, woody and reproductive parts and other genera) are considered, in the 15-year-old forest only 'other genera' were structured in space, while in the 100-year-old all of the components were.

Table 8.2. Distance at which spatial independence is attained (range) for litter components and soil nutrient variables in different successional stages. The range was calculated based on model variograms fitted to the data (modified from Negrete-Yankelevich *et al.*, 2006).

	Range (m)[a]			
	15-year-old	45-year-old	75-year-old	100-year-old
Litter components				
Total litter mass (g)	n.a.	n.a.	n.a.	>25
Unidentified (g)	n.a.	>25	16.52	>25
Quercus (g)	–	14.62	17.61	>25
Pinus (g)	>25	28.09	439.41	–
Other genera (g)	55.60	12.09	7.74	n.a.
Lauraceae (g)	–	–	n.a.	11.98
Woody and reproductive (g)	n.a.	>25	n.a.	>25
Soil chemistry				
Total carbon (%)	>25	–	–	–
Ln P (cmol kg^{-1})	>25	0.12	n.a.	>25
Mg^{2+} (cmol kg^{-1})	>25	33.08	n.a.	29.30
Na^{+} (cmol kg^{-1})	18.33	>25	n.a.	>25
K^{+} (cmol kg^{-1})	n.a.	n.a.	>25	–
Ca^{2+} (cmol kg^{-1})	15.07	n.a.	79.98	>25

[a]The range has been indicated as >25 m for variables that do not show a tendency to spatial independence within the studied distance (linear models).
n.a. indicates those variables that had no autocorrelation at the distances studied; – indicates variables that had a variation coefficient smaller than 0.25.

There are two main pathways by which a single tree can influence the nutrient content in the soil around itself. First, litter from different tree species varies in quantity, nutrient quality and litter fall timing (Negrete-Yankelevich, 2004), which are determinant factors for the nutrient release to the soil (Swift *et al.*, 1979). Second, trees regulate the proliferation of fine roots according to nutrient availability in different areas (Burton *et al.*, 2000), consuming and competing with other plants for nutrients in a spatially heterogeneous manner (Day *et al.*, 2003). In Juquila, pine and oak trees seem to be influencing the spatial distribution of nutrients in the soil through the first mechanisms. Pine litter was found to be relatively scarce in Mg, and oak litter in P (Table 8.1). Accordingly, patches of *Quercus* litter in late succession and of *Pinus* litter in early succession appear to coincide with the areas of P and Mg deficit in the soil respectively (Fig. 8.4a–d). Further, it was found that in the 100-year-old forest, the decomposer organisms resident under oak canopies were able to decompose oak leaves incorporating less P from external sources than were the decomposers under the canopies of other tree genera (Negrete-Yankelevich, 2004). In this late successional stage, where the dominance of oak has diminished considerably and the remaining trees from this genus are presumably old, there has been a long time to develop a zone of influence in the soil around individual oaks. This zone of influence might have promoted a decomposer food-web specialized in a more efficient decomposition of phosphorus-poor oak leaves. Therefore, the composition of both the above-ground and below-ground communities may depend on the nutrient environment (in litter and soil) that develops through the long-term interaction between the two communities themselves.

The distribution of macroinvertebrate taxa also suggests the development through succession of a spatial relationship between the above- and below-ground communities. In Juquila, the increase in tree co-dominance through secondary succession was accompanied by an increase in macroinvertebrate mean community diversity (Fig. 8.3a and b). Simultaneously, the members of the macroinvertebrate community seemed to become more frequently aggregated at a 5–25 m scale in the oldest successional stage (Table 8.3). The development of a patchy litter layer by a diverse tree community can promote spatial structuring and high diversity in the invertebrate community through resource partitioning, reducing some competitive pressure (Amarasekare, 2003). In the soil community the mechanism could be mediated by the development of a patchy nutrient environment in the soil through the production of litter of different qualities. In this study the matching spatial distributions of oak litter, phosphorus concentration and Coleoptera larvae in the soil of the old-growth forest give some support to this hypothesis (Fig. 8.4c–e). Therefore evidence suggests that above-ground and below-ground biodiversities develop close interdependence in their composition, spatial structure and function. The homogenization of the organic matter input to the soil that follows the colonization of pines in disturbed forests may threaten the conservation of the spatial structuring of above- and below-ground communities and possibly the conservation of their high biodiversity.

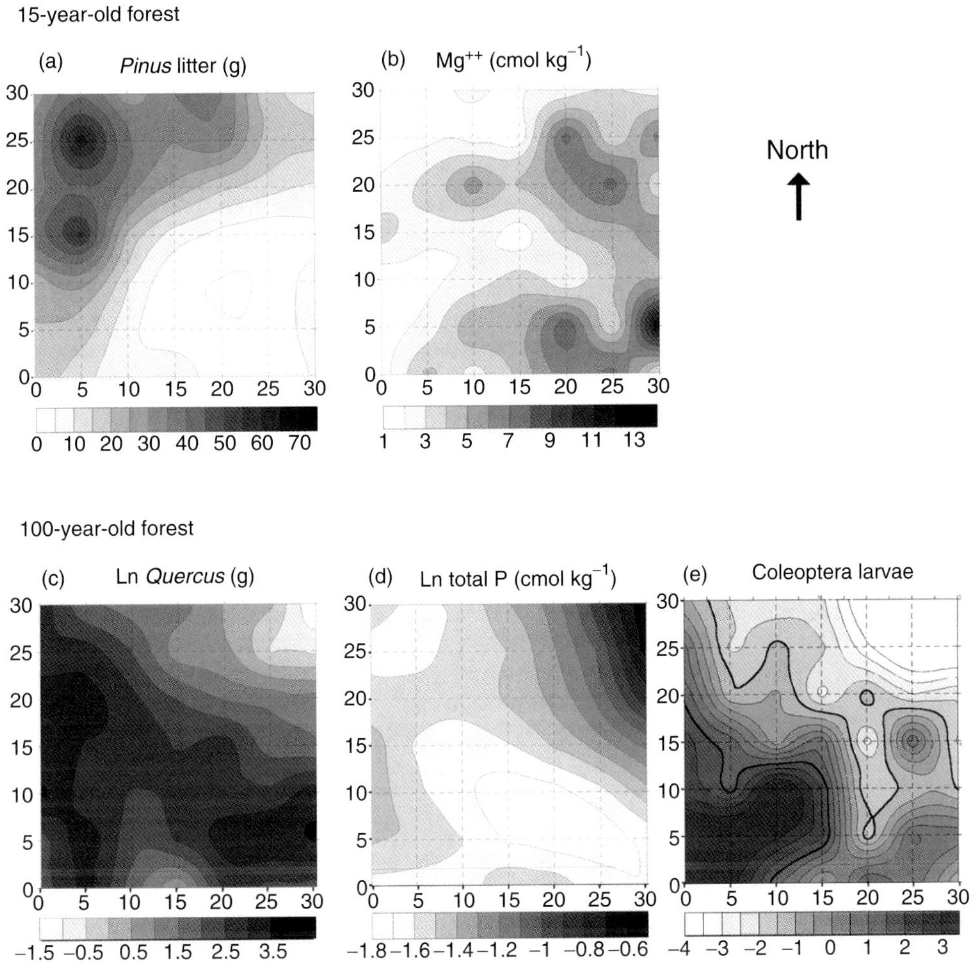

Fig. 8.4. Distribution maps of some of the litter components, nutrients and Coleoptera larvae in the 15- and 100-year-old forests. Litter and nutrient maps are Kriged contours and the Coleoptera larvae map is contours based on υ values. In the υ map dark contours enclose areas where there is more than 50% aggregation ($\upsilon \geq 1.5$) or emptiness ($\upsilon \geq -1.5$) than expected at random. The grid scale is expressed in metres from the coordinate (0,0). Contour maps were filled with a gradient of greys where the minimum value in the map is represented by white and the maximum by black. The names of those variables that have been log-transformed to meet distributional assumptions are preceded by Ln (modified from Negrete-Yankelevich *et al.*, 2006).

Conclusion

Results from the below-ground study in El Rincón should be useful for decision making in the conservation of the TMCF in Oaxaca. Low intensity selective logging (sometimes followed by a few years of agriculture) compromises the compositional and spatial components of biodiversity

Table 8.3. Per cent of aggregation of macroinvertebrate taxa in the soil and litter of different successional stages.

	Aggregation (%)[a]			
	15-year-old	45-year-old	75-year-old	100-year-old
Litter				
Chilopoda	n.a.	n.a.	n.a.	n.a.
Diplopoda	n.a.	n.a.	n.a.	n.a.
Coleoptera	n.a.	n.a.	n.a.	68
Formicidae	n.a.	n.a.	n.a.	n.a.
Diplura	n.a.	n.a.	n.a.	n.a.
Coleop. larvae	n.a.	94	n.a.	48
Other larvae and pupae	n.a.	n.a.	n.a.	n.a.
Soil				
Chilopoda	n.a.	n.a.	64	12
Diplopoda	n.a.	n.a.	n.a.	82
Coleoptera	n.a.	n.a.	n.a.	51
Formicidae	n.a.	n.a.	n.a.	21
Diplura	n.a.	n.a.	n.a.	n.a.
Coleop. larvae	n.a.	n.a.	n.a.	23
Other larvae and pupae	n.a.	n.a.	n.a.	n.a.

[a]The per cent of aggregation is based on the value of the SADIE Index of Aggregation (I_a) or Index of Clustering (J_a) when these were significant after a randomization test. See Negrete-Yankelevich *et al.* (2006) for details.
n.a. = not significantly aggregated.

above and below ground and full recovery may take more than 100 years. The macroinvertebrate community composition in both recently logged sites and pristine forests are distinct compared to secondary successional stages. The complex spatial structures developed in old forests may be essential for the maintenance of a fully functional and diverse soil system. Therefore, the homogenization of the canopy is likely to be a threat to the conservation of these forests and all efforts should be put in place to control sustained disturbance, even if it is of low intensity. Of particular importance was the presence of pine trees. Even if they did not always dominate the basal area of the forest, pine litter conspicuously dominated early and mid-successional stages. The distribution of pine litter in early succession, as well as that of oak litter in old-growth forests, seems to be correlated with the spatial distribution of nutrients in the soil. Finally, this study has succeeded in directing future research towards vulnerable taxa and processes that may be crucial to sustain the diverse, slow-growing and nutrient-poor character of the TMCF in Mexico. In the absence of further information, conservation measures should pay particular attention to the native species of the worm *R. willsoni* that seems not to be able to re-establish in secondary forest, and to the release of P to the soil by continuous disturbance. The persistence of late successional tropical trees in these forests may partly be associated with their ability to compete in the noticeably phosphorus-poor soils of mature forests.

References

Amarasekare, P. (2003) Competitive coexistence in spatially structured environments: a synthesis. *Ecology Letters* 6, 1109–1122.

Bautista-Cruz, A. and del Castillo, F. (2005) Soil changes during secondary succession in a tropical montane cloud forest area. *Soil Science Society of America Journal* 69, 906–914.

Bengtsson, J., Persson, T. and Lundkvist, H. (1997) Long-term effects of logging residue addition and removal on macroarthropods and enchytraeids. *Journal of Applied Ecology* 34, 1014–1022.

Benjamini, Y. and Hochberg, Y. (1995) Controlling the false discovery rate: a practical and powerful approach to multiple testing. *Journal of Royal Statistical Society B* 57, 289–300.

Bezkorovainaya, I.N. and Yashikhin, G.I. (2003) Effects of soil hydrothermal conditions on the complexes of soil invertebrates in coniferous and deciduous forest cultures. *Russian Journal of Ecology* 34, 52–58.

Blanco-Macías, A.M. (2001) Análisis sucesional del bosque mesófilo de montaña en El Rincón, Sierra Norte de Oaxaca. Licenciatura thesis. Facultad de Estudios Superiores Iztacala, Universidad Nacional Autónoma de México (UNAM), Mexico City, Mexico.

Brown, G.G., Fragoso, C., Barois, I., Rojas, P., Patrón, J.C., Bueno, J., Moreno, A.G., Lavelle, P., Ordaz, V. and Rodríguez, C. (2001) Diversidad y rol funcional de la macrofauna edáfica en los ecosistemas tropicales mexicanos. *Acta Zoologica Mexicana (Xalapa, Mexico) N*, spec 1, 79–110.

Bruijnzeel, L.A. and Proctor, J. (1995) Hydrology and biochemistry of tropical montane cloud forest: what do we really know? In: Hamilton, L.S., Juvik, J.O. and Scatena, F.N. (eds) *Tropical Montane Cloud Forest*. Springer, New York, pp. 38–78.

Bruijnzeel, L.A. and Veneklaas, E.J. (1998) Climatic conditions and tropical montane forest productivity: the fog has not lifted yet. *Ecology* 79, 3–9.

Burton, J.A., Pregitzer, K.S. and Hendrick, R.L. (2000) Relationship between fine root dynamics and nitrogen availability in Michigan northern hardwood forests. *Oecologia* 125, 389–399.

Challenger, A. (1998) *Utilización y Conservación de los Ecosistemas Terrestres de México. Pasado, Presente y Futuro*. Comisión Nacional para el Conocimiento y Uso de la Biodiversidad (CONABIO), Instituto de Biología, Universidad Nacional Autónoma de México (UNAM), Mexico City, Mexico.

Cordova, J. and del Castillo, R.F. (2001) Changes in epiphyte cover in three chronosequences in a tropical montane cloud forest in Mexico. In: Gottsberger, G. and Liede, S. (eds) *Life Forms and Dynamics in Tropical Forests*. J. Cramer in der Gebrüder Borntraeger Verlagsbuchhandlung, Berlin–Stuttgart, Germany, pp. 1–16.

Davies, R.G., Eggleton, P., Dibog, L., Lawton, J.H., Bignell, D.E., Brauman, A., Hartmann, C., Nunes, L., Holt, J. and Rouland, C. (1999) Successional response of a tropical forest termite assemblage to experimental habitat perturbation. *Journal of Applied Ecology* 36, 946–962.

Day, K.J., Hutchings, M.J. and John, E.A. (2003) The effect of spatial pattern of nutrient supply on the early stages of growth in plant populations. *Journal of Ecology* 91, 305–315.

Finér, L., Mannerkoski, H., Piirainen, S. and Starr, M. (2003) Carbon and nitrogen pools in an old growth Norway spruce mixed forest in eastern Finland and changes associated with clear-cutting. *Forest Ecology and Management* 174, 51–63.

Fragoso, C. and Reynolds, J.W. (1997) On some earthworms from central and southern Mexican mountains, including two new species of the genus *Dichogaster* (Dichogastrini). *Megadrilogica* 7, 9–19.

Galindo-Jaimes, L., González-Espinosa, M., Quintana-Ascencio, P. and García-Barrios, L. (2002) Tree composition and structure in disturbed stands with varying dominance by *Pinus* spp. in the highlands of Chiapas, México. *Plant Ecology* 162, 259–272.

González-Espinosa, M., Quintana-Ascencio, P., Ramírez-Marcial, N. and Gaytán-Guzmán, P. (1991) Secondary succession in disturbed *Pinus–Quercus* forests in the highlands of Chiapas, Mexico. *Journal of Vegetation Science* 2, 351–360.

Gross, K.L., Pregitzer, K.S. and Burton, J.A. (1995) Spatial variation in nitrogen availability in three successional plant communities. *Journal of Ecology* 83, 357–367.

Hendrick, R.L. and Pregitzer, K.S. (1996) Temporal and depth-related patterns of fine root dynamics in northern hardwood forest. *Journal of Ecology* 84, 167–176.

Hobbie, S.E. (1992) Effects of plant species on nutrient cycling. *Trends in Ecology and Evolution* 7, 336–339.

Lavelle, P. (2000) Ecological challenges for soil science. *Soil Science* 165, 73–86.

Magurran, A.E. (1996) *Ecological Diversity and its Measurement*. Chapman and Hall, London, UK.

Negrete-Yankelevich, S. (2004) Integrating soil macroinvertebrate diversity, litter decomposition and secondary succession in a tropical montane cloud forest in México. PhD thesis. University of Edinburgh, Edinburgh, UK. Available at: http://www.era.lib.ed.ac.uk/handle/1842/592

Negrete-Yankelevich, S., Fragoso, C., Newton, A.C., Russell, G. and Heal, O.W. (2006) Spatial patchiness of litter, nutrients and macroinvertebrates during secondary succession in a tropical montane cloud forest in Mexico. *Plant and Soil* 286, 123–139.

Negrete-Yankelevich, S., Fragoso, C., Newton, A.C. and Heal, O.W. (2007) Successional changes in soil, litter and macroinvertebrate parameters following selective logging in a Mexican cloud forest. *Applied Soil Ecology* 35, 340–355.

Nilsson, L.O., Huttl, R.F. and Johansson, U.T. (1995) *Nutrient Uptake and Cycling in Forest Ecosystems*. Kluwer Academic, Dordrecht, The Netherlands.

Olsson, B.A., Bengtsson, J. and Lundkvist, H. (1996a) Effect of different forest harvest intensities on the pools of exchangeable cations in coniferous forest soils. *Forest Ecology and Management* 84, 135–147.

Olsson, B.A., Staaf, H., Lundkvist, H., Bengtsson, J. and Rosén, K. (1996b) Carbon and nitrogen in coniferous forests soils after clear-felling and harvests of different intensity. *Forest Ecology and Management* 82, 19–32.

Perry, J.N. (1998) Measures of spatial patterns for counts. *Ecology* 79, 1008–1017.

Pietikäinen, J., Haimi, J. and Siitonen, J. (2003) Short-term responses of soil macroarthropod community to clear felling and alternative forest regeneration methods. *Forest Ecology and Management* 172, 339–353.

Quintana-Ascencio, P. and González-Espinosa, M. (1993) Afinidad fitogeográfica y papel sucesional de la flora leñosa de los bosques de pino-encino de los altos de chiapas, Mexico. *Acta Botánica Mexicana* 21, 43–57.

Ramírez-Marcial, N., González-Espinosa, M. and Williams-Linera, G. (2001) Anthropogenic disturbance and tree diversity in montane rain forest in Chiapas, Mexico. *Forest Ecology and Management* 154, 311–326.

Römbke, J. (2003) The role of Gilberto Rhighi in the development of tropical Microdrile taxonomy. *Pedobiologia* 47, 405–412.

Romero-Nájera, I. (2000) Estructura y condiciones microambientales en bosques perturbados de los altos de Chiapas, Mexico. Undergraduate thesis. Universidad Nacional Autónoma de México (UNAM), Mexico City, Mexico.

Rossi, R.E., Mulla, D.J., Journel, A.G. and Franz, E.H. (1992) Geostatistical tools for modeling and interpreting spatial dependence. *Ecological Monographs* 62, 277–314.

Siira-Pietikäinen, A., Pietikäinen, J., Fritze, H. and Haimi, J. (2001) Short-term responses of soil decomposer communities to forest management: clear felling versus alternative forest harvesting. *Canadian Journal of Forest Research* 31, 88–99.

Swift, M.J., Heal, O.W. and Anderson, J.M. (1979) *Decomposition in Terrestrial Ecosystems*. Blackwell, Oxford, UK.

Switzer, G.L. and Shelton, M.G. (1979) Successional development of the forest floor and soil surface on upland sites of the east gulf coastal plain. *Ecology* 60, 1162–1171.
Tanner, E.V.J., Vitousek, P.M. and Cuevas, E. (1998) Experimental investigation of nutrient limitation of forest growth on wet tropical mountains. *Ecology* 79, 10–22.
Vitousek, P.M. (1984) Litterfall, nutrient cycling, and nutrient limitation in tropical forests. *Ecology* 65, 285–298.
Walker, L.R., Zimmerman, J.K., Lodge, D.J. and Guzmán-Grajales, S. (1996) An altitudinal comparison of growth and species composition in hurricane-damaged forests in Puerto Rico. *Journal of Ecology* 84, 877–889.
Wardle, D.A. (1992) A comparative assessment of factors which influence microbial biomass carbon and nitrogen levels in the soil. *Biological Review* 67, 321–358.
Wardle, D.A., Verhoef, H.A. and Clarholm, M. (1998) Trophic relationships in the soil microfood-web: predicting the responses to a changing global environment. *Global Change Biology* 4, 713–727.
Webster, R. and Oliver, M.A. (1990) *Statistical Methods for Land Resource Survey*. Oxford University Press, Oxford, UK.
Zaitsev, A.S., Chauvat, M., Pflug, A. and Wolters, V. (2002) Oribatid mite diversity and community dynamics in a spruce chronosequence. *Soil Biology and Biochemistry* 35, 1919–1927.

9 Applying Succession Models to the Conservation of Tropical Montane Forest

D. Golicher and A.C. Newton

Tropical montane forest in the Highlands of Chiapas, Mexico. Photo: Luis Cayuela

Summary

Forest succession models have been widely used to simulate long-term vegetation dynamics and to provide insights into successional processes. However, they have rarely been used to address questions relating specifically to forest conservation. In this chapter we briefly discuss the history and application of models of forest succession. We then demonstrate how a forest succession model can be developed from quite simple principles and equations, and applied to modelling of complex forest dynamics. The application of an individual tree-based gap model is illustrated with reference to the specific case of montane forests in Chiapas, Mexico. Results from this modelling exercise suggested that the recovery rate of tropical montane forest following disturbance is likely to be very low, even when a source of colonists is assumed to exist nearby. Given their relatively low rates of colonization and growth, results suggest that it may take several hundred years to re-establish a canopy dominated by shade-tolerant tree species. The model also provides insights into the factors influencing the relative dominance of pine and oak in montane forests of Chiapas, an issue of conservation concern. Model simulations indicated that the relative abundance of pine and oak within forest stands can be understood as a function of the anthropogenic disturbance regime, enabling theories of recent pine colonization to be refined. Results highlight the sensitivity of tropical montane forest to anthropogenic disturbance, and highlight the urgent need to conserve those remaining forest fragments that are relatively undisturbed.

Introduction

Computer models provide a structure through which data, observations and assumptions can be combined and explored. As they can add considerable value to existing information, computer models have become important tools used in forest research. They can be used to provide more accurate or more detailed predictions, or more reliable insights into the consequences of assumptions regarding how a forest system functions. Two main types of forest model may be differentiated: growth and yield models, and ecological models (Newton, 2007). Growth and yield models are generally produced by empirically deriving equations that describe the relationships between stand density, stem diameter (dbh) and tree height using standard statistical procedures such as regression (Vanclay, 1995). Ecological models investigate forest dynamics from an ecological perspective, or simulate ecological processes or characteristics of forests (Newton, 2007). Such models may be classified in a variety of different ways (see, for example, Liu and Ashton, 1995; Shugart, 1998; Porte and Bartelink, 2002). Forest succession models can be considered as a sub-class of ecological models that are designed to simulate long-term vegetation dynamics. They are usually applied in a semi-theoretical context with the aim of providing greater insights into successional processes and their influence on forest structure and composition, suggesting new hypotheses for investigation or providing new ways of exploring existing data.

Practical forest conservation involves identifying priorities for action, and identifying potential trade-offs between intervention options. These pragmatic issues sometimes appear far removed from the technical concerns of forest modellers. However, planning the conservation of a naturally forested landscape requires knowledge that can often be derived from models of successional processes. Decision makers may require tools that can help to conceptualize the complex ecology of natural forests. An important concern is the relationship between human activity and biological diversity. Under what circumstances might it be desirable to prevent or modify human activities that affect the functioning of a forest as an ecological system? Is it possible to find ways in which the spatial and temporal characteristics of human disturbances match those produced by natural processes? Succession models can potentially help to answer these questions by providing a source of evidence to inform the debate, although in practice such models have very rarely been applied specifically to questions relating to forest conservation (Newton, 2007). Despite their promise for this type of application, it is important to remember that evidence derived from models must also be combined with empirical observations wherever possible.

In this chapter we briefly discuss the history and application of models of forest succession. We then demonstrate how a forest succession model can be developed from quite simple principles and equations, and applied to modelling complex forest dynamics. The application of an individual tree-based gap model is illustrated with reference to the specific case of montane forests in Chiapas, Mexico. Gap models can be applied to the prediction of forest yield under different silvicultural treatments over comparatively short (<30 years) periods (Vanclay, 1995). For such an application, highly accurate modelling of individual tree growth, as determined by inherent growth potential and position in the canopy, is needed. The field data to achieve this are usually derived from long-term monitoring of forest plots. However, gap models can also be used to understand the long-term consequences of disturbance. In this case more general simulations are appropriate, making quantitative precision, while desirable, rather less important. The model we present was built and parameterized in this context.

History of Models of Forest Succession

Understanding forest change involves confronting a unique set of challenges associated with describing complex, often non-linear, dynamics. The search for conceptual models that predict successional trends has a very long history. Clements (1916) suggested fundamental predictability from simple holistic principles. This was influential in forming the vocabulary used to describe vegetation processes. However, Clements' concepts were criticized from their inception as leading to over-generalized statements that failed to emphasize linkages between pattern and process (Watt, 1947). Detailed, species- and site-specific information seems to be necessary in order to build effective predictive models representing the behaviour of complex vegetation systems (Gleason, 1917, 1939; Tansley, 1935). Commentators and reviewers have continued to raise fundamental objections to a simple and orderly concept of successional change (Drury and Nisbet, 1973; Picket, 1976; Connell and Slatyer, 1977; McIntosh, 1981; Peet and Christensen, 1988). The legacy of this debate has been that early terms such as *climax* communities fell into disuse, although descriptors such as *primary* and *secondary* forest, which are associated with a linear view of succession and a particular type of disturbance, continue to be freely used in the literature, even though they tend to oversimplify a complex system.

When vegetation is viewed as a dynamic system, the artificial division between individualistic and holistic views is replaced by a more coherent synthesis. In effect, computer simulation modelling helped convert Gleason's individualistic perspective into an operational tool (Shugart and O'Neill, 1979; Huston, 1994; Acevedo *et al.*, 1995; Bazzaz, 1996; Shugart, 1998; Urban *et al.*, 1999). Models can be used to extract generality from case studies, and to find common patterns that can be linked to repeated processes (Levin, 1992; Shugart, 1998). The aim is to identify simplifications that reduce complexity to tractable levels (Botkin, 1993b).

Individual-based forest simulators known as 'gap models' have proven to be very influential in the development of contemporary views of forest dynamics. The earliest such model to successfully capture interactions in a mixed forest was the JABOWA gap model (Botkin *et al.*, 1972a, b), although Ek and Monserud's (1974) FOREST model was developed simultaneously and in many respects foreshadowed the development of more complex spatially explicit simulators. The term individual-based model (IBM) was formalized by Huston *et al.* (1988). Individual-based modelling acknowledges two fundamental biological principles. The first is that individual organisms are all potentially distinct owing to genetic or environmental influences. The second is that interactions between individuals are inherently local. Sedentary organisms such as trees are influenced mainly by other nearby sedentary organisms. IBMs can be contrasted with some other detailed forest simulation models in which the numbers of trees in size classes is used as a state variable (Bossel and Krieger, 1994; Vanclay, 1994).

The simulation of many individual organisms places considerable demands on computational resources (Bugmann, 1996). It also leads to detailed output, which requires additional routines to produce automated summaries. However, the equations used by IBMs can be relatively simple. This simplicity arises from the fact that individuals can usually be represented as having a limited set of key properties that determine the outcome of a limited set of key processes (Judson, 1994). The link between processes and properties can often be expressed as mathematical equations with a small number of terms, or as logical rules that apply to specific situations arising during the lifetime of an individual organism. Although the formal representations of processes that change an individual's properties may be intuitive, when they are applied iteratively to many individuals over time even apparently simple IBMs can generate phenomenologically realistic and often complex behaviour (DeAngelis *et al.*, 1984; Huston *et al.*, 1988). Highly detailed individual tree models are not usually IBMs *sensu* Huston *et al.* (1988). Forest IBMs are *based* on individuals. They are not models *of* individuals (see Deutschman *et al.*, 1995).

The Study Area

The Highlands of Chiapas are a biologically diverse region extending over 11,000 km^2 that include 30% of some 9000 vascular plant species that are native to Chiapas (Breedlove, 1981). Several forest types have been identified, including oak, pine–oak, pine and evergreen cloud forests (Miranda, 1952; Breedlove, 1981; González-Espinosa *et al.*, 1991; Rzedowski, 1991). Traditional agricultural practices have produced a mosaic landscape of forest fragments embedded in a matrix of secondary vegetation and crop fields (Ramírez-Marcial *et al.*, 2001; Galindo-Jaimes *et al.*, 2002). Changes in land use, particularly during the past three decades, have accelerated deforestation and disturbance (Ochoa-Gaona and González-Espinosa, 2000; Cayuela *et al.*, 2006a). Almost all forest in the region has arisen from natural regeneration following long-term anthropogenic disturbance at a range of scales, with intact undisturbed mature forest being restricted to a few inaccessible fragments. There are only a few, very small, localized examples of forest plantations. Recent studies of the forests of the region are beginning to tease apart the complex way in which human disturbance, natural disturbance, soils and climate interact to shape forest composition (see Chapters 3 and 10). Gap models can support this research by forming a point of reference for theories regarding the ecological impacts of human disturbance.

Model Structure

In regions such as the Highlands of Chiapas, high-quality measurements for parameterizing complex physiologically based models have not been made to date. The long-term data sets from permanent plots that are needed in order to verify forest yield models are also not yet available. Fortunately, classic gap models can be parameterized using relatively simple measurements and observations. We matched model structure with available data by programming our own model based on the JABOWA–FORET class of gap models in the open-source R language (see The R Foundation for Statistical Computing, http://www.r-project.org/). R is a programming environment that has many advantages for modelling. Although it is a high-level language, R code that takes a fully vector-based approach can run quickly by using calls to underlying functions written in C. R is also a powerful environment for statistical analysis and graphics. This means that many tools for analysing and visualizing model output are available within the environment in which the model runs.

In common with many other individual-based forest stand simulators, our model uses a species-specific function that predicts the expected diameter increment for a tree of a given diameter under optimal growth conditions. The model follows JABOWA–FORET in using the fundamental growth equation given by Botkin (1993a) as:

$$\delta D = \frac{GD(1-(D(137+b_2D-b_3D^2)/D_{max}H_{max}))}{274+3b_2-4b_3D^2} \quad (9.1)$$

where D is diameter at a height of 137 cm (breast height), G is a species-specific constant, H_{max} is the maximum height in cm the species reaches and D_{max} is its maximum diameter, and b_2 and b_3 are allometric constants linking diameter with height. Modelled individuals do not grow at this optimum, owing to constraints imposed by shading, temperature, water or nutrient availability.

It is assumed that any tree growing without appreciable competition will reach the maximum observed diameter increment for its species δD_{max} at some point during its lifetime. The value for the growth parameter G used in the JABOWA–FORET models may be calculated using the approximation

$$G \cong 5H_{max}\left(\frac{\delta D_{max}}{D_{max}}\right) \tag{9.2}$$

where H_{max} = the maximum height obtained during the tree's lifetime, D_{max} = the maximum observed diameter and δD_{max} = the width of the widest annual diameter increment if δt is set to 1 year. Botkin (1993a) comments that this formulation is preferable because it is linked to observation. The use of these two very easily obtained measurements has practical advantages for building forest growth simulators in areas with limited data availability.

Trees typically interact with their neighbours through competition for light. Much debate regarding forest models has revolved around whether the precise spatial position of an individual tree must be known in order to produce realistic behaviour (Pacala and Deutschman, 1995; Deutschman *et al.*, 1997). Descriptions of forests as a mosaic of gaps and non-gaps (Watt, 1925, 1947; Shugart, 1984; Hubbell and Foster, 1986; Whitmore, 1989) suggest a natural framework for modelling localized interactions. A gap or patch model assumes that, although spatial heterogeneity is important in structuring forests, sufficient detail can be captured by dividing the stand into arbitrarily small units within which the position of the modelled individual is unimportant (Botkin *et al.*, 1972a, b; Shugart and West, 1980; Shugart, 1984; Urban *et al.*, 1991; Solomon and Cramer, 1993). The term patch model is in many ways a more appropriate description of the JABOWA–FORET class of models, but the use of the description 'gap model' is now so well established that it is retained here. Gap models can be contrasted with a more detailed form of spatial representation in which the precise position of every tree is known. The best-known recent model of this type is SORTIE (Pacala *et al.*, 1996), but explicit tree positions have also been used in models by Luan (1994) and Young (1998), among others.

Gap models use a simple allometric relationship for estimating the shading effect of individual tree canopies. It is assumed that total leaf area is proportional to the diameter squared of the tree (Shugart, 1984):

$$La = C_{leaf}D^2 \tag{9.3}$$

where La is total leaf area, C_{leaf} is a constant of proportionality and D is diameter at breast height. C_{leaf} can be estimated from data on crown dimensions and leaf area index (L). If leaf area index is assumed to be a species-specific constant, Eqn 9.3 becomes:

$$La = C_{canopy}LD^2 \tag{9.4}$$

where C_{canopy} is also some constant. These simplifications ignore a great deal of the complexity of tree form.

Canopy light transmission is typically calculated in a gap model using the Monsi–Saeki equation:

$$I_h = I_0 \exp(-kL_h) \tag{9.5}$$

where I_h is the light at some height h in the canopy, L_h is the projected leaf area index for all canopy elements above height h and I_0 is the incident light at the top of the canopy. Most gap models make the simplifying assumption that k is constant for all canopies (Shugart, 1984; Urban and Shugart, 1992). However, as Urban and Shugart (1992) point out, 'A smaller extinction coefficient makes the canopy leaky and reduces the asymmetry of light competition, while a larger coefficient strengthens this asymmetry. Total stand level productivity is also sensitive to this coefficient.' Direct measurement of the canopy transmission coefficient for a mixed forest is challenging and it is this element of the traditional formulation of gap models that has been most criticized. The algorithm used in our model places the trees in order of height within each patch and calculates the leaf area for each tree using Eqn 9.4. A cumulative summation of the leaf area is then used to calculate the relative proportion of available light received by each individual tree from Eqn 9.5. This is based on the assumption that all leaf area is concentrated at the top of the tree. This is obviously unrealistic as a model of competition for light alone, but seems to reproduce more complex features of competition, perhaps being a case of 'the right results for the wrong reasons' (Pacala *et al.*, 1994).

A continuum of whole tree growth response to light availability can be modelled using the equation

$$f(A) = c_1 \left(1 - e^{-c_2(A - c_3)}\right) \tag{9.6}$$

where $f(A)$ is the light response function, A is the available light and c_1, c_2 and c_3 are constants. This poses a further challenge for fully empirical parameterization. Light response curves for whole trees are not well known. Gap models such as FORET have been parameterized with reference to 'the foresters concept of tolerance' (Shugart, 1984), assuming that species' empirical response to light under field conditions are understood by experienced silviculturalists (Oliver and Larson, 1996). For our model we interpreted the results of field and greenhouse trials (see Chapter 15) to produce the assumed growth response with respect to overall light availability shown in Fig. 9.1.

Shading not only leads to poor growth, but is also assumed to be a cause of mortality in gap models. A very useful study of comparative rates of sapling mortality under naturally occurring conditions was carried out by Kobe (1996). Such studies show that the trade-off between fast growth under high light conditions and mortality in shade may be a key factor shaping undisturbed forest dynamics. Few saplings of shade-intolerant trees survive over 5 years if growing at below 20% of their maximum potential. Saplings of shade-tolerant trees do, however, survive almost indefinitely in

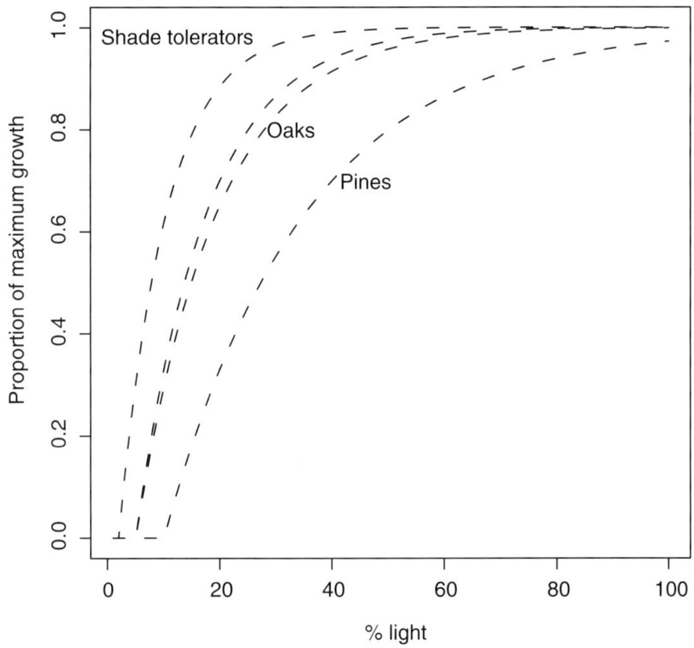

Fig. 9.1. Light response curves for the groups of species used in the gap model.

deep shade, though without achieving a substantial amount of growth. The rule for density-dependent mortality incorporated in this model was:

$$\text{p(mortality} \mid \text{diameter increment} < \text{critical increment)} = 0.15 \tag{9.7}$$

where the critical increment was the width of the smallest growth ring found. Assuming a rule-based critical increment model has the advantage of being a simple, easily measured parameter for most species that form annual rings and relatively easily estimated for those that do not.

The parameter, AGEMAX, represents the maximum expected age of a tree. This is used to estimate non-density-dependent mortality by assuming a small proportion (1%) of trees reach this age. From Shugart (1998) this gives:

$$Pm = 1 - \mathrm{e}^{\frac{-4.605}{\mathrm{AGEMAX}}} \tag{9.8}$$

Establishment can be modelled in many different ways within the gap model paradigm. A typical simple approach is to allow the same number of new stems of each species to attempt to establish in a patch each year. Actual establishment is then determined by the conditions within the patch. Facilitation (Clements, 1928; Connell and Slatyer, 1977) can also be incorporated into successional models and has been defined by Glenn-Lewin and van der Maarel (1992) as describing a 'situation in which one or more species enable the growth or development of other species'. This definition recognizes that the processes involved may be extremely complex and cannot be easily summarized in a single term (Walker and Chapin, 1987; Bazzaz, 1996). Field observations and experiments (see Chapters 10 and 15) in the region for which the model has been designed have shown clearly that completely unshaded conditions are extremely stressful for juveniles of many tree species, mainly due to hydric and thermal stress during the dry, sunny months between November and March. Frost is also common in open areas during this part of the year. This led us to incorporate a simple establishment rule into the model. The rule is based on the proportion of ambient light received at ground level in the model. Species are placed into three classes with minimum and maximum requirements for light at ground level. Light at ground level is considered as a surrogate for a range of factors involving canopy development. A class of heliophiles such as pines and light-demanding shrubs and small trees can establish at light availabilities of between 35% and 100%. Slightly more exposure-sensitive species such as oaks are assumed to establish at light availabilities between 15% and 95%. Shade-tolerant species of the forest interior are assumed to be able to establish at light availabilities of between 10% and 40%.

It is clear from this description that a classic gap model such as ours does not represent the full complexity involved in competition between trees. Some arbitrary decisions must be made regarding their structure and parameterization that can alter long-term model behaviour. Nevertheless, exploration of such a model can provide useful lessons for conservation as we illustrate below through a series of simulations.

Simulations

In order to use our model, we set the parameters from measurements derived at a study site in the east of the region (16° 31′ N, 92° 00′ W) where human disturbance appeared to have resulted in a shift from a diverse broadleaved forest to a structurally complex, yet biologically less diverse pine–oak forest. We took field measurements to obtain parameters of two species of pine, *Pinus maximinoi* and *Pinus oocarpa*, and two species of oak, *Quercus segoviensis* and *Quercus cripipilis*. In addition, we added three idealized functional groups in order to represent the more diverse elements of mixed forests. These were: (i) heliophilic shrubs such as *Crataegus pubescens* and *Baccharis vaccinioides*; (ii) understorey trees such as *Cornus* species; and (iii) shade-tolerant and shade-forming late-successional species such as *Magnolia* species. The simulations are meant to be applicable to a wider region in which many forests are in transition between evergreen cloud forest and pine–oak forest. Table 9.1 presents the basic parameters used in the simulations.

We created three simulations using this parameter set, running models for 200 years, beginning with a completely clear area of potential forest of 1 ha divided into 100 independent 100 m^2 patches. This level of internal replication was sufficient for the model to produce very consistent behaviour. No barriers to seed dispersal were assumed and it was also assumed that forest clearing did not lead to any degradation of soil properties or other barriers to forest re-establishment.

Table 9.1. Parameters used in model simulations.

Name	Dmax	Hmax	AgeMax	IncMax	LAI	CLeaf	C1	C2	C3	CritIncrement	Establishment	Colonization
Pinus oocarpa	100	3400	110	1.34	3	0.02	1	0.1	4	0.2	1	30
Pinus maximinoi	100	3200	120	1.8	3	0.02	1	0.1	4	0.2	1	30
Quercus segoviensis	80	2200	200	0.67	4	0.032	1	0.05	7	0.05	2	20
Quercus crispipilis	90	2800	220	0.74	4	0.03	1	0.05	8	0.05	2	20
Heliophilic shrubs	5	500	10	2.34	3	0.02	1.5	0.25	4	0.1	1	100
Understorey	20	1000	40	0.51	5	0.02	1	0.02	12	0.1	3	20
Generic broadleaf	100	2000	200	0.51	6	0.03	1	0.02	12	0.1	3	10

Simulation 1. The first simulation included no external disturbance. Gaps are created in the model when trees die from natural causes. Figure 9.2 summarizes the results. Maximum basal area of 30 m^2 ha^{-1} occurred after around 50 years. At this stage pines contributed most of the basal area, although at this point there were more oak stems than pine stems, as oak had become the major component of the understorey. After 100 years of forest succession, pine-dominated forest was replaced by a more stable oak-dominated forest. Other large shade-tolerant broadleaved species slowly began to contribute to total basal area during this phase. In addition, shorter 'red' oaks such as *Quercus segoviensis* were gradually replaced by taller 'white' oaks such as *Quercus crispipilis*. A three-dimensional picture of the forest structure after 200 years of succession is shown in Fig. 9.3.

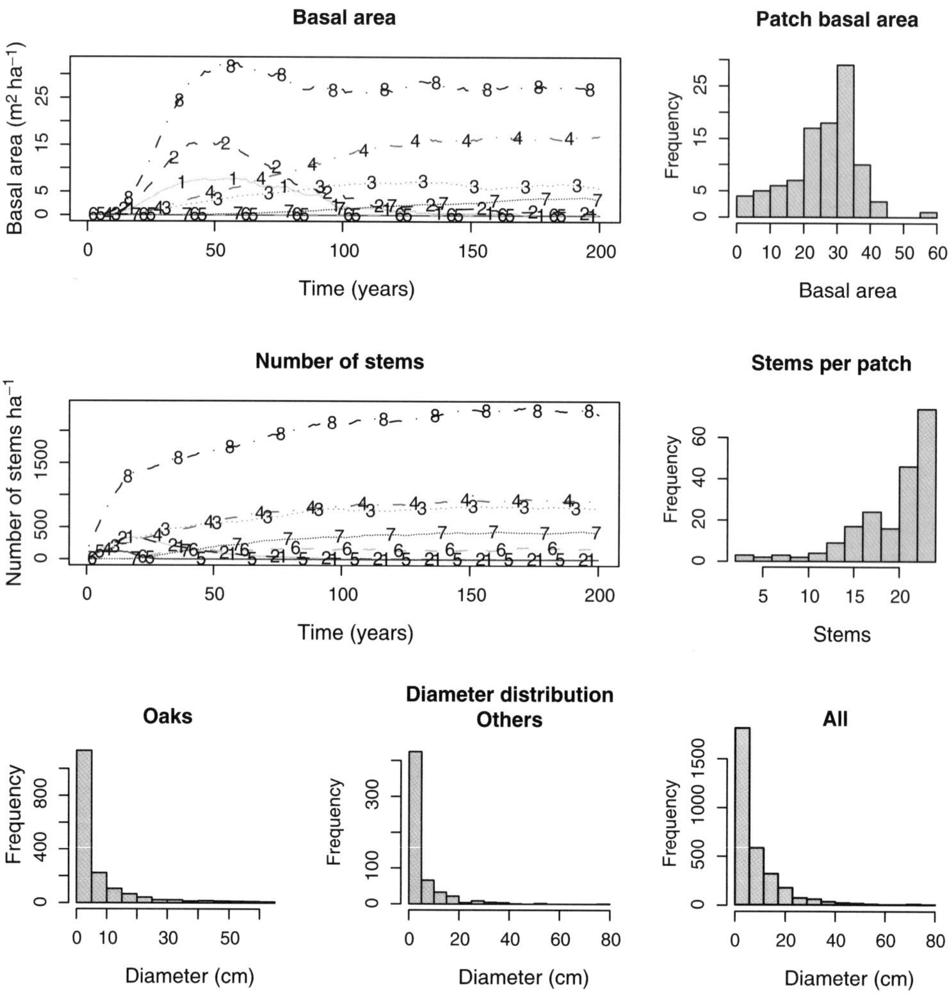

Fig. 9.2. Dynamics and final forest structure produced under a simulation with no disturbance (simulation 1). 1, *Pinus oocarpa*; 2, *Pinus maximinoi*; 3, *Quercus segoviensis*; 4, *Quercus crispipilis*; 5, heliophilic shrubs; 6, understorey; 7, generic broadleaf; 8, total.

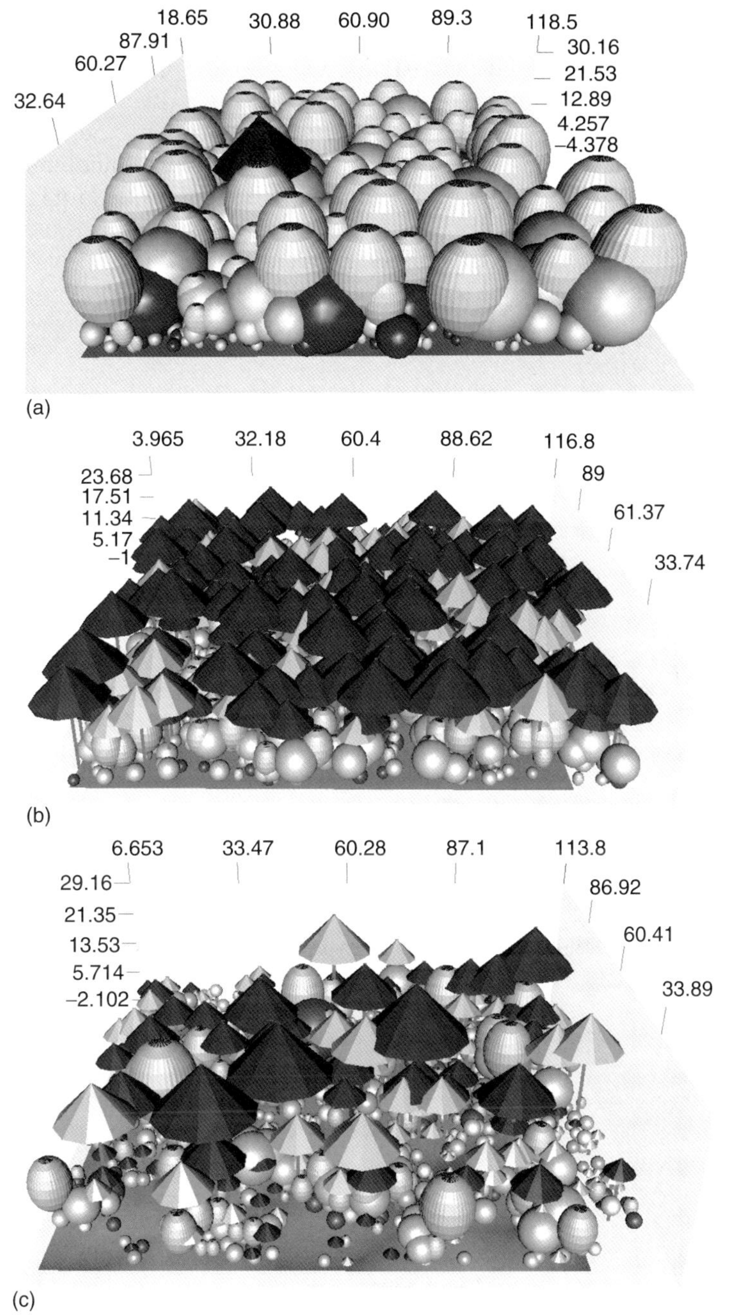

Fig. 9.3. Caricature profile of a pine–oak forest after 200 years of succession. Medium grey ellipsoids represent white oak species such as *Quercus crispipilis*. Light grey spheres are red oak species such as *Quercus segoviensis*. Dark grey spheres represent late-successional species. (a) Undisturbed simulation. (b) Simulation with clear cuts every 40 years. (c) Simulated clearance of small patches of forest at random intervals of time.

An interesting conclusion from this model is that pines either become extinct in undisturbed forest or perhaps survive as 'fugitive species' dependent on occasional larger scale disturbances (Fig. 9.4).

Simulation 2. In the second simulation all trees were subjected to a clear cut at regular intervals of 40 years. In this case the model suggested that a rather uniform, even-aged overstorey of pines would develop following each disturbance. The diameter distribution of pines has a unimodal distribution centred around 25 cm, with maximum diameters of just over 50 cm. The oak understorey has an 'inverse-j' shaped distribution of diameters with stems of between 1 and 20 cm. This type of forest structure is quite commonly encountered in the highlands of Chiapas. Where reliable disturbance history can be obtained, forests with this type of structure appear to have resulted from former slash-and-burn clearance of patches of around 1 ha for timber extraction combined with short-term rotational planting of maize. In Chiapas, the introduction of fertilizers

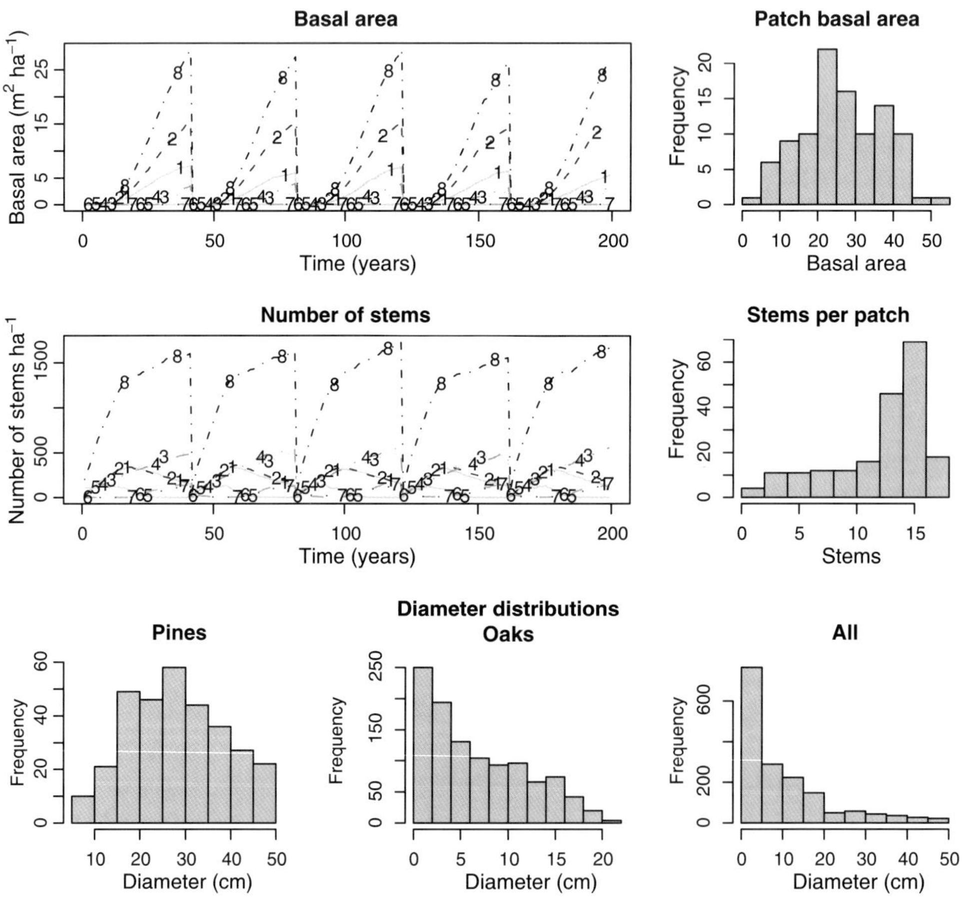

Fig. 9.4. Dynamics and final forest structure produced under a simulation with clear cuts every 40 years followed by unlimited regeneration (simulation 2). 1, *Pinus oocarpa*; 2, *Pinus maximinoi*; 3, *Quercus segoviensis*; 4, *Quercus crispipilis*; 5, heliophilic shrubs; 6, understorey; 7, generic broadleaf; 8, total.

allowed intensification of maize production on permanent sites around 30 years ago. This change led to a decline in this pattern of disturbance, but the impacts of slash-and-burn can still often be perceived in forest structure (Fig. 9.5).

Simulation 3. In the third simulation we simulated clearance of small patches of forest at random intervals of time. The probability that any patch would be cleared in any given year was set to 3/100. In other words, around 300 m² of forested area was cut each year from the total area of 10,000 m². This low-level disturbance is typical of areas that are used by indigenous communities for timber and fuelwood production. This simulation results in a form of 'dynamic equilibrium' being established after around 25 years of succession. Both pines and oaks constantly regenerate under this regime, which allows light to reach the forest floor. The total number of stems and total

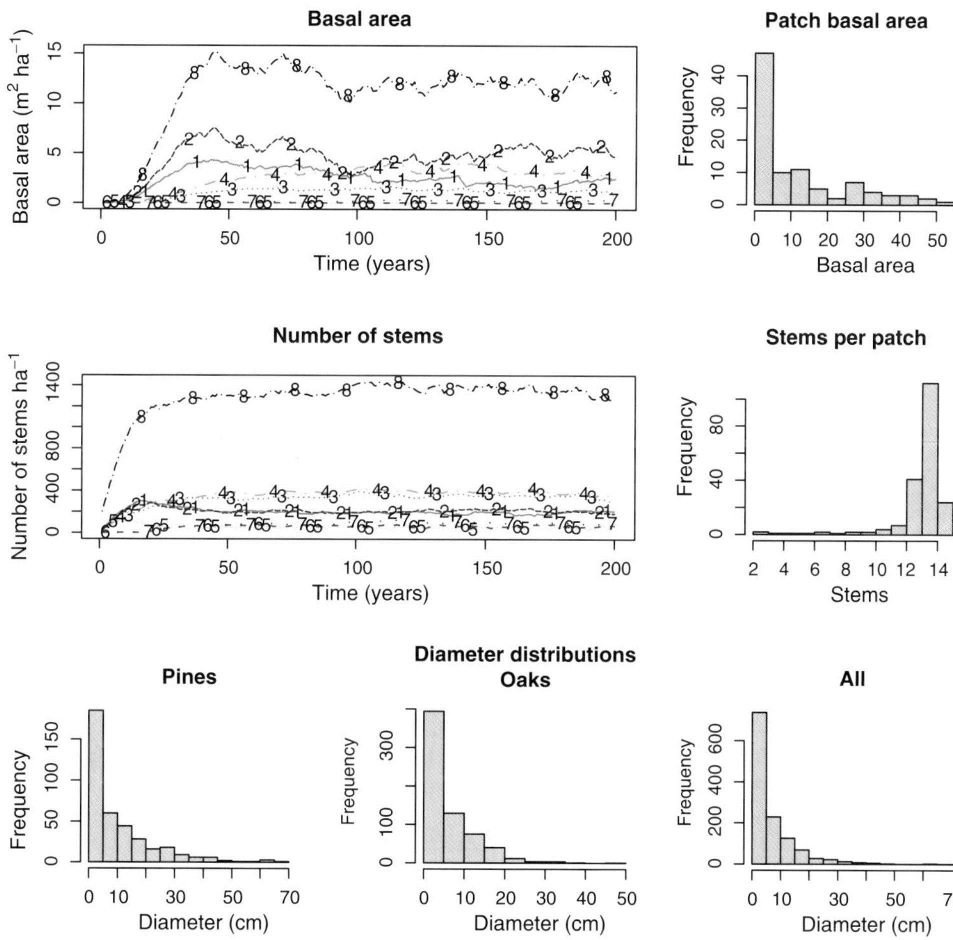

Fig. 9.5. Dynamics and final forest structure produced under a simulation with clearance of small patches of forest at random intervals of time; the probability that any patch would be cleared in any given year was set to 3/100 (simulation 3). 1, *Pinus oocarpa*; 2, *Pinus maximinoi*; 3, *Quercus segoviensis*; 4, *Quercus crispipilis*; 5, heliophilic shrubs; 6, understorey; 7, generic broadleaf; 8, total.

basal area show only minor fluctuations over time. Under this rather low level of disturbance a few patches begin to develop a rather more mature structure within the heterogeneous mosaic. Some of the more shade-tolerant broadleaved species can establish under this regime of low-level disturbance, providing an external seed source is assumed to exist.

Discussion

One of the main challenges facing the use of models to explore long-term forest dynamics is the difficulty of testing the predictions made against observations (Newton, 2007). Clear chronosequences are not easily distinguished in the highlands of Chiapas, but it is informative to compare the model simulations presented here with the investigation of successional chronosequences in Oaxaca described in Chapter 7. While the general characteristics of the forest communities analysed in the two investigations are broadly similar, there are some important differences. In Oaxaca, early successional stands are dominated by a single pine species, *Pinus chiapensis*. In Chiapas, this species is relatively rare, and in the study area two other pine species (*P. oocarpa* and *P. maximinoi*) predominate, and are considered by the model. These species differ in their ecological characteristics; for example, *P. oocarpa* is known to resprout after fire, a trait that has not been observed in *P. chiapensis* (Keeley and Zedler, 1998). Another key difference is the role of oaks (*Quercus* spp.) in the two areas; whereas in Chiapas oaks may dominate the canopy, in Oaxaca they tend to be co-dominant with a range of other broadleaved species. Despite these differences, the results from the two investigations are broadly similar. In model simulation 1, basal area of pines peaked at around 40–50 years, and tree density at 20–30 years, closely approximating the values observed in Oaxaca.

Observations made in Oaxaca indicated that relatively shade-tolerant broadleaved tree species, which dominate the canopy in late-successional or undisturbed tropical montane forest, were found to be restricted to stands aged 75 or more years since disturbance. This is again supported by model results from simulation 1, where basal area of such species remained close to zero for the first 70 years after disturbance, then increased gradually thereafter. The presence of such shade-tolerant species is of particular importance from a conservation standpoint, as it is within this functional group that most of the tree species richness of this forest type resides. Conversion of relatively undisturbed tropical montane forest to secondary forest dominated by relatively shade-intolerant species is one of the main conservation issues in the region (Chapter 3), and therefore the extent to which forests can recover from disturbance has important conservation implications. For example, if the degraded forest communities are to be restored (see Chapter 15), information is needed on the rate at which shade-tolerant tree species are able to recolonize and re-establish themselves in the forest canopy.

Results from this modelling exercise suggested that the recovery rate of tropical montane forest following disturbance is likely to be very low, even when a source of colonists is assumed to exist nearby. Given their relatively low rates of colonization and growth, results suggest that it may take several hundred years to re-establish a canopy dominated by shade-tolerant tree species. Again, this supports results obtained from successional chronosequences in Oaxaca. An important issue in this context is the apparent dominance of oak in the Chiapas model projections; for example, in simulation 1, basal area reached a plateau value at around 130 years following clearcutting, and remained stable thereafter. The replacement of oaks in the forest canopy by shade-tolerant broadleaved species is likely to occur only after the death and senescence of the cohort of oaks that established following the disturbance event. This may take at least 200 years, based on the available estimates of oak longevity.

It should be noted that oak species themselves differ in their ecological characteristics. The two oak species considered in Chiapas differ in both leaf morphology and architecture. Observations both within the study area and elsewhere in Chiapas suggest that *Q. segoviensis* may be rather more abundant on shallower rocky soils (Alvarez-Moctezuma *et al.*, 1999). Perhaps reflecting its shorter stature, *Q. segoviensis* tends to be associated with more xeric environments, where it presumably suffers less competition with *Q. crispipilis* for light. This suggests that the two oaks will tend to have differing distributions, perhaps determined by changes in moisture availability along an altitudinal gradient. Gap model simulations showed that *Q. crispipilis*, rather than *Q. segoviensis*, tends to be more dominant in terms of basal area, particularly when disturbance is relatively infrequent (simulation 1). However, field observations also suggest that *Q. segoviensis* can occur within a pine-dominated community as an understorey tree beneath pine canopies. This was also shown by other simulations with the same model framework (Golicher, 2001).

The model also provides insights into the factors influencing the relative dominance of pine and oak in montane forests of Chiapas. The key to understanding forest structure is an understanding of the historical usage pattern. From the perspective of the rural population of forest users, pine–oak forests have three principal roles, each of which is associated with a particular form of anthropogenic impact. The forests provide timber, fuelwood and ecological services. Species composition is the critical factor that determines the ability of pine–oak forest to meet the human demands placed upon it. Pine timber is used locally for construction and carpentry. Sale of pine timber is one of the few available sources of income for some rural communities. The rural population relies exclusively on fuelwood for cooking and heating (Gonzalez-Espinosa *et al.*, 1995). Oaks provide dense slow-burning wood and are therefore an essential fuel resource for all subsistence farming communities. Although resinous pine wood is used for starting fires, pine burns rapidly and is not a preferred domestic fuel. Cutting of oak for fuel can be an important cause of deforestation (Montoya-Gomez, 1995a). The most important ecological services provided by the forest are hydrological buffering and nutrient cycling. These forest services have been particularly important in sustaining subsistence agriculture. Slash-and-burn maize farming, known as *milpa*, has traditionally been a cyclical activity that has disturbed the forest and initiated secondary stand development (Collier, 1975; Pool-Novelo, 1997). Under this system, trees played an essential role in restoring the productive potential of sites following temporary maize cultivation. The recent increase in human population density has led to intensification of agriculture. Contemporary slash-and-burn is now more commonly associated with permanent deforestation, particularly when *milpa* is combined with grazing.

In addition to the disturbance associated with these three principal types of usage, unplanned fires (see Chapter 13) and the chronic disturbance caused by browsing must be included in the list of factors responsible for forest change. Yet, despite such intense human impacts, these forests have maintained a greater degree of structural and functional diversity than many comparable temperate systems (Breedlove, 1973). This has occurred without documented formal management. However, social and economic forces are now rapidly altering the pattern of land use. The long-term consequences of changing patterns of forest use are unknown. Succession models, such as that presented here, can help predict the impacts of such change.

In recent years, tree cover in this area has declined and become increasingly fragmented (González-Espinosa *et al.*, 1991, 1995; De Jong *et al.*, 1999; Ochoa-Gaona and González-Espinosa, 2000; Chapters 2 and 3), primarily as a consequence of permanent clearance for agriculture rather than cyclical slash-and-burn. Remaining forests are vulnerable to degradation and alteration as a result of human activities such as timber extraction, fuelwood collection and browsing by livestock. As our simulations show, changes in usage patterns result in changes in both forest structure and composition (González-Espinosa *et al.*, 1991; Quintana-Ascencio and González-Espinosa, 1993; Ramírez-Marcial and García-Moya, 1996). Concern has been expressed that an extensive and irreversible spread of pines within these montane forests may be taking place, as a result of the current anthropogenic disturbance regime (González-Espinosa *et al.*, 1997). Because pine-dominated forests contain fewer understorey species than oak forests (Rzedowski, 1991; González-Espinosa *et al.*, 1997), the change could be threatening the rich regional floristic diversity of the region (González-Espinosa *et al.*, 1995; Ramírez-Marcial and García-Moya, 1996). While increased pine dominance is a conservation issue in some native forests, in others the long-term future of pine timber production is under threat. Sustainable forestry relies completely on the natural regeneration of existing stocks, which may be impaired by human activities such as over-extraction of timber and livestock husbandry. Although regulations are in place to prevent over-extraction, concern has been expressed regarding both the efficacy of current legal restrictions and the extent to which they are respected (Montoya-Gómez, 1995a, b). Timber companies have traditionally paid small stumpage fees to indigenous communities in return for permission to cut pine timber. Initiatives to encourage community forestry have only recently begun to offer alternatives to unsustainable use.

The model simulations indicate that the relative abundance of pine and oak within forest stands can be understood as a function of the disturbance regime relating to the pattern of use. Following harvesting or forest clearance, pines will initially tend to dominate the forest canopy, but in the absence of any further disturbance, oaks will tend to dominate after a period of approximately 70 years (simulation 1). However, if subjected to recurrent disturbance equivalent to timber cutting, pine dominance within the stands (as indicated by basal area values) may continue indefinitely (simulation 2). A regime of less intensive use, involving disturbance of small patches, creates a forest that is heterogeneous in structure in which either pines or oaks may dominate.

Exploration of the model, supported by results from field surveys and more recent analyses of forest composition and structure at a regional level (Cayuela *et al.*, 2006b), enables the operational theory first proposed for the dynamics of disturbed montane forest in Chiapas by González-Espinosa *et al.* (1991) to be further refined. The original postulate that pine invasion is occurring as a result of human disturbance (González-Espinosa *et al.*, 1991) can be modified by a series of statements, listed below. These have been derived both from the results presented here and more extensive simulation studies using the model (Golicher, 2001):

- In the absence of anthropogenic disturbance, pines are only dominant where edaphic conditions are unsuitable for oaks. Autogenic gap-phase disturbance is insufficient to permit dense pine populations to develop within an oak-dominated matrix (simulation 1).
- Rotational slash-and-burn cultivation transforms oak-dominated forest to pine–oak forest (simulations 2 or 3). The historical importance of this form of agriculture is a partial explanation for the widespread occurrence of mixed woodland in the landscape of the highlands of Chiapas.
- Pure pine stands (as opposed to mixed pine–oak) are not usually derived from former oak woodland through rotational slash-and-burn. Either permanent land conversion followed by abandonment, catastrophic stand destruction or chronic degradation through browsing is probably needed in order to convert oak woodland into a pure pine stand, as neither simulation 2 nor simulation 3 displayed a dynamic towards pure pine stands.
- Pine–oak forests are in dynamic equilibrium only if a degree of anthropogenic disturbance continues (simulation 3).
- Removal of disturbance from pine–oak systems leads to a reversion to oak domination over a period of around 100 years (simulation 1).
- Mature mixed broadleaf forest is highly vulnerable to long-term change through disturbance by slash-and-burn farming and shows extremely slow recovery rates (simulation 1). Note that this forest type appears to be restricted to relatively moist sites (Chapter 10).
- Chronic stress caused by grazing is likely to lead to permanent deforestation, especially when combined with fire, rotational slash-and-burn farming or logging (Golicher, 2001).

These rules are summarized in Fig. 9.6. Although expressed in terms of single outcomes, a probabilistic interpretation of these relationships would be much more appropriate. The pathways illustrated thus represent the most likely directions of change. This revised theory requires considerable further testing and validation against time-series and regional-scale data.

In conservation terms, the most important result of this modelling exercise is the clear need to conserve the remaining undisturbed forest fragments in the highlands in an intact state. The threat arises not only from clearance, but also from low-level chronic disturbance that is expected to continue to erode regional floristic diversity. Disturbed, species-rich montane broad-leaved forests cannot be expected to return to their original state within a time frame that would be considered acceptable under any planned form of land management through purely natural successional processes. The modelling exercise suggests that this is the case even when seed sources are available from relictual trees. The increase in pine in disturbed forests provides locally acting incentives for conservation of the forest as a productive system. In some circumstances this can even allow forests to develop a more mature physiognomy than would occur in the absence of pine. However, restoration of compositionally degraded cloud forest will require carefully planned interventions that should be informed by evidence provided by empirical observation, experimentation and simulation.

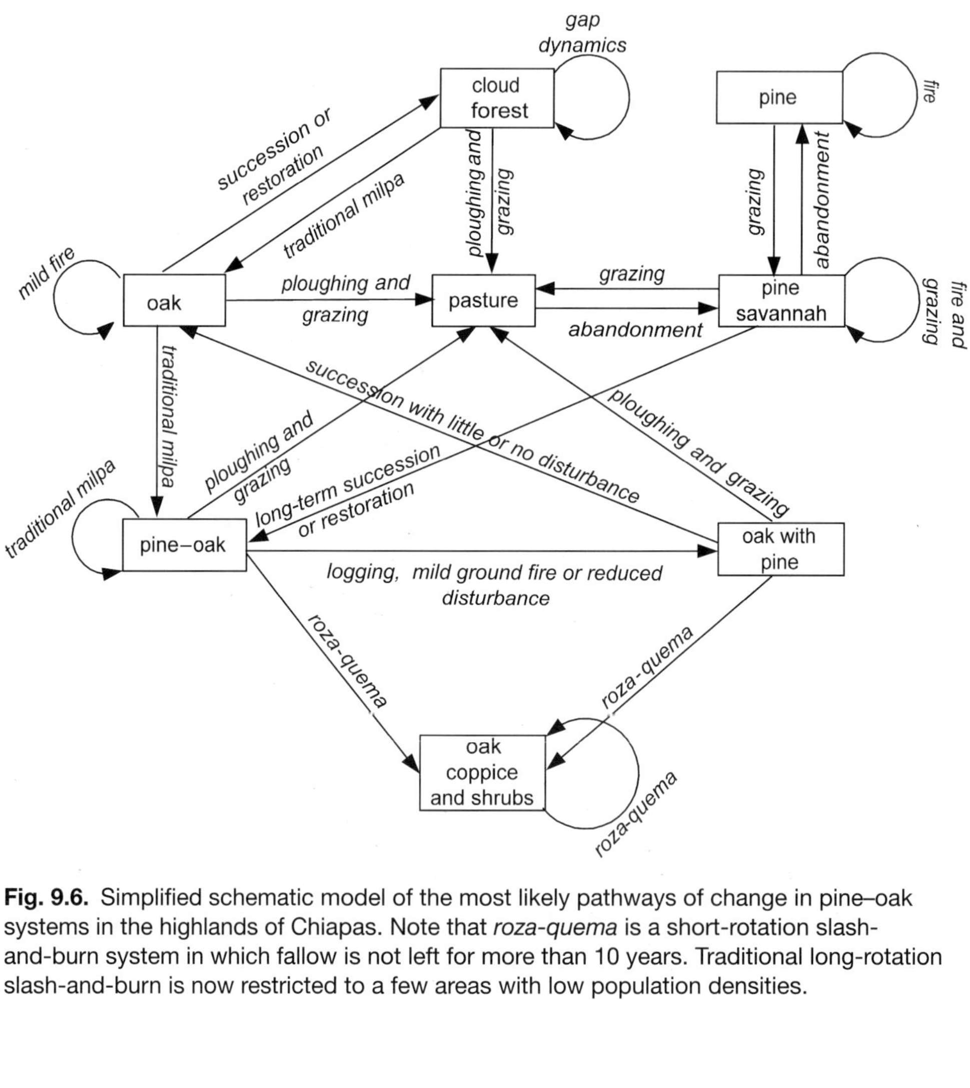

Fig. 9.6. Simplified schematic model of the most likely pathways of change in pine–oak systems in the highlands of Chiapas. Note that *roza-quema* is a short-rotation slash-and-burn system in which fallow is not left for more than 10 years. Traditional long-rotation slash-and-burn is now restricted to a few areas with low population densities.

References

Acevedo, M.F., Urban, D.L. and Ablan, M. (1995) Transition and gap models of forests dynamics. *Ecological Applications* 5, 1040–1055.

Alvarez-Moctezuma, J.G., Ochoa-Gaona, S., De Jong, B.H.J. and Soto-Pinto, M.L. (1999) Habitat and distribution of five *Quercus* (Fagaceae) species in the Chiapas Central Plateau, Mexico. *Revista de Biologia Tropical* 47, 351–358.

Bazzaz, F.A. (1996) *Plants in Changing Environments: Linking Physiological, Population and Community Ecology*. Cambridge University Press, Cambridge, UK.

Bossel, H. and Krieger, H. (1994) Simulation of multi-species tropical forest dynamics using a vertically and horizontally structured model. *Forest Ecology and Management* 69, 123–144.

Botkin, D.B. (1993a) *Forest Dynamics: An Ecological Model*. Oxford University Press, Oxford, UK.

Botkin, D.B. (1993b) *JABOWA-II: A Computer Model of Forest Growth*. Oxford University Press, Oxford, UK.

Botkin, D.B., Janak, J.F. and Wallis, J.R. (1972a) Rationale, limitations and assumptions of a northeastern forest growth simulator. *IBM Journal of Research and Development* 16, 101–116.

Botkin, D.B., Janak, J.F. and Levitan, R.E. (1972b) Some ecological consequences of a computer model of forest growth. *Journal of Ecology* 60, 849–872.

Breedlove, D.E. (1973) The phytogeography and vegetation of Chiapas, Mexico. In: Graham, A. (ed.) *Vegetation and Vegetational History of Northern Latin America*. Elsevier, Amsterdam, The Netherlands, pp. 149–165.

Breedlove, D. (1981) *Flora of Chiapas. Part I: Introduction to the Flora of Chiapas*. California Academy of Sciences, San Francisco, California.

Bugmann, H.K.M. (1996) A simplified forest model to study species composition along climate gradients. *Ecology* 77, 2055–2074.

Cayuela, L., Golicher, J.D. and Rey-Benayas, J.M. (2006a) The extent, distribution and fragmentation of vanishing montane cloud forest in the highlands of Chiapas, Mexico. *Biotrópica* 38, 544–554.

Cayuela, L., Golicher, D., Rey-Benayas, J.M., González-Espinosa, M. and Ramírez-Marcial, N. (2006b) Fragmentation, disturbance and tree diversity conservation in tropical montane forests. *Journal of Applied Ecology* 43, 1172–1181.

Clements, F.E. (1916) *Plant Succession: An Analysis of the Development of Vegetation*. Carnegie Institute Publ. 242, Washington, DC.

Clements, F.E. (1928) *Plant Succession and Indicators*. Wilson, New York.

Collier, G.A. (1975) *Fields of the Tzotzil: The Ecological Bases of Tradition in Highland Chiapas*. The University of Texas Press, Austin, Texas.

Connell, J.H. and Slatyer, R.O. (1977) Mechanisms of succession in natural communities and their role in community stability and organization. *American Naturalist* 111, 1119–1144.

De Jong, B.H.J., Cairns, M.A., Haggerty, P.K., Ramirez-Marcial, N., Ochoa-Gaona, S., Mendoza-Vega, J., Gonzalez-Espinosa, M. and March-Mifsut, I. (1999) Land-use change and carbon flux between 1970s and 1990s in central highlands of Chiapas, Mexico. *Environmental Management* 23, 373–385.

DeAngelis, D.L., Allen, T.H.F. and Starr, T.B. (1984) Hierarchy-perspectives for ecological complexity. *Bioscience* 34, 264.

Deutschman, D.H., Levin, S.A. and Pacala, S.W. (1995) Seeing the forest for the trees: community-wide predictions of a spatially explicit, individual-based mode of forest dynamics are insensitive to detail at the tree level. *Bulletin of the Ecological Society of America* 76, 320.

Deutschman, D.H., Levin, S.A., Devine, C. and Buttel, L.A. (1997) Scaling from trees to forests: analysis of a complex simulation model. *Science* 277, 1688.

Drury, W.H. and Nisbet, I.C.T. (1973) Succession. *Journal of the Arnold Arboretum* 54, 331–368.

Ek, A.R. and Monserud, R.A. (1974) *Forest: A Computer Model for Simulating the Growth and Reproduction of Mixed Species Forest Stands*. Research Report No. R263. School of Natural Resources, University of Wisconsin, Madison, Wisconsin.

Galindo-Jaimes, L., González-Espinosa, M., Quintana-Ascencio, P.F. and García-Barrios, L.E. (2002) Tree composition and structure in disturbed stands with varying dominance by *Pinus* spp. in the highlands of Chiapas, Mexico. *Plant Ecology* 162, 259–272.

Gleason, H.A. (1917) The structure and development of the plant association. *Bulletin of the Torrey Botanical Club* 44, 463–481.

Gleason, H.A. (1939) The individualistic concept of the plant association. *American Midland Naturalist* 21, 92–110.

Glenn-Lewin, D.C. and van der Maarel, E. (1992) Patterns and processes of vegetation dynamics. In: Glenn-Lewin, D.C., Peet, R.K. and Veblen, T.T. (eds) *Plant Succession: Theory and Prediction*. Chapman and Hall, London, UK, pp. 11–44.

Golicher, J.D. (2001) The dynamics of disturbed Mexican pine–oak forest: a modelling approach. PhD thesis. University of Edinburgh, Edinburgh, UK.

González-Espinosa, M., Quintana-Ascencio, P.F., Ramírez-Marcial, N. and Gaytan-Guzman, P. (1991) Secondary succession in disturbed *Pinus–Quercus* forests in the highlands of Chiapas, Mexico. *Journal of Vegetation Science* 2, 351–360.

González-Espinosa, M., Ramírez-Marcial, N., Quintana-Ascencio, P.F. and Martinez-Icó, M. (1995) La utilización de los encinos y la conservación de la biodiversidad en los altos de Chiapas. *Memorias del III Seminario Nacional sobre Utilización de Encinos*. Facultad de Ciencias Forestales Universidad Autonoma de Nuevo Leon, Linares N.L., Mexico, pp. 183–197.

González-Espinosa, M., Ochoa-Gaona, S., Ramírez-Marcial, N. and Quintana-Ascencio, P.F. (1997) Contexto vegetacional y florístico de la agricultura. In: Parra-Vázquez, M.R. and Díaz-Hernández, B.M. (eds) *Los Altos de Chiapas. Agricultura y Crisis Rural. Tomo I. Los Recursos Naturales*. El Colegio de la Frontera Sur, San Cristóbal de Las Casas, Chiapas, Mexico, pp. 85–117.

Hubbell, S.P. and Foster, R.B. (1986) Canopy gaps and the dynamics of a neotropical forest. In: Crawley, M.J. (ed.) *Plant Ecology*. Blackwell Scientific, Oxford, UK, pp. 77–97.

Huston, M.A. (1994) *Biological Diversity: The Coexistence of Species on Changing Landscapes*. Cambridge University Press, Cambridge, UK.

Huston, M.A., DeAngelis, D. and Post, W. (1988) New computer models unify ecological theory. *BioScience* 38, 682–691.

Judson, O.P. (1994) The rise of the individual-based model in ecology. *Trends in Ecology and Evolution* 9, 9–14.

Keeley, J.E. and Zedler, P.H. (1998) Evolution of life histories in *Pinus*. In: Richardson, D.M. (ed.) *Ecology and Biogeography of Pinus*. Cambridge University Press, Cambridge, UK, pp. 219–250.

Kobe, R.K. (1996) Intraspecific variation in sapling mortality and growth predicts geographic variation in forest composition. *Ecological Monographs* 66, 181–201.

Levin, S.A. (1992) The problem of pattern and scale in ecology. *Ecology* 73, 1943–1967.

Liu, J.G. and Ashton, P.S. (1995) Individual-based simulation-models for forest succession and management. *Forest Ecology and Management* 73, 157–175.

Luan, J. (1994) Simulation of forest ecosystem dynamics, with respect to the problem of hierarchy. PhD thesis. University of Edinburgh, Edinburgh, UK.

McIntosh, R.P. (1981) Succession and ecological theory. In: West, D.C., Shugart, H.H. and Botkin, D.B. (eds) *Forest Succession: Concepts and Application*. Springer, New York, pp. 10–23.

Miranda, F. (1952) *La Vegetación de Chiapas, Primera Parte*. Ediciones del Gobierno del Estado, Tuxtla Gutiérrez, Chiapas, Mexico.
Montoya-Gómez, G. (1995a) El subsector forestal en los altos de Chiapas: frontera de recursos en vías de extinción. In: Parra, M.R. and Díaz-Hernández, B.M. (eds) *Los Altos de Chiapas: Agricultura y Crisis Rural. Tomo II*. ECOSUR, San Cristóbal de las Casas, Chiapas, Mexico.
Montoya-Gómez, G. (1995b) La explotación maderera en la subregion San Cristobal y las reformas al Articulo 27 Constitucional. In: Miranda-Ocampo, R. (ed.) *Chiapas: El Regreso a la Utopía*. Universidad Autónoma de Guerrero, Mexico, pp. 33–43.
Newton, A.C. (2007) *Forest Ecology and Conservation. A Handbook of Techniques*. Oxford University Press, Oxford, UK.
Ochoa-Gaona, S. and González-Espinosa, M. (2000) Land use patterns and deforestation in the highlands of Chiapas, Mexico. *Applied Geography* 20, 17–42.
Oliver, C.D. and Larson, B.C. (1996) *Forest and Stand Dynamics*. Wiley, New York.
Pacala, S.W. and Deutschman, D.H. (1995) Details that matter: the spatial distribution of individual trees maintains forest ecosystem function. *Oikos* 74, 357–365.
Pacala, S.W., Canham, C.D., Silander, J.A. and Kobe, R.K. (1994) Sapling growth as a function of resources in a north temperate forest. *Canadian Journal of Forest* 24, 2172–2183.
Pacala, S.W., Canham, C.D., Silander, J.A.J., Kobe, R.K. and Ribbens, E. (1996) Forest models defined by field measurements: estimation, error analysis and dynamics. *Ecological Monographs* 66, 1–43.
Peet, R.K. and Christensen, N.L. (1988) Changes in species diversity during secondary forest succession on the North Carolina piedmont. In: During, H.J., Werger, M.J.A. and Willems, J.H. (eds) *Diversity and Pattern in Plant Communities*. SPB Academic Publishing, The Hague, The Netherlands, pp. 233–245.
Picket, S.T.A. (1976) Succession: an evolutionary interpretation. *American Naturalist* 110, 107–119.
Pool-Novelo, L. (1997) Intensificación de la agricultura tradicional y cambios de uso del suelo. In: Parra-Vázquez, M.R. and Díaz-Hernández, B.M. (eds) *Los Altos de Chiapas: Agricultura y Crisis Rural. Tomo I. Los Recursos Naturales*. El Colegio de la Frontera Sur, San Cristóbal de Las Casas, Chiapas, Mexico, pp. 1–22.
Porte, A. and Bartelink, H.H. (2002) Modelling mixed forest growth: a review of models for forest management. *Ecological Modelling* 150, 141–188.
Quintana-Ascencio, P.F. and González-Espinosa, M. (1993) Afinidad fitogeográfica y papel sucesional de la flora leñosa de los bosques de pino-encino de los altos de Chiapas, México. *Acta Botánica Mexicana* 21, 43–57.
Ramírez-Marcial, N. and García-Moya, E. (1996) Establecimiento de *Pinus* spp y *Quercus* spp en matorrales y pastizales de los altos de Chiapas. *Agociencia* 30, 249–257.
Ramírez-Marcial, N., González-Espinosa, M. and Williams-Linera, G. (2001) Anthropogenic disturbance and tree diversity in montane rain forests in Chiapas, Mexico. *Forest Ecology and Management* 154, 311–326.
Rzedowski, J. (1991) Análisis preliminar de la flora vascular de los bosques mésofilos de montaña de México. *Acta Botánica Mexicana* 35, 25–44.
Shugart, H.H. (1984) *A Theory of Forest Dynamics*. Springer, New York.
Shugart, H.E. (1998) *Terrestrial Ecosystems in Changing Environments*. Cambridge University Press, Cambridge, UK.
Shugart, H.H. and O'Neill, R.V. (1979) *Systems Ecology*. Dowden, Hutchinson and Ross, Stroudsburg, Pennsylvania.
Shugart, H.H. and West, D.C. (1980) Forest succession models. *BioScience* 30, 308–313.
Solomon, A.M. and Cramer, W. (1993) Biospheric implications of global change. In: Solomon, M. and Shugart, H.H. (eds) *Vegetation Dynamics and Global Change*. Chapman and Hall, New York, pp. 25–52.

Tansley, A.G. (1935) The use and abuse of vegetational concepts and terms. *Ecology* 16, 284–307.
Urban, D.L. and Shugart, H.H. (1992) Individual-based models of forest succession. In: Glenn-Lewin, D.C., Peet, R.K. and Veblen, T.T. (eds) *Plant Succession: Theory and Prediction*. Chapman and Hall, London, UK, pp. 249–292.
Urban, D.L., Bonan, G.B., Smith, T.M. and Shugart, H.H. (1991) Spatial applications of gap models. *Forest Ecology and Management* 42, 95–110.
Urban, D.L., Acevedo, M.F. and Garman, S.L. (1999) Scaling fine-scale processes to large-scale patterns using models derived from models: meta-models. In: Mladenoff, D.J. and Baker, W.L. (eds) *Spatial Modeling of Forest Landscape Change: Approaches and Applications*. Cambridge University Press, Cambridge, UK, pp. 70–98.
Vanclay, J.K. (1994) *Modelling Forest Growth and Yield: Applications to Mixed Tropical Forests*. CAB International, Wallingford, UK.
Vanclay, J.K. (1995) Growth models for tropical forests: a synthesis of models and methods. *Forest Science* 41, 7–42.
Walker, R.L. and Chapin, F.S. (1987) Interaction among processes controlling successional change. *Oikos* 50, 131–137.
Watt, A.S. (1925) On the ecology of British beech woods with special reference to their regeneration. II. The development and structure of beech communities on the Sussex Downs. *Journal of Ecology* 13, 27–73.
Watt, A.S. (1947) Pattern and process in the plant community. *Journal of Ecology* 35, 1–22.
Whitmore, T.C. (1989) Canopy gaps and the two major groups of forest trees. *Ecology* 70, 536–538.
Young, A.C. (1998) A framework for modelling tropical forest dynamics. PhD thesis. University of Edinburgh, Edinburgh, UK.

10 Models of Regional and Local Stand Composition and Dynamics of Pine–Oak Forests in the Central Highlands of Chiapas (Mexico): Theoretical and Management Implications

M.A. Zavala, L. Galindo-Jaimes and M. González-Espinosa

Open canopy of pine–oak forest in the Highlands of Chiapas, Mexico. Photo: Mario González-Espinosa

Summary

A sound analysis of the long-term implications of fragmentation and habitat loss for forest biodiversity requires the identification of the mechanisms underlying forest structure and composition. In this chapter we compile evidence from several multi-scale empirical and theoretical studies conducted in the Central Highlands of Chiapas to propose possible mechanisms underlying observed tree species richness patterns. In agreement with a niche-based perspective, tree segregation and coexistence patterns at regional scales provide partial evidence of niche differentiation along environmental gradients. Also recruitment patterns and Markovian models of stand composition parameterized at local scales suggest differential regeneration niches and a predictable successional dynamics with convergence towards a forest dominated by a broadleaved canopy. However, both the scales of unaccounted spatial variation in regional models of stand composition and the existence of predictable local successional dynamics associated with guild groups rather than species (pines, oaks, canopy broadleaves and understorey broadleaves) support the idea of neutral processes operating within guilds or functional groups. These results suggest that from a theoretical point of view pine–oak forests of the Highlands of Chiapas are an intermediate stage between highly diverse humid tropical forests and more simplified pine–oak temperate forests, with segregation and coexistence among a given number of functional groups or guilds, but also with neutral mechanisms driving community composition at more local scales. From a management perspective, in addition to the preservation of large fragments (as implied from neutral theories), the forest area preserved or to be restored should include a representative array of forest habitat types (as implied from a niche-based perspective) arranged so that spatial proximity among fragments and connectivity is maximized.

Introduction

A sound analysis of the long-term implications of fragmentation and habitat loss for forest biodiversity requires the identification of the mechanisms underlying forest structure and composition (e.g. Chave and Norden, 2007). In recent decades, much of the debate has focused on two seemingly opposed views regarding the mechanisms maintaining biodiversity in plant communities (e.g. Chave, 2004; Purves and Pacala, 2005). On one side, a number of widely accepted models of plant community assembly have been inspired by the niche concept. These attempt to explain species segregation or coexistence in terms of differences in species responses to environmental heterogeneity and habitat spatial structure. Examples include, among others, models considering partitioning of resource heterogeneity (Pacala and Tilman, 1994) or environmental variability (Chesson, 2000), trade-offs between competitive and colonization ability (Levins and Culver, 1971), and Janzen–Connell density-dependent effects based on a hypothetical advantage of rare species to pest damage (Janzen, 1970). Alternatively, the so-called neutral theories assume that all individuals in a community are equivalent with respect to their prospects of survival and reproduction (Hubbell, 2001, 2005). Stochastic individual-level processes result in emergent community-level patterns that are in close agreement with patterns of community structure found in many species-rich communities.

Lack of consensus regarding the fundamental mechanisms underlying the maintenance of forest community structure largely restricts our capability to evaluate current threats to biodiversity in the face of global change. The reconciliation of these two seemingly contrasting viewpoints is in part hampered by mathematical and methodological challenges (Chave, 2004) but, more importantly, because of a lack of integration between current theoretical models and experimental and field observations. There is therefore a critical need for studies conducted in specific forest ecosystems that allow us to test predictions from theory and the underlying assumptions of such theory.

Tropical montane mixed pine–oak forests of the Central Highlands of Chiapas (southern Mexico) and Guatemala exhibit remarkable species richness in relation to temperate pine–oak ecosystems in the northern latitudes. In the Highlands of Chiapas some 350–400 tree species have been recorded around an area of approximately 11,000 km^2, including 11 pine and 23 oak species (Alba-López *et al.*, 2003; González-Espinosa *et al.*, 2006). This situation contrasts with other pine–oak-dominated regions in Europe, North America and Asia, where the number of coexisting oak and pine species is much lower. Area, water and energy availability are considered major determinants of species biodiversity worldwide (Currie and Paquin, 1987; Adams and Woodward, 1989). Nevertheless, biogeographical mechanisms associated with post-glacial migration and species colonization potential also now appear as critical determinants of biodiversity loss in temperate regions, particularly in Europe (Bennett *et al.*, 1991; Hawkins and Porter, 2003). In contrast, in the Central Highlands of Chiapas, as in much of Mexico, high biodiversity may be explained by the confluence of the Neotropical and the Holarctic biogeographical regions, and by the absence of significant barriers to latitudinal tree migration (Breedlove, 1981; Rzedowski, 1993). This suggests that stochastic-driven processes may operate alongside niche-based mechanisms to maintain species richness in this highly diverse region. Under the current conditions of high rates of habitat fragmentation and loss, understanding the mechanisms underlying these patterns becomes a critical issue (Ochoa-Gaona and González-Espinosa, 2000; Chave and Norden, 2007).

In this chapter we compile evidence from several ongoing empirical and theoretical studies to support or reject evidence of mechanisms supporting neutral theories of forest community structure; we also propose possible mechanisms underlying observed patterns of segregation and coexistence of pine and oak species. First, we investigate whether, as expected from a niche-based perspective, pine and oak species segregate along environmental gradients, and whether environment–species dependencies can account for observed spatial aggregation patterns. Second, we test for functional equivalence among selected pine and oak species in the regeneration niche by examining recruitment patterns under different forest microsites and canopy tree species; we also evaluate the dynamical consequences of observed recruitment patterns on long-term forest stand composition with simple Markovian stochastic models (Horn, 1975; Usher, 1981, 1992). Finally, we discuss the theoretical implications of these findings in relation to niche and drift-based viewpoints, as well as potential implications for forest management and conservation.

Regional Patterns of Pine–Oak Distributions

Correlational studies among community structure and environment (e.g. direct gradient analyses) have been the most widespread approach to investigating patterns in plant community structure worldwide (Whittaker, 1975). If patterns of community structure repeat in relation to a given environmental gradient independently of the geographical locality, this would suggest that common niche-based underlying mechanisms may be responsible (Tilman, 1988). Conversely, lack of consistent patterns may reflect a predominant role of random non-deterministic processes in shaping community structure.

The distribution patterns of selected pine and oak species in the Highlands of Chiapas can be explored through a simple non-linear regression model (logistic regression) in which the probability of finding a given species at a point is governed by a binomial process. We can then investigate the likelihood of various models in which the probability of success (finding a given species) is a function of the environment. In this way we can test the likelihood of environmental dependency in species distributions in relation to randomly generated patterns.

Forest inventory plots in the Highlands of Chiapas are still limited in number and are inadequate for developing an accurate cartographic description of current tree distributions. To circumvent this problem we interpolated a probability distribution surface for each species based on a limited number of floristic inventories (a total of 666 plots obtained during 1995–2004). First we parameterized the logistic models (modelling species probability of occurrences as a function of environmental variables) and we computed semi-variograms of model residuals to investigate the scale of unexplained spatial variation. The combination of maps of predicted probabilities produced by the regression model and spatial variability allowed us to develop continuous models of the probability of encountering a given species at a point (Figs 10.1 and 10.2). For simplicity we targeted 12 pine and oak species that are major structural components of forest communities in this region: *Pinus ayacahuite* var. *ayacahuite*, *P. devoniana*, *P. montezumae*, *P. oocarpa* var. *oocarpa*, *P. pseudostrobus*, *P. tecunumanii*, *Quercus candicans*, *Q. crassifolia*, *Q. crispipilis*, *Q. laurina*, *Q. rugosa* and *Q. segoviensis*. We used elevation and rainfall in January (dry season) as environmental drivers (INEGI, 1984a, b, 1985).

The results from the model show that seven out of the 12 species considered were concentrated in the Central Highlands of Chiapas: *P. ayacahuite*, *P. montezumae*, *P. pseudostrobus*, *P. tecunumanii*, *Q. crassifolia*, *Q. laurina* and *Q. rugosa*. These species, along with *Q. crispipilis*, which appears widely distributed across the study area, were often found in mixed stands for altitudes above 2000 m. *P. devoniana*, *P. oocarpa* and *Q. segoviensis* occupied the south-east of the studied area, associated with sites lower in elevation (1500–2000 m) and with a lower precipitation. Finally, *Q. candicans* was found in the north-west region, occupying lower (1500–2000 m) and moister sites because of the interception of moist north winds (Figs 10.1 and 10.2).

Patterns of Recruitment in Relation to Forest Type and Tree Cover

A complementary approach to investigating differential niche regeneration strategies is to examine patterns of establishment and sapling abundance under different microsites in the field. Both seedling and sapling stages have been shown to have a disproportionate effect on the stand dynamics and composition of temperate deciduous (Kobe *et al.*, 1995) and other pine–oak forests (Zavala, 1999; Zavala and Zea, 2004). Specifically, we classified forest fragments

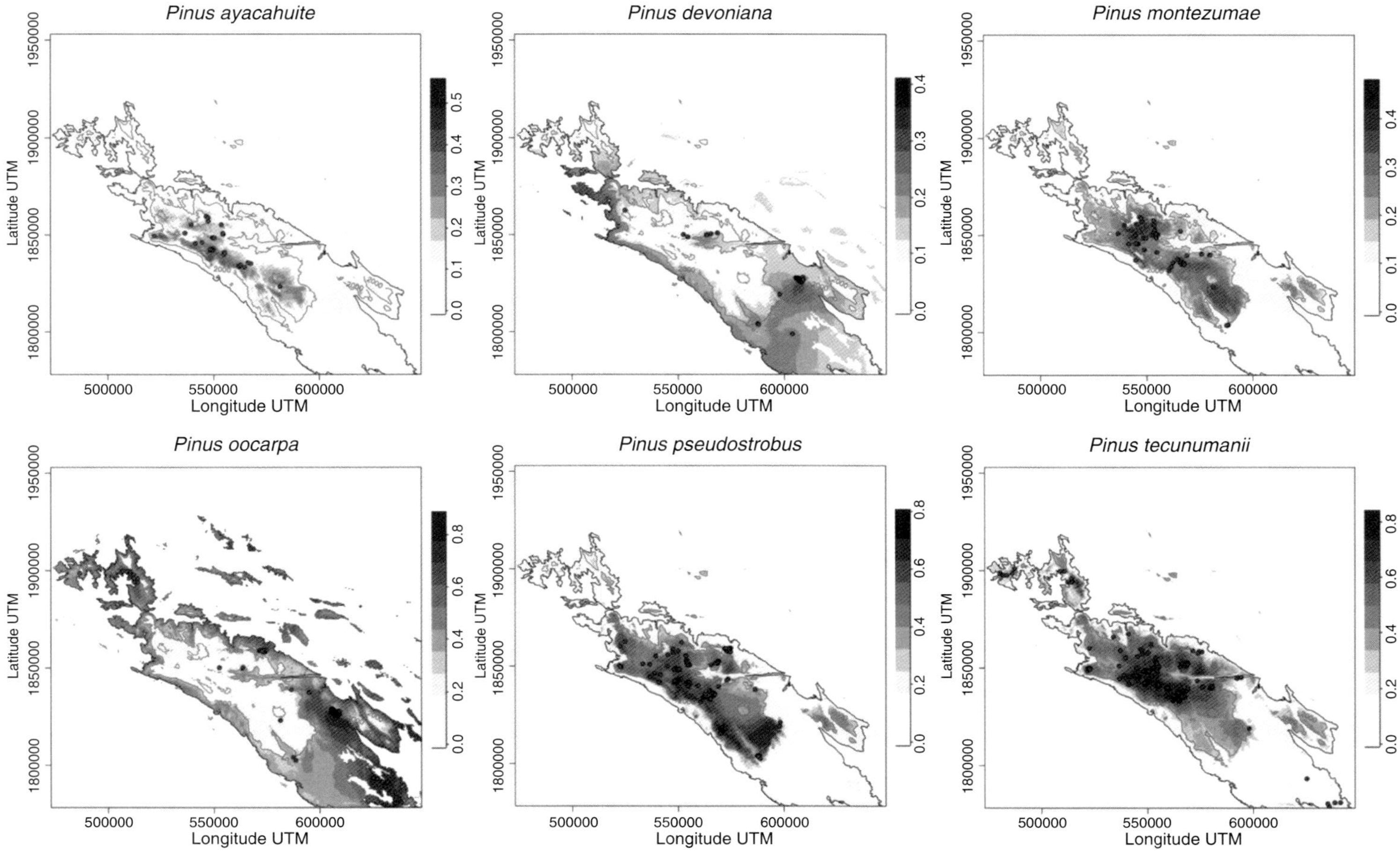

Fig. 10.1. Spatialized predictions of probability of occurrence for *Pinus* spp. in the study area. Solid points show floristic inventories where a given species is present. Grey tones indicate probability of occurrence according to the predictions of the logistic model (tone scale is shown along the vertical right axis).

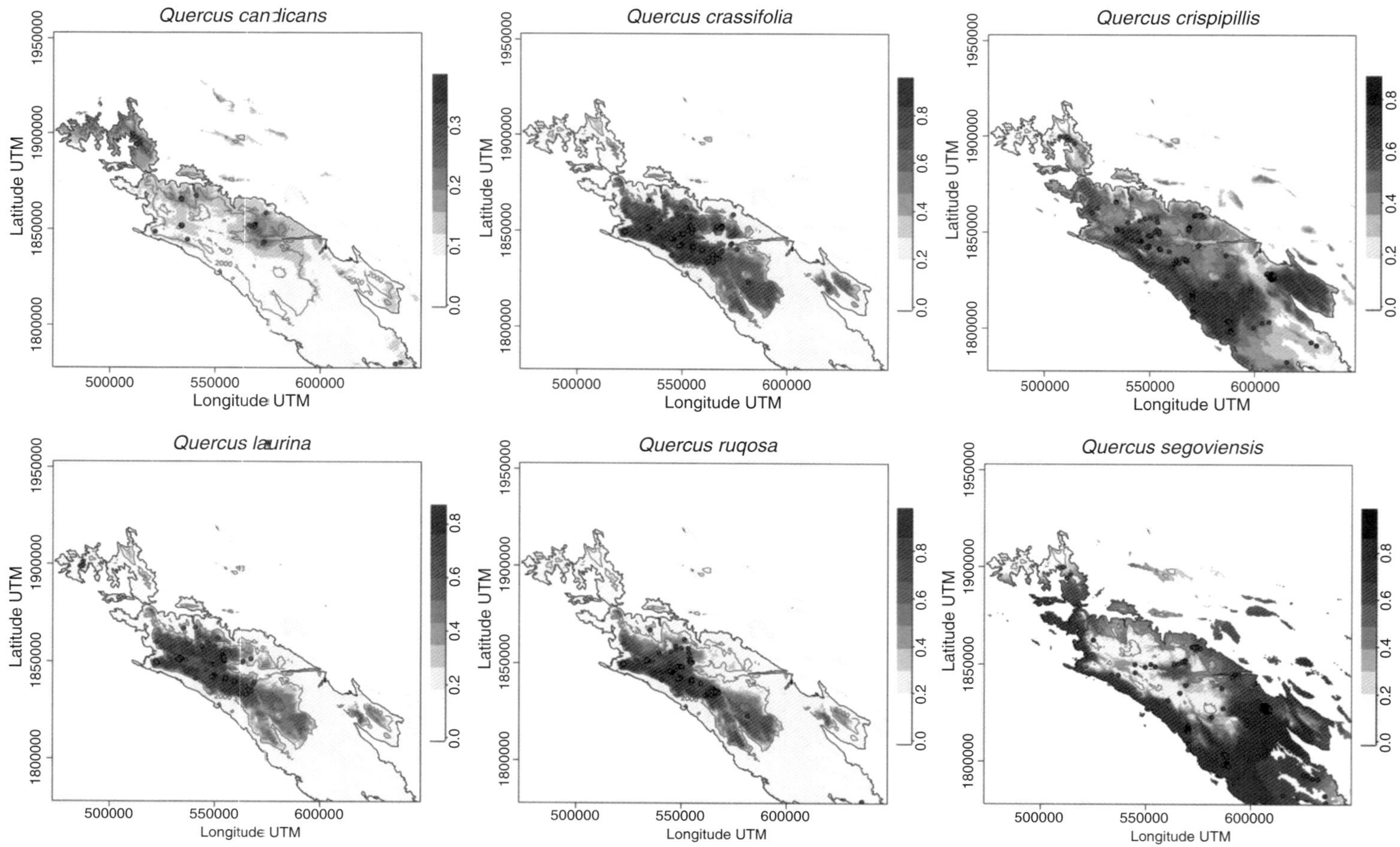

Fig. 10.2. Spatialized predictions of probability of occurrence for *Quercus* spp. in the study area. Solid points show floristic inventories where a given species is present. Grey tones indicate probability of occurrence according to the predictions of the logistic model (tone scale is shown along the vertical right axis).

according to pine–oak dominance as pine-dominated (P), mixed pine–oak (PO) and oak-dominated (O) stands; we grouped species as pines (P), oaks (O), broadleaved canopy dominant (CB; height >20 m), and other understorey broadleaved species (UB; height < 20 m). We set 60 transects (50 m long) across fragments (4–5 transects separated 10 m from each other in each forest type), and recorded seedlings (height < 50 cm) and saplings (height >50 cm and dbh < 5 cm) of each species. We also selected canopy trees (dbh > 20 cm) around 1 m at each side of the transect, and established a 3-m area of influence around each of them under which we estimated seedling and sapling density.

Pines only regenerated in pine-dominated fragments (P), but oaks (O) and understorey broadleaved species (UB) also showed a high regeneration in pine-dominated stands (P). Oak-dominated stands (O) supported regeneration for both canopy (CB) and understorey (UB) broadleaved species. The best setting for CB and UB regeneration was in protected sites under their own crowns, where they exhibited a higher density of juveniles and tended to monopolize regeneration (Fig. 10.3). When forest type was considered, we observed that understorey broadleaved species (UB) could regenerate in either forest type (although with variation in composition), while broadleaved species (CB) preferentially regenerated in oak-dominated stands (Fig. 10.4).

These trends were statistically tested with a log-linear analysis that examined which factors influence seedling and sapling occurrence for a given group (pines, oaks, and canopy or understorey broadleaved species). Categorical factors in the analyses included forest type (F), canopy species above the seedling or sapling (C), age or size (E) (seedling or sapling) and species (G) (of the seedling or sapling individual). The best fit to the data (best model, FCE,FCDG,EG, $\chi^2 = 25.11$, df = 33, $P = 0.84$) suggests that regeneration composition is a function of the stage (seedling or sapling) considered, but that this relationship is also a function of both forest type (F) and canopy species (C). This analysis also confirms that the presence of either seedlings or saplings of a particular species appears clearly differentiated in each forest type (pine, $\chi^2 = 23.54$; pine–oak, $\chi^2 = 37.15$ and oak, $\chi^2 = 138.11$) and canopy species (pine, $\chi^2 = 50.81$; oak, $\chi^2 = 58.57$; canopy broadleaved, $\chi^2 = 39.51$ and understorey broadleaved, $\chi^2 = 56.95$). Therefore, occurrence of both pine and oak seedlings and saplings takes place in pine forests and under pine individuals, while broadleaved juveniles tend to be found in oak- and other broadleaved-dominated forests (Fig. 10.5).

Recruitment Patterns and Stand Dynamics

The dynamical consequences of recruitment patterns for stand dynamics and composition can be easily explored with a Markov model, which assumes that the probability of replacement of one canopy-dominant species by another is proportional to the density of juvenile individuals of the latter (Horn, 1975). The transition matrix that evaluates the probability of replacement across species combinations is then multiplied by a vector that represents initial stand composition, t (in our case based on species-specific basal area; Galindo-Jaimes

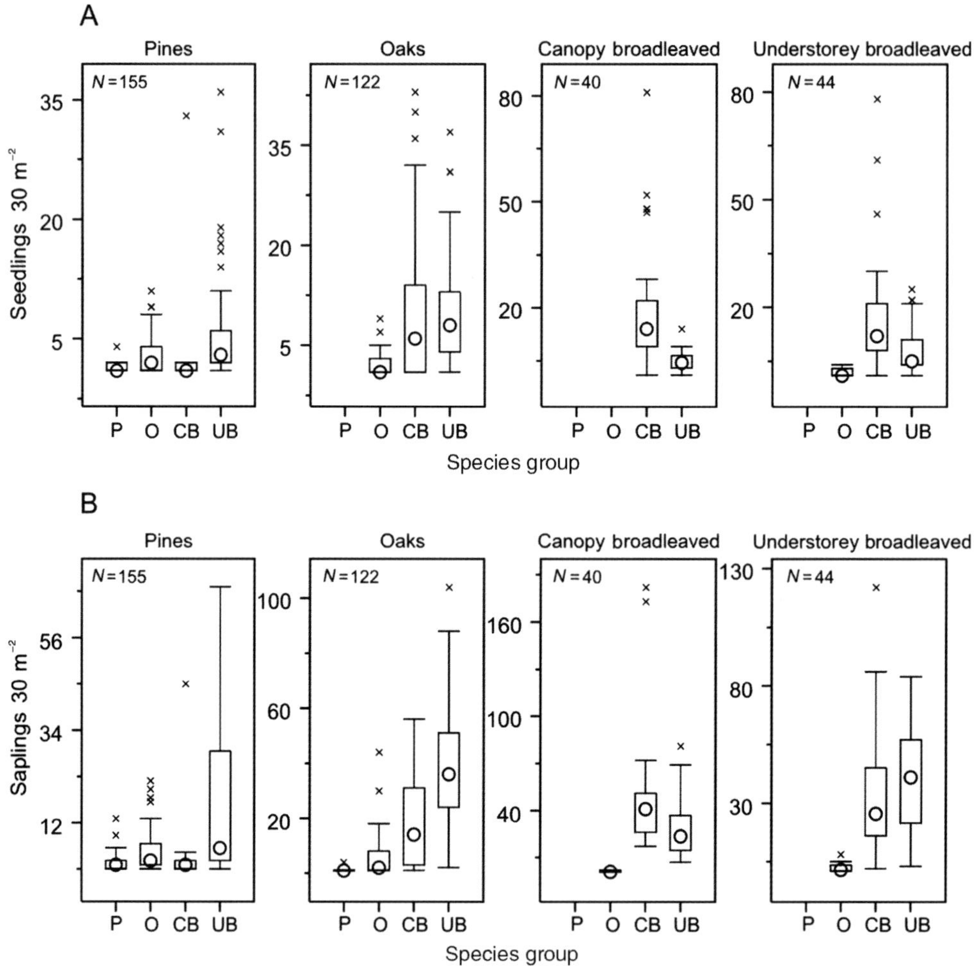

Fig. 10.3. Seedling (A) and sapling (B) density for four functional groups of tree species (P = pine species, O = oak species, CB = canopy broadleaved species, UB = understorey broadleaved species) found under different types of canopy dominant individuals (pines, oaks, canopy broadleaved and understorey broadleaved), *N* = number of trees sampled (30 m^2).

et al., 2002). The resulting product is another vector representing the expected stand composition at the next generation *t*+1 (the generation time is the time that a tree can remain as a canopy dominant, or maximal age, which we assume to be 100–120 years; González-Espinosa *et al.*, 1991). Successive multiplication of this vector by the transition matrix allows us to investigate the existence of an equilibrium or 'climax' state. The direction and intensity of the changes can also be evaluated analytically by inspection of the dominant eigenvalue (this type of probability matrix necessarily includes one as a dominant eigenvalue). If this value appears more than once, it indicates the existence of several absorbent states, or, in biological terms, multidirectional succession or polyclimax (Horn, 1975; Usher, 1981, 1992).

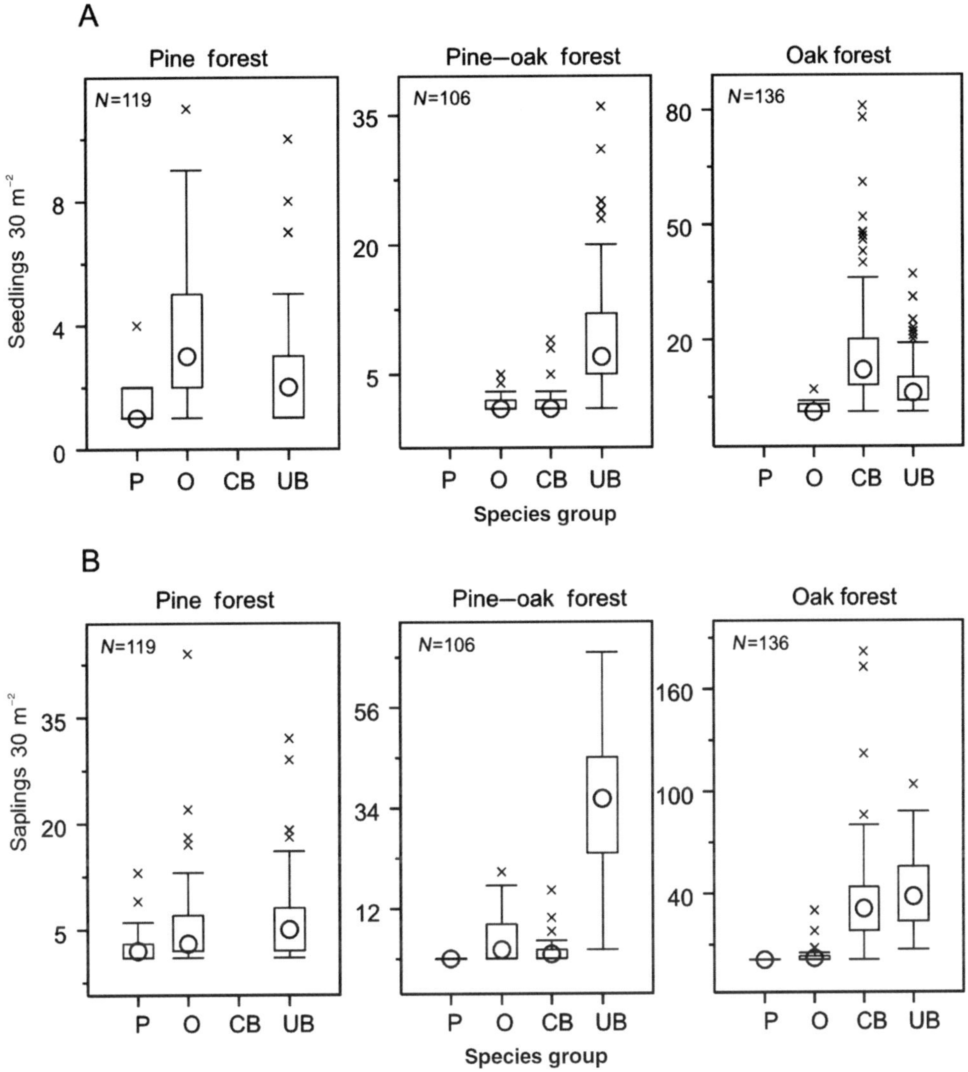

Fig. 10.4. Seedling (A) and sapling (B) density for four groups of tree species (P = pine species, O = oak species, CB = canopy broadleaved species, UB = understorey broadleaved species) found under different forest types according to relative pine–oak dominance (pine forest, mixed pine–oak forest, oak forest), *N* = number of trees sampled (30 m²).

We used the same data set described in the previous section to parameterize a transition matrix for these stands. Specifically, we counted all the seedlings and saplings beneath a 3-m area of influence of selected canopy trees (dbh > 20 cm) located around 1 m at each side of the transect. We constructed a matrix in which each element indicates the proportion of saplings of each species (rows) under each species (columns). According to the matrix, seedlings and saplings more frequently found under oak forests were *Persea americana*, *Styrax argenteus*, *Oreopanax xalapensis* and *Prunus rhamnoides*. *O. xalapensis*,

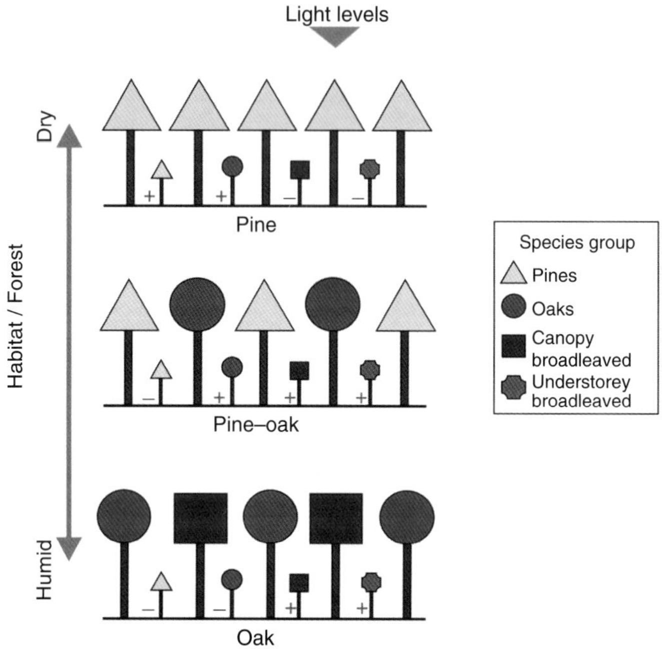

Fig. 10.5. Variation in forest regeneration patterns as a function of canopy stand structure along a water availability–land-use gradient (see Romero-Najera, 2000; Galindo-Jaimes *et al.*, 2002). Small trees represent seedlings and saplings and larger trees indicate canopy dominant trees. Groups of species are represented by different canopy symbols (triangle, pines; circle, oaks; square, canopy broadleaved; cross, understorey broadleaved).

Rapanea juergensenii, Ternstroemia lineata and *Q. laurina* were the species most frequently found in the understorey of pine–oak forests. Finally, regeneration in pine forests was dominated by seedlings and saplings of *Q. laurina, Q. crassifolia, R. juergensenii* and *T. lineada*. The probability of occurrence (transition matrix) for juveniles of 20 tree species is shown in Table 10.1. Self-replacement probability is low within the genus *Pinus* (< 0.15), as well as replacement of pines by broadleaves other than *Quercus* (<0.17). In contrast, *Quercus* seedlings and saplings exhibit a 0.67 probability of replacing pines. Probability of replacement of *Quercus* species by *Pinus* is very low (0.02) and *Quercus* self-replacement probabilities were also rather low (<0.17). Broadleaved tree species other than *Quercus* with highest probabilities to recruit under pines and oaks were *P. americana* and *S. argenteus* (0.14–0.45), and self-replacement probabilities or probabilities of replacement by other broadleaved trees were relatively high (0.20–0.80). Understorey broadleaved species exhibit both low self-replacement probabilities and low frequencies under oaks.

Projections of both the seedling and sapling matrices (only data from the sapling matrix are shown) predict long-term competitive exclusion of *Pinus, Quercus* and other broadleaved trees species such as *Alnus acuminata* and

Table 10.1. Transition matrix for seedlings (A) and juveniles (B). Matrix coefficients represent the probability of replacement for each canopy tree species (the first 12 species) and understorey species (the last eight species). Species are given their Latin names (first column) and Latin name abbreviation (first row). The bold figures highlight self-replacement values for each species (matrix diagonal).

	A. Seedlings																			
	PAY	PMO	PPS	PTE	QAC	QCR	QLA	QRU	CMA	PAM	SAR	OBE	MSH	CTH	CDI	RSH	PRH	ZME	AAC	CGU
Pinus ayacahuite	**0.00**	0.00	0.00	0.00	0.00	0.00	0.50	0.50	0.00	0.00	0.00	0.00	0.00	0.00	0.00	0.00	0.00	0.00	0.00	0.00
Pinus montezumae	0.00	**0.00**	0.00	0.00	0.09	0.34	0.32	0.00	0.00	0.11	0.00	0.00	0.00	0.00	0.00	0.07	0.05	0.02	0.00	0.00
Pinus pseudostrobus	0.00	0.00	**0.07**	0.00	0.01	0.33	0.11	0.17	0.06	0.17	0.03	0.00	0.00	0.00	0.02	0.02	0.00	0.01	0.00	0.00
Pinus tecunumanii	0.00	0.00	0.00	**0.03**	0.01	0.30	0.30	0.07	0.00	0.07	0.02	0.00	0.01	0.04	0.02	0.05	0.04	0.01	0.00	0.00
Quercus acatenangensis	0.00	0.00	0.00	0.00	**0.41**	0.00	0.00	0.00	0.33	0.07	0.00	0.00	0.00	0.04	0.00	0.11	0.00	0.04	0.00	0.00
Quercus crassifolia	0.00	0.00	0.00	0.00	0.00	**0.01**	0.13	0.00	0.00	0.53	0.07	0.00	0.01	0.01	0.00	0.07	0.18	0.00	0.00	0.00
Quercus laurina	0.00	0.00	0.00	0.00	0.00	0.00	**0.05**	0.00	0.02	0.63	0.11	0.00	0.00	0.00	0.01	0.04	0.12	0.00	0.00	0.00
Quercus rugosa	0.00	0.00	0.00	0.00	0.00	0.00	0.18	**0.00**	0.05	0.59	0.00	0.00	0.00	0.05	0.00	0.05	0.09	0.00	0.00	0.00
Clethra macrophylla	0.00	0.00	0.00	0.00	0.00	0.00	0.00	0.00	**0.02**	0.88	0.07	0.00	0.01	0.00	0.01	0.00	0.01	0.00	0.00	0.00
Persea americana	0.00	0.00	0.00	0.00	0.00	0.00	0.00	0.00	0.02	**0.82**	0.12	0.00	0.01	0.00	0.01	0.01	0.02	0.00	0.00	0.00
Styrax argenteus	0.00	0.00	0.00	0.00	0.00	0.00	0.00	0.00	0.10	0.65	**0.19**	0.00	0.00	0.00	0.00	0.00	0.06	0.00	0.00	0.00
Olmediella betschleriana	0.00	0.00	0.00	0.00	0.00	0.00	0.00	0.00	0.06	0.52	0.36	**0.00**	0.00	0.00	0.00	0.00	0.06	0.00	0.00	0.00
Magnolia sharpii	0.00	0.00	0.00	0.00	0.00	0.00	0.01	0.00	0.01	0.77	0.16	0.01	**0.00**	0.00	0.01	0.01	0.04	0.00	0.00	0.00
Cleyera theaeoides	0.00	0.00	0.00	0.00	0.00	0.00	0.00	0.00	0.01	0.79	0.10	0.00	0.00	**0.00**	0.02	0.04	0.04	0.00	0.00	0.00
Cornus disciflora	0.00	0.00	0.00	0.00	0.00	0.00	0.05	0.00	0.00	0.58	0.22	0.00	0.01	0.00	**0.01**	0.02	0.12	0.00	0.00	0.00
Rhamnus sharpii	0.00	0.00	0.00	0.00	0.00	0.00	0.00	0.00	0.50	0.50	0.00	0.00	0.00	0.00	0.00	**0.00**	0.00	0.00	0.00	0.00
Prunus rhamnoides	0.00	0.00	0.00	0.00	0.00	0.00	0.00	0.00	0.00	0.82	0.14	0.00	0.00	0.00	0.00	0.00	**0.05**	0.00	0.00	0.00
Zanthoxylum melanostictum	0.00	0.00	0.00	0.00	0.00	0.00	0.00	0.00	0.00	0.90	0.06	0.00	0.00	0.00	0.00	0.00	0.03	**0.00**	0.00	0.00
Alnus acuminata	0.00	0.00	0.00	0.00	0.00	0.00	0.50	0.00	0.00	0.17	0.00	0.00	0.00	0.00	0.00	0.17	0.00	0.00	**0.17**	0.00
Citharexylum donnell-smithii	0.00	0.00	0.00	0.00	0.00	0.00	0.00	0.00	0.17	0.83	0.00	0.00	0.00	0.00	0.00	0.00	0.00	0.00	0.00	**0.00**

Continued

Table 10.1. *Continued*

	B. Saplings																			
	PAY	PMO	PPS	PTE	QAC	QCR	QLA	QRU	CMA	PAM	SAR	OBE	MSH	CTH	CDI	RSH	PRH	ZME	AAC	CGU
Pinus ayacahuite	**0.13**	0.00	0.03	0.00	0.00	0.00	0.67	0.03	0.00	0.00	0.00	0.00	0.00	0.13	0.00	0.00	0.00	0.00	0.00	0.00
Pinus montezumae	0.00	**0.00**	0.01	0.03	0.39	0.13	0.14	0.01	0.00	0.07	0.01	0.00	0.02	0.03	0.06	0.05	0.05	0.02	0.00	0.00
Pinus pseudostrobus	0.02	0.00	**0.15**	0.05	0.05	0.15	0.24	0.08	0.02	0.14	0.02	0.00	0.00	0.03	0.02	0.02	0.00	0.01	0.00	0.00
Pinus tecunumanii	0.02	0.00	0.00	**0.03**	0.01	0.09	0.38	0.03	0.00	0.05	0.02	0.00	0.02	0.04	0.06	0.17	0.04	0.01	0.01	0.00
Quercus acatenangensis	0.00	0.01	0.00	0.00	**0.36**	0.00	0.02	0.00	0.11	0.03	0.00	0.00	0.01	0.07	0.00	0.02	0.02	0.35	0.00	0.00
Quercus crassifolia	0.00	0.00	0.00	0.00	0.00	**0.00**	0.15	0.00	0.01	0.37	0.08	0.00	0.05	0.04	0.06	0.11	0.11	0.01	0.00	0.00
Quercus laurina	0.00	0.00	0.00	0.00	0.00	0.00	**0.07**	0.00	0.06	0.45	0.13	0.00	0.03	0.02	0.03	0.04	0.13	0.02	0.00	0.00
Quercus rugosa	0.02	0.00	0.00	0.01	0.00	0.01	0.17	**0.02**	0.03	0.33	0.10	0.00	0.00	0.08	0.00	0.12	0.06	0.04	0.00	0.00
Clethra macrophylla	0.00	0.00	0.00	0.00	0.00	0.00	0.00	0.00	**0.08**	0.55	0.23	0.00	0.02	0.01	0.03	0.01	0.06	0.01	0.00	0.00
Persea americana	0.00	0.00	0.00	0.00	0.00	0.00	0.00	0.00	0.04	**0.64**	0.24	0.00	0.01	0.00	0.01	0.01	0.04	0.01	0.00	0.00
Styrax argenteus	0.00	0.00	0.00	0.00	0.00	0.00	0.00	0.00	0.02	0.54	**0.26**	0.00	0.03	0.00	0.02	0.01	0.12	0.01	0.00	0.01
Olmediella betschleriana	0.00	0.00	0.00	0.00	0.00	0.00	0.00	0.00	0.08	0.58	0.14	**0.06**	0.00	0.02	0.06	0.00	0.00	0.06	0.00	0.00
Magnolia sharpii	0.00	0.00	0.00	0.00	0.00	0.00	0.01	0.00	0.01	0.59	0.18	0.00	**0.05**	0.01	0.03	0.01	0.10	0.00	0.00	0.00
Cleyera theaeoides	0.00	0.00	0.00	0.00	0.00	0.00	0.01	0.00	0.03	0.59	0.20	0.00	0.04	**0.02**	0.04	0.01	0.05	0.01	0.00	0.00
Cornus disciflora	0.00	0.00	0.00	0.00	0.00	0.00	0.02	0.00	0.02	0.48	0.16	0.00	0.04	0.02	**0.04**	0.02	0.18	0.01	0.00	0.00
Rhamnus sharpii	0.00	0.00	0.00	0.00	0.00	0.00	0.00	0.00	0.09	0.53	0.09	0.00	0.00	0.06	0.00	**0.00**	0.24	0.00	0.00	0.00
Prunus rhamnoides	0.00	0.00	0.00	0.00	0.00	0.00	0.00	0.00	0.00	0.59	0.23	0.00	0.03	0.00	0.08	0.00	**0.08**	0.00	0.00	0.00
Zanthoxylum melanostictum	0.00	0.00	0.00	0.00	0.00	0.00	0.00	0.00	0.00	0.55	0.35	0.00	0.01	0.00	0.01	0.00	0.08	**0.00**	0.00	0.00
Alnus acuminata	0.00	0.00	0.00	0.00	0.00	0.08	0.67	0.00	0.00	0.00	0.00	0.00	0.00	0.00	0.00	0.08	0.08	0.00	**0.08**	0.00
Citharexylum donnell-smithii	0.00	0.00	0.00	0.00	0.00	0.00	0.00	0.00	0.03	0.36	0.24	0.00	0.06	0.00	0.03	0.00	0.27	0.00	0.00	**0.00**

Olmediella betschleriana by canopy broadleaved trees such as *P. americana* and *S. argenteus*, which show the highest values at equilibrium (25–60%; Fig. 10.6). In contrast, understorey broadleaved species that are not completely excluded, such as *Cleyera theaeoides* and *Citharexylum donnell-smithii*, remain in the community, but with very low abundances. Other understorey broadleaves, such as *Magnolia sharpii*, *Cornus disciflora*, *P. rhamnoides*, *Ramnus sharpii*,

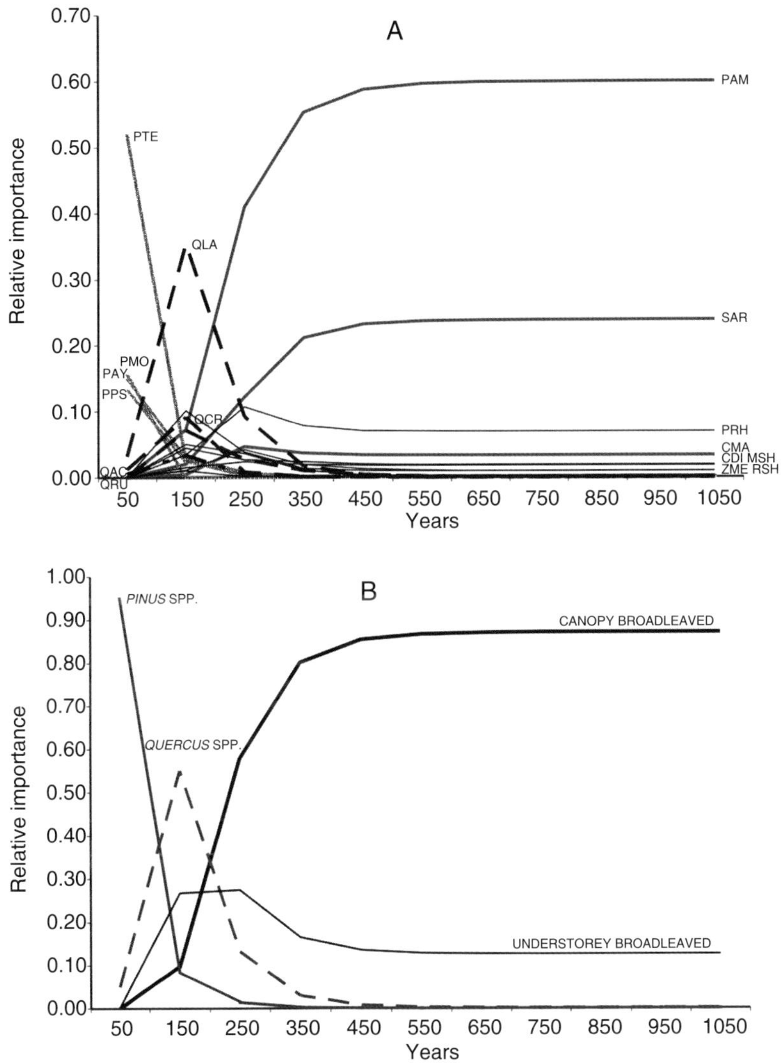

Fig. 10.6. Projections of the transition matrix over time (100 years per generation) until equilibrium is reached. (A) Dynamics of 20 canopy and understorey species. Species names are given by their Latin name abbreviation. (B) Dynamics of the four species groups as defined in the text (*Pinus* spp., *Quercus* spp., canopy broadleaved species, understorey broadleaved species).

Zanthoxylum melanostictum and canopy broadleaved *Clethra macrophylla* show higher abundances but without achieving dominance (individually these can reach values below 0.1 and altogether can comprise up to 15% of stand composition; Fig. 10.6).

Predicted equilibrium or climax state matches the stand structure and composition observed in remnant old-growth forest stands in mesic sites of the study region. This suggests that in some mesic localities, and with long periods without major disturbances, current mixed pine–oak forests could be eliminated and replaced by broadleaved canopy dominant species. Yet these mesic sites are rather scarce in the study region and seem to be concentrated in the northern highland region. There, species of *Persea, Styrax, Clethra, Olmediella, Podocarpus, Weinmannia,* and even *Quercus acatenangensis* and *Q. laurina* can dominate the canopy (Ramírez-Marcial *et al.*, 2001).

Challenges for the Development of a Mechanistic Theory of Forest Dynamics

Large-scale analyses and local analyses of recruitment patterns as well as predictions of stand dynamics offer some insights into possible mechanisms contributing to the maintenance of tree species richness in this region. Species segregation and coexistence patterns in the Central Highlands of Chiapas provide partial evidence of niche differentiation along environmental gradients. Pine and oak species of other regions can segregate along altitudinal gradients that often correlate with the variability in resources (e.g. water availability) or environmental conditions (e.g. temperature or radiation) (Barton, 1993; Zavala *et al.*, 2000). Segregation effects are particularly strong in species such as *P. oocarpa, Q. candicans, Q. segoviensis* or *Q. crispilis,* which appear to be associated with specific conditions (e.g. relatively dry sites). In contrast, there is also significant overlapping in the distributions of other species, such as *P. ayacahuite, P. montezumae, P. pseudostrobus, P. tecunumanii, Q. crassifolia, Q. laurina* and *Q. rugosa,* which exhibit similar distributional areas and are often found in mixed stands. Spatial autocorrelation operates in model residuals at scales around 30–40 km, suggesting that possible unaccounted for environmental effects that operate at these scales would in fact segregate species. Alternatively, historical contingencies and chance may interact with ecologically driven mechanisms to produce these patterns. Thus, at regional scales the *Pinus* and *Quercus* segregation patterns show some evidence of niche-based mechanisms (e.g. habitat preferences through seedling establishment) that tend to segregate species along environmental gradients (e.g. associated with altitude and winter rainfall). Spatial scales of aggregation, in turn, suggest that microclimatic or environmental factors not considered in this study may be operating, or that, as suggested by neutral theory, stand composition at intermediate scales may be dominated by stochastic events (Hubbell, 2001).

At a local scale, recruitment patterns depart from randomness, and suggest a strong bias towards habitat specificity and differential regeneration niches for pines, oaks and broadleaved species. Self-replacement by pines is rather low and pine recruitment is almost absent from other forest types. This agrees with the idea of pines as colonizer species that can colonize open, recently disturbed sites (González-Espinosa *et al.*, 1991; Richardson and Bond, 1991), given their dispersal capability and growth patterns (Rejmánek and Richardson, 1996; Richardson, 1998). In contrast, oak juveniles are more frequently found under pine cover (Galindo-Jaimes *et al.*, 2002) and several studies have emphasized the positive effects of intermediate closed canopies on the germination and establishment of *Q. crassifolia* (Ramírez-Marcial *et al.*, 1996), *Q. rugosa* (López-Barrera and González-Espinosa, 2001) and *Q. laurina* (Camacho-Cruz *et al.*, 2000). Finally, broadleaved species find a favourable habitat for effective establishment in oak-dominated stands or under themselves in the case of *C. macrophylla*, *P. americana* and *S. argenteus*, and are absent from pine stands. This has been interpreted as a result of the relatively more humid and temperate conditions encountered under oak stands (Camacho-Cruz *et al.*, 2000; Romero-Nájera, 2000; Galindo-Jaimes *et al.*, 2002).

Previous studies have suggested that the structure and dynamics of pine–oak forests are primarily the result of land-use shifts and individual species responses to resource heterogeneity and disturbances (Hong *et al.*, 1995; Vetaas, 1997; Ramírez-Marcial *et al.*, 2001; Galindo-Jaimes *et al.*, 2002). Differential establishment patterns among these three groups (pines, oaks and broadleaved species) support the idea of niche-based mechanisms operating during the regeneration process (i.e. regeneration niches *sensu* Grubb, 1977) along a light- and possibly a water-stress gradient. Species microhabitat preferences seem to indicate the existence of facilitation mechanisms underlying secondary forest succession with pine facilitating oaks and oaks facilitating broadleaved species. These mechanisms result, according to the Markovian model, in predictable successional patterns driven by facilitation and competition and tend to converge towards a forest dominated by a broadleaved canopy.

Some studies, however, suggest that pines and oaks either tend to exhibit mutual exclusion or tend to coexist depending on disturbance (particularly chronic human disturbance) and climatic conditions (González-Espinosa *et al.*, 1991; Ramírez-Marcial *et al.*, 2001; Galindo-Jaimes *et al.*, 2002). Particularly in drier or highly disturbed sites, the positive effects of pine cover on oak seedling establishment may not be enough to facilitate oak regeneration and stand composition may converge towards a pine-dominated equilibrium driven by autosuccessional dynamics (Zavala, 1999). The rate of recovery and transition, if any, from pine to oaks in these stands may be rather slow and depend both on rainfall variability and proximity to seed sources (Zavala and Zea, 2004). As a result, pines, oaks and canopy broadleaves would conform to a compositional gradient across the landscape associated with water availability, time since last disturbance and history.

Relative seedling or sapling density under canopy-dominant trees, however, may be a poor predictor of the tree-by-tree replacement process. A more detailed account of critical population stages and species differential performance (e.g. germination, establishment and seedling and sapling growth and mortality) along resource and disturbance gradients may be required to properly describe stand dynamics (see Pacala *et al.*, 1994; Kobe *et al.*, 1995). Specifically, trade-offs in species' ability to tolerate shade, drought and repeated disturbances can, along with dispersal, explain pine and oak coexistence in heterogeneous landscapes by means of niche-based mechanisms (Zavala *et al.*, 2000; Zavala and Zea, 2004). The existence of niche-based mechanisms operating at the regeneration stage may partially explain the segregation of species along a successional gradient, while species-specific habitat requirements could help explain species segregation along altitudinal and rainfall gradients. It is unlikely, however, that these mechanisms can fully explain the existence of roughly 11 pine and 23 oak species over an area of approximately 11,000 km^2, and the maintenance of mixed stands across the region (Alba-López *et al.*, 2003; González-Espinosa *et al.*, 2006). The scales of unaccounted spatial variation and the existence of predictable successional dynamics associated with guild groups rather than species (pines, oaks, canopy broadleaves and understorey broadleaves) also supports the idea of neutral processes operating within guilds or functional groups. Therefore, as suggested by Purves and Pacala (2005), the balance between neutrality and niche structuring mechanisms may depend on the dimensionality of the niche structure being low compared to the number of species in the community. That is, when within-guild diversity is sufficiently large, some functional equivalence in the community may occur. According to our results, mixed pine–oak forests of the Highlands of Chiapas are intermediate between highly diverse humid tropical forests and more simplified pine–oak temperate forests, with contrasting segregation and coexistence among a given number of functional groups or guilds, but also with neutral mechanisms driving community composition at more local scales.

Conservation and Restoration Implications of Biological Diversity in Fragmented Forest Landscapes

Current rates of deforestation, land-use changes and forest fragmentation are high in the study region (Chapters 2 and 3) and in other highly populated neotropical montane areas (Ochoa-Gaona and González-Espinosa, 2000; Kappelle, 2004, 2006). Yet, even in those cases where forest fragments are large enough to be considered under some category of 'forest cover' in broad-scale inventories (e.g. Palacio-Prieto *et al.*, 2000), detailed site-based studies indicate severe modifications in structure and floristic composition of the remaining forest patches (González-Espinosa *et al.*, 1995; Ramírez-Marcial *et al.*, 2001; Galindo-Jaimes *et al.*, 2002; Ochoa-Gaona *et al.*, 2004).

On a more theoretical level, studies have focused on the mechanisms involved in environmental degradation at the local community scale (e.g. Pacala and Tilman, 1994). However, both biogeography and evolutionary theory indicate that processes of species coexistence may also be driven by factors acting at a regional scale, not solely at the local scale (Ricklefs, 1987, 2004). Also, the impact of fragmentation on community dynamics may be rather slow and its effects may only become detectable after decades (Laurance *et al.*, 2001). Therefore, there is a need for the development of management tools that link broad-scale patterns of species diversity driven by regional ecological gradients and socio-economic forces (Román-Cuesta *et al.*, 2003) with local processes that can reflect planning and assessment of forest dynamics at the stand level. Predictive models of possible responses of forest ecosystems to human-induced environmental change seem promising (e.g. see Chapters 9 and 11), especially for assessing the long-term impact of fragmentation and forest clearance on species diversity.

The richness of the floristic pool involved in conservation and restoration of neotropical montane forests is considerable (Ramírez-Marcial *et al.*, 2005, 2006; González-Espinosa *et al.*, 2006). Yet it is very difficult (perhaps impossible) to parameterize ecological models that maximize generality, realism and precision simultaneously (Lawton, 1999; Purves *et al.*, 2007). Some previous studies take approaches that focus on the interaction of a single species with its environment, moving away from the biological processes that operate at the community level. Here we model stand-level dynamics with a simple model that seeks to predict system responses considering four basic functional groups: pines, oaks, canopy broadleaved and understorey broadleaved trees. The groups, particularly the latter two, may be further broken down in order to provide more realism for predictions according to different regional habitat conditions and local site potential. Ongoing experimental, observational and modelling work will provide more detailed analysis of the behaviour of forest dynamics and tree diversity along gradients of chronic human disturbance.

Notwithstanding its relative simplicity, some advantages of restricting ourselves to considering these four basic groups might become apparent if the model can successfully be used to mimic some of the major forest development policies that may be adopted in the region in the near future. For example, models of the kind we have developed here might help to assess the impacts of widespread pine plantations or an induced increase in the abundance of pine following traditional forest use. These forest use options are becoming increasingly widespread at the regional level and are driving the replacement of stands of native tree species in other areas of Latin America (e.g. Guatemala, Honduras, Colombia, Chile), and there is a need for predicting mid- and long-term consequences of such increases in pine dominance in forested landscapes. On the other hand, the dynamics of two major functional groups that are explicitly considered in the model, oak species and understorey broadleaved trees, seems crucial to predicting habitat availability for other species. These two groups of trees are very diverse and seem to be related to novel forest use options based on non-conventional timber and non-timber uses, including fuelwood from broadleaved hardwoods (Marshall *et al.*, 2006). For example, epiphyte load and therefore its potential commercial harvesting are highly dependent on the canopy and understorey cover provided by oak species (e.g. Wolf and Konings, 2001; Wolf and Flamenco, 2003, 2005). There is therefore a need to predict oak dominance with greater detail at both the local and landscape spatial scales.

Recent advances in models of community dynamics show evidence that the interplay between regional and local scales in a spatial context strongly influence community responses to fragmentation (Solé *et al.*, 2004; Chave and Norden,

2007). Although the gap between theory and application is still considerable (Simberloff, 2004), these studies indicate that a deep understanding of the mechanisms underlying the maintenance of biological diversity is key for properly assessing the potential effects of fragmentation on biodiversity. If, as neutral theory and our results suggest, there is some functional equivalence among species ('guilds') and intra-guild diversity is driven by drift, then the maintenance of species richness may be critically related to area (Hubbell, 2001). Accordingly, for the purpose of biodiversity conservation, a single large reserve may be better than smaller multiple ones (e.g. Zavala and Burkey, 1997). On the other hand, if – as also suggested by our results – there is also evidence of niche structure at least among guilds, then maintenance of diversity (which requires inter-guild diversity) may require the maintenance of habitat heterogeneity. That is, the forest area preserved or to be restored should include a representative array of forest habitat types (e.g. as suggested by the filter approach, Zavala and Oria de Rueda, 1995). This would lend support to a conservation and restoration strategy based on establishing or maintaining forest corridors and diversity maintenance through natural dispersal from scattered patches of variable size (Damschen *et al.*, 2006). In this regard, an essential effect of fragmentation on species richness results from the inability of the species confined to patches to colonize a previously degraded habitat owing to their limited dispersal ability. This suggests that, in addition to preserving large fragments of the most important habitats, effective conservation and forest restoration practices may require a geometrical array of these fragments that maximizes spatial proximity among fragments and connectivity (Levey *et al.*, 2005; Damschen *et al.*, 2006; Purves *et al.*, 2007).

Acknowledgements

This work was financed by the European Commission, BIOCORES Project, INCO (Rey Benayas Jose M) IV Contract ICA4-CT-2001-10095. We thank Duncan Golicher and Luis Cayuela for their assistance with logistic models and spatial statistics and Pedro Quintana-Ascencio for his advice on logit analyses and project design.

References

Adams, J.M. and Woodward, F.I. (1989) Patterns in tree species richness as a test of the glacial extinction hypothesis. *Nature* 339, 699–701.

Alba-López, M.P., González-Espinosa, M., Ramírez-Marcial, N. and Castillo-Santiago, M.A. (2003) Determinantes de la distribución de *Pinus* spp. en la Altiplanicie Central de Chiapas, México. *Boletín de la Sociedad Botánica de México* 73, 7–15.

Barton, A.M. (1993) Factors controlling plant distributions: drought, competition, and fire in montane pines. *Ecological Monographs* 63, 367–397.

Bennett, K.D., Tzedakis, P.C. and Willis, K.J. (1991) Quaternary refugia of north European trees. *Journal of Biogeography* 18, 103–115.

Breedlove, D. (1981) *Flora of Chiapas. Part I: Introduction to the Flora of Chiapas*. California Academy of Sciences, San Francisco, California.

Camacho-Cruz, A., González-Espinosa, M., Wolf, J.H.D. and De Jong, B.H.J. (2000) Germination and survival of tree species in disturbed forests of the highlands of Chiapas, Mexico. *Canadian Journal of Botany* 78, 1309–1318.

Chave, J. (2004) Neutral theory and community ecology. *Ecology Letters* 7, 241–253.

Chave, J. and Norden, N. (2007) Changes of species diversity in a simulated fragmented neutral landscape. *Ecological Modelling* (in press).

Chesson, P. (2000) Mechanisms of maintenance of species diversity. *Annual Review of Ecology and Systematics* 31, 343–366.

Currie, D.J. and Paquin, V. (1987) Large-scale biogeographical patterns of species richness of trees. *Nature* 329, 326–327.

Damschen, E.I., Haddad, N.M., Orrock, J.L., Tewksbury, J.J. and Levey, D.J. (2006) Corridors increase plant species richness at large scales. *Science* 313, 1284–1286.

Galindo-Jaimes, L., González-Espinosa, M., Quintana-Ascencio, P.F. and García-Barrios, L.E. (2002) Tree composition and structure in disturbed stands with varying dominance by *Pinus* spp. in the highlands of Chiapas, Mexico. *Plant Ecology* 162, 259–272.

González-Espinosa, M., Quintana-Ascencio, P.F., Ramírez-Marcial, N. and Gaytán-Guzmán, P. (1991) Secondary succession in disturbed *Pinus–Quercus* forests in the highlands of Chiapas, México. *Journal of Vegetation Science* 2, 351–360.

González-Espinosa, M., Ochoa-Gaona, S., Ramírez-Marcial, N. and Quintana-Ascencio, P.F. (1995) Current land-use trends and conservation of old-growth forest habitats in the highlands of Chiapas, Mexico. In: Wilson, M.H. and Sader, S.A. (eds) *Conservation of Neotropical Migratory Birds in Mexico*. Maine Agriculture and Forest Experiment Station, Miscellaneous Publication 727, Orono, Maine, pp. 190–198.

González-Espinosa, M., Ramírez-Marcial, N. and Galindo-Jaimes, L. (2006) Secondary succession in montane pine–oak forests of Chiapas, México. In: Kappelle, M. (ed.) *Ecology and Conservation of Neotropical Oak Forests. Ecological Studies 185*. Springer, Berlin, Germany, pp. 209–221.

Grubb, P.J. (1977) The maintenance of species-richness in plant communities: the importance of the regeneration niche. *Biological Reviews* 52, 107–145.

Hawkins, B.A. and Porter, E.E. (2003) Relative influences of current and historical factors on mammal and bird diversity patterns in deglaciated North America. *Global Ecology and Biogeography* 12, 475–481.

Hong, S.-K., Nakagoshi, N. and Kamada, M. (1995) Human impacts on pine-dominated vegetation in rural landscapes in Korea and western Japan. *Vegetatio* 116, 161–172.

Horn, H. (1975) Markovian properties of forest succession. In: Cody, M.L. and Diamond, J. (eds) *Ecology and Evolution of Communities*. Belknap Press, Cambridge, Massachusetts, pp. 196–211.

Hubbell, S.P. (2001) *A Unified Neutral Theory of Biodiversity and Biogeography*. Princeton University Press, Princeton, New Jersey.

Hubbell, S.P. (2005) Neutral theory in community ecology and the hypothesis of functional equivalence. *Functional Ecology* 19, 166–172.

INEGI (1984a) Carta topográfica, E15DG2 (San Cristóbal de Las Casas), escala 1:50,000. SPP/INEGI Instituto Nacional de Estadística, Geografía y Informática, Mexico City, Mexico.

INEGI (1984b) Carta de efectos climáticos regionales, E15–11 (Tuxtla Gutiérrez), escala 1:250,000. SPP/INEGI, Instituto Nacional de Estadística, Geografía y Informática, Mexico City, Mexico.

INEGI (1985) Carta edafológica, E15–11 (Tuxtla Gutiérrez), escala 1:250,000. SPP/INEGI, Instituto Nacional de Estadística, Geografía y Informática, Mexico City, Mexico.

Janzen, D.H. (1970) Herbivores and the number of tree species in tropical forests. *American Naturalist* 104, 501–508.

Kappelle, M. (2004) Tropical montane forests. In: Burley, J., Evans, J. and Youngquist, J.A. (eds) *Encyclopedia of Forest Sciences,* Volume 4. Elsevier, Oxford, UK, pp. 1782–1793.

Kappelle, M. (ed.) (2006) *Ecology and Conservation of Neotropical Montane Oak Forests. Ecological Studies 185*. Springer, Berlin, Germany.
Kobe, R.K., Pacala, S.W., Silander, J.A. and Canham, C.D. (1995) Juvenile tree survivorship as a component of shade tolerance. *Ecological Applications* 5, 517–532.
Laurance, W.F., Cochrane, M.A., Bergen, S., Fearnside, P.M., Delamônica, P., Barber, C., D'Angelo, S. and Fernandes, T. (2001) The future of the Brazilian Amazon. *Science* 291, 438–439.
Lawton, J.H. (1999) Are there general laws in ecology? *Oikos* 84, 177–192.
Levey, D.J., Bolker, B.M., Tewksbury, J.J., Sargent, S. and Haddad, N.M. (2005) Effects of landscape corridors on seed dispersal by birds. *Science* 309, 146–148.
Levins, R. and Culver, D. (1971) Regional coexistence of species and competition between rare species. *Proceedings of National Academy of Sciences, USA* 68, 1246–1248.
López-Barrera, F. and González-Espinosa, M. (2001) Influence of litter on emergence and early growth of *Quercus rugosa*: a laboratory study. *New Forests* 21, 59–70.
Marshall, E., Schreckenberg, K. and Newton A.C. (2006) *Commercialization of Non-Timber Forest Products. Factors Influencing Success: Lessons Learned from Mexico and Bolivia and Policy Implications for Decision-Makers*. United Nations Environmental Programme World Conservation Monitoring Centre, Cambridge, UK.
Ochoa-Gaona, S. and González-Espinosa, M. (2000) Land use deforestation in the highlands of Chiapas, Mexico. *Applied Geography* 20, 17–42.
Ochoa-Gaona, S., González-Espinosa, M., Meave, J.A. and Sorani-dal Bon, V. (2004) Effect of forest fragmentation on the woody flora of the highlands of Chiapas, Mexico. *Biodiversity and Conservation* 13, 867–884.
Pacala, S.W. and Tilman, D. (1994) Limiting similarity in mechanistic and spatial models of plant competition in heterogeneous environments. *American Naturalist* 143, 222–257.
Pacala, S.W., Canham, C.D., Silander, J.A. and Kobe, R.K. (1994) Sapling growth as a function of resources in a north temperate forest. *Canadian Journal of Forest Research* 24, 2172–2183.
Palacio-Prieto, J.L., Bocco, G., Velásquez, A., Mas, J.-F., Takaki-Takaki, F., Victoria, A., Luna-González, L., Gómez-Rodríguez, G., López-García, J., Palma-Muñoz, M., Trejo-Vázquez, I., Peralta-Higuera, A., Prado-Molina, J., Rodríguez-Aguilar, A., Mayorga-Saucedo, R. and González-Medrano, F. (2000) La condición actual de los recursos forestales en México: resultados del Inventario Forestal Nacional 2000. *Investigaciones Geográficas* 43, 183–203.
Purves, D.W. and Pacala, S.W. (2005) Ecological drift in niche-structured communities: neutral pattern does not imply neutral process. In: Burslem, D., Pinard, M. and Hartley, S. (eds) *Biotic Interactions in the Tropics*. Cambridge University Press, Cambridge, UK.
Purves, D., Zavala, M.A., Ogle, K., Prieto, F. and Rey-Benayas, J.M. (2007) Distribution of three *Quercus* species in a Mediterranean landscape: environmental forcing, dispersal and metapopulation dynamics. *Ecological Monographs* (in press).
Ramírez-Marcial, N., González-Espinosa, M. and García-Moya, E. (1996) Establecimiento de *Pinus* spp. y *Quercus* spp. en matorrales y pastizales de los altos de chiapas, México. *Agrociencia*, 30, 249–257.
Ramírez-Marcial, N., González-Espinosa, M. and Williams-Linera, G. (2001) Anthropogenic disturbance and tree diversity in montane rain forests in Chiapas, Mexico. *Forest Ecology and Management* 154, 311–326.
Ramírez-Marcial, N., Camacho-Cruz, A. and González-Espinosa, M. (2005) Potencial florístico para la restauración de bosques en los altos y las montañas del norte de Chiapas. In: González-Espinosa, M., Ramírez-Marcial, N. and Ruiz-Montoya, L. (eds) *Diversidad Biológica en Chiapas*. Plaza y Valdés, Mexico City, Mexico, pp. 325–363.
Ramírez-Marcial, N., Camacho-Cruz, A., González-Espinosa, M. and López-Barrera, F. (2006) Establishment, survival and growth of tree seedlings under successional montane oak forests in Chiapas, Mexico. In: Kappelle, M. (ed.) *Ecology and Conservation of Neotropical Montane Oak Forests, Ecological Studies 185*. Springer, Berlin, Germany, pp. 177–189.

Rejmánek, M. and Richardson, D.M. (1996) What attributes make some plants species more invasive? *Ecology* 77, 1655–1661.
Richardson, D.M. (1998) *Ecology and Biogeography of Pinus*. Cambridge University Press, Cambridge, UK.
Richardson, D.M. and Bond, W.J. (1991) Determinants of plant distribution: evidence from pine invasions. *The American Naturalist* 137, 639–668.
Ricklefs, R.E. (1987) Community diversity: relative roles of local and regional processes. *Science* 235, 167–171.
Ricklefs, R.E. (2004) A comprehensive framework for global patterns in biodiversity. *Ecology Letters* 7, 1–15.
Román-Cuesta, R.M., Gracia, M. and Retana, J. (2003) Environmental and human factors influencing fire trends in ENSO and non-ENSO years in tropical Mexico. *Ecological Applications* 13, 1177–1192.
Romero-Nájera, I. (2000) Estructura y condiciones microambientales en bosques perturbados de los altos de Chiapas, México. BSc thesis. Universidad Nacional Autónoma de México, DF, Mexico.
Rzedowski, J. (1993) Diversity and origins of the phanerogamic flora of Mexico. In: Ramamoorthy, T.P., Bye, R., Lot, A. and Fa, J. (eds) *Biological Diversity of Mexico: Origins and Distribution*. Oxford University Press, New York, pp. 129–144.
Simberloff, D. (2004) Community ecology: is it time to move on? *American Naturalist* 163, 787–799.
Solé, R., Alonso, D. and Saldaña, J. (2004) Habitat fragmentation and biodiversity collapse in neutral communities. *Ecological Complexity* 1, 65–75.
Tilman, D. (1988) *Plant Strategies and the Dynamics and Structure of Plant Communities. Monographs in Population*. Princeton University Press, Princeton, New Jersey.
Usher, M.B. (1981) Modeling ecological succession with particular reference to Markovian models. *Vegetatio* 46, 11–18.
Usher, M.B. (1992) Statistical models of succession. In: Glenn-Lewin, D.C., Peet, R.K. and Veblen, T.T. (eds) *Plant Succession – Theory and Prediction*. Chapman and Hall, London, UK, pp. 215–248.
Vetaas, O.R. (1997) The effect of canopy disturbance on species richness in central Himalayan oak forests. *Plant Ecology* 132, 29–38.
Whittaker, R.H. (1975) *Communities and Ecosystems*. Macmillan, New York.
Wolf, J.H.D. and Flamenco, A. (2003) Patterns in species richness and distribution of vascular epiphytes in Chiapas, México. *Journal of Biogeography* 30, 1689–1707.
Wolf, J.H.D. and Flamenco, A. (2005) Distribución y riqueza de epífitas de Chiapas. In: González-Espinosa, M., Ramírez-Marcial, N. and Ruiz-Montoya, L. (eds) *Diversidad Biológica en Chiapas*. Plaza y Valdés, Mexico City, Mexico, pp. 127–162.
Wolf, J.H.D. and Konings, C.J.F. (2001) Toward the sustainable harvesting of epiphytic bromeliads: a pilot study from the highlands of Chiapas, Mexico. *Biological Conservation* 101, 23–31.
Zavala, M. (1999) A model of stand dynamics for holm oak–Aleppo pine forests. In: Rodà, F., Retana, J., Gracia, C. and Bellot, J. (eds) *Ecology of Mediterranean Evergreen Oak Forests*. Springer, Berlin, Germany, pp. 105–117.
Zavala, M.A. and Burkey, T.V. (1997) Application of ecological models to landscape planning: the case of the Mediterranean basin. *Landscape and Urban Planning* 38, 213–227.
Zavala, M.A. and Oria de Rueda, J.A. (1995) Preserving biological diversity in managed forests: a meeting point for forestry and ecology. *Landscape and Urban Planning* 31, 363–378.
Zavala, M.A. and Zea, E. (2004) Mechanisms maintaining biodiversity in Mediterranean pine–oak forests: insights from a spatial simulation model. *Plant Ecology* 171, 197–207.
Zavala, M.A., Espelta, J.M. and Retana, J. (2000) Constraints and trade-offs in Mediterranean plant communities: the case of mixed holm oak–Aleppo pine forests. *Botanical Review* 66, 119–149.

11 Process-based Modelling of Regeneration Dynamics and Sustainable Use in Species-rich Rainforests

N. Rüger, J.J. Armesto, A.G. Gutiérrez,
G. Williams-Linera and A. Huth

Old-growth forest of *Nothofagus dombeyi* (in the background) surrounded by secondary forest of *Drimys winteri* that originated as a result of logging of large trees for fuelwood in Chiloé Island, X Region, Chile. Photo: Cristian Echeverría

Summary

Sustainable use of species-rich moist forests needs an understanding of forest dynamics and the response of the forest to wood harvesting events. This chapter studies natural forest dynamics, explores the productivity of native managed forests and quantifies the ecological impacts of different management strategies. The process-based forest growth model FORMIND is applied to study natural forest succession and to assess long-term ecological implications of fuelwood extraction for tropical montane cloud forest in central Veracruz, Mexico, as well as to compare potential harvesting strategies for Valdivian temperate rainforest in northern Chiloé Island, Chile, regarding forest productivity and ecological consequences.

Simulation results show that both forest types have a relatively high potential for wood production. As wood extraction increases, the forest structure becomes simplified because large old trees disappear from the forest. The species composition shifts to tree species that are favoured by the respective harvesting strategy. The overall ecological impact increases linearly with the amount of extracted wood. Simulation results allow management strategies to be defined that balance conservation and production objectives, promote the regeneration of desired tree species, or minimize shifts in the species composition of the forest. Process-based forest models enhance our understanding of the dynamics of species-rich moist forests and are indispensable tools to assess long-term implications of anthropogenic impacts on forest ecosystems. They can thereby contribute to the conservation and sustainable use of native forests outside protected areas.

Introduction

Due to massive deforestation and forest fragmentation in many regions of the world, conservation, sustainable management and restoration of native forests have become major goals for numerous governmental and non-governmental organizations. The tropical montane forests in Mexico and temperate forests in southern South America, which were the focus of the research described here, have traditionally received less scientific and public attention than tropical lowland rainforests, yet they provide important ecosystem goods and services at global, regional and local scales (Chapter 1).

Apart from protection within national parks or reserves, ecologically appropriate management of forests can contribute to the conservation of native biodiversity and ecosystem services (e.g. Lindenmayer and Franklin, 2002; Fredericksen and Putz, 2003). To determine which types of management are appropriate and sustainable, information on long-term forest dynamics under different disturbance or management regimes is required. An understanding of long-term forest dynamics is often lacking owing to the long timescales of forest development and the lack of long-term experience with forest management. Even the global standards for certification of sustainable forest management defined by the Forest Stewardship Council (FSC, 2004) only demand a 'rationale for rate of annual harvest and species selection' because quantitative tools for the determination of sustainable cutting limits or the estimation of ecological consequences of different management options are in most cases unavailable.

This chapter aims to contribute to an ecologically appropriate use of species-rich moist forests by addressing three general objectives: (i) to gain a better understanding of natural forest dynamics; (ii) to explore the productivity of the native forests under different management scenarios; and (iii) to quantify ecological impacts of these anthropogenic disturbances. The chapter focuses on two study regions: tropical montane cloud forest (TMCF) in central Veracruz, Mexico, and Valdivian temperate rainforest (VTRF) in northern Chiloé Island, Chile. The specific objectives reflect the different socio-economic context in the two study regions.

In central Veracruz, Mexico, land use is highly diverse and fragmented. Agricultural fields, pastures and shade-coffee plantations are intermingled with old-growth TMCF forest fragments and secondary forests that are re-growing after abandonment of previous land uses such as cultivation of coffee or cattle grazing (Williams-Linera *et al.*, 2002). Until the present time, most fuelwood consumed for cooking and heating in the region has been obtained from old-growth TMCF fragments, where individual people regularly cut large living trees for their own needs or to supply local markets. This type of wood extraction has a long tradition. However, it is unclear what ecological consequences such extraction has for forest structure and composition in the long term. Hence, the specific objectives regarding TMCF in central Veracruz, Mexico, are to simulate natural forest succession and to investigate long-term impacts of repeated low-intensity selective logging on forest structure and composition.

The case of VTRF in southern Chile is different. Dynamics of these forests are complex and not well known. The forests on the study site are apparently not in equilibrium, as there is no regeneration of the relatively shade-intolerant *Eucryphia cordifolia*, which is present almost exclusively as large, mature individuals. Furthermore, pristine native forests are severely threatened by conversion to pure plantations of exotic species, because there is very little experience with their management, and because they are considered to be too complex to be managed. Therefore, the specific objectives regarding VTRF in southern Chile are to study long-term forest dynamics under different disturbance regimes as well as to show that the native forests have a silvicultural potential and to explore different management strategies as regards their productivity and ecological impacts.

There are two potential approaches to address these questions. First, one could design and conduct experiments combined with long-term monitoring of forest response to different silvicultural treatments. However, the design, execution and monitoring of large silvicultural experiments are costly and operationally difficult. Therefore modelling approaches that are complementary to experimental studies are needed to assess the long-term consequences of different management options and to provide guidelines for forest managers and planners aiming at reconciling conservation and production objectives (e.g. Lindenmayer and Franklin, 2002).

We apply the process-based forest model FORMIND to address the questions posed. The individual-oriented approach of FORMIND allows for a detailed incorporation of different logging scenarios. FORMIND was developed in the late 1990s at the Center for Environmental Systems Research of the University of Kassel, Germany. Its relatively easy parameterization has allowed a successful application to tropical lowland rainforest in several regions of the world and made it one of the most widely applied models of species-rich forests. It has been used to study forest dynamics and effects of logging, fragmentation and climate change in Malaysia (Köhler *et al.*, 2001; Huth *et al.*, 2004, 2005; Köhler and Huth, 2004), sustainable timber harvesting in Venezuela and Paraguay (Kammesheidt *et al.*, 2001, 2002), and fragmentation effects in French Guyana (Köhler *et al.*, 2003).

The simulations described in this chapter were performed to enhance our understanding of the dynamics of species-rich rainforests. Simulation results contribute to the conservation and use of native biodiversity outside protected areas by providing guidelines for future sustainable management and highlighting their potential for provision of ecosystem services (Franklin, 1993; Armesto *et al.*, 1998).

Methods

Study areas

Central Veracruz

TMCF in central Veracruz, Mexico (19° 30′ N, 96° 54′ W) occurs at an altitude between 1200 and 2000 m. The climate is mild and humid throughout the year, with three seasons. A relatively dry-cool season extends from November to March, a dry-warm season from April to May and a wet-warm season from June to October. Annual precipitation varies between 1350 and 2200 mm; mean annual temperature is between 12 and 18°C. The soil has been classified as andosol. Dominant tree species include *Carpinus caroliniana, Clethra mexicana, Fagus grandifolia, Liquidambar styraciflua, Quercus germana, Quercus leiophylla, Quercus xalapensis* and *Turpinia insignis* (Williams-Linera, 2002).

Five old-growth forest fragments were selected as reference sites for this study. To enable an individual-based simulation of forest dynamics, the 58 native tree species that occur at the study sites were grouped into plant functional types (PFTs). Criteria for classification into PFTs were light demand and maximum attainable height. Three levels of shade tolerance were distinguished: shade-intolerant (i), intermediate (m) and shade-tolerant (t). Three height groups were considered: small trees (≤ 15 m tall, ≤ 35 cm dbh), canopy trees (≤ 25 m tall, ≤ 80 cm dbh) and emergent trees (≤ 35 m tall, ≤ 100 cm dbh). This classification resulted in six PFTs, because some of the combinations are rare (Table 11.1).

Chiloé

The study site used as a reference for the simulations was a large remnant (200 ha) of VTRF located in Guabún, Chiloé Island, Chile (41° 50′ S), about 30 km north-west of Ancud. The prevailing climate is wet-temperate, with a strong oceanic influence (Di Castri and Hajek, 1976). Rainfall occurs throughout the year. The nearest meteorological station in Punta Corona (41° 47′ S, 73° 52′ W) has an annual average of 2444 mm of rainfall and a mean annual temperature of 10.7°C. Mean maximum and minimum monthly temperatures are 13.8°C (January) and 8.3°C (July). Floristically, this forest type is dominated

Table 11.1. Definition of plant functional types (PFTs) according to shade tolerance (T) and maximum attainable height (H_{max}). Three levels of shade tolerance are distinguished: i, shade-intolerant; m, intermediate; t, shade-tolerant. The successional status refers to the stage of succession in which a PFT attains maximum basal area values.

Plant functional type	PFT	T	H_{max}	Examples
Early successional small trees	1	i	15 m	*Heliocarpus*, *Myrsine*
Mid-successional small trees	2	m	15 m	*Miconia*, *Oreopanax*
Late successional small trees	3	t	15 m	*Cinnamomum*, *Ilex*
Mid-successional canopy trees	4	m	25 m	Some *Quercus* spp.
Late successional canopy trees	5	t	25 m	*Magnolia*, *Beilschmiedia*
Emergents	6	m	35 m	*Liquidambar*, *Clethra*

by *Eucryphia cordifolia* (Eucryphiaceae), *Aextoxicon punctatum* (only member of the endemic Aextoxicaceae), *Laureliopsis philippiana* (Atherospermataceae) and several tree species of the Myrtaceae family.

In this study we focused on four PFTs, three of which each correspond to a single species and one to a species group. *E. cordifolia* is a canopy-emergent species (up to 40 m in height and 2 m in diameter). It is considered light-demanding and requires medium to large-scale disturbances for establishment (Donoso *et al.*, 1985; Veblen, 1985). *A. punctatum* and *L. philippiana* are shade-tolerant species occurring in the main canopy of the forest (e.g. Donoso *et al.*, 1999). They reach heights of 30 m and diameters of up to 1 m. Finally, five tree species in the Myrtaceae family (*Amomyrtus luma, Amomyrtus meli, Luma apiculata, Myrceugenia ovata* and *Myrceugenia planipes*) were combined in one species group because of their similar ecological characteristics. They are shade-tolerant species with maximum heights of 15–20 m, which often dominate the lower canopy (e.g. Donoso *et al.*, 1999). A few other tree species (e.g. *Drimys winteri, Pseudopanax laetevirens*) occur at the study site, but they are relatively rare and were not included in the simulations.

The process-based forest growth model FORMIND

The individual-oriented forest growth model FORMIND simulates the spatial and temporal dynamics of uneven-aged mixed species forest stands (e.g. Köhler and Huth, 1998; Kammesheidt *et al.*, 2001; Köhler *et al.*, 2001, 2003). The model simulates a forest (in annual time steps) as a mosaic of interacting grid cells with a size of 20 m × 20 m, which is the approximate crown size of a large mature tree. It is assumed that light availability is the main driver of individual tree growth and forest succession. Within each grid cell all trees compete for light and space following the gap model approach (Shugart, 1984). For the explicit modelling of the competition for light, each grid cell is divided into horizontal layers. In each height layer the leaf area is summed up and the light climate in the forest interior is calculated via an extinction law. The carbon balance of each individual tree is modelled explicitly, including the main physiological processes (photosynthesis, respiration) and litter fall. Growth process equations are modified from the models FORMIX3 and FORMIX3-Q (Ditzer *et al.*, 2000; Huth and Ditzer, 2000, 2001). Allometric functions relate above-ground biomass, stem diameter, tree height, crown diameter and stem volume. Tree mortality can occur either through self-thinning in densely populated grid cells, senescence, gap formation by large falling trees, or medium-scale windthrows (800–1600 m^2) in the case of Chiloé. Gap formation links neighbouring grid cells. Tree regeneration rates are effective rates of recruitment of small trees at a diameter at breast height (dbh) threshold of 1 cm, with seed loss through predation and seedling mortality already being implicitly incorporated. Water and nutrient availability are assumed to be homogeneous and there is no inter-annual variability of climatic conditions in the model.

A detailed model description that follows a standard protocol for describing individual- and agent-based models can be found in the online appendix of Grimm *et al.* (2006) and in Rüger (2006). Tables with model parameters for the two study regions are given in the Appendix (see also Rüger, 2006).

The model was tested by comparing long-term characteristics (steady state) of a simulated forest with field data from old-growth forests including maximum diameter increment for each PFT, diameter distribution and stem number and basal area for each PFT. Results showed that the model was able to reproduce these different field observations (Rüger *et al.*, 2007a, c). Additionally, independent data from a chronosequence approach (Muñiz-Castro *et al.*, 2006) could be used to validate model predictions regarding the regeneration of TMCF (Rüger *et al.*, 2007b).

Forest regeneration

To study the successional dynamics of the forests we simulated the regeneration of Mexican TMCF and Chilean VTRF after large-scale disturbance, such as clearcutting. We assumed that seed input is not limited. In the case of TMCF, we assumed that no further disturbances – other than gap creation by falling trees – occur during the course of succession. In the case of VTRF, we simulated two different disturbance scenarios – with and without occasional windthrow events which create canopy gaps of 800–1600 m^2 – to study the dependence of the current forest composition on natural medium-scale disturbances. The assumed probability that such a disturbance occurs is 0.008 ha $year^{-1}$. Ten simulations were carried out for a simulation area of 1 ha and 400 years (TMCF) or 1500 years (VTRF), and mean and standard deviation of the basal area of the different PFTs were determined.

Logging scenarios

Central Veracruz

We simulated selective logging scenarios of old-growth TMCF by varying the extracted stem volume. There are few data available on actual wood extraction. For this reason, we varied the logging intensity in a broad range

(5–100 m^3 ha^{-1} with a logging cycle of 10 years). Total standing wood volume of an undisturbed old-growth forest is *c.*500 m^3 ha^{-1}, and 1–20% of total wood volume is extracted every 10 years by the logging scenarios. We used inventory data from the study sites as initial condition and then simulated forest dynamics over a 100-year period to allow the model to establish a steady-state old-growth forest. Selective logging scenarios were then applied over a simulation time of 400 years (i.e. time steps 100–500 in the model).

Four selective logging scenarios were simulated (Table 11.2). In the first two scenarios (*S1, S2*), only trees of PFTs 4 and 6 were logged, because preferred tree species for fuelwood (e.g. *Quercus* spp., *L. styraciflua, C. caroliniana* and *C. mexicana*) were mainly classified as PFTs 4 and 6. Scenarios *S3* and *S4* allowed logging of all canopy species. In scenarios *S1* and *S3*, logging concentrated on medium-sized trees with a dbh of 40–60 cm which are preferentially cut in the study area for fuelwood and charcoal production for local market supply (G. Williams-Linera, personal observation). Scenarios *S2* and *S4* allowed cutting of all trees >40 cm dbh. Within the range of allowed dbh values, the largest trees were logged first. If at a given time step the stem volume of all harvestable trees in the simulation area did not reach the volume value aimed for by the logging scenario, the respective logging operation was omitted. This was done to keep logging scenarios comparable and to clearly reveal the limits of a sustained fuelwood extraction. Felled trees were directed to already existing gaps if possible. Apart from trees that were destroyed by the falling tree, no additional logging damage was considered because wood extraction in the study area is carried out without heavy machinery but with the help of pack animals.

Chiloé

We simulated three logging strategies (selective logging with and without retention of large old trees and logging in bands), which either resemble current logging practices or which have been suggested as suitable options for management of VTRF (Donoso, 1989; Armesto *et al.*, 1999a). Within each strategy, different scenarios were simulated which varied the extracted wood volume and logging cycle (in the case of selective logging) and the logging cycle (in the case of band logging). The model was initialized with inventory data from an old-growth forest stand. Logging cycles were repeated over 400 years.

Table 11.2. Logged plant functional types (PFTs) and diameter classes used in simulations of selective logging scenarios. PFT 4, mid-successional canopy trees (e.g. *Carpinus caroliniana*); PFT 5, late successional canopy trees (e.g. *Quercus xalapensis*); PFT 6, mid-successional emergent trees (e.g. *Liquidambar styraciflua*).

Scenario	Logged PFT	Diameter range (cm)
S1	4, 6	40–60
S2	4, 6	>40
S3	4, 5, 6	40–60
S4	4, 5, 6	>40

Selective logging in this case refers to the extraction of trees with a dbh of 50–100 cm. The two selective logging strategies simulated here differ in the way large old and probably senescent trees are treated. In the first case (with retention of large trees), trees >1 m dbh are left standing, because they often exhibit heart rot, do not provide valuable timber, but provide habitat and resources for birds and other species (e.g. Díaz *et al.*, 2005). In the second case of selective logging (without retention of large trees), all trees >1 m dbh are removed prior to the simulation of logging scenarios to enhance growth of potential future crop trees by reducing shading. We varied the time between two sequential harvestings (logging cycle) from 10 to 50 years. For each logging cycle we also varied the volume of harvested wood (harvest aim) such that on an annual basis 1–10 m^3 ha^{-1} were harvested. For a logging cycle of 10 years this corresponds to harvesting 10–100 m^3 ha^{-1}, and for a logging cycle of 50 years to 50–500 m^3 ha^{-1}. When the harvestable volume was lower than the harvest aim, the logging operation was omitted. Within the diameter range of 50–100 cm, the largest trees were always logged first. Logging damage to the remaining trees was simulated as direct damage by the falling tree and additional damage due to skidding. We assumed reduced-impact logging where falling trees are directed to existing gaps if possible. No damage occurred to trees >50 cm dbh. Skidding damage was assumed to increase from 6% of the remaining vegetation when 10 m^3 ha^{-1} were harvested to 50% when 500 m^3 ha^{-1} were harvested (Rüger *et al.*, 2007c).

Logging in bands was simulated with clearcutting trees in 20 m wide bands. The size of the created gap was 0.2 ha ha^{-1}. The return time to each band was varied from 50 to 150 years. Skidding damage was assumed to be only 10%, because logging bands can be used to extract trees from the stand. No damage occurred to trees >50 cm dbh.

Assessment of logging scenarios

We evaluated the economic and ecological consequences of a given logging scenario on four variables, namely mean annual harvest, forest structure, forest composition and an overall ecological index, which measures the similarity of a logged forest to undisturbed old-growth forest.

Mean annual harvest was calculated for the logging period of 400 years under each logging scenario. To assess changes in forest structure, stem numbers in five (TMCF) or three (VTRF) diameter classes were analysed. In TMCF, diameter classes are 5–20 cm, 20–40 cm, 40–60 cm, 60–80 cm and 80–100 cm dbh. In VTRF, diameter classes are 5–50 cm, 50–100 cm and 100–200 cm dbh. The species composition was evaluated based on importance values (*IV*) of the different PFTs, calculated as

$$IV_i = \frac{1}{2}\left(\frac{ba_i}{ba_{total}} + \frac{n_i}{n_{total}}\right)$$

i.e. the normalized sum of relative basal area (*ba*, m^2 ha^{-1}) and relative density (*n*, trees ha^{-1}) of the focal PFT in relation to all PFTs (*total*). Both forest structure and composition were averaged over the last 100 years of the simulation time to exclude transient states of the forest from the analysis.

To contrast economic benefit and ecological impact of a logging scenario, we calculated an ecological index (*EI*), which measures the similarity of a logged forest to undisturbed old-growth forest (Rüger *et al.*, 2007c). *EI* takes into account two aggregate measures of the similarity of the structure and composition of the simulated forest to an unlogged control forest (index of structural change, *ISC* and index of compositional change, *ICC*) and the stand leaf area index as an indicator of erosion risk (*LAI*). In the case of VTRF, additionally the number of old trees (>1 m dbh, *OLD*) was included as an indicator for old-growth conditions and amount of habitat for species that depend on large old trees. The values of *ISC, ICC, LAI* and *OLD* were determined for each logging scenario and divided by the maximum value obtained from all logging scenarios (ISC_{max}, ICC_{max}, LAI_{max}, OLD_{max}), respectively, and summed up as follows:

$$EI = \frac{1}{4}\left(\left(1 - \frac{ISC}{ISC_{max}}\right) + \left(1 - \frac{ICC}{ICC_{max}}\right) + \frac{LAI}{LAI_{max}} + \frac{OLD}{OLD_{max}}\right)$$

Results

Forest regeneration

Central Veracruz

Figure 11.1 shows the course of succession over 400 years, starting from a treeless area. Total basal area reached a steady state after approximately 80–90 years at about 44 m^2 ha^{-1}. During the first 20 years, pioneer species (PFT 1) accounted for most of the stand's basal area, due to their fast growth. Then they were rapidly replaced by PFTs with intermediate shade tolerance (PFTs 2, 4, 6), which reached their maximum basal area after approximately 50 years. PFT 5, the slow-growing shade-tolerant canopy species, was the last in arriving at its steady-state basal area after approximately 300 years. Its increase in basal area was accompanied by a decrease of PFTs 4 and 6.

Chiloé

The simulation of long-term forest dynamics over a 1500-year period without external disturbances is shown in Fig. 11.2A. Total basal area reached a steady state after 300 years of simulation at approximately 95 m^2 ha^{-1}. The first 400 years of succession were dominated by *E. cordifolia*, which was then replaced by the shade-tolerant species and tended to disappear from the forest after approximately 800 years. In the long run, the myrtaceous species accounted for the highest basal area, followed by *L. philippiana* and *A. punctatum*. Incorporating natural medium-scale disturbances (e.g. multiple tree falls) into the model changed long-term forest dynamics (Fig. 11.2B). This is similar to simulating forest dynamics on a larger spatial scale, where forest patches representing different successional stages occur side by side. Again, at the beginning of stand succession, the forest

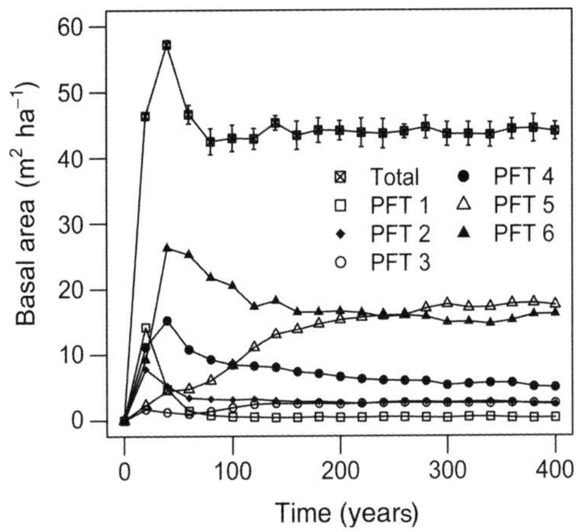

Fig. 11.1. Simulation of regeneration of tropical montane cloud forest in central Veracruz, Mexico, after large-scale disturbance. Basal area values of the different PFTs are means of ten simulations for 1 ha and 400 years (individuals ≥ 5 cm dbh). Standard deviation is shown for total basal area (from Rüger *et al.*, 2007a).

was dominated by *E. cordifolia* and was then gradually replaced by the shade-tolerant species. But in contrast to forest dynamics without medium-scale disturbances, *E. cordifolia* now persisted in the forest, with a few large *E. cordifolia* trees accounting for a large proportion of the stand's basal area. A steady state of the basal area of the three species and one species group was only reached after 1000 years. As a consequence of the medium-scale disturbances, fluctuations of basal area were stronger than in the simulations without disturbances.

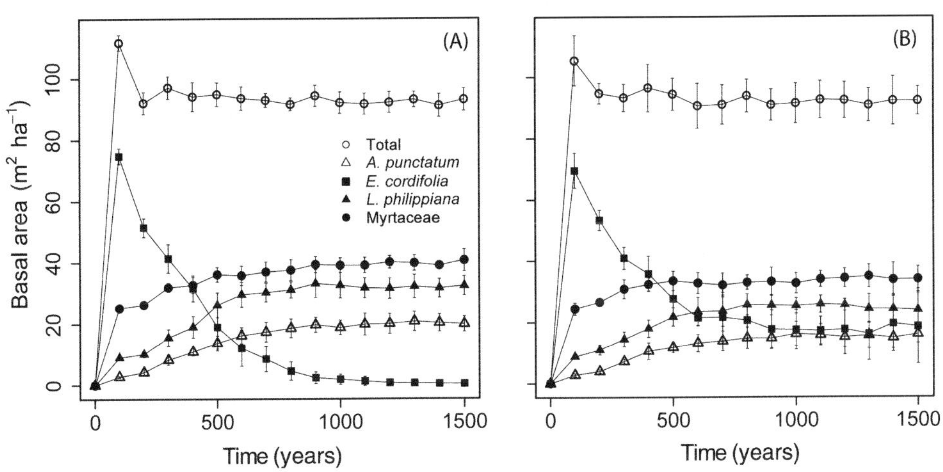

Fig. 11.2. Simulation of regeneration of Valdivian temperate rainforest in northern Chiloé Island, Chile, after large-scale disturbance. Basal area values of the different PFTs (A) without and (B) with medium-sized disturbances (windthrows). Mean and standard deviation of basal area for ten simulations for 1 ha and 1500 years (individuals ≥ 5 cm dbh) (from Rüger *et al.*, 2007c).

Logging scenarios

Central Veracruz

MEAN ANNUAL HARVEST Mean annual harvest over the 400-year period of logging is shown in Fig. 11.3. Only in the scenario where all canopy species >40 cm dbh were allowed to be harvested (*S4*) could the logging target always be met and up to 120 m^3 ha^{-1} could be harvested from the forest every 10 years (results >100 m^3 ha^{-1} not shown). Thus, mean annual harvest increased linearly up to 12 m^3 ha^{-1} $year^{-1}$. All other scenarios reached a limit of wood extraction. Restricting the diameter range of harvestable trees to 40–60 cm dbh (*S3*), this limit was at about 60 m^3 ha^{-1}. Beyond this threshold logging operations had to be omitted because there were not enough harvestable trees and mean annual harvest saturated at about 7 m^3 ha^{-1} $year^{-1}$. When only trees of PFT 4 and 6 were logged (scenarios *S1* and *S2*), only 20 and 30 m^3 ha^{-1} could be harvested from the forest every 10 years, respectively. If volumes >30 m^3 ha^{-1} were to be logged under scenario *S2*, the time lags between two logging events had to be prolonged, and mean annual harvest saturated at about 3.5 m^3 ha^{-1} $year^{-1}$. When the diameter range of logged trees was restricted to 40–60 cm dbh (*S1*), volumes higher than 50 m^3 ha^{-1} could never be achieved, and therefore no logging took place in these scenarios. Maximum mean annual harvest was only 2.5 m^3 ha^{-1} $year^{-1}$ under scenario *S1*.

FOREST STRUCTURE Figure 11.4 shows the changes of stem numbers in five diameter classes for the four logging scenarios. Stem numbers in the two smallest diameter classes (5–20 cm and 20–40 cm dbh) increased with increasing wood extraction for all scenarios. The intermediate diameter class (40–60 cm dbh) was the only diameter class directly affected by logging under scenarios *S1* and *S3*. Here, a decline of stem numbers was observed for scenarios *S1* and *S3*. For *S2*, stem numbers remained constant for low levels of wood extraction and slightly decreased for higher levels. For *S4* stem numbers increased up to a mean annual harvest of *c.*7.5 m^3 ha^{-1} $year^{-1}$ and then sharply declined. Stem numbers in the

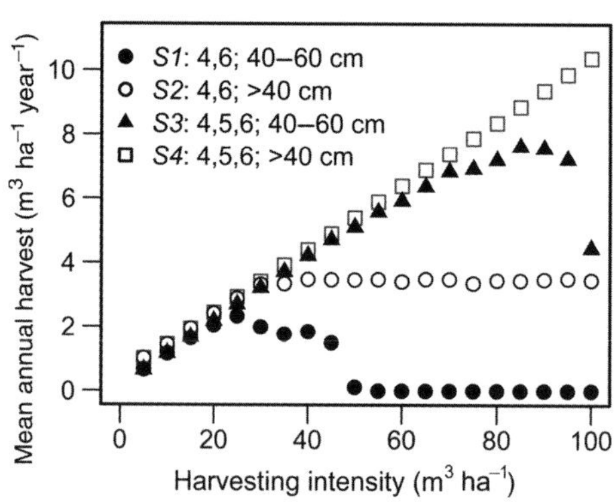

Fig. 11.3. Mean annual harvest for four selective logging scenarios (*S1–S4*) of tropical montane cloud forest in central Veracruz, Mexico. Harvesting intensity varies from 5 to 100 m^3 ha^{-1} that are harvested every 10 years for a simulation period of 400 years (from Rüger *et al.*, 2007b).

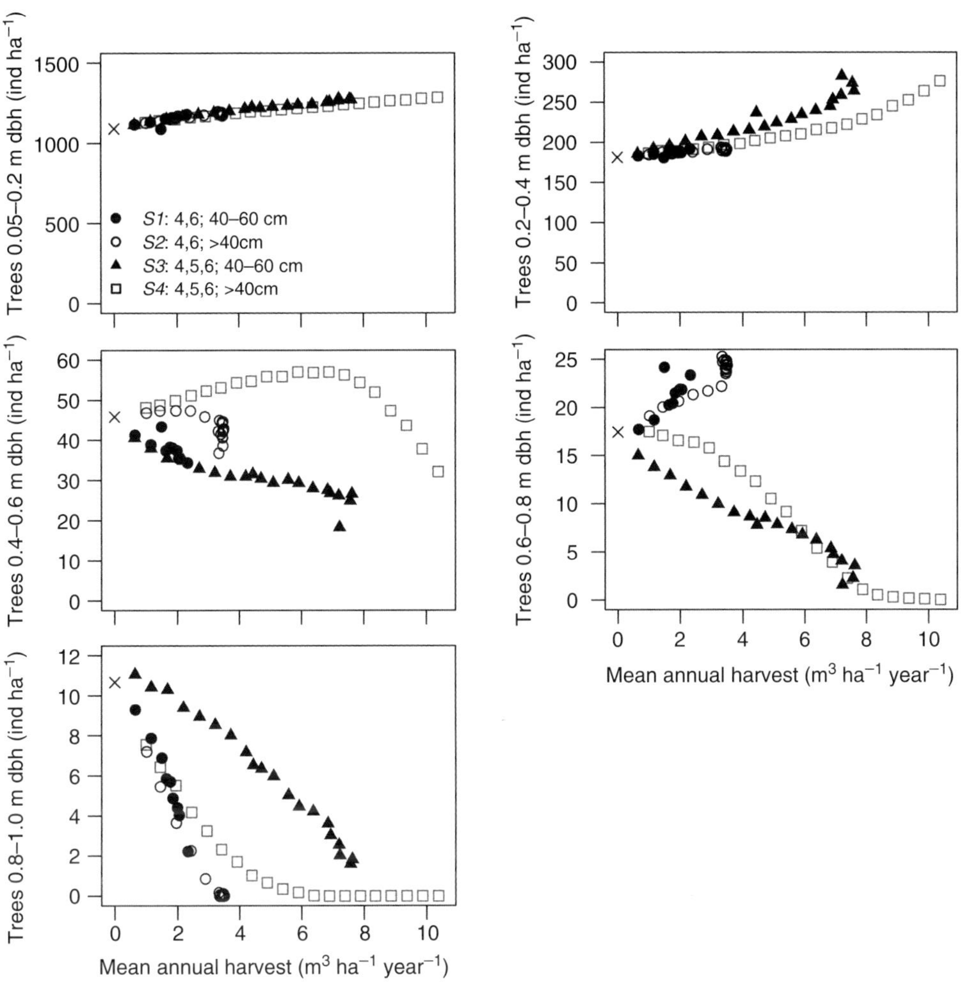

Fig. 11.4. Mean number of trees in five diameter classes for simulation time 400–500 years for four selective logging scenarios (*S1–S4*) of tropical montane cloud forest in central Veracruz, Mexico. Mean values for undisturbed old-growth forest are displayed for comparison (×) (from Rüger *et al.*, 2007b).

60–80 cm diameter class increased for scenarios *S1* and *S2* because they benefited from the decrease of emergent trees in the largest diameter class. For scenarios *S3* and *S4* they declined, in the case of *S4* to very low numbers for high levels of wood extraction. When *S4* was simulated with wood extraction levels >9.5 m^3 ha^{-1} year^{-1}, no trees >60 cm dbh remained in the forest. The strongest impact of logging scenarios was observed for the largest diameter class, which only contained emergent trees of PFT 6 (80–100 cm dbh). Even at low levels of wood extraction, stem numbers in this diameter class decreased sharply for all scenarios. For scenario *S3*, the decline occurred slightly more slowly with increasing wood extraction and under this scenario more large trees were maintained in the forest compared to the other scenarios.

FOREST COMPOSITION Impacts of logging scenarios on forest composition were measured by importance values of the 6 PFTs, which are based on relative stem numbers and basal area. The detailed impact of the logging scenarios on the species composition is shown in Fig. 11.5. Importance values for PFTs 1–4 were consistently altered by all the logging scenarios. Whereas importance values for PFT 1 remained constant, they increased slightly for PFTs 2, 3 and 4. Scenarios *S1* and *S2*, which only targeted PFTs 4 and 6, had the strongest impact on the forest composition. Importance of PFT 5 increased under these scenarios at the expense of PFT 6. Under scenarios *S3* and *S4* the importance of PFTs 5 and 6 decreased steadily for increasing levels of wood extraction.

ECOLOGICAL INTEGRITY VERSUS HARVEST With increasing mean annual harvest, similarity to undisturbed old-growth forest (*EI*) decreased almost linearly for all logging scenarios (Fig. 11.6). Hence, every surplus in harvest is accompanied by an increase of ecological impact. For scenarios *S1* and *S2*, the decrease was more pronounced than for *S3* and *S4*, due to their stronger impact on the species composition.

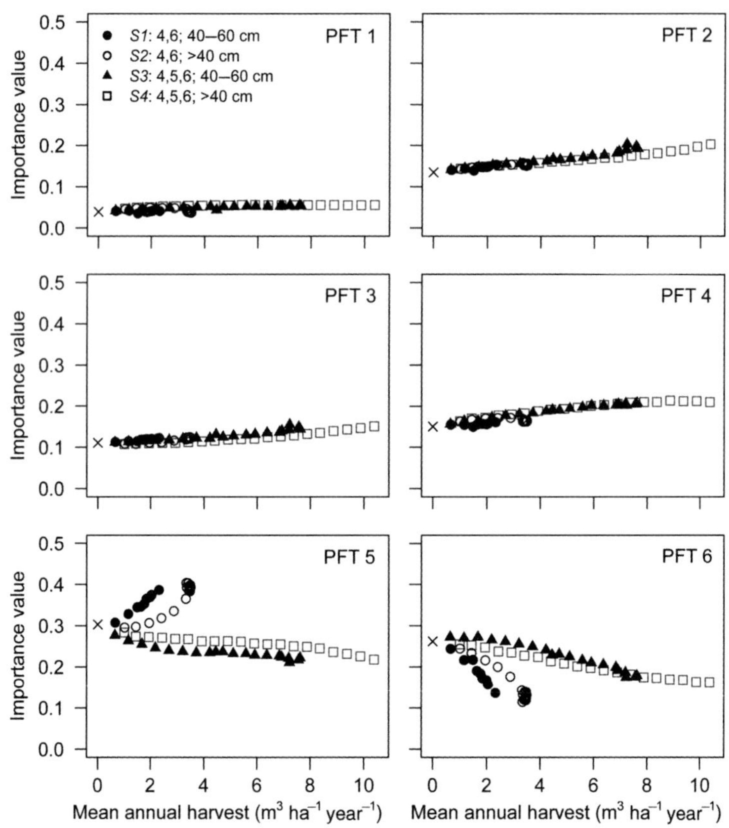

Fig. 11.5. Importance values as a measure of dominance of six PFTs for four selective logging scenarios (*S1–S4*) of tropical montane cloud forest in central Veracruz, Mexico. Importance values of the undisturbed old-growth forest are displayed for comparison (×) (from Rüger *et al.*, 2007b).

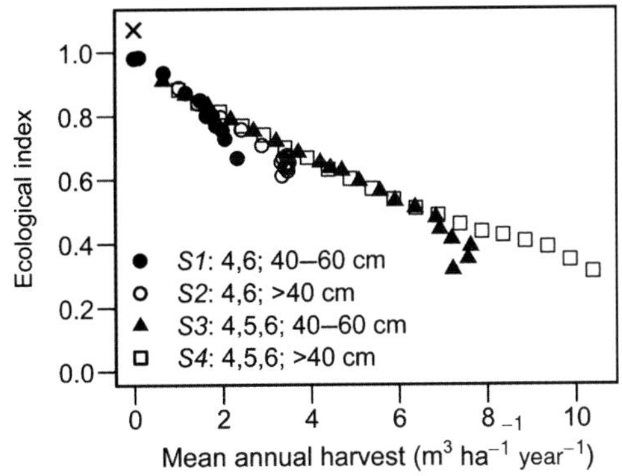

Fig. 11.6. Ecological index versus mean annual harvest for four selective logging scenarios (*S1–S4*) of tropical montane cloud forest in central Veracruz, Mexico (from Rüger *et al*., 2007b).

Chiloé

MEAN ANNUAL HARVEST We simulated wood extraction for three logging strategies (selective logging with and without retention of large old trees and logging in bands). For the selective logging scenarios, we varied the logging cycle from 10 to 50 years and the harvest aim from 1 to 10 m^3 ha^{-1} on an annual basis (Fig. 11.7). Up to about 4 m^3 ha^{-1} $year^{-1}$, the harvest aim could be met by all scenarios. Selective logging scenarios with large tree retention reached a limit of sustainable wood extraction at 6.5 m^3 ha^{-1} $year^{-1}$. When large trees were removed prior to the simulation of logging scenarios, up to 8 m^3 ha^{-1} $year^{-1}$ could be harvested. Logging in bands achieved a higher annual harvest that ranged from 6 m^3 ha^{-1} $year^{-1}$ for a logging cycle of 150 years to 13.4 m^3 ha^{-1} $year^{-1}$ for a logging cycle of 60 years.

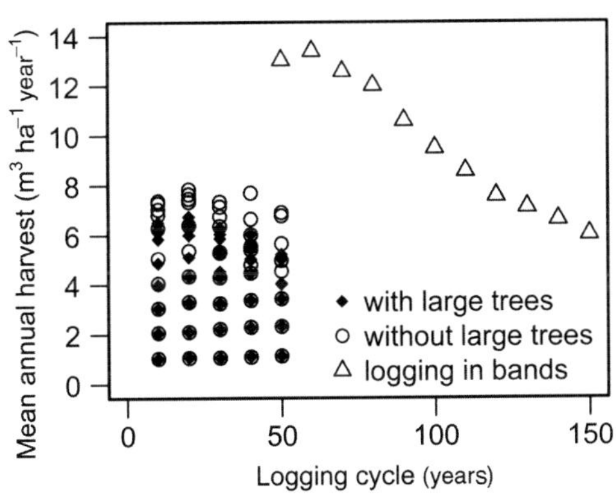

Fig. 11.7. Mean annual harvest for three logging strategies (selective logging with and without large tree retention, logging in bands) of Valdivian temperate rainforest in northern Chiloé Island, Chile (from Rüger *et al*., 2007c). For selective logging, logging cycle varied from 10 to 50 years, harvest aim (i.e. amount of extracted wood aimed at by the logging scenario) varied from 10 to 500 m^3 ha^{-1}, depending on the logging cycle. Converted to an annual basis, harvest aim ranged between 1 and 10 m^3 ha^{-1} $year^{-1}$. For logging in bands the logging cycle was varied from 50 to 150 years.

FOREST STRUCTURE To study the impact of logging scenarios on forest structure, we distinguish three diameter classes (5–50, 50–100, 100–200 cm dbh). The number of small trees (5–50 cm dbh) increased for increasing levels of wood extraction (Fig. 11.8). The number of large trees (50–100 cm dbh) remained stable for low levels of wood extraction (up to 5 m^3 ha^{-1} year^{-1}), but sharply decreased for higher levels of wood extraction. For logging in bands, the decrease of the number of large trees occurred at higher levels of wood extraction (8–14 m^3 ha^{-1} year^{-1}). The number of old trees (>1 m dbh) decreased linearly up to a mean annual harvest of 8 m^3 ha^{-1} year^{-1}. Beyond

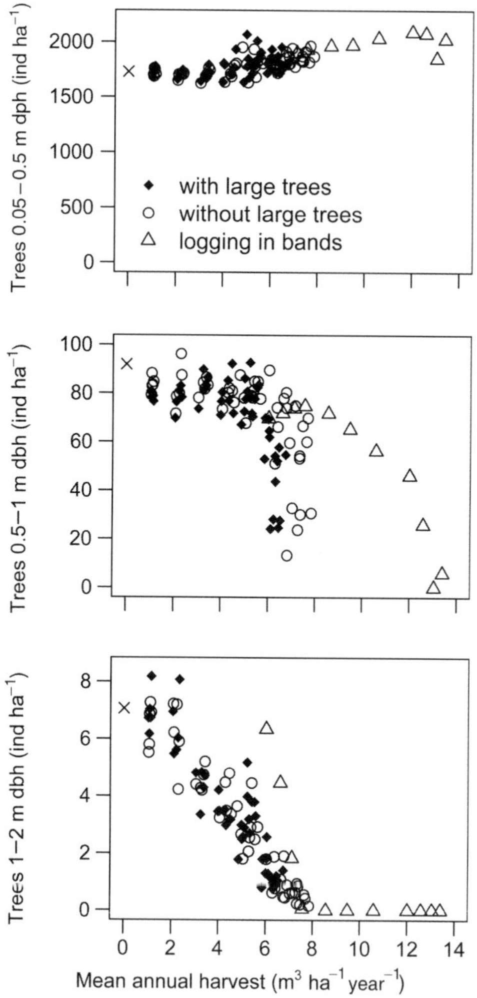

Fig. 11.8. Mean number of trees in three diameter classes for simulation time 400–500 years for three logging strategies (selective logging with and without large tree retention, logging in bands) of Valdivian temperate rainforest in northern Chiloé Island, Chile. Mean values for undisturbed old-growth forest are displayed for comparison (×) (from Rüger *et al.*, 2007c).

that threshold, no old trees remained in the forests in the long term, because large trees were harvested before they attained a dbh of 1 m.

FOREST COMPOSITION Logging scenarios mainly had an effect on importance values of *E. cordifolia* and *L. philippiana* (Fig. 11.9). Importance values of *E. cordifolia* were more than twice as high in the logging in bands scenarios compared to the selective logging scenarios. This increase occurred at the expense of *L. philippiana*, for which importance values in the logging in bands scenarios halved compared to selective logging. The inverse pattern was observed within the selective logging scenarios for increasing levels of wood extraction. While *E. cordifolia*'s importance values decreased, importance values of *L. philippiana* increased. Importance values of *A. punctatum* remained relatively stable under the different logging scenarios. The myrtaceous species showed the same trends as *L. philippiana*, but to a lesser extent.

ECOLOGICAL INTEGRITY VERSUS HARVEST As in the case of selective logging of TMCF, the ecological index (*EI*) that describes overall similarity of the logged forest to undisturbed old-growth forest decreased almost linearly with increasing harvesting intensity of selective logging scenarios for VTRF (Fig. 11.10). *EI* was lowest under the logging in bands scenarios.

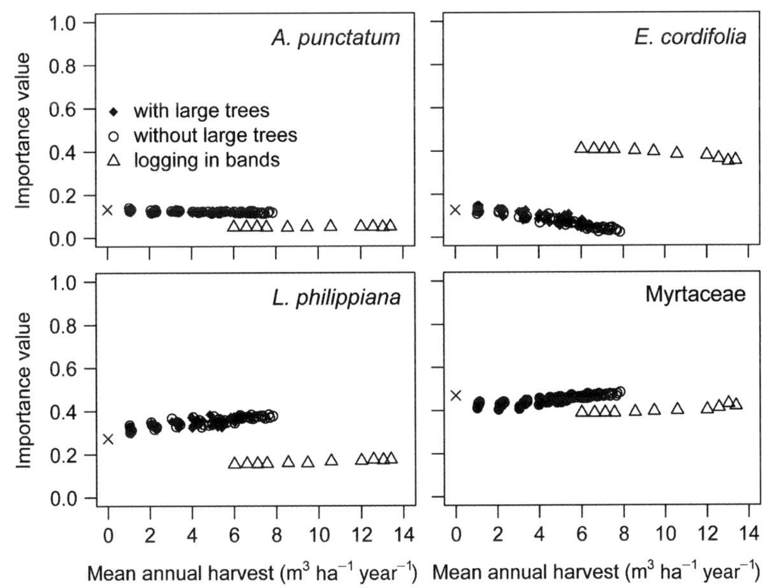

Fig. 11.9. Importance values as a measure of dominance of three species and one species group for three logging strategies (selective logging with and without large tree retention, logging in bands) of Valdivian temperate rainforest in northern Chiloé Island, Chile. Importance values of the undisturbed old-growth forest are displayed for comparison (×) (from Rüger *et al*., 2007c).

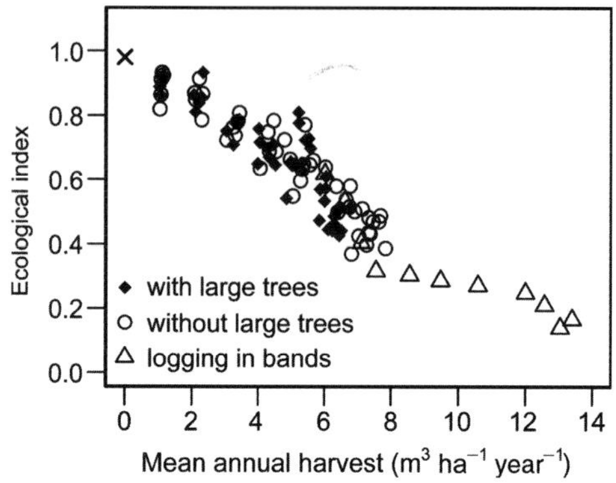

Fig. 11.10. Ecological index versus mean annual harvest for three logging strategies (selective logging with and without large tree retention, logging in bands) of Valdivian temperate rainforest in northern Chiloé Island, Chile (from Rüger *et al.*, 2007c).

Discussion

Comparison of study regions

The forest growth model FORMIND has been applied to study natural forest succession, productivity and ecological impacts of different management scenarios on native species-rich tropical montane cloud forest in central Veracruz, Mexico, and Valdivian temperate rainforest in northern Chiloé Island, Chile. The first part of the discussion is aimed at summarizing and comparing the results for the two study regions as well as drawing general conclusions for a sustainable use of species-rich moist forests.

Forest dynamics

The most conspicuous difference between tropical montane cloud forest (TMCF) in central Veracruz, Mexico, and Valdivian temperate rainforest in southern Chile (VTRF) is their tree species richness. In TMCF more than 100 tree species have been counted (G. Williams-Linera, unpublished data), whereas in VTRF about 15 tree species occur (Donoso, 1993). The forests also differ largely in their structure and dynamics due to differences in life-history traits of the tree species and the disturbance regime in the study regions.

In TMCF, the trees grow relatively rapidly (up to 2 cm year^{-1} in stem diameter, Williams-Linera, 1996). Tree lifespans seem to be short, and large old trees rarely exceed a maximum diameter of 1 m. The model estimates an annual turnover rate of 5% (trees ≥ 5 cm dbh), which corresponds to short-term observations from the study site (Williams-Linera, 2002). In VTRF, the trees grow more slowly (up to 1 cm year^{-1}, A. Gutiérrez, unpublished data). They reach maximum diameters of up to 2 m (especially *Eucryphia cordifolia*) and have longer lifespans (e.g. Lusk and del Pozo, 2002). The model suggests that annual turnover rates are as low as 2% (trees ≥ 5 cm dbh).

According to simulation results, TMCF regenerates rapidly after disturbance, and field data confirm this (Muñiz-Castro *et al.*, 2006). After a few decades, aggregated forest characteristics such as density, basal area and leaf area index (LAI) have recovered (Rüger *et al.*, 2007a, b). The successional dynamics of TMCF correspond to the typical temporal pattern: an initial stage dominated by pioneer species is followed by an intermediate stage where species with intermediate shade tolerance gain the highest basal area values, and finally a climax stage where shade-tolerant species attain their maximum share. Forest structure and species dominance in terms of basal area of different plant functional types (PFTs) reach a steady state and old-growth conditions within 300 years of a large-scale disturbance.

As in TMCF, aggregated characteristics of VTRF recover rapidly, but the temporal pattern of the succession is different and the dynamics are slower. Following a large-scale disturbance, the light-demanding *E. cordifolia* dominates in terms of basal area for about 400 years. Then *E. cordifolia* is slowly replaced by shade-tolerant species and tends to disappear from the forest after about 800 years if no medium to large-sized disturbances occur. The proportion of basal area of the different species reaches a steady state after approximately 1000 years.

With a basal area of about $45\,m^2\,ha^{-1}$ and an above-ground biomass of $480\,Mg\,ha^{-1}$, TMCF in central Veracruz stores more biomass than Amazonian tropical lowland rainforests, where basal area values of $25–30\,m^2\,ha^{-1}$ were measured and estimated biomass ranged between 220 and $340\,Mg\,ha^{-1}$ (Baker *et al.*, 2004). With a density of about 1800 individuals ha^{-1} ($\geq 5\,cm$ dbh, Gutiérrez *et al.*, unpublished data) and a basal area of nearly $100\,m^2\,ha^{-1}$, VTRF belongs to the densest forests recorded, with the highest stem volume (up to $1000\,m^3\,ha^{-1}$) (Armesto *et al.*, 1999b) and biomass ($800\,Mg\,ha^{-1}$) in the world. Volume increments of TMCF and VTRF are similar due to the rapid growth of TMCF and the large amount of biomass of VTRF.

The differences in forest dynamics and structure are possibly due to differences in the disturbance regime in both regions. In central Veracruz (TMCF), the prevalent disturbance regime is gap creation on a small spatial scale, because most of the dying trees fall over and only a small portion remains standing (Williams-Linera, 2002). Natural large-scale disturbances such as hurricanes, landslides, fire or floods are rare and negligible for forest dynamics (G. Williams-Linera, unpublished data). In northern Chiloé Island (VTRF), the disturbance regime seems to be composed of single and multiple tree falls, which apparently occur with lower frequency than in TMCF, and infrequent windthrow events, which open much larger gaps of up to several hectares in size (A. Lara, Instituto de Silvicultura, Universidad Austral de Chile, personal communication).

Tree species that dominate early successional phases of VTRF, such as *E. cordifolia*, *Drimys winteri* or *Embothrium coccineum*, are not able to successfully establish in small canopy gaps but require large canopy openings, whereas, in TMCF, pioneer species and species with intermediate shade tolerance are able to establish in single tree-fall gaps. In old-growth VTRF, LAI is very high and light availabilities at the forest floor are low (Saldaña and Lusk, 2003). *E. cordifolia* as a species can only survive in the forest due to its long lifespan (Lusk and del Pozo, 2002) that allows it to persist in the forest until a new large disturbance occurs, and its emergent stature that ensures a wide distribution of the wind-dispersed seeds. Hence, *E. cordifolia* is an example of the relatively rare long-lived pioneer species (e.g. Loehle, 1988; Lusk, 1999). On the contrary, the majority of tree species of VTRF are adapted to low light levels and can establish and persist underneath a closed canopy (cf. Figueroa and Lusk, 2001; Lusk and del Pozo, 2002).

The occurrence of large-scale disturbances together with long tree lifespans also causes high spatial heterogeneity of VTRF. Gaps of different size, young dense patches, old-growth forest with emerging *E. cordifolia*, and forest patches where *E. cordifolia* is lacking and shade-tolerant species dominate occur side by side on a large spatial scale. This spatial heterogeneity is difficult to assess with conventional inventory data in small sample plots. Thus, in terms of forest structure, VTRF is more heterogeneous than TMCF and can be regarded to be 'in equilibrium' only on very large temporal and spatial scales.

Forest productivity

On a global scale, and in the context of increasingly globalized wood fibre production, sustainable management of native forests in Mexico and Chile is not likely to be competitive from the economic point of view (Franklin, 2003). Quantitatively, annual wood volume increments of up to 12 m^3 ha^{-1} in Mexican TMCF and up to 13 m^3 ha^{-1} in Chilean VTRF fall well short of growth rates of plantations of *Eucalyptus* spp. or *Pinus radiata*, which reach mean annual volume increments of 40 and 30 m^3 ha^{-1}, respectively (Ugalde and Pérez, 2001). In Mexico, the current mean annual yield from the management of native forests is as low as 1.2 m^3 ha^{-1}, and in a sustainable development scenario this value is envisioned to rise to 1.8 m^3 ha^{-1} until the year 2025 (Torres-Rojo, 2004). The simulated maximum sustainable harvest suggests that Mexican TMCF has a much higher potential for wood production, although simulated wood extraction rates refer to gross stem volume values and not to net commercial volume. Simulated annual volume increments are also much higher than those predicted for various tropical lowland forests, ranging between 1 and 4 m^3 ha^{-1} (Huth and Ditzer, 2001; Kammesheidt *et al.*, 2002; van Gardingen *et al.*, 2003). However, Silva *et al.* (1995) measured an annual volume increment of 6 m^3 ha^{-1} directly after the logging of rainforest in the Brazilian Amazon.

Selective logging of Chilean VTRF favoured shade-tolerant species, whereas *E. cordifolia* regeneration was promoted by clearcutting in bands. Clearcutting in bands creates larger gaps which are more suitable for the establishment of *E. cordifolia*. The same rationale applies to other shade-intolerant commercial tree species (e.g. mahogany (*Swietenia macrophylla*)), which do not sufficiently regenerate after selective logging to sustain desired yields (e.g. Fredericksen and Putz, 2003, and references therein). Thus, successful management of native species-rich forests requires an adaptation of management practices to the ecological properties of the target species.

Compared with plantations of non-native species, sustainable management of the native forests can provide a continuous supply of timber and fuelwood and has ecological and economic advantages that might offset the lower growth rates under certain circumstances (Franklin, 2003). Economic advantages include lower management costs for small owners or local communities, the proximity of wood production to local markets and a higher timber quality. From the ecological perspective, the management of native forests improves the quality of land use. The conservation of native biodiversity ensures the maintenance of mutualistic interactions (Armesto *et al.*, 1999b; Smith-Ramírez *et al.*, 2005), the regulation of hydrological cycles (e.g. Iroumé and Huber, 2002), the storage of higher amounts of carbon (e.g. Chen *et al.*, 2005), and the supply of non-timber forest products.

Ecological impacts of logging and implications for conservation

A widely adopted definition of ecologically sustainable forest management is given by Lindenmayer and Recher (1998): 'Ecologically sustainable forest management perpetuates ecosystem integrity while continuing to provide wood and non-wood values; where ecosystem integrity means the maintenance of forest structure, species composition, and the rate of ecological processes and functions with[in] the bands of normal disturbance regimes.' However, every anthropogenic intervention in the form of wood extraction, even at a very low intensity, has an ecological impact on the forest. The above definition is, in a strict sense, impossible to fulfil. Tree felling is an additional disturbance which increases mortality, and this increased mortality has an effect on the forest structure and composition. The crucial point is rather: how severe are the ecological impacts? In all logging scenarios, the overall ecological impact in terms of the aggregated ecological index increased linearly with the amount of extracted wood.

The most notable effect of wood extraction on the forest structure was the loss of large old trees from the forest. Every kind of management that does not explicitly retain a number of large old trees leads in the long term to the loss of those trees, whereas the number of small trees increases. Consequently, the forest structure becomes more simplified, and the forests become younger and more homogeneous. These changes in the forest structure can take between a few decades and more than 100 years and can therefore hardly be observed directly (Rüger *et al.*, 2007b).

Management decisions are based on a multitude of different criteria that are given different priority by different stakeholders. Simulation results serve to define a type of management that balances conservation and production objectives according to these preferences. Apart from adjusting harvesting method and intensity, variable retention systems provide a flexible means of combining different management objectives (e.g. Lindenmayer and Franklin, 2002). Variable retention systems allow a certain amount of forest structures (e.g. large living trees, dead trees, undisturbed forest floor, patches of understorey shrubs and herbs, or groups of juvenile trees in a forest gap) to be left untouched, facilitating the recovery of biodiversity and ecosystem processes, as well as ensuring the maintenance of islands of original habitat and landscape connectivity (Armesto *et al.*, 1999a). In this way, the loss of large old trees can be partially compensated. Areas with different amounts of retained elements and of varying extension can be combined to ensure a spatially diverse forest structure.

Evaluation of the process-based modelling approach

FORMIND was developed for the simulation of the dynamics of species-rich moist forests where competition for light is the main driver for forest dynamics. It includes key processes such as recruitment, growth, mortality, competition for light and space, gap creation through falling dead trees and external disturbances. The main focus of the model is on the response of the forest to natural and anthropogenic disturbances at the stand level and on the temporal scale of decades to hundreds of years. To cope with the high species richness of tropical forests, tree species are grouped into plant functional types (PFTs) according to their maximum height and light demand. FORMIND differs from most other models of mixed-species forests in the calculation of single-tree growth as growth is calculated on the basis of carbon balance, including photosynthesis and respiration. The individual-oriented approach allows for a model evaluation on different levels (e.g. trees, PFTs, entire tree community).

The parameters used in FORMIND can be divided into environmental parameters (e.g. average light intensity above the forest, light extinction coefficient), parameters describing allometric relations (e.g. between stem diameter, height, crown diameter, crown depth, form factor), physiological parameters (e.g. maximum rate of photosynthesis, slope of light–response curve, respiration parameters), and demographic parameters (recruitment and mortality rates).

The data basis for parameter estimation is usually very heterogeneous. Environmental parameters and allometric relations of tree geometry are relatively easy to obtain from field measurements. Measurements of physiological parameters, on the other hand, are often not available, especially measurements of respiration parameters. However, field data on diameter increment of single trees, either from growth measurements over several years or from dendrochronological analyses, are often available. Physiological parameters can be adjusted in such a way that observed growth characteristics are reproduced (cf. Rüger *et al.*, 2007a, c).

Apart from these methods to derive model parameters from field data that to some extent ensure that single processes produce realistic outcomes, independent field data were used to validate overall model results (Rüger *et al.*, 2007a, c). Simulated diameter distributions corresponded to field data from the study site (Rüger *et al.*, 2007a). Furthermore, the results of a chronosequence study that covered 0.5–80-year-old secondary TMCF in central Veracruz could be used to validate model predictions regarding the regeneration of TMCF (Muñiz-Castro *et al.* 2006; Rüger *et al.*, 2007a). The qualitative development of simulated forest regeneration corresponded to the field data. However, the model slightly overestimated the velocity of forest recovery due to a higher recruitment rate during the first decade and a slightly overestimated tree growth (Rüger *et al.*, 2007a). Simulation results for harvesting scenarios of TMCF in central Veracruz therefore seem to be reliable, although sustainable harvesting rates could be slightly overestimated due to the overestimated recruitment rates and tree growth.

The high spatial heterogeneity of VTRF made an evaluation of model performance for this forest type difficult. Simulation results regarding dynamics and sustainable use of VTRF should therefore be interpreted carefully. They should be regarded as possible scenarios derived from currently available information on the forest rather than predictions of forest development. To improve the data basis for model parameterization and evaluation, it would be desirable to obtain inventory data from larger areas, from secondary forests of different ages, as well as information about mortality rates and the frequency and extent of large-scale disturbances.

Conclusions and Outlook

Simulation results showed that both forest types have a high potential for wood production. However, every anthropogenic intervention in the form of wood extraction, even at very low levels, has an ecological impact on the forests. Comparing all logging scenarios, the overall ecological impact increased linearly with the amount of extracted wood. The developed ecological index, which integrates several ecological criteria, provides a first approach for the determination of management strategies serving multiple purposes. These may include economic income from wood production and relative maintenance of forest structure and composition to ensure the protection of non-economic ecosystem services from the native forests, e.g. soil protection, water capture, biodiversity conservation, cultural and recreation values. Moreover, simulation results serve to design management strategies that promote the regeneration of desired tree species and/or minimize shifts in the species composition of the forest.

However, an enhancement of economic aspects is necessary if model results are to be valuable for decision makers and stakeholders that depend on the forest as a source of income. Economic extensions could include an incorporation of logging costs for different management strategies and the consideration of economic concepts such as discounting and price development. Integrated sustainability indices should be developed incorporating economic and ecological criteria which can be weighted according to preferences of stakeholders. A particular focus should be on the management of secondary forests as their area is increasing, their growth rates are relatively high, and their structure and species composition are less vulnerable to tree harvesting than old-growth forests.

The new graphical user interface of the model provides the opportunity to use the model in workshops with decision makers, stakeholders or students to explore implications of alternative management scenarios and to raise awareness and understanding of the underlying ecological processes of forest dynamics. Simulation exercises can support the education of forestry students with respect to the management of native forests, an issue that is often neglected in conventional curricula.

Models such as FORMIND enhance our understanding of the dynamics of species-rich moist forests and are indispensable tools to assess long-term implications of anthropogenic disturbances for forest ecosystems. Together with empirical studies, simulation approaches contribute substantially to the conservation and sustainable use of native species-rich forests outside protected areas by providing guidelines for ecologically sound management and highlighting their potential for provision of ecosystem services.

Appendix

Table A11.1. Parameters of FORMIND2.3 for tropical montane cloud forest in central Veracruz, Mexico.v

Parameter	Description	Unit	PFT 1	PFT 2	PFT 3	PFT 4	PFT 5	PFT 6	Reference
Environmental parameters									
k	Light extinction coefficient	m^2 ground m^{-2} leaf			0.5				Hafkenscheid (2000)
I_0	Average irradiance above canopy	µmol(photons) m^{-2} s^{-1}			600				G. Williams-Linera, unpublished data
S_d	Mean sunshine hours per day	h day^{-1}			12				Estimated
Recruitment parameters									
D_S	Diameter of ingrowing trees	m			0.01				Technical parameter
I_{min}	Minimum light intensity for establishment	% of I_0	10	3	1	3	1	3	Estimated
I_{max}	Maximum light intensity for establishment	% of I_0			100				Estimated
N_{max}	Ingrowth rates of small trees	ha $year^{-1}$	1000	400	250	400	250	400	Fitted using inventory data
Mortality parameters									
m_B	Basic mortality	$year^{-1}$	0.05	0.015	0.015	0.01	0.008	0.01	Estimated
m_{max}	Maximum mortality of small trees	$year^{-1}$			0.25				Estimated
D_{mort}	Diameter up to which mortality is increased	m			0.1				Estimated

D_{fall}	Minimum diameter of falling trees	m				0.35				Arriaga (1987, 2000), Williams-Linera (2002)
p_{fall}	Probability of large dying trees to fall	%				80				Williams-Linera (2002)
Tree geometry parameters										
h_0	Parameter of diameter–height relationship	cm m^{-1}	2.24	2.24	2.24		2.15	2.15	2.1	Aguilar-Rodríguez *et al*. (2001)
h_1	Parameter of diameter–height relationship	m	18.55	18.55	18.55		29.26	29.26	42	
f	Form factor	–				0.5				Köhler (2000)
cd	Parameter of diameter–crown diameter relationship	m cm^{-1}				0.2				G. Williams-Linera (unpublished data)
H_{max}	Maximum height	m	15	15	15		25	25	35	G. Williams-Linera (personal observation)
D_{max}	Maximum diameter	m	0.35	0.35	0.35		0.8	0.8	1.0	
c	Crown length fraction	–				0.1				Estimated
L_{max}	Maximum leaf area index per tree	m^2 leaf m^{-2} ground				2				Estimated
sw	Fraction of stem wood biomass to total biomass	–				0.7				Köhler (2000)
Biomass production parameters										
p_{max}	Maximum rate of photosynthesis	µmol(CO_2) m^{-2} s^{-1}	20	16	10		16	10	16	Ellis *et al*. (2000), Dillenburg *et al*. (1995)

Continued

Table A11.1. *Continued*

Parameter	Description	Unit	PFT 1	PFT 2	PFT 3	PFT 4	PFT 5	PFT 6	Reference
α	Slope of light–response curve	$\mu mol(CO_2)\ \mu mol\ (photons)^{-1}$	0.15	0.2	0.25	0.2	0.25	0.2	Estimated
ρ	Wood density	$t\ m^{-3}$	0.55	0.65	0.7	0.65	0.7	0.65	Bárcenas *et al.* (1998), Aguilar-Rodríguez *et al.* (2001)
r_g	Parameter of growth respiration	–	0.2						Ryan (1991)
r_0	Parameter of maintenance respiration	–	0.79	0.59	0.41	0.3	0.19	0.23	Fitted using diameter increment data (Williams-Linera, 1996)
r_1	Parameter of maintenance respiration	–	1.2	1.2	1.2	1.07	1.1	1.02	
m	Transmission coefficient of leaves	–	0.1						Larcher (2001)
codm	Parameter for conversion in organic dry matter	$t\ \mu mol(CO_2)^{-1}$	$0.63\ 44e^{-12}$						Larcher (2001)
Technical parameters									
a	Patch size	m^2	400						Technical parameter
Δh	Step width of vertical discretization	m	0.5						Technical parameter

Table A11.2. Parameters of FORMIND2.3 for Valdivian temperate evergreen rainforest in northern Chiloé Island, Chile. AP, *Aextoxicon punctatum*; EC, *Eucryphia cordifolia*; LP, *Laureliopsis philippiana*; MY, myrtaceous species.

Parameter	Description	Unit	AP	EC	LP	MY	Reference
Environmental parameters							
k	Light extinction coefficient	m^2 ground m^{-2} leaf	0.5				Estimated[a]
I_0	Average irradiance above canopy	μmol(photons) m^{-2} s^{-1}	700				C. Lovengreen (unpublished data)
s_d	Mean sunshine hours per day	h day^{-1}	12				Estimated
Recruitment parameters							
D_s	Diameter of ingrowing trees	m	0.01				Technical parameter
I_{min}	Minimum light intensity for establishment	% of I_0	1	70	3	1	Estimated[b]
I_{max}	Maximum light intensity for establishment	% of I_0	90	100	90	100	Estimated[b]
N_{max}	Maximum recruitment rates of small trees	ha $year^{-1}$	50	100	150	250	Estimated[b]
Mortality parameters							
m_B	Basic mortality	year	0.01	0.006	0.006	0.004	Estimated[c]
m_{max}	Maximum mortality of small trees	year	0.12				Estimated[c]
D_{mort}	Diameter up to which mortality is increased	m	0.1				Estimated[c]
D_{fall}	Minimum diameter of falling trees	m	0.45				Estimated[c]
p_{fall}	Probability of dying trees to fall	%	30				Estimated[c]

Continued

Table A11.2. *Continued*

Parameter	Description	Unit	AP		EC	LP	MY	Reference
Tree geometry parameters								
h_0	Parameter of diameter–height relationship	cm m^{-1}		1.2				Brun (1969)[d]
h_1	Parameter of diameter–height relationship	m	41.6		48.7	40.1	27.7	Brun (1969)[d]
f	Form factor	–	0.4		0.4	0.4	0.35	Estimated[e]
cd	Parameter of diameter–crown diameter relationship	m cm^{-1}		0.12				
H_{max}	Maximum height	m	30		40	30	20	e.g. Brun (1969), Lusk and del Pozo (2002)
D_{max}	Maximum diameter	m	1		2	1	0.7	
c	Crown depth fraction	–		0.25				Estimated
L_{max}	Maximum leaf area index per tree	m^2 leaf m^{-2} ground		4				Saldaña and Lusk (2003)
sw	Fraction of stem wood biomass to total biomass	–		0.7				Köhler (2000)
Biomass production parameters								
p_{max}	Maximum rate of photosynthesis	µmol(CO_2) m^{-2} s^{-1}	5.6		10	6.4	7	Lusk *et al.* (2003)
α	Slope of light–response curve	µmol(CO_2) µmol(photons)$^{-1}$	0.25		0.2	0.2	0.35	Estimated
ρ	Wood density	t m^{-3}	0.59		0.72	0.55	1.15	Pérez-Galaz (1983), Diaz-vaz *et al.* (2002)

r_g	Parameter of growth respiration	–		0.2				Ryan (1991)
r_0	Parameter of maintenance respiration	–	0.1		0.11	0.13	0.1	Estimated using diameter increment data (Gutiérrez *et al.*, in preparation)
r_1	Parameter of maintenance respiration	–	0.0		0.0	0.0008	0.003	Estimated[f]
r_2	Parameter of maintenance respiration	–	0.0001		0.0	0.0	0.0	Estimated[f]
m	Transmission coefficient of leaves	–		0.1				Larcher (2001)
codm	Parameter for conversion in organic dry matter	t μmol$(CO_2)^{-1}$		0.63 $44e^{-12}$				Larcher (2001)
Technical parameters								
a	Patch size	m^2		400				Technical parameter
Δh	Step width of vertical discretization	m		0.5				Technical parameter

[a]Pierce and Running (1988), Brown and Parker (1994).
[b]Estimated based on inventory data (A. Gutiérrez, unpublished).
[c]Estimated based on Lusk (2002), Lusk and del Pozo (2002), Coomes *et al.* (2003) and inventory data (A. Gutiérrez, unpublished).
[d]C. Echeverría (unpublished data), A. Gutiérrez (unpublished data).
[e]Estimated based on Emanuelli and Pancel (1999), Salas (2002).
[f]Estimated using diameter increment data (Gutiérrez *et al.*, in preparation).

References

Aguilar-Rodríguez, S., Abundiz-Bonilla, L. and Barajas-Morales J. (2001) Comparación de la gravedad específica y características anatómicas de la madera de dos comunidades vegetales en México. *Anales del Instituto de Biología, Universidad Autónoma de México, Serie Botánica* 72, 171–185.

Armesto, J.J., Rozzi, R., Smith-Ramírez, C. and Arroyo, M.T.K. (1998) Conservation targets in South American temperate forests. *Science* 282, 1271–1272.

Armesto, J.J., Franklin, J.F., Arroyo, M.T.K. and Smith-Ramírez, C. (1999a) El sistema de cosecha con 'retención variable': una alternativa de manejo para conciliar los objetivos de conservación y producción en los bosques nativos chilenos. In: Donoso, C. and Lara, A. (eds) *Silvicultura de Los Bosques Nativos de Chile*. Editorial Universitaria, Santiago, Chile, pp. 69–94.

Armesto, J.J., Lobos, P.L. and Arroyo, M.K. (1999b) Los bosques templados del sur de Chile y Argentina: una isla biogeográfica. In: Armesto, J.J., Villagrán, C. and Arroyo, M.T.K. (eds) *Ecología de Los Bosques Nativos de Chile*, 3rd edn. Editorial Universitaria, Santiago, Chile, pp. 23–28.

Arriaga, L. (1987) Perturbaciones naturales por la caída de árboles. In: Puig, H. and Bracho, R. (eds) *El Bosque Mesófilo de Montaña de Tamaulipas*. Instituto de Ecología, Mexico, pp. 133–152.

Arriaga, L. (2000) Types and causes of tree mortality in a tropical montane cloud forest of Tamaulipas, Mexico. *Journal of Tropical Ecology* 16, 623–636.

Baker, T.R., Phillips, O.L., Malhi, Y., Almeida, S., Arroyo, L., Di Fiore, A., Erwin, T., Killeen, T.J., Laurance, S.G., Laurance, W.F., Lewis, S.L., Lloyd, J., Monteagudo, A., Neill, D.A., Patino, S., Pitman, N.C.A., Silva, J.N.M. and Vásquez-Martínez, R. (2004) Variation in wood density determines spatial patterns in Amazonian forest biomass. *Global Change Biology* 10, 545–562.

Bárcenas, G., Dávalos, R. and Enríquez, M. (1998) Banco de información sobre las características tecnológicas de maderas mexicanas. *Memories del Segundo Congreso Mexicano de Tecnología de Productos Forestales*. 25–27 November, Morelia, Michigan.

Brown, M.J. and Parker, G.G. (1994) Canopy light transmittance in a chronosequence of mixed-species deciduous forests. *Canadian Journal of Forest Research* 24, 1694–1703.

Brun, R. (1969) Strukturstudien im gemäßigten Regenwald Südchiles als Grundlage für Zustandserhebungen und Forstbetriebsplanung. PhD thesis. Albert-Ludwig-Universität, Freiburg i. Br., Germany.

Chen, G.S., Yang, Y.S., Xie, J.S., Guo, J.F., Gao, R. and Qian, W. (2005) Conversion of a natural broad-leafed evergreen forest into pure plantation forests in a subtropical area: effects on carbon storage. *Annals of Forest Science* 62, 659–668.

Coomes, D.A., Duncan, R.P., Allen, R.B. and Truscott, J. (2003) Disturbances prevent stem-size density distributions in natural forests from following scaling relationships. *Ecology Letters* 6, 980–989.

Di Castri, F. and Hajek, E. (1976) *Bioclimatología de Chile*. Universidad Católica de Chile, Santiago, Chile.

Díaz, I., Armesto, J.J., Reid, S., Sieving, K.E. and Willson, M.F. (2005) Linking forest structure and composition: avian diversity in successional forests of Chiloe Island, Chile. *Biological Conservation* 123, 91–101.

Diaz-vaz, J.E., Poblete, H., Juacida, R. and Devlieger, F. (2002) *Maderas Comerciales de Chile*, 3rd edn. Ed. Marisa Cuneo, Valdivia, Chile.

Dillenburg, L.R., Teramura, A.H., Forseth, I.N. and Whigham, D.F. (1995) Photosynthetic and biomass allocation responses of *Liquidambar styraciflua* (Hamamelidaceae) to vine competition. *American Journal of Botany* 82, 454–461.

Ditzer, T., Glauner, R., Förster, M., Köhler, P. and Huth, A. (2000) The process-based stand growth model FORMIX3-Q applied in a GIS environment for growth and yield analysis in a tropical rain forest. *Tree Physiology* 20, 367–381.

Donoso, C. (1989) Antecedentes básicos para la silvicultura del tipo forestal siempreverde. *Bosque* 10, 37–53.

Donoso, C. (1993) *Bosques Templados de Chile y Argentina. Variación, Estructura y Dinámica*. Editorial Universitaria, Santiago, Chile.

Donoso, C., Escobar, B. and Urrutia, J. (1985) Estructura y estrategias regenerativas de un bosque virgen de ulmo (*Eucryphia cordifolia* Cav.) tepa (*Laurelia philippiana* Phil.) Looser en Chiloé, Chile. *Revista Chilena de Historia Natural* 58, 171–186.

Donoso, C., Donoso, P., González, M. and Sandoval, V. (1999) Los bosques siempreverdes. In: Donoso, C. and Lara, A. (eds) *Silvicultura de los Bosques Nativos de Chile*. Editorial Universitaria, Santiago, Chile, pp. 297–339.

Ellis, A.R., Hubbell, S.P. and Potvin, C. (2000) In situ field measurements of photosynthetic rates of tropical tree species: a test of the functional group hypothesis. *Canadian Journal of Botany* 78, 1336–1347.

Emanuelli, P. and Pancel, L. (1999) Funciones de volumen para la Reserva Nacional Valdivia. Documento de trabajo, Proyecto Manejo Sustentable del Bosque Natiro, CONAF-GTZ.

Figueroa, J.A. and Lusk, C.H. (2001) Germination requirements and seedling shade tolerance are not correlated in a Chilean temperate rain forest. *New Phytologist* 152, 483–489.

Franklin, J.F. (1993) Preserving biodiversity: species, ecosystems, or landscapes? *Ecological Applications* 3, 202–205.

Franklin, J.F. (2003) Challenges to temperate forest stewardship – focusing on the future. In: Lindenmayer, D.B. and Franklin, J.F. (eds) *Toward Forest Sustainability*. Island Press, Washington, DC, pp. 1–13.

Fredericksen, T.S. and Putz, F.E. (2003) Silvicultural intensification for tropical forest conservation. *Biodiversity and Conservation* 12, 1445–1453.

FSC (2004) *Principles and Criteria for Forest Stewardship*. Forest Stewardship Council A.C., Bonn, Germany.

Grimm, V., Berger, U., Bastiansen, F., Eliassen, S., Ginot, V., Giske, J., Goss-Custard, J., Grand, T., Heinz, S., Huse, G., Huth, A., Jepsen, J.U., Jørgensen, C., Mooij, W.M., Müller, B., Robbins, A.M., Robbins, M.M., Rossmanith, E., Rüger, N., Pe'er, G., Piou, C., Railsback, S.F., Strand, E., Souissi, S., Stillmann, R., Vabø, R., Visser, U. and DeAngelis, D.L. (2006) A standard protocol for describing individual-based and agent-based models. *Ecological Modelling* 198, 115–126.

Hafkenscheid, R. (2000) Hydrology and biogeochemistry of tropical montane rain forests of contrasting stature in the Blue Mountains, Jamaica. PhD thesis. Vrije Universiteit Amsterdam, The Netherlands.

Huth, A. and Ditzer, T. (2000) Simulation of the growth of a lowland Dipterocarp rain forest with FORMIX3. *Ecological Modelling* 134, 1–25.

Huth, A. and Ditzer, T. (2001) Long-term impacts of logging in a tropical rain forest – a simulation study. *Forest Ecology and Management* 142, 33–51.

Huth, A., Drechsler, M. and Köhler, P. (2004) Multicriteria evaluation of simulated logging scenarios in a tropical rain forest. *Journal of Environmental Management* 71, 321–333.

Huth, A., Drechsler, M. and Köhler, P. (2005) Using multicriteria decision analysis and a forest growth model to assess impacts of tree harvesting in Dipterocarp lowland rain forests. *Forest Ecology and Management* 207, 215–232.

Iroumé, A. and Huber, A. (2002) Comparison of interception losses in a broadleaved native forest and a *Pseudotsuga menziesii* (Douglas fir) plantation in the Andes Mountains of southern Chile. *Hydrological Processes* 16, 2347–2361.

Kammesheidt, L., Köhler, P. and Huth, A. (2001) Sustainable timber harvesting in Venezuela: a modelling approach. *Journal of Applied Ecology* 38, 756–770.
Kammesheidt, L., Köhler, P. and Huth, A. (2002) Simulating logging scenarios in secondary forest embedded in a fragmented neotropical landscape. *Forest Ecology and Management* 170, 89–105.
Köhler, P. (2000) Modelling anthropogenic impacts on the growth of tropical rain forests. PhD thesis. University of Kassel, Germany. Der Andere Verlag, Osnabrück, Germany.
Köhler, P. and Huth, A. (1998) The effect of tree species grouping in tropical rain forest modelling – simulation with the individual based model FORMIND. *Ecological Modelling* 109, 301–321.
Köhler, P. and Huth, A. (2004) Simulating growth dynamics in a South-East Asian rain forest threatened by recruitment shortage and tree harvesting. *Climatic Change* 67, 95–117.
Köhler, P., Ditzer, T., Ong, R.C. and Huth, A. (2001) Comparison of measured and modelled growth on permanent plots in Sabahs rain forests. *Forest Ecology and Management* 144, 101–111.
Köhler, P., Chave, J., Riera, B. and Huth, A. (2003) Simulating long-term response of tropical wet forests to fragmentation. *Ecosystems* 6, 114–128.
Larcher, W. (2001) *Ökophysiologie der Pflanzen*, 6th edn. Eugen Ullmer, Stuttgart, Germany.
Lindenmayer, D.B. and Franklin, J.F. (2002) *Conserving Forest Biodiversity: A Comprehensive Multiscaled Approach*. Island Press, Washington, DC.
Lindenmayer, D.B. and Recher, H.F. (1998) Aspects of ecologically sustainable forestry in temperate eucalypt forests – beyond an expanded reserve system. *Pacific Conservation Biology* 4, 4–10.
Loehle, C. (1988) Tree life histories: the role of defenses. *Canadian Journal of Forest Research* 18, 209–222.
Lusk, C.H. (1999) Long-lived light-demanding emergents in southern temperate forests: the case of *Weinmannia trichosperma* (Cunoniaceae) in Chile. *Plant Ecology* 140, 111–115.
Lusk, C.H. (2002) Leaf area accumulation helps juvenile evergreen trees tolerate shade in a temperate rain forest. *Oecologia* 132, 188–196.
Lusk, C.H. and del Pozo, A. (2002) Survival and growth of seedlings of 12 Chilean rainforest trees in two light environments: gas exchange and biomass distribution correlates. *Austral Ecology* 27, 173–182.
Lusk, C.H. and Kelly, C.K. (2003) Interspecific variation in seed size and safe sites in a temperate rain forest. *New Phytologist* 158, 535–542.
Lusk, C.H., Wright, I. and Reich, P.B. (2003) Photosynthetic differences contribute to competitive advantage of evergreen angiosperm trees over evergreen conifers in productive habitats. *New Phytologist* 160, 329–336.
Muñiz-Castro, M.A., Williams-Linera, G. and Rey-Benayas, J.M. (2006) Distance effect from cloud forest fragments on plant community structure in abandoned pastures in Veracruz, Mexico. *Journal of Tropical Ecology* 22, 431–440.
Pérez-Galaz, V.A. (1983) *Manual de propiedades físicas y mecánicas de maderas chilenas*. Documento de trabajo N° 47, Investigación y Desarrollo Forestal CONAF/FAO, Santiago de Chile, Chile.
Pierce, L.L. and Running, S.W. (1988) Rapid estimation of coniferous forest leaf area index using a portable integrating radiometer. *Ecology* 69, 1762–1767.
Rüger, N. (2006) Dynamics and sustainable use of species-rich moist forests – a process-based modelling approach. PhD thesis. University of Osnabrück, Germany.
Rüger, N., Williams-Linera, G. and Huth, A. (2007a) Long-term dynamics of secondary tropical montane cloud forests in central Veracruz, Mexico. (Submitted.)
Rüger, N., Williams-Linera, G., Kissling, W.D. and Huth, A. (2007b) 'Tala hormiga' – Simulating long-term impacts of fuelwood extraction on a Mexican cloud forest. (Submitted.)

Rüger, N., Gutiérrez, A.G., Kissling, W.D., Armesto, J.J. and Huth, A. (2007c) Ecological impacts of different harvesting options for temperate evergreen rain forest in southern Chile – a simulation experiment. (In press.)

Ryan, M.G. (1991) Effects of climate change on plant respiration. *Ecological Applications* 1, 157–167.

Salas, C. (2002) Ajuste y validación de ecuaciones de volumen para un relicto del bosque de Roble-Laurel-Lingue. *Bosque* 23, 81–92.

Saldaña, A. and Lusk, C.H. (2003) Influencia de las especies del dosel en la disponibilidad de recursos y regeneración avanzada en un bosque templado lluvioso del sur de Chile. *Revista Chilena de Historia Natural* 76, 639–650.

Shugart, H.H. (1984) *A Theory of Forest Dynamics: The Ecological Implications of Forest Succession Models*. Springer, New York.

Silva, J.N.M., de Carvalho, J.O.P., Lopes, J. do C.A., de Almeida, B.F., Costa, D.H.M., de Oliveira, L.C., Vanclay, J.K. and Skovsgaard, J.P. (1995) Growth and yield of a tropical rain forest in the Brazilian Amazon 13 years after logging. *Forest Ecology and Management* 71, 267–274.

Smith-Ramírez, C., Martínez, P., Núñez, M., González, C. and Armesto, J.J. (2005) Diversity, flower visitation frequency and generalism of pollinators in temperate rain forests of Chiloé Island, Chile. *Botanical Journal of the Linnean Society* 147, 399–416.

Torres-Rojo, J.M. (2004) *Latin American Forestry Sector Outlook Study Working Paper*. Informe Nacional – México. FAO, Rome, Italy.

Ugalde, L. and Pérez, O. (2001) *Mean Annual Volume Increment of Selected Industrial Forest Plantation Species by Forest Plantation Thematic Papers, Working Paper 1*. Forest Resources Development Service, Forest Resources Division, FAO, Rome, Italy.

van Gardingen, P.R., McLeish, M.J., Phillips, P.D., Fadilah, D., Tyrie, G. and Yasman, I. (2003) Financial and ecological analysis of management options for logged-over Dipterocarp forests in Indonesian Borneo. *Forest Ecology and Management* 183, 1–29.

Veblen, T.T. (1985) Forest development in tree-fall gaps in the temperate rain forests of Chile. *National Geographic Research* 1, 161–184.

Williams-Linera, G. (1996) Crecimiento diamétrico de árboles caducifolios y perennifolios del bosque mesófilo de montaña en los alrededores de Xalapa. *Madera y Bosques* 2, 53–65.

Williams-Linera, G. (2002) Tree species richness complementarity, disturbance and fragmentation in a Mexican tropical montane cloud forest. *Biodiversity and Conservation* 11, 1825–1843.

Williams-Linera, G., Manson, R.H. and Isunza-Vera, E. (2002) La fragmentación del bosque mesófilo de montaña y patrones de uso del suelo en la región oeste de Xalapa, Veracruz, México. *Madera y Bosques* 8, 73–89.

12 Testing Forest Biodiversity Indicators by Assessing Anthropogenic Impacts along Disturbance Gradients

A.C. Newton, C. Echeverría, M. González-Espinosa, G. Williams-Linera, N. Ramírez-Marcial, O. Thiers, J.J. Armesto, J.C. Aravena and A. Lara

Relatively pristine south temperate rainforest, in the vicinity of San Pablo de Tregua, Chile. Photo: Adrian Newton

Summary

Current efforts at sustainable forest management depend on the development of appropriate and effective indicators of forest biodiversity. While many such indicators have been proposed, few have been rigorously tested. To address this, forest structure, composition and soil characteristics were assessed along anthropogenic disturbance gradients in four study areas, namely Highland Chiapas and Central Veracruz (Mexico), and San Pablo de Tregua and Chiloé Island (Chile). Indicators selected for analysis included soil pH, organic matter content and bulk density; species richness; and the species composition of forest stands. Results indicated highly contrasting responses to disturbance in the four study areas. For example, in the Highlands of Chiapas, highly significant ($P<0.001$) correlations were obtained between basal area and both soil organic matter content and bulk density. A highly significant relationship between basal area and bulk density was also recorded in San Pablo de Tregua ($P=0.001$), but in this case the relationship was negative. No significant correlations between basal area and soil characteristics were recorded in the other study areas. With respect to species richness, the only significant correlation was recorded in Chiloé, which was positive and highly significant ($P<0.001$). With respect to tree composition, in the Highlands of Chiapas basal area was found to be positively correlated with the relative abundance of shade-tolerant mature trees ($P=0.037$), and the density of both mature and juvenile shade-tolerant trees ($P=0.015$ and 0.021, respectively). Similar results were obtained in Chiloé, but in San Pablo de Tregua no significant correlations were recorded. The lack of consistent responses suggests that none of the indicators appears to be applicable over a broad geographical area. In many cases, the indicators failed to be sensitive to variation in disturbance, and therefore appear to have low value for monitoring forest condition. This implies that individual indicator sets may need to be developed for each individual forest area of interest, and the goal of a set of easily measured, generally applicable indicators may be difficult to achieve.

Introduction

It is now widely recognized that understanding the response of forest ecosystems to disturbance is of crucial importance for developing management approaches that are genuinely sustainable (Lindenmayer and Franklin, 2002). Traditionally, ecological researchers have devoted greater attention to investigating natural disturbance regimes than the impacts of human activities. As a result, substantial progress has been made in understanding the influence of natural disturbance regimes on forest structure and composition (Pickett and White, 1985; Attiwill, 1994). In contrast, the impacts of anthropogenic disturbance remain less well understood. Interest in this issue has increased in recent years, partly in response to global policy initiatives aimed at supporting sustainable forest management. In particular, following the development of the Forest Principles and Chapter 11 of Agenda 21 at UNCED in 1992, a large number of national and international initiatives have developed criteria and indicators (C&I) for monitoring progress towards sustainable forest management objectives (Castañeda, 2001; Higman *et al.*, 2005). The effectiveness of such indicators depends to a large extent on understanding the environmental impacts of forest management interventions.

Although large numbers of indicators have been developed, many have proved to be impractical for implementation at the scale of the forest management unit (FMU) (Franc *et al.*, 2001; Angelstam and Dönz-Breuss, 2004). This is especially the case for those indicators developed to monitor forest biodiversity (Stork *et al.*, 1997). This partly reflects the early emphasis on development of C&I for use at the national level, which produced indicators that are not sufficiently sensitive to be useful at the FMU level (Raison *et al.*, 2001). Yet many of the important changes in biodiversity that occur as a result of human activity can only be detected at the local scale. Many of the forest biodiversity indicators that have been proposed have been poorly tested and require rigorous validation in order to be interpreted with confidence (Noss, 1999). Ideally, the relationship between proposed indicators and the variables of interest ('end points') should be determined using appropriate statistical approaches (Hyman and Leibowitz, 2001), but this is rarely achieved in practice. Newton and Kapos (2002) reviewed the forest biodiversity indicator sets developed by the various international C&I processes, and found that the following variables are commonly included: the area of forest under different successional stages, the area and percentage of forests affected by anthropogenic disturbance, and the complexity and heterogeneity of forest structure. This further emphasizes the need to quantify the impact of anthropogenic disturbance on forests.

What are the key characteristics that determine whether a proposed indicator is suitable or appropriate? Noss (1990) suggested that indicators of biodiversity should ideally be: (i) sufficiently sensitive to provide an early warning of change; (ii) widely applicable over a broad geographical area; (iii) capable of providing a continuous assessment over a wide range of disturbance; (iv) relatively independent of sample size; (v) easy and cost-effective to measure; (vi) able to differentiate between natural and anthropogenic impacts or disturbance; and (vii) relevant to ecologically significant phenomena (such as key ecological processes). Noss (1990) also noted that a single indicator will seldom suffice, and therefore a suite of indicators will usually be required.

How might the suitability or effectiveness of a set of indicators therefore be tested? One potential approach is to compare forest stands subjected to different anthropogenic disturbance regimes (Angelstam and Dönz-Breuss, 2004). This would allow the criteria identified by Noss (1990) to be rigorously tested, for example by identifying the relationships between selected indicators and the intensity of disturbance, and by assessing whether these relationships are consistent between different geographical areas. In fact, this approach has been relatively little used by researchers. Examples include: Liow *et al.* (2001), who examined bee diversity along a disturbance gradient in tropical lowland forests in South-east Asia; Jones *et al.* (2003), who described the collapse of termite assemblages along a land-use intensification gradient in lowland central Sumatra, Indonesia; and research by Eggleton *et al.* (1995, 1996) on species richness of termites under different levels of forest disturbance in Cameroon. Lawton *et al.* (1998) describe extension of the latter investigation to include other species groups. Relatively little work of this nature appears to have been performed on temperate rainforest or tropical montane forest; Estrada and Fernandez (1999) provide an example of ant diversity studied along a successional gradient in a Colombian cloud forest.

This chapter describes results obtained from four parallel investigations, performed in different study areas, designed to examine anthropogenic impacts on native forests. Detailed results from three of these investigations, undertaken in Chiloé Island (Chile), Highland Chiapas and Xalapa, Veracruz (Mexico), have been presented previously by Aravena *et al.* (2002), Ramírez-Marcial *et al.* (2001) and Williams-Linera (2002), respectively. Data from a fourth investigation, performed in San Pablo de Tregua (Chile), have not been published elsewhere. Each of these investigations involved assessment of forest structure and composition along gradients of anthropogenic disturbance. Here, a set of easily measured indicators were selected for analysis using data obtained from these studies, including soil pH, organic matter content and bulk density; species richness; and the composition of forest stands, analysed in terms of the relative abundance of different functional groups of tree species. The objective of the research presented here was to test these indicators, by examining their relationship with the intensity of disturbance, using basal area as a proxy for the latter. Our initial hypothesis was that the impacts of anthropogenic disturbance on successional processes in these forests are likely to be broadly similar. If this is the case, then the indicators should demonstrate consistent responses across the four study areas.

Description of the Study Areas

Parallel investigations were undertaken in four study areas, described below.

Highlands of Chiapas, Mexico

Plots were established at eight localities, in forest fragments varying from 0.4 to 1.8 ha in size, at elevations of between 1700 and 2300 masl in the northern Highlands of Chiapas (17° 08′–17° 14′ N; 92° 52′–92° 52′ W). Slopes ranged between 3 and 60°, more commonly in the range 20–45°. The stands included a wide variety of successional stages, from relatively undisturbed old-growth forests to heavily disturbed and open stands. Annual rainfall is 1400–2000 mm, and heavy fog occurs for 4–8 h each day during 8–10 months of the year in the higher locations. Soils are typically developed on karst. The principal form of anthropogenic disturbance in the area is conversion of forest into *milpa* fields under the traditional slash-and-burn system of agriculture. The remaining forest fragments are subjected to firewood and timber extraction, as well as extensive livestock browsing. Survey plots were each 0.1 ha in area, and were circular. At least six plots were evaluated at each locality, and the data pooled for analysis.

San Pablo de Tregua, Chile

The study area was San Pablo de Tregua, an experimental area owned by Universidad Austral de Chile, Valdivia. The site is located in the Chilean

Andes (72° 02′–72° 09′ W, 39° 30′–39° 38′ S) and is 2200 ha in extent. Altitude is 620–840 masl and precipitation 4000–5000 mm year^{-1}. Soils are derived from volcanic ash and pumitic material, overlying andesitic and basaltic lavas and fluvio-glacial sediments. The site is covered by mature forest of Coigüe-Raulí-Tepa (see Chapter 14), which has been subjected to different degrees of anthropogenic disturbance, including selective logging and browsing by livestock. The area has also been subjected to natural disturbance such as earthquakes, which has resulted in landslips. A series of ten permanent plots of 1000 m^2 were established in areas that had experienced different intensities of disturbance, based on several aerial photographs (1:10,000) and historical information. Plots were located longitudinally perpendicular to the slope of the site.

Central Veracruz, Mexico

Study sites were located between 1250 and 1875 m altitude in the eastern Sierra Madre mountains in central Veracruz, in an area extending from 19° 30′ 1.03″ N to 19° 36′ 5.87″ N and 96° 54′ 14.20″ W to 97° 2′ 43.11″ W. Total annual precipitation varies between 1300 and 2200 mm, while mean annual temperature is between 12 and 18°C. Soils are andosols. In the study area, tropical montane cloud forest (TMCF) was originally the dominant land cover type. Tree species dominance changes from one forest fragment to the next, but, in general, composition is similar in the studied fragments of central Veracruz. Seven forest fragments were selected for detailed analysis separated by an average distance of around 4 km, except two sites which were 15 and 40 km away from the other sites. Forest fragments were surrounded by a diversity of other land cover uses (pastures, row crops, coffee plantations and secondary vegetation), which are representative of those found in the region. Forest fragments are subject to ongoing disturbance involving harvesting and removal of forest products, particularly firewood, and impacts of browsing animals.

Chiloé Island (Region X), Chile

The study area comprised approximately 400 km^2 in the north-eastern corner of Chiloé Island, about 20 km north of the city of Ancud (41° 50′ S–73° 40′ W). The landscape can be characterized as rolling hills with altitudes ranging from 50 to 100 m. Soils are generally thin (<1 m), originating from Pleistocenic moraine fields and glacial outwash plains, often with poor drainage. Lowland forests in the area have been logged since the early 1800s, but land clearing became more intense in the second half of the 20th century. The present-day rural landscape is characterized by a mosaic of remnant forest fragments, woodlands and grazing pastures. The major forms of human impact on forests during the last century have been selective logging of valuable timber trees, widespread use of fire to clear land for pastures and increasing patch fragmentation (Willson and Armesto, 1996). The prevailing climate is described as wet temperate with a strong oceanic influence (Di-Castri and Hajek, 1976). Annual rainfall is approximately 2000 mm, with a mean annual temperature of 12°C. Rainfall occurs throughout the year, but 64% of the precipitation is concentrated from April to September.

Nine stands were selected, representing a broad successional chronosequence, including forests that were disturbed in recent decades, fast-growing mid-successional stands, late successional and old-growth forests. Initial stand selection was based on visual assessment of the present condition of forests. Studies were conducted in forest patches varying from 10 ha to larger than 100 ha in size, separated from one another by areas of open pastures, but connected to each other by secondary scrubland vegetation. All forests, including late successional stands, have been subjected to occasional grazing by cattle, mainly along the margins, and some canopy trees have been selectively harvested. Early and mid-successional stands were recovering from non-catastrophic anthropogenic fire and subsequent timber extraction, as indicated by the local presence of charcoal, woody detritus and stumps.

Methods

Full details of the methods employed are provided by Ramírez-Marcial *et al.* (2001), Aravena *et al.* (2002) and Williams-Linera (2002). A brief summary is provided here.

In each forest stand, a survey plot was established at least 30 m from any forest edge. Plots were each 1 ha in size (100 m × 100 m) (unless stated otherwise above). In each plot, structural, floristic and environmental variables were recorded. Forest structure was characterized in terms of density and basal area of woody plants. All mature individuals (≥5 cm dbh) were censused, either in the entire plot, or (in Veracruz) in ten 50 m × 2 m parallel plots randomly located along the slope. The diameter of all trees ≥5 cm dbh was measured and identified to species. Juvenile individuals (<5 cm dbh and ≥1.3 m tall) were also measured, typically in 5 m × 5 m subplots randomly located within the larger plot.

Replicate soil samples were collected at 0–10 cm depths and mixed into a composite sample per site to estimate bulk density gravimetrically (e.g. clod method; Blake, 1979), pH (soil–water mixture of 1:2.5), and organic matter content (e.g. using the Walkley–Black method; Walkley and Black, 1934).

To quantify differences in the representation of functional groups of tree species, trees were classified according to their shade tolerance, based on previously published studies and personal observations. Shade-intolerant tree species are defined here as species that are always found as adults in open successional areas and are known to regenerate after catastrophic stand disturbance (Veblen and Alaback, 1996). Shade-tolerant trees, in turn, are generally absent as adults from open successional sites, but often form dense sapling 'banks' in the shaded understorey of second-growth and old-growth forests. They regenerate within small canopy openings originated by tree falls (Armesto and Figueroa, 1987; Veblen and Alaback, 1996). A third group of trees was classified as intermediate in shade tolerance because they are capable of regenerating in medium-sized tree-fall gaps as well as in large openings (Armesto and Fuentes, 1988), but they rarely form sapling banks.

Results were analysed by performing Spearman rank correlations between variables, using SPSS v. 11 (SPSS Inc.).

Results

The results indicate highly contrasting responses to anthropogenic disturbance in the four study areas. A number of significant relationships of stand structure variables with soil characteristics were recorded (Table 12.1). In the Highlands of Chiapas, highly significant ($P<0.001$) correlations were obtained between basal area and both soil organic matter content and bulk density. A highly significant relationship between basal area and bulk density was also recorded in San Pablo de Tregua ($P=0.001$), but in this case the relationship was negative (i.e. bulk density declined with increasing basal area). However, no significant correlations between basal area and soil characteristics were recorded in the other two study areas. With respect to species richness, the only significant correlation was recorded in Chiloé, which was positive and highly significant ($P<0.001$). This result was not obtained in any of the other study areas.

In order to examine the impact of disturbance on tree species composition, basal area was correlated with the relative abundance of different functional groups of tree species, defined in terms of their shade tolerance. This

Table 12.1. Correlations between forest stand structure variables, soil characteristics and species richness along gradients of anthropogenic disturbance in four study areas.

Variable		BA (%)	BA	Dbh (cm)	pH	Org (%)	BD	SR
(a) Highland Chiapas, Mexico								
BA(%)	*r*	–	1.000	0.476	−0.602	**0.952**	**0.952**	0.643
	P	–	–	0.233	0.114	**0.000**	**0.000**	0.086
BA	*r*	1.000	–	0.476	−0.602	**0.952**	**0.952**	0.643
	P	–	–	0.233	0.114	**0.000**	**0.000**	0.086
Dbh (cm)	*r*	0.476	0.476	–	0.193	0.476	0.476	0.500
	P	0.233	0.233	–	0.647	0.233	0.233	0.207
pH	*r*	−0.602	−0.602	0.193	–	−0.675	−0.675	−0.193
	P	0.114	0.114	0.647	–	0.066	0.066	0.647
Org (%)	*r*	0.952	0.952	0.476	−0.675	–	1.000	0.619
	P	0.000	0.000	0.233	0.066	–	–	0.102
BD	*r*	**0.952**	**0.952**	0.476	−0.675	1.000	–	0.619
	P	**0.000**	**0.000**	0.233	0.066	–	–	0.102
SR	*r*	0.643	0.643	0.500	−0.193	0.619	0.619	–
	P	0.086	0.086	0.207	0.647	0.102	0.102	–
(b) San Pablo de Tregua, Chile								
BA (%)	*r*	–	1.000	**0.857**	−0.612	0.188	**−0.872**	0.425
	P	–	–	**0.002**	0.060	0.603	**0.001**	0.221

Continued

Table 12.1. *Continued*

Variable		BA (%)	BA	Dbh (cm)	pH	Org (%)	BD	SR
BA	*r*	1.000	–	**0.857**	–0.612	0.188	**–0.872**	0.425
	P	–	–	**0.002**	0.060	0.603	**0.001**	0.221
Dbh (cm)	*r*	**0.857**	**0.857**	–	–0.274	0.353	**–0.820**	0.539
	P	**0.002**	**0.002**	–	0.444	0.318	**0.004**	0.108
pH	*r*	–0.612	–0.612	–0.274	–	0.139	0.604	0.163
	P	0.060	0.060	0.444	–	0.701	0.065	0.654
Org (%)	*r*	0.188	0.188	0.353	0.139	–	–0.250	0.388
	P	0.603	0.603	0.318	0.701	–	0.486	0.268
BD	*r*	**–0.872**	**–0.872**	**–0.820**	0.604	–0.250	–	–0.258
	P	**0.001**	**0.001**	**0.004**	0.065	0.486	–	0.472
SR	*r*	0.425	0.425	0.539	0.163	0.388	–0.258	–
	P	0.221	0.221	0.108	0.654	0.268	0.472	–
(c) Central Veracruz, Mexico								
BA (%)	*r*	–	1.000	0.214	–0.631	0.536	–0.371	–0.143
	P	–	–	0.645	0.129	0.215	0.468	0.760
BA	*r*	1.000	–	0.214	–0.631	0.536	–0.371	–0.143
	P	–	–	0.645	0.129	0.215	0.468	0.760
Dbh (cm)	*r*	0.214	0.214	–	–0.234	–0.536	–0.371	–0.679
	P	0.645	0.645	–	0.613	0.215	0.468	0.094
pH	*r*	–0.631	–0.631	–0.234	–	–0.180	0.638	–0.342
	P	0.129	0.129	0.613	–	0.699	0.173	0.452
Org (%)	*r*	0.536	0.536	–0.536	–0.180	–	–0.086	0.250
	P	0.215	0.215	0.215	0.699	–	0.872	0.589
BD	*r*	–0.371	–0.371	–0.371	0.638	–0.086	–	–0.257
	P	0.468	0.468	0.468	0.173	0.872	–	0.623
SR	*r*	–0.143	–0.143	–0.679	–0.342	0.250	–0.257	–
	P	0.760	0.760	0.094	0.452	0.589	0.623	–
(d) Chiloé Island, Chile								
BA (%)	*r*	–	1.000	0.333	0.077	0.067	0.218	**0.953**
	P	–	–	0.381	0.844	0.864	0.574	**0.000**
BA	*r*	1.000	–	0.333	0.077	0.067	0.218	**0.953**
	P	–	–	0.381	0.844	0.864	0.574	**0.000**
Dbh (cm)	*r*	0.333	0.333	–	0.530	–0.577	–0.122	0.271
	P	0.381	0.381	–	0.142	0.104	0.755	0.481
pH	*r*	0.077	0.077	0.530	–	–0.571	–0.455	0.049
	P	0.844	0.844	0.142	–	0.108	0.218	0.900
Org (%)	*r*	0.067	0.067	–0.577	–0.571	–	–0.127	–0.061
	P	0.864	0.864	0.104	0.108	–	0.745	0.875
BD	*r*	0.218	0.218	–0.122	–0.455	–0.127	–	0.315
	P	0.574	0.574	0.755	0.218	0.745	–	0.409
SR	*r*	**0.953**	**0.953**	0.271	0.049	–0.061	0.315	–
	P	**0.000**	**0.000**	0.481	0.900	0.875	0.409	–

Abbreviations: *r*, correlation coefficient; *P*, significance value (result of Spearman rank correlation analysis); BA(%), basal area expressed as a percentage of maximum value recorded across the disturbance gradient; BA, basal area of the forest stand; Dbh, mean stem diameter of mature trees (at breast height); pH, soil pH; Org., organic matter content of soil (%); BD, bulk density of soil; SR, species richness. Significant correlations ($P < 0.05$) are highlighted in bold.

was achieved by calculating the percentages of total basal area comprised of shade-tolerant and shade-intolerant tree species (considering mature trees only, >5 cm dbh). The relative abundance of these two functional groups was further examined by calculating the percentage of the total number of tree stems comprised of shade-tolerant and shade-intolerant tree species, considering mature trees (>5 cm dbh) and juveniles (<5 cm dbh) separately. Each of these variables was correlated with total basal area in separate correlation analyses. Again, contrasting results were obtained in the different study areas. In the Highlands of Chiapas, basal area was found to be significantly, positively correlated with the relative abundance of shade-tolerant mature trees expressed in terms of basal area ($P = 0.037$), and the density of both mature and juvenile shade-tolerant trees ($P = 0.015$ and 0.021, respectively). Conversely, significant negative relationships were recorded between basal area and relative abundance of shade-intolerant mature trees expressed in terms of basal area, and the density of both mature and juvenile shade-intolerant trees (Table 12.2). In central Veracruz, the only significant relationship recorded was a negative correlation between basal area and the density of shade-tolerant juvenile trees ($P = 0.036$), contrasting with the results from Chiapas, where a positive correlation between these variables was recorded. In Chiloé, two significant correlations were obtained, which were consistent with results from Chiapas. Significant negative correlations were recorded between basal area and relative abundance of shade-intolerant trees expressed in terms of basal area ($P = 0.01$), and between basal area and the density of shade-intolerant juveniles ($P = 0.036$). No significant correlations were recorded in San Pablo de Tregua.

Table 12.2. Correlations between forest stand composition variables and stand structure along gradients of anthropogenic disturbance in four study areas.

		BA tol	BA intol	D tol	D intol	D tol juv	D intol juv
Highland Chiapas	*r*	**0.738**	**−0.886**	**0.810**	**−0.910**	**0.786**	**−0.952**
	P	**0.037**	**0.003**	**0.015**	**0.002**	**0.021**	**0.000**
San Pablo de Tregua	*r*	−0.356	0.467	−0.239	0.467	0.350	−0.350
	P	0.313	0.173	0.506	0.173	0.321	0.321
Central Veracruz	*r*	−0.464	−0.020	−0.714	0.222	**−0.786**	0.535
	P	0.294	0.967	0.071	0.632	**0.036**	0.216
Chiloé Island	*r*	0.410	**−0.800**	0.343	−0.259	0.533	**−0.700**
	P	0.273	**0.010**	0.366	0.574	0.139	**0.036**

Abbreviations: *r*, correlation coefficient; *P*, significance value (result of Spearman rank correlation analysis). Significant correlations ($P < 0.05$) are highlighted in bold. Correlations were between BA(%), basal area expressed as a percentage of maximum value recorded across the disturbance gradient, and the following variables describing stand composition: BA tol, the percentage of total basal area comprised of shade-tolerant species; BA intol, the percentage of total basal area comprised of shade-intolerant species; D tol, the percentage of stem density of mature trees (>5 cm dbh) comprised of shade-tolerant species; D intol, the percentage of stem density of mature trees (>5 cm dbh) comprised of shade-intolerant species; D tol juv, the percentage of stem density of juvenile trees (<5 cm dbh) comprised of shade-tolerant species; D intol juv, the percentage of stem density of juvenile trees (<5 cm dbh) comprised of shade-intolerant species.

The correlation analyses were tested for linear relationships between variables (see Fig. 12.1). The data collected provided little evidence for any non-linear relationships between basal area and the other variables measured. However, in Chiapas, pH showed a peaked (Gaussian) distribution, with the highest values occurring at intermediate values of basal area (data not illustrated).

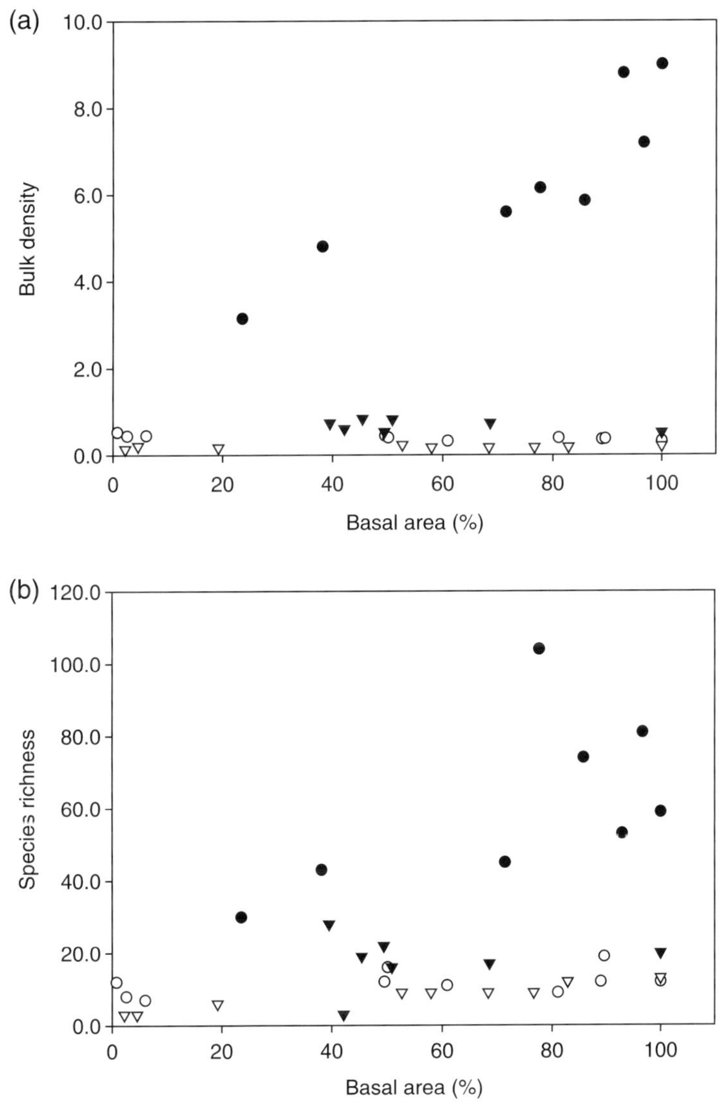

Fig. 12.1. Relationship between basal area of forest stands along gradients of anthropogenic disturbance in four study areas, and (a) soil bulk density, (b) species richness, (*Continued*)

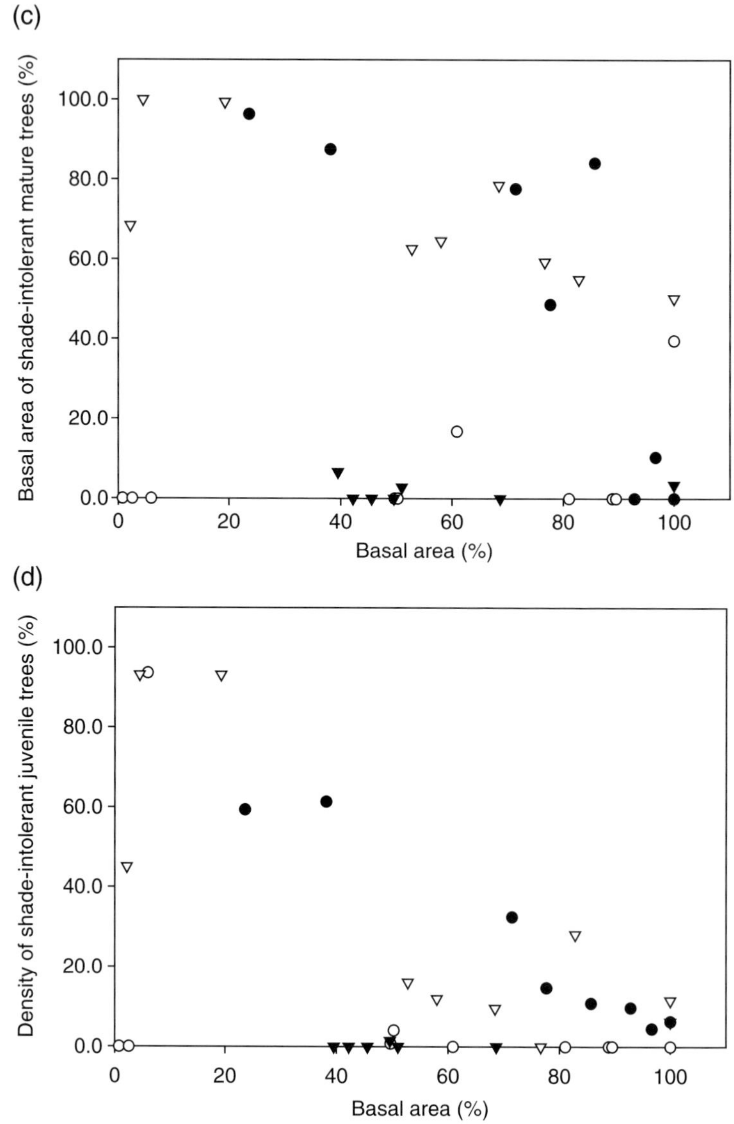

Fig. 12.1. *Continued* (c) basal area of shade-intolerant mature trees (expressed as a percentage of total basal area) and (d) density of shade-intolerant juvenile trees (expressed as a percentage of total density of juvenile trees). In these graphs, basal area is expressed as a percentage of the maximum value recorded along the disturbance gradient, to facilitate comparison between different study areas. Solid circles, Highlands of Chiapas; open circles, San Pablo de Tregua, Chile; open triangles, Central Veracruz, Mexico; solid triangles, Chiloé Island, Chile.

Discussion

These results indicate that none of the indicators selected displayed a consistent response across all four study areas. Rather, the relationships detected were highly idiosyncratic, and differed markedly between the four study areas. Overall, relatively few relationships were detected between the selected indicators and basal area. The only relationship that appeared somewhat consistent across sites was the negative correlation between basal area and the abundance of shade-intolerant species, recorded in both Chiloé Island and Highland Chiapas.

The lack of a consistent response may partly reflect the methodological difficulties of obtaining genuinely comparable data sets from different areas. While the studies were planned from the outset to generate comparable data, by employing a shared set of methods, there were inevitable differences between the data sets simply as a result of the contrasting characteristics of the forests concerned. A more general problem relates to assessment of disturbance. Here, within each study area, basal area was used as a proxy for the intensity of disturbance, but it is recognized that this is not ideal: the two variables may not always be closely correlated. Other methods were used by the individual studies to characterize disturbance regimes; for example Williams-Linera (2002) assessed the presence of the number of cut stumps and the amount of woody debris, whereas Ramírez-Marcial *et al.* (2001) evaluated the frequency and intensity of harvesting and fire by assessing the number of cut and burnt stumps, respectively, and livestock browsing using a categorical variable based on observations of stock density and abundance of faeces. However, such assessments proved difficult to apply consistently in different study areas; for example, Williams-Linera (2002) originally planned to estimate browsing impacts by surveying faeces, but in practice very few were observed during the survey, despite browsing animals being widespread. Characterization of a forest disturbance regime presents substantial practical difficulties (Newton, 2007); ideally, detailed information on disturbance events would be collected over a prolonged time period. Another problem is the validity of assuming that differences between sites can be attributed solely to disturbance history (Pickett, 1989), although attempts were made to address this by selecting sites that were floristically and edaphically very similar (Aravena *et al.*, 2002). The preferred approach would be to perform experimental manipulations of the forests and examine their responses, although this is seldom achievable in practice, particularly when the need to replicate treatments across several sites is considered.

Given these limitations, these results should clearly be viewed with caution. Intriguingly, however, some significant relationships were identified. The most striking of these were those recorded in Chiapas, where significant correlations were obtained between basal area and soil organic matter content and bulk density, as well as the relative abundance of both shade-tolerant and shade-intolerant tree species, both juvenile and mature. The linear trend in bulk density recorded in Chiapas is particularly noteworthy (Fig. 12.1a), meeting the Noss (1990) criterion of varying continuously over a wide range of disturbance.

The reasons why these relationships were not reproduced across the other study areas remain an enigma, but, as a consequence, another of the criteria listed by Noss (1990) is clearly not met: none of the indicators appears to be applicable over a broad geographical area. In fact, in many cases, the indicators selected here also failed to be sensitive to variation in disturbance, another of the criteria listed by Noss (1990), and therefore appear to have low value for monitoring forest condition, at least in some areas. It is therefore perhaps a matter of concern that these indicators are widely incorporated into C&I sets used to assess sustainable forest management (Newton and Kapos, 2002).

Angelstam and Dönz-Breuss (2004) describe a detailed evaluation of forest biodiversity indicators across a range of sites in Europe with contrasting histories. The method employed multiple sample plots to assess variation along gradients of 'naturalness' within each of five sites. In all study areas, basal area increased consistently with increasing naturalness, supporting its use as a proxy for disturbance, as in the current investigation. Of nine biodiversity indicators that were selected for study, two appeared to give consistent results across all sites: the amount of dead wood and the frequency of occurrence of uprooted trees. The response of the other indicators, however, varied between sites and, as a result, some (such as presence of particular tree species or threatened lichen species) appeared to have limited value. The contrasting behaviour of different indicators on different sites, as recorded in the current investigation, therefore has some support from another study. If many indicators have limited general applicability, then individual indicator sets may need to be developed for each individual forest area of interest.

Angelstam and Dönz-Breuss (2004) highlight the need for studies at multiple spatial scales of how biodiversity indicators behave along gradients of land-use change, and how indicators are related to each other. While the current investigation was designed to address precisely this need, results suggest that the goal of a set of easily measured, generally applicable indicators may be difficult to achieve. A thorough assessment of anthropogenic impacts on forest biodiversity would ideally involve the simultaneous assessment of a wide range of taxa, but, as noted by Lawton *et al.* (1998), the resources required to achieve this are beyond the reach of what is currently available, at least for species-rich forests. A practical alternative may be to use structural characteristics of forest stands, which are relatively easy to measure, and are widely believed to relate closely to the requirements of a range of species (Noss, 1990, 1999; Ferris and Humphrey, 1999). McElhinny *et al.* (2005) review the structural indices that have been used previously by researchers, but note the need for demonstrating an association between the measures made and the elements of biodiversity that are of interest. Again, this is often lacking.

Forest structure is another of the eight generalized indicators identified by Newton and Kapos (2002) as common to many C&I processes. Other indicators considered by these authors include the degree of fragmentation of forest types, and rate of conversion of forest cover (by type) to other land cover types. These can be assessed using remote sensing and GIS techniques (Chapter 2), and collection of appropriate field data can be used to identify relationships between forest area and fragmentation metrics and the distribution of different species (Chapter 3). These approaches may offer particular advantages for assessing biodiversity indicators, as they can readily be applied in a standard way to different study areas, and future research might therefore usefully focus on further development and application of these methods.

References

Angelstam, P. and Dönz-Breuss, M. (2004) Measuring forest biodiversity at the stand scale – an evaluation of indicators in European forest history gradients. *Ecological Bulletins* 51, 305–332.

Aravena, J.C., Carmona, M.R., Perez, C.A. and Armesto, J.J. (2002) Changes in tree species richness, stand structure and soil properties in a successional chronosequence in northern Chiloe Island, Chile. *Revista Chilena de Historia Natural* 75, 339–360.

Armesto, J.J. and Figueroa, J. (1987) Stand structure and dynamics in the rain forest of Chiloé Archipelago, Chile. *Journal of Biogeography* 14, 367–376.

Armesto, J.J. and Fuentes, E.R. (1988) Tree species regeneration in a mid-elevation temperate forest in Isla Chiloé, Chile. *Vegetatio* 74, 151–159.

Attiwill, P.M. (1994) The disturbance of forest ecosystems: the ecological basis for conservative management. *Forest Ecology and Management* 63, 247–300.

Blake, G.R. (1979) Bulk density. In: Black, C.A., Evans, D.D., White, J.L., Ensminger, L.E. and Clark, F.E. (eds) *Methods of Soil Analysis, Part 1*. American Society of Agronomy, Inc., Madison, Wisconsin, pp. 374–390.

Castañeda, F. (2001) Collaborative action and technology transfer as means of strengthening the implementation of national-level criteria and indicators. In: Raison, R.J., Brown, A.G. and Flinn, D.W. (eds) *Criteria and Indicators for Sustainable Forest Management*. IUFRO Research Series No. 7. CAB International, Wallingford, UK, pp. 145–163.

Di-Castri, F. and Hajek, E.R. (1976) *Bioclimatología de Chile*. Vicerrectoría de Comunicaciones, Universidad Católica de Chile, Santiago, Chile.

Eggleton, P., Bignell, D.E., Sands, W.A., Waite, B., Wood, T.G. and Lawton, J.H. (1995) The species richness of termites (Isoptera) under differing levels of forest disturbance in the Mbalmayo Forest Reserve, southern Cameroon. *Journal of Tropical Ecology* 11, 1–14.

Eggleton, P., Bignell, D.E., Sands, W.A., Mawdsley, N.A., Lawton, J.H. and Bignell, N.C. (1996) The diversity, abundance and biomass of termites under differing levels of disturbance in the Mbalmayo Forest Reserve, southern Cameroon. *Philosophical Transactions of the Royal Society of London Series B: Biological Sciences* 351, 51–68.

Estrada, C. and Fernandez, F. (1999) Diversity of ants (Hymenoptera: Formicidae) in a successional gradient of a cloud forest (Narino, Colombia). *Revista de Biología Tropical* 47, 189–201.

Ferris, R. and Humphrey, J.W. (1999) A review of potential biodiversity indicators for application in British forests. *Forestry* 72, 313–328.

Franc, A., Laroussinie, O. and Karjalainen, T. (2001) *Criteria and Indicators for Sustainable Forest Management at the Forest Management Unit Level*. European Forest Institute, Proc. 38, Gummerus Printing, Saarjärvi, Finland.

Higman, S., Mayers, J., Bass, S., Judd, N. and Nussbaum, R. (2005) *Sustainable Forestry Handbook*, 2nd edn. Earthscan, London, UK.

Hyman, J.B. and Leibowitz, S.G. (2001) JSEM: a framework for identifying and evaluating indicators. *Environmental Monitoring and Assessment* 66, 207–232.

Jones, D.T., Susilo, F.X., Bignell, D.E., Hardiwinoto, S., Gillison, A.N. and Eggleton, P. (2003) Termite assemblage collapse along a land-use intensification gradient in lowland central Sumatra, Indonesia. *Journal of Applied Ecology* 40, 380–391.

Lawton, J.H., Bignell, D.E., Bolton, B., Bloemers, G.F., Eggleton, P., Hammond, P.M., Hodda, M., Holt, R.D., Larsen, T.B., Mawdsley, N.A., Stork, N.E., Srivastava, D.S. and Watt, A.D. (1998) Biodiversity inventories, indicator taxa and effects of habitat modification in tropical forest. *Nature* 391, 72–76.

Lindenmayer, D.B. and Franklin, J.F. (2002) *Conserving Forest Biodiversity. A Comprehensive Multiscaled Approach*. Island Press, Washington, DC.

Liow, L.H., Sodhi, N.S. and Elmqvist, T.H. (2001) Bee diversity along a disturbance gradient in tropical lowland forests of south-east Asia. *Journal of Applied Ecology* 38, 180–192.

McElhinny, C., Gibbons, P., Brack, C. and Bauhus, J. (2005) Forest and woodland stand structural complexity: its definition and measurement. *Forest Ecology and Management* 218, 1–24.

Newton, A.C. (2007) *Forest Ecology and Conservation. A Handbook of Techniques*. Oxford University Press, Oxford, UK.

Newton, A.C. and Kapos, V. (2002) Biodiversity indicators in national forest inventories. *Unasylva* 53, 56–64.

Noss, R.F. (1990) Indicators for monitoring biodiversity: a hierarchical approach. *Conservation Biology* 4, 355–364.

Noss, R.F. (1999) Assessing and monitoring forest biodiversity: a suggested framework and indicators. *Forest Ecology and Management* 115, 135–146.

Pickett, S.T.A. (1989) Space-for-time substitution as an alternative to long-term studies. In: Likens, G.E. (ed.) *Long-Term Studies in Ecology: Approaches and Alternatives*. Springer, New York, pp. 110–135.

Pickett, S.T.A. and White, P.S. (1985) *The Ecology of Natural Disturbance and Patch Dynamics*. Academic Press, New York.

Raison, R.J., Flinn, D.W. and Brown, A.G. (2001) Application of criteria and indicators to support sustainable forest management: some key issues. In: Raison, R.J., Brown, A.G. and Flinn, D.W. (eds) *Criteria and Indicators for Sustainable Forest Management*. IUFRO Research Series No. 7. CAB International, Wallingford, UK, pp. 5–18.

Ramírez-Marcial, N., González-Espinosa, M. and Williams-Linera, G. (2001) Anthropogenic disturbance and tree diversity in montane rain forests in Chiapas, Mexico. *Forest Ecology and Management* 154, 311–326.

Stork, N.E., Boyle, T.J.B., Dale, V., Eeley, H., Finegan, B., Lawes, M., Manokaran, N., Prabhu, R. and Soberon, J. (1997) *Criteria and Indicators for Assessing the Sustainability of Forest Management: Conservation of Biodiversity*. CIFOR Working Paper No. 17. CIFOR, Jakarta, Indonesia.

Veblen, T.T. and Alaback, P.B. (1996) A comparative review of forest dynamics and disturbance in the temperate rain forests of North and South America. In: Lawford, R.G., Alaback, P.B. and Fuentes, E. (eds) *High Latitude Rain Forests and Associated Ecosystems of the West Coast of the Americas: Climate, Hydrology, Ecology and Conservation*. Springer, New York, pp. 173–213.

Walkley, O. and Black, I.A. (1934) An examination of the Degtjareff method for determining soil organic matter, and a proposed modification of the chromic acid titration method. *Soil Science* 37, 29–38.

Williams-Linera, G. (2002) Tree species richness complementarity, disturbance and fragmentation in a Mexican tropical montane cloud forest. *Biodiversity and Conservation* 11, 1825–1843.

Willson, M.F. and Armesto, J.J. (1996) The natural history of Chiloé: on Darwin's trail. *Revista Chilena de Historia Natural* 69, 149–161.

13 Fire Challenges to Conserving Tropical Ecosystems: the Case Study of Chiapas

R.M. Román-Cuesta, J. Retana and M. Gracia

Fragmented forest landscape in the Highlands of Chiapas, Mexico. Photo: Mario González-Espinosa

Summary

Among the threats that jeopardize conservation efforts in forest ecosystems, fire has become one of the most important. Chiapas represents an informative case study as it displays the same structural fire causes common to other tropical areas: presence of fire spreading from farming sites, fragmented landscapes and a high sensitivity of fuels to ENSO-driven droughts. Since 1984, one million hectares have burned in the state of Chiapas, with peaks recorded in the ENSO events in 1986–1987, 1997–1998 and 2003. Chiapas ecosystems include fire-adapted communities such as pine stands and subdeciduous forests, as well as fire-sensitive ecosystems such as evergreen rainforests and montane cloud forests, which are both suffering from an excess of fire. While fire-adapted ecosystems have higher fire incidence and larger burned areas under normal climatic conditions (i.e. non-ENSO), the presence of ENSO droughts alters this pattern, resulting in higher burned areas in fire-sensitive ecosystems (e.g. montane cloud forests of Chimalapas, evergreen broadleaved ecosystems of the Lacandon and El Ocote). This partly relates to the presence of continuous fuel layers in such systems. Even though the natural fire regimes of these different ecosystems are still to be defined, the current fire situation in Chiapas has reshaped fire regimes throughout the state, reducing the intervals without fire and jeopardizing the stability of highly diverse and valuable ecosystems. A particular issue of concern relates to the presence of fire in protected areas. Results suggest that protected areas are not being effective in mitigating fire impacts within their boundaries, and peaks of burned area are common inside parks in years of severe conditions (e.g. 50% of the total burned area in 2003 occurred within El Ocote Biosphere Reserve). Accumulated fuel loads from previous fire disturbances, severe climatic conditions and strong human pressures currently combine to degrade some of the last remaining well-preserved areas in the state. Conservation alternatives and compensations are urgently required in this highly populated region, with decreased use of fire, fire management programmes, fuel treatment, environmental education and economic incentives among the key aspects to consider.

Introduction

Most of today's fires occur in tropical and subtropical areas (Dwyer *et al.*, 1999). In these regions, fire has been traditionally used as a land management tool to favour forest conversions to agrarian uses, grassland regeneration, disposal of farming residues, enhanced used of secondary forest products, etc. (Goldammer, 1993; Rodríguez, 1996; Fulé and Covington, 1997). In the history of the seasonal tropics, fire has contributed considerably to reshaping forest ecosystems and savannas (e.g. pine, pine–oak forests, tropical deciduous forests (*sensu* Leopold, 1950)) (Goldammer and Peñafiel, 1990). The influence of fire over millennia has favoured tropical vegetation communities that are considered stable ecosystems over the long term (i.e. fire-dependent ecosystems). The extreme case of this fire dependence is presented by induced savannas or grasslands that are maintained by fire and that would return to seasonal forests if fire were excluded (Mueller-Dombois and Goldammer, 1990). At the other end of the gradient are non-seasonal fire-sensitive ecosystems, which do not present fire adaptation traits because fire has not been a co-evolving agent, although it might have been present as a sporadic disturbance factor (Sanford *et al.*, 1985) (e.g. tropical evergreen forests or cloud forests (*sensu* Leopold, 1950)) (Fig. 13.1).

In recent decades, particular political and socio-economic conditions, environmental pressures and changes in the traditional uses of fire have

(a) (b)

Fig. 13.1. View from a street in San Cristóbal de las Casas, Chiapas, during the dry (fire) season of ENSO 2003 (a) versus the wet season (b), and the visibility and respiratory problems deriving from it. The Huitepec Volcano is not observable due to high aerosol contents at the end of the fire season in 2003 (a). (Photo: R.M. Román-Cuesta 2003.)

resulted in altered tropical fire regimes, where both too little and too much fire represent a threat to biodiversity. Although there are tropical areas where too little fire is a problem, the increasing size and frequency of fires in tropical forests appear to be more of a concern (Kinnaird and O'Brien, 1998; Cochrane, 2003; Du Toit *et al.*, 2004). These fires are threatening fire-sensitive and poorly fire-adapted ecosystems (e.g. moist broadleaf forests) while promoting the spread of fire-dependent stands (e.g. pines) outside their natural areas of occurrence (Goldammer and Peñafiel, 1990; Ramírez-Marcial *et al.*, 2001; García-Barrios and González-Espinosa, 2004). The virtues of fire as a traditional farming tool are therefore being undermined because of its destructive power, which is globally exacerbated by ENSO drought-related episodes. For example, in the 1997–1998 fire season, 10 million ha burned in Indonesia (Siegert *et al.*, 2001), 4 million ha in Brazil (Lindenmayer *et al.*, 2004), 3 million ha in Bolivia (Cochrane, 2003) and 2.5 million ha in Central America (Cochrane, 2002). In this single season, Mexico experienced 14,445 wildfires spreading over 849,632 hectares – the largest area ever burned in Mexico in a single season (Cedeño, 2001). Chiapas and Oaxaca, the most biodiverse Mexican states (Deininger and Minten, 2002), each contributed up to *c*.20% of the total Mexican burned area in 1998 (198,000 and 210,000ha, respectively) (Román-Cuesta *et al*., 2003; Asbjorsen *et al*., 2005).

In the case of tropical Mexico, and in Chiapas in particular, in the 1998 El Niño season most protected areas were affected (Montebello, El Ocote, Montes Azules, Naha, Sepultura, Cañón del Sumidero, Encrucijada, Triunfo), severely damaging fire-sensitive ecosystems (i.e. cloud montane forests in Chimalapas, evergreen rainforests of the Lacandon). In spite of this, the fact that fire could be a problem was new to most people. Deeply rooted in the farmers' minds, fire has been used in Chiapas as a land management tool for centuries. The idea of uncontrolled fire as a severe agent of degradation was, therefore, astonishing. One could argue, however, that this astonishment was more a reflection of detailed *media coverage* that fire received that year, in a place where worse fire conflagrations had occurred in the past without raising much public debate other than the voices of a few local conservationists such as Gertrude Blom (Harris and Sartor, 1984) and Miguel Alvárez del Toro (Alvarez-del Toro, 1985):

> I am writing this at the end of April 1983 and I would like to state, once more, that a vast expanse of the Lacandon jungle is burning and one of Mexico's greatest riches is being destroyed forever – lost in suffocating smoke and asphyxiant heat. In the face of the general indiference towards this situation, it is heartbreaking to see the impotence of those who care.
>
> (Gertrude Duby Blom. Arde la Selva, arde)

While 'fire' was one of the most uttered words in the autumm of 1998, there seemed to be a general ignorance of 'what had burned', 'where', 'how', 'how much' and 'how often' that vegetation had burned in the past. As a result, the authors began investigating the role of fire in Chiapas' forest conservation, mainly concentrating on four aspects that will be addressed in this chapter: (i) characterization of fire regimes in Chiapas; (ii) the role of climatic disturbances such as the El Niño-Southern Oscillation (ENSO) on fire patterns in Chiapas; (iii) the effectiveness of protected areas in mitigating fire within their boundaries; and (iv) the development of fire management plans for the most fire-impacted protected areas.

Fire Regime Characterization in Chiapas

The goal of characterizing the fire regime in any ecosystem is to identify the ecological role of fire, often with the ultimate objective of guiding conservation management strategies. Knowing 'how, why, when, what, where, how much and how often should the vegetation be burnt' depends on quantifying the variables that define a fire regime, such as: fire frequency, fire return intervals, extent of annual burned areas, number–size relationships of fires, affected ecosystems, seasonality of fire, fire causality, severity of fire impacts and types of fire (e.g. ground, surface or crown fires). Several major problems arise when attempting to answer these questions, some of which are particularly relevant to tropical countries:

1. *The selection of the appropriate temporal scale for identifying 'natural' fire regimes.* The importance of learning about 'natural' fire regimes lies in their offering clues about the ecologically appropriate levels of fire for each vegetation community, and the understanding of the role of fire in tropical ecosystems. However, in practice, there is not a clear definition of what a 'natural' fire regime means (i.e. pre-industrial? pre-human?), or how useful this concept would be to guide current ecological conservation strategies, given that contemporary management approaches need to take into account a full range of social and economic factors in addition to ecological ones (Goldammer, 1993; Covington and Fulé, 1999; Román-Cuesta and Martínez-Vilalta, 2006). Therefore, the ecologically appropriate levels of fire depend on contrasting and frequently inaccessible issues such as the knowledge of 'natural' fire regimes versus 'acceptable' fire regimes (based on socio-economic needs).

Also included in the selection of the appropriate temporal scale is the presence of inter-annual or inter-decadal disturbances, such as the ENSO phenomenon (Veblen *et al.*, 1999; Kitzberger *et al.*, 2001). Thus, in addition to long-term trends in mean climatic conditions, multi-decadal scale changes in year-to-year variability need to be considered in assessments of the potential influence of climatic change on fire regime (Kitzberger *et al.*, 2001). In Chiapas, the presence of the ENSO reshapes fire regimes in all ecosystems (i.e. from small, surface, low-intensity fires in evergreen rainforests in non-ENSO years, to large, crown, high-intensity fires in severe ENSO years). Choosing the right temporal and spatial scale of analysis is, therefore, an important issue (Román-Cuesta *et al.*, 2003).

2. *Lack of fire data in many tropical countries,* owing to the lack of governmental agencies responsible for forest fire management, insufficient staff to attend and register all fire emergencies, remoteness and inaccessibility of many tropical fires, or political-administrative issues regarding the public access to fire data. In most cases, however, the ultimate reason for this lack of fire data relates to lack of funding. The cost of an efficient fire service is a luxury that many tropical countries cannot afford. When in existence, prevention and suppression activities have priority. Mexico, and Chiapas in particular, is a positive example of how much can be achieved, even with financial and personnel restrictions. Initially SEMARNAT and, latterly, CONAFOR (National Forest Commission) have taken institutional responsibility for fire management (monitoring, prevention, supression, fire database development), with weekly data now available via their web site (http://www.conafor.gob.mx/portal/), and more detailed data available upon request.

As an alternative to existing ground fire detection efforts, or in addition to them, some tropical countries are reinforcing their fire monitoring programmes through remote sensing techniques. In Mexico, since the severe 1998 fire season, the Mexican National Commission for the Knowledge and Use of Biodiversity (CONABIO) developed a remote sensing fire-monitoring programme, which has provided helpful support for ground detection programmes (Pedro Martínez, personal communication). Freely accessible archives, with georeferenced polygons for all fires detected in Mexico, can be downloaded from their website (CONABIO, http://www.conabio.gob.mx/).

Some care is required, however, when using remotely sensed fire data, as each sensor and fire algorithm has different strengths and limitations for fire detection (Pereira and Setzer, 1996; Fuller and Fulk, 2000).

3. *Knowledge of the spatial distribution of major ecosystem types and their sensitivity to fire.* From a conservation point of view it is interesting to know the specificities of fire with respect to individual vegetation communities, and even the individual responses of plant species to fire. However, few tropical countries have updated vegetation or land cover maps, and even fewer possess data on the ecological responses of vegetation to fire (e.g. fire avoidance, fire resistance and fire tolerance). As a way of simplifying these requirements, vegetation is frequently categorized into fire-dependent, fire-influenced and fire-sensitive ecosystems, in a decreasing order of fire co-evolution and fire adaptability (Myers, 2006) (Fig. 13.3, further in the text).

So what is currently known about fire in Chiapas? First, we present information for the period 1984–2006, indicating that more than 1 million ha have burned in Chiapas, with a quarter of this affecting forestland, especially in the years 1986–1987, 1991–1992, 1997–1998 and 2003 (Table 13.1). With the exception of 1998, these severe fire seasons were not matched by the rest of the Mexican Republic, reflecting a higher sensitivity of tropical ecosystems either to ENSO phenomena, or to local socio-economic conditions prevalent within Chiapas.

What are the characteristics of fires in Chiapas?

Based on fire data for 1993–2003 (Román-Cuesta *et al.*, 2004), in Chiapas fire incidence is concentrated in small fires. Thus, fires between 1 and 250 ha in extent accounted for 80% of the total number of fires, but were responsible for only 22% of the total area burned. Fires larger than 500 ha represented a small number (8% of the incidents) but 62% of the total area burned, with fires larger than 5000 ha – just 0.3% – responsible for 22% of the total affected area (Fig. 13.2).

Regarding types of fires, surface fires were the most frequent, accounting for 83% of the incidences and 62% of the total affected area. There were also mixed fires, in which surface fires combined with torching episodes or with isolated crown fires (15% of the incidences, 31% of the total area). Stand-replacement fires (crown fires) were not frequent in Chiapas in this period, but severe droughts in recent years are making them a much more frequent phenomenon, especially in fire-sensitive ecosystems such as evergreen rainforests or cloud montane forests.

When do fires occur?

Chiapas has marked rainfall seasonality, with an official rainy season starting in May and ending in October, and a dry season starting in November and

Table 13.1. Fire statistics for the state of Chiapas for 1984–2006. Data were obtained from SEMARNAT (1984–1999) and CONAFOR (2000–2006).

	Chiapas							Mexico	
Years	Number fires	Herbaceous (ha)	Shrubland (ha)	Regrowth (ha)	Non-arboreal layers (ha)	Arboreal layers (ha)	Total affected area (ha)	Number of fires	Total affected area (ha)
1984	426	26,201	3,358	7,623	37,182	13,024	50,206	6,120	236,032
1985	584	15,792	2,262	1,803	19,857	5,615	25,472	4,386	152,224
1986	530	169,000	5,480	6,265	180,745	21,859	202,604	8,482	290,815
1987	646	43,815	8,250	9,203	61,268	72,255	133,523	9,263	287,347
1988	444	26,490	3,573	4,253	34,316	30,152	64,468	10,492	518,286
1989	317	9,112	2,241	770	12,123	6,008	18,131	9,946	507,471
1990	161	2,858	955	456	4,269	4,148	8,417	3,443	80,400
1991	234	8,984	2,861	1,387	13,232	12,181	25,413	8,621	269,266
1992	121	12,658	530	148	13,336	4,234	17,570	2,829	44,401
1993	127	7,791	2,201	179	10,171	4,165	14,336	10,251	235,020
1994	121	9,519	1,977	876	12,372	4,301	16,673	7,830	141,502
1995	150	11,008	3,776	1,403	16,187	8,006	24,193	7,850	309,097
1996	197	7,496	4,070	1,268	12,834	5,740	18,574	9,256	248,765
1997	181	23,851	10,759	2,309	36,919	11,195	48,114	5,163	107,845
1998	405	85,335	47,590	20,112	153,037	45,771	198,808	14,445	850,000
1999	203	5,051.5	1,369	141	6,561.5	274	6,835.5	7,880	244,000
2000	307	18,013	7,598	841	26,452	2,170	28,623	8,534	236,120
2001	442	22,074	8,122	1,392	31,588	1,928	33,515	6,304	256,266
2002	461	19,062	4,664	997	24,723	990	25,712	8,147	79,288
2003	493	29,019	22,107	1,780	52,906	14,248	67,153	8,106	315,230
2004	289	8,394	1,991	273	10,658	371	11,028	6,190	197,311
2005	461	15,690	4,854	411	20,955	2,554	23,508	9,491	136,084
2006	329	8,030	1,828	175	10,033	451	10,485	8,585	223,872
Total	7,607	585,244	152,416	64,065	801,725	271,640	1,073,362	181,614	5,966,642

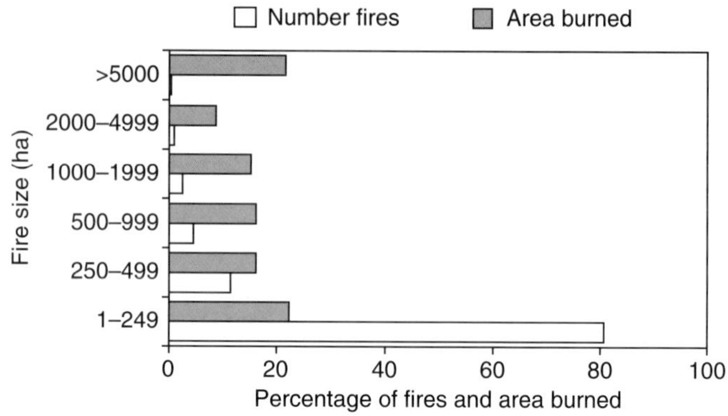

Fig. 13.2. Percentage of fires and area burned for different fire sizes, in Chiapas, for 1993–1999. (Adapted from Román-Cuesta *et al.*, 2004.)

ending in April. However, at least in the last decade, May holds a significant percentage of each year's fires (up to 37% in dry years) with an increasing number of fire days (Fig. 13.3). There are also records of fire in June, for almost all years with fire data (1993–2005). Both 1993 and 2005 show the maximum number of fire incidences in June, with 5% of the total fires during that month.

What are the causes of fire?

Three components may be differentiated:

1. *Anthropogenic component*: As is the case for the rest of Mexico, and many other tropical countries, farming negligence and deliberate burning are the most important causes of fire (51% and 33% of the incidences; 57% and 21%

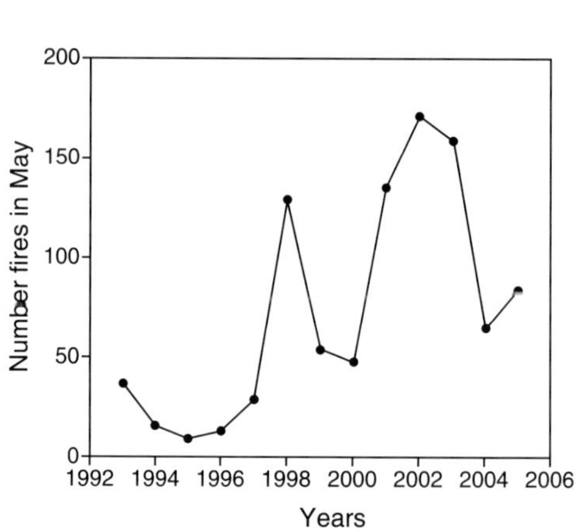

Fig. 13.3. Number of fires occurring in May, in Chiapas, for 1993–2005. (Source: CONAFOR's fire database.)

of the total area burned, respectively). The proportion of fires caused by negligence and deliberate burning are not related to inter-annual variations of fire incidence, as their combined percentage has remained stable over time (70–80% of the total number of fires), including the 1998 ENSO year (Román-Cuesta *et al.*, 2004).

2. *Climatic component*: The role of climate and particularly the presence of ENSO-related droughts are major conservation issues in Chiapas, and strongly influence the distribution and character of their fire seasons (Román-Cuesta *et al.*, 2003). This issue is discussed in more detail in the following sections.

3. *Vegetation component*: Another structural factor determining Chiapas' fire patterns refers to vegetation types and their flammability. The diversity of climates and soils in Chiapas favours a wide range of vegetation communities that have different responses and levels of adaptation to the presence of fire. Figure 13.4 shows the distribution of fuel types based on Mexican vegetation and land-use cover in 2000 (Mas *et al.*, 2004). The evergreen rainforests, montane cloud forests and firs are relatively uninflammable forest types, and pine–oak the most inflammable (Fulé and Covington, 1997). In Chiapas, for the period 1993–1999,

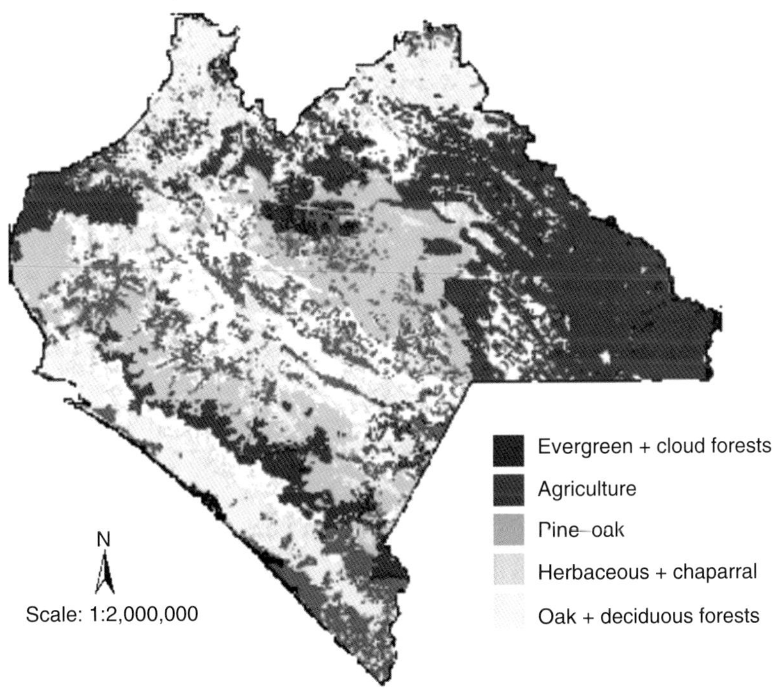

Fig. 13.4. Fuel distribution in Chiapas, based on flammability properties of vegetation types. Black areas correspond to fire-sensitive ecosystems, shades of grey to fire-influenced and fire-dependent ecosystems. One could argue that oaks and deciduous forests are also fire-dependent ecosystems. However, in their more mesophitic areas, the ecological role of fire is still to be determined. (Source: National Forest Inventory, 2000.)

the most affected vegetation communities were pine–oak forests (85% of fire incidence, and 65% of total area burned) (Román-Cuesta *et al.*, 2004). Inspection of the distribution of CONAFOR-reported fires in Fig. 13.5b, compared with the distribution of pine–oaks in Fig. 13.4, confirms this trend. However, the situation changes with ENSO conditions. The most affected communities during 1998 were rainforests (55% of the total area burned), with a decreased contribution of 1.3% in the non-ENSO years (Román-Cuesta *et al.*, 2003). In contrast, in 1998, pine–oak communities were comparatively less affected than in non-ENSO years (18% of the total area in 1998 versus 87% in the non-ENSO years). This suggests a dual vegetation impact, depending on whether it is an ENSO or a non-ENSO year, raising the interesting question of whether fires in Chiapas are climate-driven or fuel-driven, and how this should influence forest management and landscape conservation strategies. This debate has major implications for conservation priorities, and funding investment, as explained by Minnich and Chou (1997), Keeley and Fotheringham (2001) and Westerling *et al.* (2006).

How often does vegetation burn?

Literature reviews indicate that, under undisturbed conditions, fire-sensitive communities such as evergreen rainforests and montane cloud rainforests burn with a frequency of one fire every 500–1000 years (Thonicke *et al.*, 2001; Bussmann, 2005). Fire-adapted forests such as pines and mixed pine–oak stands would burn under natural conditions every 4–40 years, depending on the pine species (Rodríguez-Trejo and Fulé, 2003), and fire-dependent ecosystems such as grasslands or induced prairies would burn annually. In Chiapas, there exists limited information on fire frequencies for different vegetation communities, mainly because of the lack of spatially accurate fire data over long intervals. However, considering the current presence of fire in all ecosystems types in Chiapas (Fig. 13.5), the above-mentioned 'natural' fire frequencies are unlikely to be met by most of Chiapas' forest ecosystems. Research by the Nature Conservancy in pine forests of La Sepultura Biosphere Reserve indicates that pines and pine–oak communities are suffering from over-burning (CONANP-SEMARNAT, 2003). In La Selva del Ocote, Martínez *et al.* (2003) indicate how the ENSO 2003 fires were affecting the evergreen and semi-deciduous rainforests that had previously burned in ENSO 1998. This corresponds to a fire frequency of 5 years, very different from the putative natural fire frequencies for these ecosystems. The same phenomenon of shortened fire intervals has been observed in Las Lagunas de Montebello National Park in its pine–oak–liquidambar forests (burned in 1998 and 2003). The enhanced damage of fire revisiting already burned areas is one of the main conservation issues that protected areas in Chiapas must now address (Román-Cuesta and Martínez-Vilalta, 2006). Goldammer (1999) and Cochrane *et al.* (1999) have reported this positive feedback phenomenon in tropical fire-sensitive ecosystems, which mainly relates to reduced humidity values, increased availability of woody material that dies and accumulates,

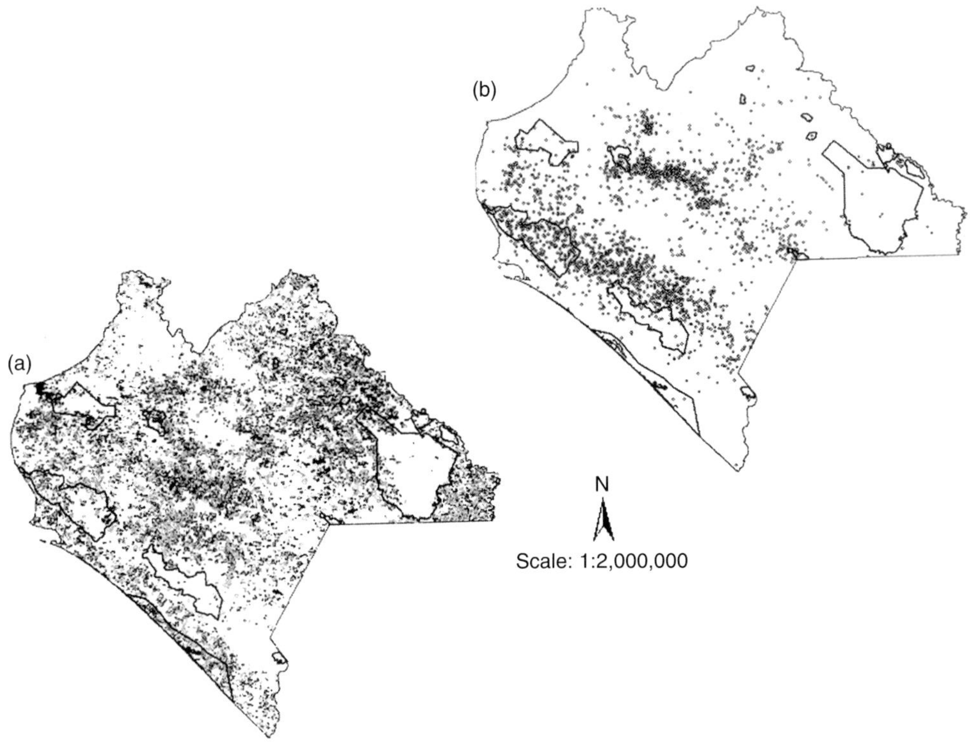

Fig. 13.5. (a) Fire hotspots as detected by MODIS for the period 2000–2005 (CONABIO). Black spots correspond to fire in ENSO 2003, and grey polygons to the remaining years. (b) Ground-detected geopositioned fire database for the same period, from CONAFOR. Black polygons correspond to Chiapas' main federal reserves. See Fig. 13.8 for the names of the reserves.

and increased inflammability of newly invading species (such as *Pteridium* ferns in Chiapas). All of these factors combine to increase the severity of the subsequent fires, in a cycle that frequently leads to severe degradation in the form of induced grasslands (Laurance and Williamson, 2001).

Where do fires concentrate?

Figure 13.5 shows fire distribution in Chiapas, from MODIS hotspots (2000–2005) on the left side, and from the CONAFOR database (1995–2005) on the right. Comparison of these two databases offers a number of interesting insights. First, there is a large difference in the total number of fires on the two images. The much lower number of fires in Fig. 13.5b mainly relates to the difference between a 'forest fire' and a 'hotspot'. Thus, CONAFOR does not consider intentional farming burns as forest fires and therefore does not register them, while the MODIS algorithm considers a 'hotspot' as anything with a temperature above 50°C (CONABIO, http://www.conabio.gob.mx/). This is particularly obvious in the lower south-eastern part of the image, the region of Marqués de Comillas, where most burnings are related to farming activities, and therefore appear in the MODIS image but are absent in CONAFOR's.

Forest fires in Chiapas are concentrated along two major montane axes: the Highlands of Chiapas (from El Cañón del Sumidero to Lagunas de Montebello park) and La Sierra Madre de Chiapas, including La Sepultura and El Triunfo parks (see Fig. 13.8 for the distribution of parks). The advance of the agrarian frontier is, therefore, concentrated on these two axes. Owing to the El Niño 2003 drought, ignitions resulted in very extensive and continuous forest fires (black spots in Fig. 13.5a), many of them affecting large areas inside protected parks (Román-Cuesta and Martínez-Vilalta, 2006) (El Ocote, Sepultura, Montes Azules, Lagunas de Montebello, Cañón del Sumidero, La Encrucijada). The 2005 fire season affected areas that had not burned in the previous 5 years, some of them expanding further within protected land (grey polygons in Fig. 13.5a) (e.g. Montes Azules, Selva El Ocote). Following Fig. 13.5, the only protected area in Chiapas where fire is not a threatening factor is the Tacana volcano and El Triunfo.

The Influence of Climatic Disturbances such as El Niño-Southern Oscillation (ENSO) on Fire Patterns in Chiapas

ENSO droughts and their consequences for forest conservation are not a new issue in Chiapas, but the level of public and media attention on these issues is. ENSO 1997–1998 was associated with the worst drought in Mexico for the past 70 years (Cedeño, 2001); however, in Chiapas, the ENSO 1986–1987 resulted in both larger arboreal and total burned areas than the 1997–1998 event (*arboreal area*: 72,255 ha versus 45,771 ha; *total burned area*: 202,604 ha versus 198,800 ha, Table 13.1). In spite of these large numbers, the 1986–1987 fire impacts were only briefly mentioned in a governmental report, pointing out the useless waste of the Pacific rainforests by fire. Chiapas gained that year (1987) the dubious honour of being the warmest place on Earth, together with Madagascar, owing to its fire situation. Another severe fire season in Chiapas was the ENSO event of 1982–1983, which was reported as an ecological fire emergency in different continents (Andreae *et al.*, 1988). The evergreen rainforests of Montes Azules (the Lacandon jungle) were among the most severe fires in Chiapas at that time (Harris and Sartor, 1984).

The inclusion of evergreen rainforests and other fire-sensitive ecosystems during ENSO droughts (Asbjornsen *et al.*, 2005) in a state where most affected communities were normally fire-dependent pine–oak forests (Román-Cuesta *et al.*, 2003) raises questions about the role of ENSO in fire trends in Chiapas. Questions can also be asked about the role of human pressures in the distribution of fires in Chiapas, especially considering that most fires in that state are human-ignited (Román-Cuesta *et al.*, 2004). To investigate these factors, an analysis was performed of their relative influence on distribution of fires and the area burned in Chiapas during 1993–1999, at a municipality level (Fig. 13.6). Moreover, since forest area affected by the ENSO 1987 and 1998 events reached

47% of the total forest area burned in the state during 1984–1999, we separated our analysis into years of normal and extreme climatic conditions (non-ENSO versus ENSO). We hypothesized that in non-ENSO years human variables would have a major importance in Chiapas because of the large role of negligent and deliberate burnings (Román-Cuesta, 2000). However, we hypothesized that in ENSO years environmental variables would play a major role because of the severe drought conditions. To test this hypothesis, we searched for causal relationships among fire, environmental and socio-economic variables in Chiapas using path analysis techniques (Román-Cuesta *et al*., 2003).

The results of this study revealed how different climate regimes (such as ENSO or non-ENSO) condition the relative influence of factors affecting

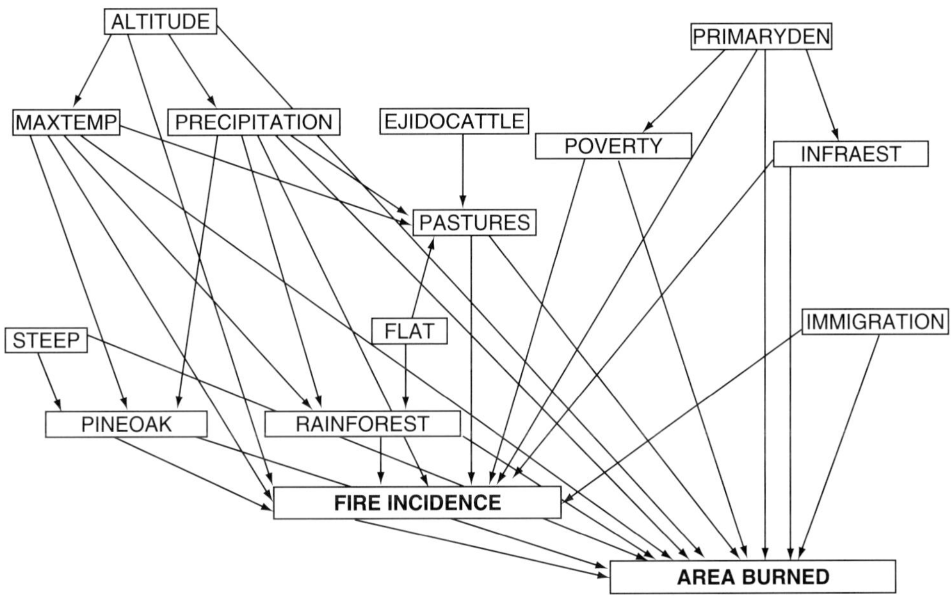

Fig. 13.6. Mixed model with environmental and human-related variables, and interactions among them, for the ENSO and non-ENSO years. Final models were obtained from this mixed model by pruning non-significant relationships. Temporal scale: 1993–1999. Spatial scale: municipalities of Chiapas ($n = 111$). Abbreviations of variables are as follow: ALTITUDE, altitude above sea level of city capitals; EJIDOCATTLE, % of ejidos with farming activities; FLAT, % of land with flat land; IMMIGRATION, % of people in each municipality coming from other states; INFRAEST, density of infrastructures; MAXTEM, maximum temperature in the dry season; PASTURES, % of land in pastures; PINEOAK, % of land in pine–oak communities; POVERTY, % of population with below minimum salary; PRECIPITATION, total rainfall in the dry season; PRIMARYDEN, % of population related to the primary sector (agriculture); RAINFOREST, % of land with rainforests; STEEP, % of land with slopes above 30%. (Source: socio-economic data belong to the Mexican 2000 Census and were obtained from the SIMBAD system, at the National Institute of Statistics and Geography. The environmental data were obtained from the Laboratory of Geographical Systems and Image treatment in ECOSUR research centre, San Cristóbal las Casas. Source: Román-Cuesta *et al*., 2003.)

fire trends (both fire incidence and area burned) in Chiapas. Contrary to our initial hypothesis, *environmental variables* played a decisive role in *non-ENSO years*, suggesting that the status of the vegetation was the main cause determining fire ignition and fire spread in these years. In contrast, the observed trends in the *ENSO year* suggested that *human-related variables* play a major role in these ENSO years, indicating that the presence of ignition agents mainly determine fire trends: the vegetation is so water-stressed that, when fire starts, everything burns independently of its flammability properties. Corroborating this idea, in the 1998 fire season in Brazil there were 1.5 million km^2 of susceptible rainforests that did not burn because of a shortage of ignition sources (Román-Cuesta *et al.*, 2003). Our research also confirmed how different climate regimes (ENSO versus non-ENSO) result in an interesting shift in the arboreal community that was more affected by fire in each period. Thus, as mentioned in the vegetation section above, during non-ENSO years the most affected communities are flammable pine–oak forests (85% of fire incidence and 65% of total area burned), while in severe ENSO years fire-sensitive water-stressed rainforests and cloud forests were the ones burning the most (55% of the total area burned) (Román-Cuesta *et al.*, 2004).

This has a number of implications for fire and forest management. In non-ENSO years, limiting the number of fires would have an effect in preserving forest resources, owing to the strong relationship between fire incidence and burned areas (most affected vegetation communities being pine–oak forests; Román-Cuesta *et al.*, 2003). However, research is required to improve knowledge of the role of fire in these communities, as fire suppression might not be a good alternative in some of the temperate ecosystems affected. In ENSO years, however, rainforests are the major concern. In these ecosystems, the low relationship between fire incidence and burned areas suggests that controlling key fires is the major priority. Thus, in these years it takes only a few fires to burn very large areas, which are out of fire-fighting control. Pastures are one of the main stressor agents in ENSO years and restricting pasture-burning might be an important measure to avoid undesired fires, although not easy to put into practice. The application of effective strategies to prevent accidental forest fires or the discontinuation of fire in land management practices is a common demand in other studies in tropical areas (Nepstad *et al.*, 1999). However, fire is the agrarian tool that fits best in the current tropical socio-economic and environmental framework, and the only alternatives that have been suggested are not realistic. This lack of alternatives and a long history of fire use are held to be responsible for the failure of fire suppression initiatives in Chiapas, despite a history of official opposition to wildfire, dating back to pre-Columbian times (Fulé and Covington, 1997).

The Effectiveness of Protected Areas in Mitigating Fire within their Boundaries

For the period 1993–1999, 7.6% of the total national land in Chiapas (state-owned) had been affected by fire, placing national landownership at the top

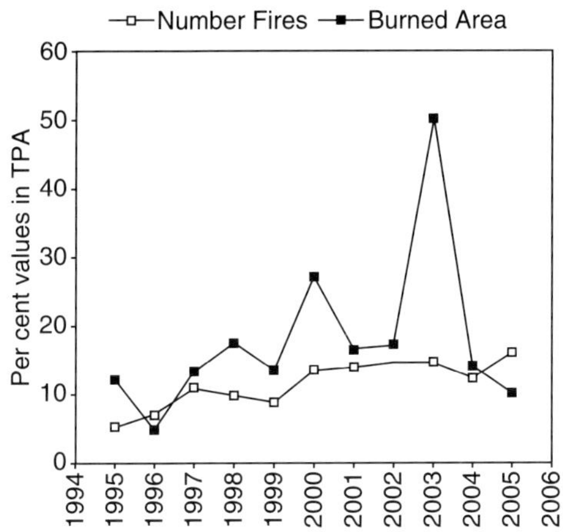

Fig. 13.7. Percentage of fire ignitions and burned area inside protected areas in Chiapas, for the period 1995–2003. (Source: Román-Cuesta and Martínez-Vilalta, 2006.)

of fire impacts (Román-Cuesta *et al.*, 2004). As a consequence of this high value, Román-Cuesta and Martínez-Vilalta (2006) investigated the percentage of total area burned in Chiapas that occurred inside protected areas (Fig. 13.7). Interestingly, in years such as the El Niño 1998, the contribution of protected areas to the total area burned (18%, 34,000 ha) was moderated by the fact that the whole state was on fire. However, during ENSO 2003, up to 50% of the total area burned was located inside protected areas (25,000 ha). Having spent the fire season of 2003 in Chiapas, the authors had first-hand experience that the severity of fire impacts was not exclusive to Chiapas' parks, but was shared by Guatemala's border reserves, where the situation in El Petén was even worse (Albacete, 2003; Ramírez, 2003; Redford *et al.*, 2006). With this in mind, the statements made by Bruner *et al.* (2001) regarding the global effectiveness of protected areas in preventing disturbances within their boundaries (e.g. fire, hunting, land-use changes), and the similar results obtained by DeFries *et al.* (2005), seem unlikely for Chiapas and, by extension, for the Mesoamerican Biological Corridor.

Based on CONAFOR's GPS georeferenced fire database for Chiapas (1995–2005), we contrasted the number of fires and burned areas in the parks with their surrounding buffers, to test if parks in Chiapas were effective in mitigating fire within their boundaries (Román-Cuesta and Martínez-Vilalta, 2006). Figure 13.8 shows the distribution of protected areas in Chiapas. Buffers were defined as concentric areas surrounding the protected boundaries, whose final areas equalled the total burnable land of each reserve. Results indicate that fire has become a first-order problem in almost all reserves in Chiapas, with both fire incidence and burned areas being significantly higher within the boundaries than outside the parks (Fig. 13.9). Parks in Chiapas, therefore, are not effective deterrents to fire. Differential fire trends between Tropical Protected

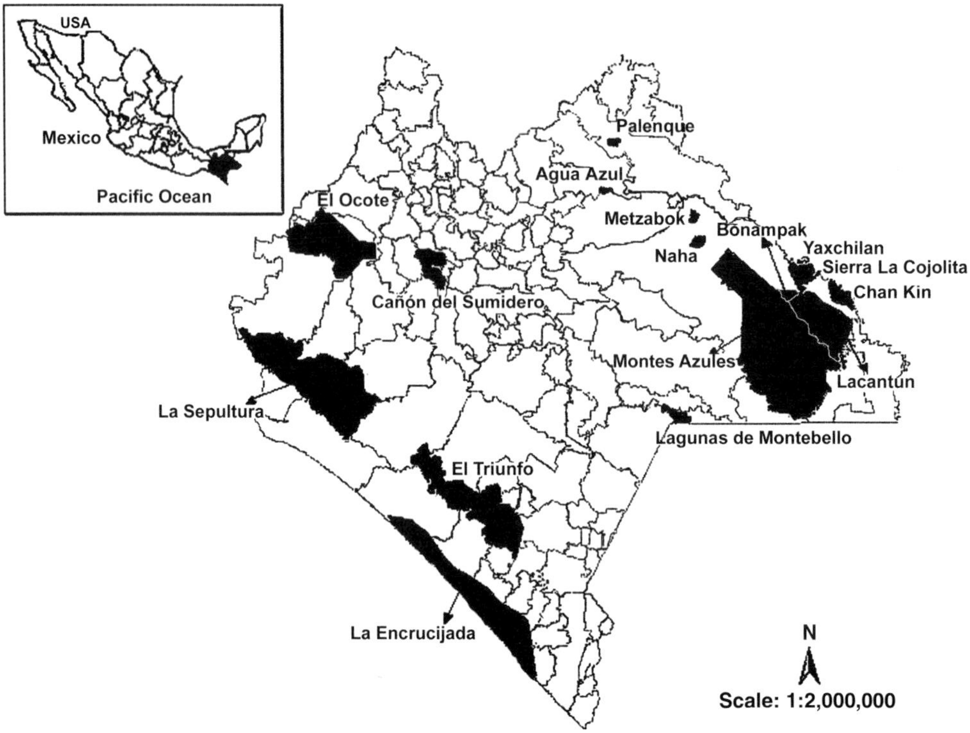

Fig. 13.8. Geographic location of Chiapas and the 15 +1 tropical protected areas (TPAs) included in this study (+1 refers to the non-federal reserve of Sierra Cojolita). Montes Azules, Bonampak, Yaxhilan, Lacantun and Sierra Cojolita were all merged to form the Lacandona TPA. (Source: Román-Cuesta and Martínez-Vilalta, 2006.)

Areas (TPAs) and their buffers were increased in ENSO years, fires being more severe inside parks than in their buffers during these years. Anthropogenic factors (agriculture and road density) played a major role in the enhanced fire incidence in TPAs, whereas natural habitat extents played a major role in the amount of area burned, favouring continuous fuel layers, which are difficult to control once the fire has started (Román-Cuesta and Martínez-Vilalta, 2006).

The ultimate importance of these fires relates to the role of parks as islands of biodiversity preservation. Among current debates are the relative merits of 'people parks' versus 'people-free parks', the role that local communities can play in terms of park conservation and the role that national governments have in terms of providing compensation and livelihood alternatives to communities living in parks (Schwartzman *et al.*, 2000; Peres and Zimmerman, 2001). If fire reduction is the goal, people-free parks (categories I–IV of IUCN (International Union for the Conservation of Nature)) represent the ideal scenario. However, in highly populated areas such as Mesoamerica, few reserves can effectively control the incursion of people, and the social problems deriving from core-area protection frequently appear in the form of fire. Fagan *et al.* (2006) suggest that categories V and VI (people parks) can

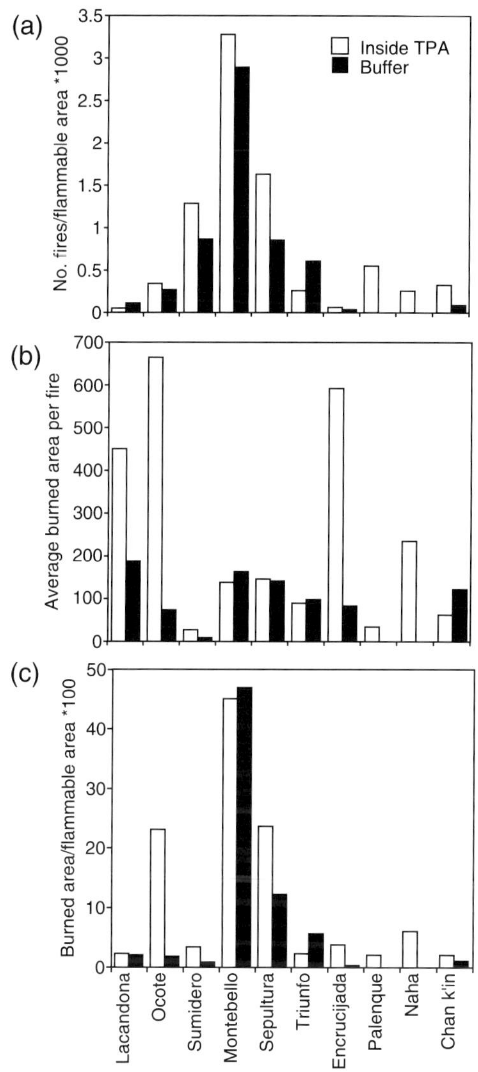

Fig. 13.9. (a) Fire incidence in tropical protected areas (TPAs) and in their buffers (number of fires/flammable area × 1000) in Chiapas from 1995 through 2005. (b) Average burned area (in hectares) for fires in TPAs and their buffers in Chiapas (1995–2005). (c) Total area affected by fires (area burned/flammable area × 100) in TPAs and their buffers in Chiapas from 1995 through 2005. (Source: Román-Cuesta and Martínez-Vilalta, 2006.)

have significant conservation values as buffers or corridors, but the ability of these buffers and corridors to sustain biodiversity as stand-alone conservation units is uncertain. The fire situation in Chiapas, and even more in Guatemala's protected areas, casts doubts on the conservation value of these categories and on the sustainability of the Mesoamerican Biological Corridor as a whole.

With respect to this situation, new approaches have been proposed to mitigate fire damages:

1. As people-free parks are not a realistic approach in Mesoamerica, projects have been developed to eliminate fire in parks and surrounding areas. The Pachuca Project, with the *mucuna*–maize green manure, is an interesting example (Eastmond and Faust, 2006).
2. Fuel management, fuel flammability and fuel continuity treatments to avoid the initiation and spread of fire. Fuel management strategies to decrease fuel loads are discussed in Agee *et al.* (2000), Gascon *et al.* (2000) and Finney (2001). Most parks in Chiapas whose fire-sensitive forests were affected in 1998 are currently accumulating large loads of dead woody biomass, substantially increasing the fire danger of new episodes (higher probabilities of ignition and higher fire severities if fire starts). Fuel management is an urgent issue in many parks. Although fire management should benefit from fuel modification, a problem appears when deciding which fuel modification practices best fit the various tropical ecosystems. Thus, in fire-intolerant humid forests, common fuel management practices, such as firebreaks or prescribed burning, can have the paradoxical effect of increasing fuel loads and fire risk (Martínez *et al.*, 2003; Barlow and Peres, 2004). The work of certain NGOs such as the Nature Conservancy, which is trying to apply fire to reduce fuels on fire-adapted ecosystems, is an innovative approach that could guide further efforts in the region.
3. Under a scenario of increased ENSO frequency (Trenberth and Hoar, 1997), fires in tropical forests may emerge as an even greater problem in the future. Fire management and fire planning will therefore become essential tools in any conservation area's management plan. Among the measures that fire management includes, fire prevention is a key factor as it concentrates on fire causality (both human and environmental). In this context, the Climate Prediction Center (NOAA, 2005) has an online ENSO–La Niña diagnosis, which helps track ENSO evolution and intensity. This service can help in making decisions on fire moratoriums, such as the one taken in Chiapas during the 2003-ENSO fire season for La Selva Zoque region (El Ocote–Chimalapas–Uxpanapa).
4. While climate variability is of great concern, the strong human influence of tropical fires necessarily requires community-based participation to decrease fire risk and fire damage. However, the population's commitment to nature conservation will not occur without external support. Governments are urged to initiate dialogue to reduce conflicts, fulfil land dispute resolutions and promises (e.g. long-lasting conflicts in Parque Nacional Lagunas Montebello), and search for economic alternatives for those communities that are living in the buffers of protected land. Thus, a well-managed local ecotourism network could clearly benefit the conservation efforts of Chiapas's parks, which are still seriously underdeveloped or are starting to be controlled by foreign leisure capital (e.g. Ecotouristic Park in El Cañón del Sumidero), quite distant from local interests and local participation. A management plan that does not include local communities as direct beneficiaries of local conservation is clearly vulnerable to failure.

Fire Management Plan for La Selva El Ocote Biosphere Reserve

Since the ENSO 1982–1983, many parks have reported recurrent fire presence in areas already affected by previous fires (e.g. Eldvidge *et al.*, 2001; Cleary and Genner, 2004). Despite increasing global awareness, when the 2003 ENSO fires affected the already burned areas of La Selva El Ocote (burned in one of the worst fire conflagrations of Chiapas in 1998; Asbjornsen *et al.*, 2005) little could be done. Thus, forests degraded by the previous ENSO season resulted in accumulated fuel loads up to 150 t ha^{-1}, which, together with dry conditions, strong winds, water inaccessibility and extremely difficult access, led to the worst fire event in Chiapas in 2003 (Martínez *et al.*, 2003). By the end of May, 20,000 ha had burned in the reserve, with severe vegetation degradation in some areas and complete vegetation removal in others. This 2003 El Ocote fire episode represented a major conservation failure in one of Mexico's most important biodiversity areas. El Ocote forms part of the biological corridor Chimalapas–Uxpanapa–El Ocote (Selva Zoque), once one of the largest pristine extensions of cloud montane rainforests in North America (Asbjornsen *et al.*, 2005).

One could argue that very little could have been done under the severe climatic conditions of 2003, together with the unattackable fuel loads, the restricted access to water points and the geological conditions of the area whose karstic material moved fire through underground tunnels. However, had the park had a fire management plan, many of the above-mentioned issues would have already been foreseen. This is because fire management plans reflect on preventive, pre-suppression and suppression strategies, which enable reduction of fire costs and damage by planning ahead and aiding fire suppression activities when they start. Martínez *et al.* (2003) provide an example of a fire management plan for a protected area, La Selva del Ocote, where there exists detailed information on each of ITTO's suggested guidelines and main issues to consider when developing a fire management plan (ITTO, 1997). Beside classical monitoring, prevention, suppression and restoration activities, strong emphasis should be placed on reinforcing institutional cooperation, strengthening local social networks and searching for economical alternatives for communities in the park, as key strategies to obtain meaningful fire management results.

Conclusions

Funding for fire-related activities has increased in Chiapas since the 1998 ENSO. However, equipment, logistics, staff training and restoration activities always require more money than the budget allows. Fire management is gradually reaching the conservation management programmes of most parks, by both reducing and increasing its presence (i.e. increased fire for prescribed burning). However, while incorporating fire management into the conservation discourse is an important first step, it is also important to frame the problem of fire within a larger socio-political context whose development decisions have a direct influence on conservation efforts by opening new areas or disturbing old ones. Because the roots of the fire problem lie within the social and macroeconomic realms (Rodríguez-Trejo and Fulé, 2003), any political decision and its derived activities will affect people's access to land and population distributions, which will in turn affect fire regimes. Protection and fire management planning, while important, will not fully address the problem of fire in tropical forests, especially not in protected areas.

Acknowledgements

We would like to express our gratitude to SEMARNAT and CONAFOR offices for their help with the fire databases. To ECOSUR staff, for their support during fieldwork and logistic arrangements. To LAIGE, for their sharing of digital data and to the European Community and its INCO-DC programme (Framework 4), funder of the SUCRE project (ERBIC-18 Oct.97-0146).

References

Agee, J., Bahro, B., Finney, M., Omi, P., Sapsis, D., Skinner, C., van Wagtendonk, J. and Weatherspoon, C. (2000) The use of fuel breaks in landscape fire management. *Forest Ecology and Management* 127, 55–66.

Albacete, C. (2003) *Guatemala: Deliberate Fires Raze Tropical Forest and Serve Logging Interests*. Bulletin 70. World Rainforest Movement, Montevideo, Uruguay.

Alvarez del Toro, M.A. (1985) *Así era Chiapas*. Publicaciones del Gobierno de Chiapas, Tuxtla Gutiérrez, Mexico.

Andreae, M., Browell, E., Garstang, M., Gregory, G., Harriss, R., Hill, G., Jacob, D., Pereira, M., Sachse, G., Setzer, A., Silva Dias, P., Talbot, R., Torres, A. and Wofsey, S. (1988) Biomass-burning emissions and associated haze layers over Amazonia. *Journal of Geophysical Research* 93, 1509–1527.

Asbjornsen, H., Velázquez-Rosas, N., García-Soriano, R. and Gallardo-Hernández, C. (2005) Deep ground fires cause massive above and belowground biomass losses in tropical montane cloud forests in Oaxaca, Mexico. *Journal of Tropical Ecology* 21, 427–434.

Barlow, J. and Peres, C. (2004) Ecological responses to El Niño-induced surface fires in central Brazilian Amazonia: management implications for flammable tropical forests. *Philosophical Transactions of the Royal Society of London B* 359, 367–380.

Bruner, A., Gullison, R., Rice, R. and da Fonseca, G. (2001) Effectiveness of parks in protecting tropical biodiversity. *Science* 291, 125–128.

Bussmann, R. (2005) Bosques Andinos del sur de Ecuador, clasificación, regeneración y uso. *Revista Peruana de Biología* 12, 1–21.

Cedeño, O. (2001) Fire management in Mexico. In: Goldammer, J.G. and Mutch, R.W. (eds) *Global Forest Fire Assessment 1990–2000*. Forest Resources Assessment – WP 55. FAO, Rome, Italy, pp. 420–437.

Cleary, D. and Genner, M. (2004) Changes in rain forest butterfly diversity following major ENSO-induced fires in Borneo. *Global Ecology and Biogeography* 13, 129–140.

Cochrane, M.A. (2002) *Spreading Like Wildfire – Tropical Forest Fires in Latin America and the Caribbean: Prevention, Assessment and Early Warning*. United Nations Environment Programme (UNEP), Mexico City, Mexico.

Cochrane, M.A. (2003) Fire science for rainforests. *Nature* 421, 913–919.
Cochrane, M.A., Alencar, A., Schulze, M.D., Souza, C.M., Nepstad, D.C., Lefebvre, P. and Davidson, E. (1999) Positive feedback in the fire dynamic of closed canopy tropical forests. *Science* 284, 1832–1835.
CONANP-SEMARNAT (2003) *Análisis de las Estadísticas de Incendios Forestales de la Reserva de la Biosfera la Sepultura, en el Periodo 1997–2003*. National Commission of Natural Protected Areas (CONANP) and Secretaría de Medio Ambiente y Recursos Naturales (SEMARNAT), Tuxtla Gutiérrez, Chiapas, Mexico.
Covington, W. and Fulé, P. (1999) Fire regime changes in la Michilia Biosphere Reserve, Durango, Mexico. *Conservation Biology* 13, 640–652.
DeFries, R., Hansen, A., Newton, A.C. and Hansen, M.C. (2005) Increasing isolation of protected areas in tropical forests over the past twenty years. *Ecological Applications* 15, 19–26.
Deininger, K. and Minten, B. (2002) Determinants of deforestation and the economics of protection: an application to Mexico. *American Journal of Agrarian Economy* 84, 943–960.
Du Toit, J., Walker, B. and Campbell, B. (2004) Conserving tropical nature: current challenges for ecologists. *Trends in Ecology and Evolution* 19, 12–17.
Dwyer, E., Pereira, J.M.C., Gregoire, J.P. and da Camara, C.C. (1999) Characterization of the spatio-temporal patterns of global fire activity using satellite imagery for the period April 1992 to March 1993. *Journal of Biogeography* 27, 57–69.
Eastmond, A. and Faust, B. (2006) Farmers, fires and forests: a green alternative to shifting cultivation for conservation of the Maya forest? *Landscape and Urban Planning* 74, 267–284.
Eldvidge, C., Hobson, V., Baugh, K., Dietz, K., Shimabukuro, Y., Krug, T., Novo, E. and Echavarria, F. (2001) DMSP-OLS estimation of tropical forest area impacted by surface fires in Roraima, Brazil: 1995 versus 1998. *International Journal of Remote Sensing* 22, 2661–2673.
Fagan, C., Peres, C. and Terborgh, J. (2006) Tropical forests: a protected-area strategy for the twenty-first century. In: Laurance, W.F. and Peres, C. (eds) *Emerging Threats to Tropical Forests*. University of Chicago Press, Chicago, Illinois, pp. 415–432.
Finney, M.A. (2001) Design of regular landscape fuel treatment patterns for modifying fire growth and behavior. *Forest Science* 47, 219–228.
Fulé, P. and Covington, W. (1997) Fire regimes and forest structure in the Sierra Madre Occidental, Durango, Mexico. *Acta Botanica Mexicana* 41, 43–79.
Fuller, D.O. and Fulk, M. (2000) Comparison of NOAA-AVHRR and DMSP-OLS for operational fire monitoring in Kalimantan, Indonesia. *International Journal of Remote Sensing* 21, 181–187.
García-Barrios, L. and González-Espinosa, M. (2004) Change in oak to pine dominance in secondary forests may reduce shifting agriculture yields: experimental evidence from Chiapas, Mexico. *Agriculture, Ecosystems and Environment* 102, 389–401.
Gascon, C., Williamson, B. and da Fonseca, G. (2000) Receding forest edges and vanishing reserves. *Science* 288, 1356–1358.
Goldammer, J.G. (1993) Historical biogeography of fire: tropical and subtropical. In: Crutzen, P. and Goldammer, J.G. (eds) *Fire in the Environment, the Ecological, Atmospheric and Climatic Importance of Vegetation Fires*. John Wiley, New York, pp. 297–315.
Goldammer, J.G. (1999) Forests on fire. *Science* 284, 1782–1783.
Goldammer, J.G. and Peñafiel, S.R. (1990) Fire in the pine–grassland biomes of tropical and subtropical Asia. In: Goldammer, J.G. (ed.) *Fire in the Tropical Biota. Ecosystems, Processes and Global Challenges*. Ecological Studies 84. Springer, Berlin, pp. 45–62.
Harris, A. and Sartor, M. (1984) *Gertrude Blom: Bearing Witness*. University of North Carolina Press, Chapel Hill, North Carolina.
ITTO (1997) *ITTO Guidelines on Fire Management in Tropical Forest*. ITTO Policy Development Series 6. ITTO, Yokohama, Japan.
Keeley, J. and Fotheringham, J. (2001) History and management of crown-fire ecosystems: a summary and response. *Conservation Biology* 15, 1561–1567.

Kinnaird, M. and O'Brien, T. (1998) Ecological effects of wildfire on lowland rainforest in Sumatra. *Conservation Biology* 12, 954–956.

Kitzberger, T., Swetnam, T.W. and Veblen, T.T. (2001) Inter-hemispheric synchrony of forest fires and the El Niño-Southern Oscillation. *Global Ecology and Biogeography* 10, 315–326.

Laurance, W.F. and Williamson, G.B. (2001) Positive feedbacks among forest fragmentation, drought and climate change in the Amazon. *Conservation Biology* 15, 1529–1535.

Leopold, A. (1950) Vegetation zones in Mexico. *Ecology* 31, 507–518.

Lindenmayer, D., Foster, D., Franklin, J., Hunter, M., Noss, R., Schmiegelow, F. and Perry, D. (2004) Salvage harvesting policies after natural disturbance. *Science* 303, 1303.

Martínez, P., Velazquez, J. and Román-Cuesta, R.M. (2003) *Versión Preliminar del Plan de Manejo del fuego Selva El Ocote, Chiapas, México*. Comisión Nacional Forestal (CONAFOR), and El Colegio de la Frontera Sur (ECOSUR), San Cristóbal de Las Casas, Chiapas, Mexico.

Mas, J.F., Velázquez, A., Reyes-Díaz-Gallegos, J., Mayorga-Saucedo, M., Alcántara, C., Bocco, G., Castro, R., Fernández, T. and Pérez-Vega, A. (2004) Assessing land use/cover changes: a nationwide multidate spatial database for Mexico. *International Journal of Applied Earth Observation and Geoinformation* 5, 249–261.

Minnich, R.A. and Chou, Y.H. (1997) Wildland fire dynamics in the chaparral of southern California and northern Baja California. *International Journal of Wildland Fire* 7, 221–248.

Mueller-Dombois, D. and Goldammer, J.G. (1990) Fire in tropical ecosystems and global environmental change: an introduction. In: Goldammer, J.G. (ed.) *Fire in the Tropical Biota. Ecosystem Processes and Global Challenges*. Springer, Berlin, Germany, pp. 1–10.

Myers, R.L. (2006) *Living with Fire: Sustaining Ecosystems and Livelihoods through Integrated Fire Management*. Global Fire Initiative, Nature Conservancy, Tallahassee, Florida.

Nepstad, D.C., Veríssimo, A., Alencar, A., Nobre, C., Lima, E., Lefebvre, P., Schlesinger, P., Potter, C.S., Moutinho, P. and Mendoza, E. (1999) Large-scale impoverishment of Amazonian forests by logging and fire. *Nature* 398, 505–508.

NOAA (2005) *El Niño/Southern Oscillation (ENSO) Diagnostic Discussion*. Climate Prediction Center, National Centers for Environmental Prediction, North Oceanic and Atlantic Administration (NOAA) Camp Springs, Maryland. Available at: http://www.cpc.ncep.noaa.gov/products/analysis_monitoring/enso_advisory/ (accessed March 2005).

Pereira, A.C. and Setzer, A.W. (1996) Comparison of fire detection in savannas using AVHRR's channel 3 and TM images. *International Journal of Remote Sensing* 17, 1925–1937.

Peres, C.A. and Zimmerman, B. (2001) Perils in parks or parks in peril? *Conservation Biology* 15, 793–797.

Ramírez, A. (2003) *Guatemala Fire Emergency*. Global Fire Monitoring Center (GFMC), Freiburg, Germany. Available at: http://www.fire.uni-freiburg.de/GFMCnew/2003/0322/20030322_gt.htm (accessed November 2006).

Ramírez-Marcial, N., González-Espinosa, M. and Williams-Linera, G. (2001) Anthropogenic disturbance and tree diversity in the montane rain forest in Chiapas, Mexico. *Forest Ecology and Management* 154, 311–326.

Redford, K., Robinson, J. and Williams, A. (2006) Parks as shibboleths. *Conservation Biology* 20, 1–2.

Rodríguez, D.A. (1996) *Los Incendios Forestales*. Universidad Autónoma de Chapingo, Mundi-Prensa México SA, Mexico.

Rodríguez-Trejo, D. and Fulé, P. (2003) Fire ecology of Mexican pines and a fire management proposal. *International Journal of Wildland Fire* 12, 23–37.

Román-Cuesta, R.M. (2000) Forest fire situation in the state of Chiapas, Mexico. In: Pugliese, J. (ed.) *Global Forest Fire Assessment 1990–2000*. Working paper 55, Forestry Department, FAO, Rome, Italy, pp. 426–437.

Román-Cuesta, R.M. and Martínez-Vilalta, J. (2006) Effectiveness of protected areas in mitigating fire within their boundaries: case study of Chiapas. *Conservation Biology* 20, 1074–1086.

Román-Cuesta, R.M., Gracia, M. and Retana, J. (2003) Environmental and human factors influencing fire trends in ENSO and non-ENSO years in tropical Mexico. *Ecological Applications* 13, 1177–1192.
Román-Cuesta, R.M., Retana, J. and Gracia, M. (2004) Fire trends in tropical Mexico: a case study of Chiapas. *Journal of Forestry* 102, 26–32.
Sanford, R., Saldarriaga, J., Clark, K., Uhl, C. and Herrera, R. (1985) Amazon rain-forest fires. *Science* 227, 53–55.
Schwartzman, S., Moreira, A. and Nepstad, D. (2000) Rethinking tropical forest conservation: perils in parks. *Conservation Biology* 14, 1351–1357.
Siegert, F., Ruecker, G., Hinrichs, A. and Hoffman, A.A. (2001) Increased damage from fires in logged forests during droughts caused by El Niño. *Nature* 414, 437–440.
Thonicke, K., Venevsky, S., Sitch, S. and Cramer, W. (2001) The role of fire disturbance for global vegetation dynamics: coupling fire into a dynamic global vegetation. *Global Ecology and Biogeography* 10, 661–678.
Trenberth, K.E. and Hoar, T.J. (1997) ENSO and climate change. *Geophysical Research Letters* 24, 3057–3060.
Veblen, T.T., Kitzberger, T., Villalba, R. and Donnegan, J. (1999) Fire history in northern Patagonia: the roles of humans and climatic variation. *Ecological Monographs* 69, 47–67.
Westerling, A., Hidalgo, H., Cayan, D. and Swetnam, T. (2006) Warming and earlier spring increases western US forest wildfire activity. *Science* 313, 940–943.

14 Identification of Priority Areas for Conservation in South-central Chile

K.A. Wilson and A.C. Newton

Temperate rainforest in the island of Chiloé, southern Chile. This forest area (Senda Darwin) has been degraded by timber extraction, fire and livestock browsing, but is now recovering gradually as the result of a forest restoration initiative. Photo: Adrian Newton

Summary

There is an increasing awareness in Chile of the need to improve the representation of the country's forest types in the national network of protected areas, and to reduce the impacts of the native forest conversion to other land uses. One promising strategy is to use the principles of systematic conservation planning to identify important areas for the conservation of biodiversity. In this chapter we use information on the vulnerability of native forest to threatening processes and information on the distribution of forest types to systematically determine priorities for biodiversity conservation in the temperate forest region of south-central Chile. We find that the existing reserve network covers approximately 12% of the study region, and that only 53% of the area of native forest estimated to be present before European settlement remained in 1997. Temperate forest is now largely restricted to the upper elevations of the Andean and coastal ranges, within a matrix of pasture, agriculture and plantations. We develop a model of the conversion of native forest to plantations and predict conversion to be more likely in warm and low rainfall areas that are close to towns and roads and on red clay and mixed alluvial soils. We argue that the priority areas for conservation should be currently unprotected areas that are vulnerable to plantation conversion and that, if lost or degraded, will result in conservation targets being compromised. We find the Evergreen forest type to be a priority for conservation owing to its lack of representation in the existing reserve network and the degree to which it has been cleared. By focusing our conservation efforts on areas with the greatest conservation value and the highest likelihood of losing significant portions of this value, we will be able to achieve maximum impact for conservation investments in south-central Chile.

Introduction

Despite the ecological importance of Chilean temperate forests (Davis *et al.*, 1994–1997; Wilcox, 1995; Olson and Dinerstein, 1998; Stattersfield *et al.*, 1998), they have experienced a long history of degradation and destruction (Veblen, 1983; CODEFF, 1992; Rozzi *et al.*, 2000; Neira *et al.*, 2002; Chapters 2 and 3) and are presently threatened with conversion to other land uses, particularly plantations of exotic species (Neira *et al.*, 2002). Between 1995 and 1998, about 6700 ha of native forest in the Los Lagos Region was replaced with plantations (CONAF-CONAMA-BIRF, 1999). Despite these threats, temperate rainforests are poorly represented in the existing protected area network (Neira *et al.*, 2002), which is biased towards the high elevation, volcanic areas of the Andes (Armesto *et al.*, 1998).

There is an increasing awareness in Chile of the need to improve the representation of the country's forest types in the reserve network (Armesto *et al.*, 1996b; Neira *et al.*, 2002) and the adequacy of coverage of the current reserve network has been questioned (Contreras *et al.*, 1979; Ormazábal 1986a, b; Armesto *et al.*, 1998; Rozzi *et al.*, 2000; Pauchard and Villarroel, 2002). For example, Armesto *et al.* (1998) illustrated that more than 90% of the reserved land in the temperate forest region is concentrated at high latitudes (greater than 43° S; specifically in regions XI and XII), outside the richest area of biodiversity (which is between 35.6° and 41.3° S), and in regions with low human population densities and few forest-related industrial developments.

Strategies are therefore required to reduce the impacts of the conversion of native forest to plantations and improve the representativeness of the reserve network within the temperate forest region of Chile. One promising strategy is to use the principles of systematic conservation planning to identify important areas for the conservation of biodiversity. Systematic conservation planning is the process of locating and designing conservation areas (ranging from strict reserves to areas that are important for off-reserve management) to promote the persistence of biodiversity *in situ* and has become the international norm for making spatially explicit decisions about reserve networks (Possingham *et al.*, 2006).

Recent advances in the field of systematic conservation planning have seen the development of principles and tools to design efficient reserve networks that meet predetermined conservation targets for the biodiversity features of interest (Margules and Pressey, 2000). Approaches to systematic conservation planning recognize that, due to constraints on the amount of land that can be set aside for biodiversity conservation, there is a need to conserve biodiversity in the most efficient manner possible (Pressey *et al.*, 1993). It is also recognized that conservation areas must be able to mitigate at least some of the processes that threaten biodiversity. However, while much attention is directed towards understanding the patterns of biodiversity, much less has been given to determining the areas of the landscape most vulnerable to threats. In this context, it is useful to assess the vulnerability of the remaining areas of native forest in south-central Chile to help identify their relative urgency for protection (Wilson *et al.*, 2005a,b).

When developing a conservation plan, vulnerable areas might be avoided so that objectives are achieved, as far as possible, in areas without liabilities for implementation and management. Considerations of defensibility, or avoiding vulnerable areas, can be especially important if resources are likely to be insufficient for effective management. When implementation of new conservation areas commences, an important consideration in scheduling their implementation will often be their relative vulnerability (Wilson *et al.*, 2005a). The more vulnerable areas might receive higher priority, especially if there are few or no alternative areas available to protect the biodiversity features they contain. This strategy can minimize the extent to which conservation objectives are compromised by threatening processes during the frequently protracted process of establishing conservation areas on the ground.

There are a variety of possible approaches available to assess the relative vulnerability of areas to threatening processes. Wilson *et al.* (2005a) reviewed methods that have been used to assess vulnerability and categorized them into four groups based mainly on the types of data employed. The first method uses information on permitted or projected land uses. The second method identifies the extent of past impacts on features and uses these data to predict future impacts on the same features. In some circumstances, the underlying spatial (e.g. proximity to cities and roads) and environmental characteristics (e.g. soil type, slope, climate) believed to have predisposed areas to threatening processes in the past are determined, and areas that are presently unaffected and share these characteristics are then identified. The third method identifies vulnerable areas as those with high concentrations of taxa with high probabilities of extinction, and the final method is based on expert knowledge. All four methods have been employed at a variety of spatial scales and resolutions in countries with differing levels of development, even in those typically regarded as data-poor. The data underpinning many of the methods are globally available and so most methods are applicable anywhere, at least at a coarse scale.

A measure of vulnerability alone is likely to be an insufficient criterion for identifying priority areas for conservation. This is because biodiversity features, such as forest types, are likely to have different spatial options available to achieve their conservation targets. The differences in the relative irreplaceability of areas can be crucial in determining the most urgent areas for conservation. Although some areas might be irreplaceable and therefore require protection in order to meet our conservation targets, a measure of conservation value alone will rarely be sufficient to define conservation priorities. This is because areas of high irreplaceability may not be threatened. Areas of the landscape that are priorities for conservation should be those that are both vulnerable to threatening processes and that, if lost or degraded, will result in targets being compromised (Margules and Pressey, 2000; Pressey and Taffs, 2001).

The objective of this chapter is to use information on the vulnerability of native forest to threatening processes and information on the distribution of forest types to systematically determine priorities for biodiversity conservation. As part of this process, areas of native forest that are both highly vulnerable and irreplaceable are identified. These areas (land costs and cultural and societal values being equal) should be the priorities for biodiversity conservation.

The study region for these analyses is located within south-central Chile and extends from approximately latitude 39.5° S to 43° S (Valdivia, Osorno and Llanquihue Provinces of Region X), and from the coastal mountain range to the Andes. This region covers approximately 4.2 million ha and contains a large proportion of the remaining temperate forest in southern Chile. The area is presently experiencing high rates of conversion to plantations (Neira *et al.*, 2002).

Historical and Current Extent of Native Forest

Prior to European settlement, native forest was estimated to cover 3.7 million ha, which corresponds to 88% of the study region (Lara *et al.*, 1999). The remainder was comprised largely of areas devoid of vegetation, such as those subject to volcanic activity. The area of native forest has subsequently been reduced and in 1997 was estimated to occupy 2 million ha (CONAF-CONAMA-BIRF, 1997). Only 53% of the area of native forest estimated to be present before European settlement remained in 1997, with 38% having been converted to pasture and agriculture and 6% converted to plantations.

Temperate forest is now largely restricted to the upper elevations of the Andean and coastal ranges. Extensive areas of native forest in the Central Valley between the coastal and Andes ranges have been converted to other land cover types (Fig. 14.1).

Prior to European settlement, the native forest in the study region was comprised of seven types (following the classification scheme of Neira *et al.*, 2002; see Appendix). Small areas of Guaitecas Cypress (52 ha) and sclerophyllous forest (1726 ha) have since established in the study region. Many of the forest types have been reduced in extent (Fig. 14.2). Prior to European settlement, the forest types that had the greatest coverage were Evergreen, Coigue–Raulí–Tepa/Roble–Raulí–Coigue and Alerce. These three types accounted for 3.25 million ha, or 87% of the native forest cover. Proportionally,

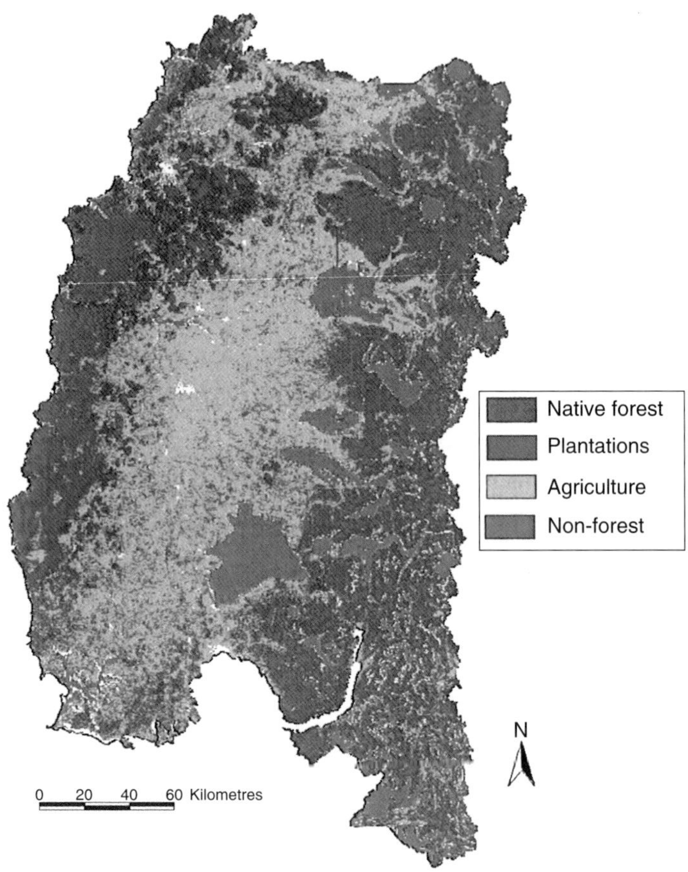

Fig. 14.1. The spatial extent of the major land cover types in 1997. Native forest is restricted to the coastal range (western portion of study region) and the Andes range (eastern portion of study region). Much of the native forest in the Central Valley between these two ranges has been converted to agriculture and plantations.

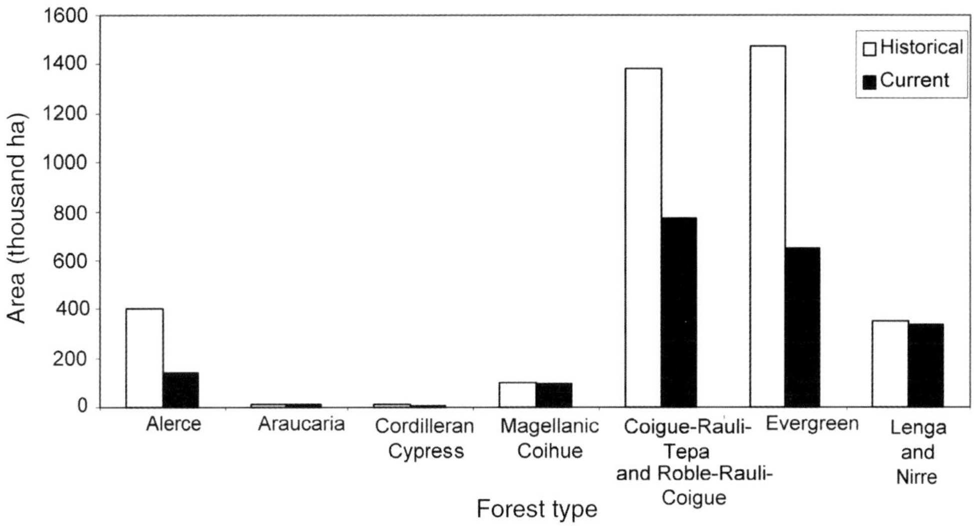

Fig. 14.2. Historical and current extent of each forest type.

these forest types remain dominant, but their area has since substantially diminished. For example, the cover of Alerce forest has been reduced by approximately 65% (Fig. 14.2).

Large proportions of the Coigue–Raulí–Tepa/Roble–Raulí–Coigue, Cordilleran Cypress, Evergreen and Alerce forest types have been converted to pasture and agriculture (Table 14.1). The Coigue–Raulí–Tepa/Roble–Raulí–Coigue and Evergreen forest types have also been converted to plantations (Table 14.1).

Table 14.1. The proportion of the original extent of each forest type that remains as native forest (of some type) or that has been converted to another type of land cover. For details of the forest types, see Appendix.

	Current land cover type					
Original forest type	Native forest	Pasture and agriculture	Wetlands	Urban areas	Plantations	Other
Alerce	81	15	1	0	0	3
Araucaria	90	4	0	0	0	6
Cordilleran Cypress	71	22	0	0	0	7
Magellanic Coihue	93	4	0	0	0	3
Coigue–Raulí–Tepa/ Roble–Raulí–Coigue	34	57	0	1	7	1
Evergreen	53	36	1	0	8	2
Lenga	85	7	0	0	0	8
Ñirre	93	2	0	0	1	4

The Existing Reserve Network

The existing reserve network covers approximately 12% of the study region and is represented by six national parks, ten private reserves, one national monument and three national reserves (Table 14.2, p.321). Using overlay analysis in geographic information systems software, it is apparent that national parks, which account for 86% of the total extent of conservation areas, are located in the Andes range, generally at elevations greater than 600 metres, and on large volcanic cones (Table 14.2). Seventy per cent (359,167 ha) of the reserve network is forested. The remaining is comprised of volcanic areas and other areas devoid of vegetation (CONAF-CONAMA-BIRF, 1997).

Our goal is to identify priority areas to protect (through either strict reserves or off-reserve management) 15% of the extent of each forest type that existed prior to European settlement. Information on the pre-European extent of each forest type was obtained from a historical land cover dataset (Lara *et al.*, 1999). Basing the target on the historical extent of each forest type ensured that larger targets were allocated to forest types that have been most reduced in extent (RACAC, 1996; Pressey, 1998; Pressey *et al.*, 2003). The conservation target allocated to each forest varies according to the proportion of the pre-European extent remaining (Table 14.3).

The representativeness of the existing reserve network was assessed by determining the number of forest types represented at the target level and the number represented above the target level. The conservation targets for the Araucaria, Magellanic Coigue and Lenga forest types are met in the existing reserve network (Table 14.4). A large proportion of the target allocated to the Alerce forest type is met in the existing reserve network (Table 14.4). In

Table 14.3. The historical and current extent of the native forest types in the study region, the % change in the extent of the forest types since European settlement and the conservation target allocated to each forest type.

Native forest type	Historical extent (ha)	Current extent (ha)	% Change in area	Target area (ha)	% of remaining extent requiring protection to meet target
Alerce	400,214	140,305	–65	60,032	42.8
Araucaria	10,292	9,587	–7	1,544	16.1
Cordilleran Cypress	10,281	6,512	–37	1,542	23.7
Magellanic Coihue	99,741	95,372	–4	14,961	15.7
Coigue–Raulí–Tepa/ Roble–Raulí–Coigue	1,381,725	770,928[a]	–44	207,259	26.9
Evergreen	1,470,945	647,277	–56	220,642	34.1
Lenga and Ñirre	353,688	338,478	–4	53,053	15.7
TOTAL	3,726,886	2,008,895		559,033	

[a]Comprising 400,748 ha of Coigue–Raulí–Tepa and 370,180 ha of Roble–Raulí–Coigue.

Table 14.2. Characteristics of the existing reserve network in the study region. The reserve network consists of National Parks, a National Monument, three National Reserves and Private Reserves. Characteristics concerning the elevation range, rainfall range and soil type of the conservation areas are provided.

Conservation area	Area (ha)	Type of reserve	Elevation range (m)	Maximum rainfall range (mm)	Soil type
PN Villarica	17,359	National Park	600–2,000	2,500–4,000	Sandy volcanic soils, recent volcanic soils (volcanic cone)
PN Hornopirén	12,141	National Park	400–1,800	5,000	Not available
PN Hornopirén	6,479	National Park	400–1,800	5,000	Not available
PN Puyehue	112,377	National Park	200–1,800	3,000–4,000	Sandy volcanic soils, recent volcanic soils (volcanic cone)
PN Vicente Pérez Rosales	249,804	National Park	0–1,800	2,500–3,500	Volcanic alluvial sands, sandy volcanic soils (volcanic cone)
PN Alerce Andino	39,882	National Park	0–1,400	2,500–4,000	Sandy volcanic soils
Subtotal National Parks	*438,042*				
San Pablo de Tregua	2,189	Private Reserve	200–1,400	2,500	Recent volcanic soils
Fundo San Julian	326	Private Reserve	200–600	2,500–3,000	Recent volcanic soils
Rodeo Grande	34	Private Reserve	0–400	2,500	Red clay
Parque Pumalin	17,200	Private Reserve	600–2,000		Not available
Campo Escuela Polincay	16	Private Reserve	0–200	2,500–4,000	Sandy volcanic soils
Parcela Lipingüe	229	Private Reserve	0–200	3,000	Recent volcanic soils
Santa Elvira	75	Private Reserve	0–200	2,500	Recent volcanic soils
Cuinco	4	Private Reserve	0–200	1,500	Red clay
Santa Anita/El Mirador	153	Private Reserve	0–200	2,000	Recent volcanic soils
Parcela Altamira – CEA	3	Private Reserve	0–200	4,000	Metamorphics
Subtotal Private Reserves	*20,229*				
Alerce Costero National Monument	2,248	National Monument	400–1,000	4,000	Metamorphics
Subtotal National Monuments	*2,248*				
RN Llanquihue	34,147	National Reserve	0–1,600	2,500	Sandy volcanic (volcanic cone)
RN Valdivia	9,789	National Reserve	0–800	4,000	Metamorphics
RN Mocho-Choshuenco	7,518	National Reserve	1,200–1,600	4,000	Sandy volcanic (volcanic cone)
Subtotal National Reserves	*51,454*				
Total extent of protected areas	***511,973***				

Table 14.4. Contribution of the existing reserve network to meeting the conservation targets for the forest types.

Native forest type	Area in existing reserve network (ha)	% of target satisfied in existing reserve network	Area of unprotected extent requiring protection to meet target (ha)	% of unprotected extent requiring protection to meet target
Alerce	34,356	57.2	25,676	18.3
Araucaria	1,702	110.2	0	0
Coigue–Raulí–Tepa/Roble–Raulí–Coigue	92,956	44.9	114,303	14.8
Cordilleran Cypress	862	55.9	680	10.4
Magellanic Coihue	54,843	366.6	0	0
Lenga and Ñirre	138,903	261.8	0	0
Evergreen	35,545	16.1	185,097	28.6
Total	*359,167*			

comparison, the Evergreen forest type is under-represented in the existing reserve network (Table 14.4).

Obtaining a Representative Reserve Network

The conservation planning decision-support tool, Marxan (Ball and Possingham, 2000), was used to determine the areas of native forest with the highest irreplaceability, or likelihood that they will require protection in order for the conservation targets to be met. The planning units employed in this analysis were the extant remnants of native forest. The following input tables for Marxan were constructed:

- Information on each planning unit, including its area and land use.
- Information on the distribution of each forest type in each planning unit.
- Information on each forest type, including its conservation target.

The simulated annealing algorithm in Marxan was used to perform the analysis. The adaptive simulated annealing schedule followed by iterative improvement was used and was configured so that the number of simulated annealing iterations was 10 million and the number of temperature decreases was 10,000. The conservation feature penalty factor was set to 1000, which ensured that all conservation targets were met. Marxan was run 100 times to produce 100 near-optimal solutions. Irreplaceable planning units were identified as those that were included in each of the 100 reserve network solutions.

Marxan is one of several tools available to perform systematic conservation planning analyses and can find good solutions to a mathematically defined optimization problem (Possingham *et al.*, 2000). Marxan underpinned the rezoning of the Great Barrier Reef and is used by over 1000 users in over 80 countries worldwide (http://www.ecology.uq.edu.au/marxan.htm). It is the primary spatial planning tool used by the Nature Conservancy (USA). While Marxan was employed in this analysis, it is recognized that many other tools have been developed, including C-Plan (NSW NPWS, 1999), ALDO (Groves, 2003), CODA (Bedward *et al.*, 1992), Diversity-ED (Faith and Walker, 1994; Margules and Redhead, 1995), ResNet/ResNet-GUI (Sarakinos *et al.*, 2001; Kelley *et al.*, 2002), Sites (Noss *et al.*, 2002; Andelman and Willig, 2003), TAMARIN (Stoms *et al.*, 2004), TARGET (Faith *et al.*, 2003) and WORLDMAP (Williams *et al.*, 2003). In addition, various commercial optimizing packages (such as LINDO, CPlex and XPRESS) have been used for conservation planning (Rodrigues *et al.*, 1999). Each tool has its relative advantages and disadvantages – we employed Marxan since we have found it to deliver efficient solutions to large and complex conservation planning problems in a timely manner.

Our analysis indicates that 12 planning units are irreplaceable (i.e. were included in each of the reserve network solutions) and will require protection in order for the conservation targets allocated to the Alerce, Cordilleran Cypress, Coigue–Raulí–Tepa/Roble–Raulí–Coigue and Evergreen forest types to be met (Table 14.5; Fig. 14.3). These planning units cover 344,499 ha. The forest occurring within the existing reserve network (359,167 ha) and the additional planning units account for approximately 35% of the remaining forested area in the study region. A large proportion of the forest in the coastal range requires protection in order for the target allocated to the Evergreen forest type to be met (Fig. 14.3).

Table 14.5. The degree to which conservation targets are met in the proposed reserve network.

Native forest type	% of target satisfied by proposed reserve network
Alerce	103
Araucaria	110
Coigue–Raulí–Tepa/ Roble–Raulí–Coigue	101
Cordilleran Cypress	116
Magellanic Coihue	367
Lenga and Ñirre	262
Evergreen	107

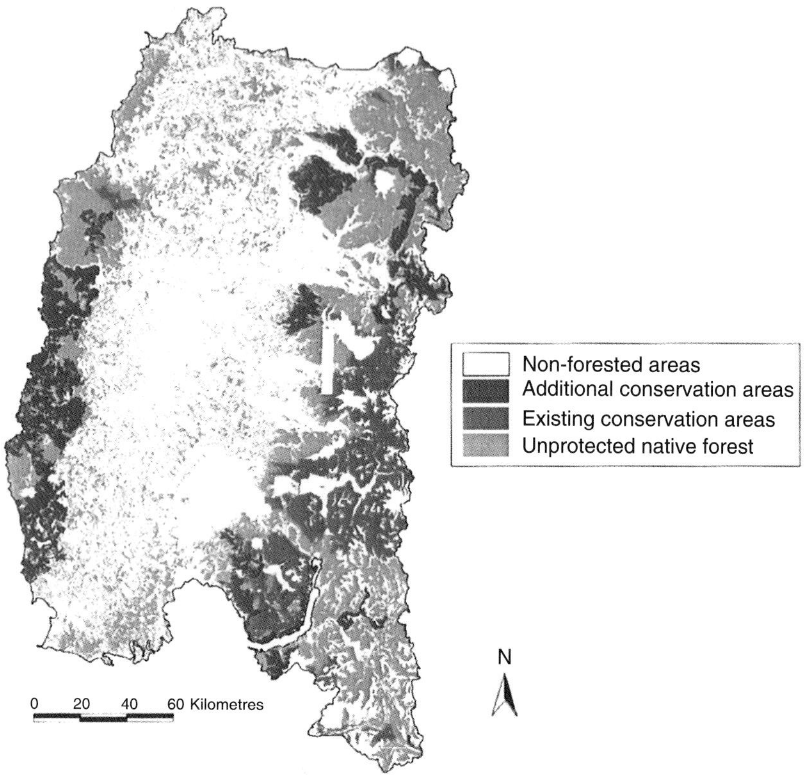

Fig. 14.3. Additional conservation areas required to meet the targets allocated to each forest type.

Vulnerability of Remaining Native Forest to Plantation Conversion

Wilson *et al.* (2005a) describe an assessment of the relative vulnerability of remaining areas of native forest to conversion to plantations. The probability of exposure of native forest to this threatening process was assessed using a quantitative method based on spatial and statistical modelling (Method 2 from the review of Wilson *et al.*, 2005a). First, a classification tree was used to identify environmental variables that may explain the spatial distribution of native forest conversion. The variables considered for inclusion in the model of native forest conversion to plantations after correlated variables were excluded were annual precipitation, latitude, soil type, slope, altitude, distance to towns and distance to roads. These variables were then further analysed using a multivariate, spatially explicit statistical model, using the statistical technique of logistic regression. The model was used to identify the variables that explain the spatial distribution of native forest conversion. The best-fit statistical model proposed that the presence of native forest conversion to plantations is a function of soil type, slope, altitude, annual precipitation, distance to roads, distance to towns and latitude (Table 14.6).

The model predicted conversion to plantations to be more likely in areas of low elevation and gentle slope. The probability of conversion was negatively related to distance to towns, distance to roads, rainfall, high altitude, steep slopes and latitude. The conversion of native forest to plantations is predicted to be more likely in relatively warm and low rainfall areas that are close to towns and roads and on red clay and mixed alluvial soils (Table 14.6). Areas of native forest with a high probability of conversion were identified in order to delineate areas of native forest highly vulnerable to plantation conversion (Fig. 14.4).

Table 14.6. The coefficients for the explanatory variables included in the model that describes the conversion of native forest to plantations.

Variable	Coefficient	SE	T-statistic	Wald statistic	*P*-value
Intercept	55.77	0.87	64.10	4,108.28	
Soil type (Marine sediments – Soil type 1)					
Sandy volcanic soils – Soil type 7	−2.18	0.16	−13.65	186.25	***
Volcanic alluvial sands – Soil type 8	−0.43	0.17	−2.53	6.40	*
Mixed alluvial – Soil type 2	0.61	0.22	2.78	7.73	**
Red clay – Soil type 3	0.62	0.14	4.50	20.21	***
Metamorphic – Soil type 4	−0.65	0.14	−4.81	23.17	***
Recent volcanic soils – Soil type 5	−0.96	0.14	−7.03	49.45	***
Salt pan – Soil type 6	−1.51	0.15	−10.31	106.20	***
Annual precipitation (mm)	−0.0002	0.00	−12.56	157.84	***
Latitude (degrees south)	−1.37	0.02	−63.40	4,019.79	***
Distance to towns (km)	−0.00003	0.00	−30.33	920.06	***
Distance to roads (km)	−0.0006	0.00	−43.30	1,874.83	***
Slope (baseline is 0–15%)					
Slope 15–30	0.09	0.02	4.41	19.48	***
Slope 30–45	−0.27	0.03	−7.92	62.71	***
Slope 45–60	−1.56	0.12	−13.10	171.67	***
Slope 60–100	−1.76	0.19	−9.49	90.11	***
Altitude (baseline is 0–200 masl)					
Altitude 200–400	0.28	0.02	13.75	189.06	***
Altitude 400–600	−0.18	0.03	−6.16	37.93	***
Altitude 600–800	−1.46	0.06	−25.02	626.01	***
Altitude 800–1000	−3.43	0.22	−15.58	242.76	***
Altitude 1000–1200	−2.96	0.27	−10.82	116.99	***
Altitude 1200–1400	−1.04	0.18	−5.74	32.92	***
Altitude 1400–1600	−3.38	0.86	−3.92	15.34	***
Altitude 1600–1800	−2.35	2.39	−0.98	0.96	0.33
Altitude 1800–2000	−5.58	34.83	−0.16	0.03	0.86

SE stands for the standard error of the estimated coefficient value. ***$P < 0.001$, **$P < 0.05$, *$P < 0.1$.

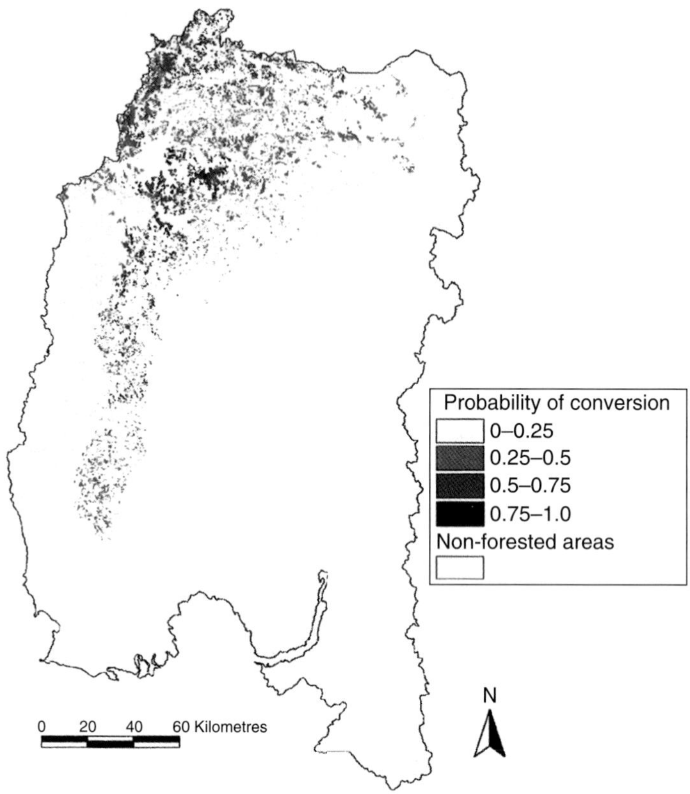

Fig. 14.4. The probability of native forest conversion to plantations. The darker areas have a higher probability of conversion (and are therefore the more vulnerable areas).

The predicted probabilities of conversion do not provide an indication of the imminence of conversion; rather they provide an estimate of how likely it is that conversion will occur in an area at some stage in the future. An exact time frame for the event is not provided. Therefore, a relative, rather than an absolute, vulnerability is predicted. Areas of native forest identified to be vulnerable to conversion are concentrated in the Central Valley region, east of the coastal range. These areas have moderate slope and elevation, mild climatic conditions and soils with minimal limitations for plant growth. Areas of low probability of conversion are concentrated in the Andes and coastal ranges: areas of steep slope, high elevation, low temperatures and high rainfall. There are some sections of the coastal range with high probabilities of conversion. These areas are situated in lower elevation areas north of the city of Valdivia and surrounding the township of Punta Falsa (south of Valdivia) (Fig. 14.4). Between 1995 and 1998, approximately 16,000 ha of native forest in the study region was degraded or converted to other land uses (Meneses, 2001). Approximately 42% (6700 ha) was converted to plantations and the majority of this conversion was in the Central Valley region (Meneses, 2001). Therefore, the areas predicted by the model to be at risk of conversion are largely associated with the location of recent activity to convert native forests.

Priority Areas for Biodiversity Conservation

Two-dimensional scatter plots (referred to henceforth as priority plots) were generated to display the irreplaceability and vulnerability values of each unreserved planning unit (Margules and Pressey, 2000). Priority plots allow the planning units that are priorities for conservation action to be visualized. Those planning units in the upper right-hand corner of the priority plots are likely to lose their conservation values and have fewest replacements. Protection of these planning units is urgent if targets are not to be compromised. The lower right-hand section of the priority plots contains planning units that are vulnerable but have more replacements, either because the forest types that occur within them are relatively common or because their targets have been partly met in existing conservation areas. These planning units could move into the upper right-hand corner, if those that are more vulnerable and have higher irreplaceability are lost. In the upper left-hand corner lie planning units with lower vulnerability but with high irreplaceability. Protection of these planning units is less urgent, but they may be used as replacements for planning units that have high irreplaceability and that are more vulnerable. In the lower left-hand corner lie planning units that do not require urgent protection, according to this analysis.

The 12 irreplaceable planning units, required to meet the conservation targets, are highlighted on the priority plot (Fig. 14.5). Eight of these are also vulnerable (have a vulnerability value of 1), with vulnerability calculated as the probability that there is conversion somewhere within each planning unit.

Low vulnerability replacements for the eight vulnerable and irreplaceable planning units were sought. Fifteen areas with high irreplaceability and low vulnerability were added to the existing reserve network instead of the eight irreplaceable and vulnerable planning units (therefore, a total of 19 planning units were added). With the less vulnerable replacements, the Coigue–Raulí–Tepa/Roble–Raulí–Coigue and Evergreen forest types could not meet their targets (Table 14.7). Therefore, whilst there is some flexibility in the reserve network solution, some of the vulnerable areas will need to be included to meet the targets for these two forest types. The planning units requiring protection to meet the targets for the Evergreen forest type are both irreplaceable and vulnerable and are largely confined to the coastal range.

The position of planning units within the priority plots is not static. Some of the vulnerable planning units are likely to be converted to plantations. As this happens, the irreplaceability of some of the remaining planning units will increase as they become more important for achieving targets for forest types that are now less extensive. Conversely, as planning units are progressively reserved, the irreplaceability of others will decrease as the features they contain approach or reach their conservation targets. The vulnerability of planning

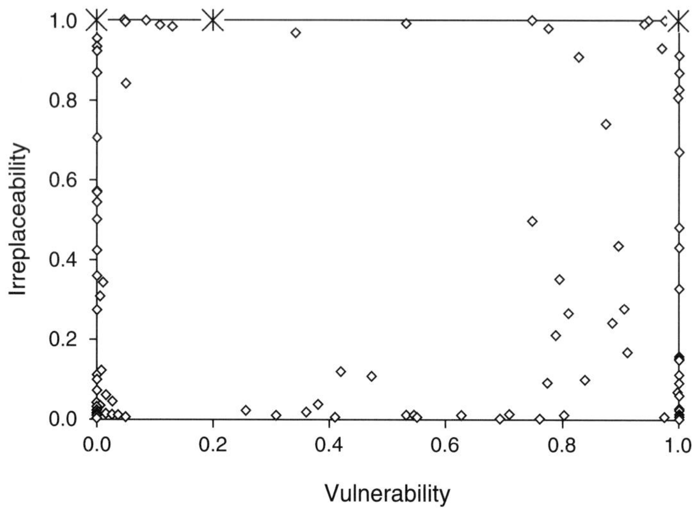

Fig. 14.5. The irreplaceability and vulnerability of the unprotected planning units. Vulnerability is measured as the probability that there is conversion somewhere within the extent of each planning unit. The planning units required to form a representative reserve network are depicted by (⋇), based on their contribution to meeting the conservation targets for the forest types occurring in the region. Possible replacement planning units are depicted by (◊).

Table 14.7. The degree to which targets are met when low vulnerability replacements for the irreplaceable and highly vulnerable planning units are employed.

Forest type	% of target satisfied in the proposed reserve network
Alerce	132.2
Araucaria	110.2
Coigue–Raulí–Tepa/ Roble–Raulí–Coigue	71.4
Cordilleran Cypress	249.6
Magellanic Coihue	366.6
Lenga	261.8
Evergreen	16.1

units is also likely to change with time, most likely by increasing, but may also decline if, for example, there is a downturn in the world woodchip or paper pulp market and the rate of conversion of native forest to plantations declines.

Discussion

The analysis of land cover change has shown that, since European settlement, a large proportion of native forest has been converted to other land uses. Large areas of the Coigue–Raulí–Tepa / Roble–Raulí–Coigue, Cordilleran Cypress, Evergreen and Alerce forest types have been converted to pasture and agriculture. Additionally, the Coigue–Raulí–Tepa / Roble–Raulí–Coigue and Evergreen forest types have been converted to plantations.

The national parks in the study region, which account for 86% of the total extent of the reserve network, are at high elevations and are situated on large volcanic cones. Much of the land in these Andean parks comprises ice and unvegetated terrain (CONAF-CONAMA-BIRF, 1997). Many of these conservation areas were reportedly chosen for their scenic or recreational value (Pauchard and Villarroel, 2002). The existing reserve network represents some forest types well above their targets (for example, Magellanic Coigue and Lenga and Ñirre) at the expense of others (for example, the Evergreen forest type). According to the forest type classification found in Gajardo (1983), Chile has a total of 85 ecosystems and vegetative subgroups, of which 19 are not represented in the reserve network. Approximately 33% of ecosystems have less than 5% of their area protected (Neira *et al.*, 2002).

The vulnerability of extant areas of native forest was predicted according to variables that limit plant growth (such as rainfall and soil type) and variables that determine the suitability of sites for the establishment of plantations (such as slope, elevation and distance to infrastructure (roads and towns)). Areas of native forest predicted to be vulnerable to conversion are concentrated in the Central Valley and western portion of the study region. These are areas of moderate slope and elevation, with mild climatic conditions, and soils with minimal limitations for plant growth. The forests are also accessible by the existing road network and are in close proximity to major towns. These vulnerable areas have been identified by others to have minimal climatic and edaphic limitations for plantation establishment (Schlatter and Gerding, 1995; Schlatter *et al.*, 1995).

In generating the vulnerability model it was assumed that the past pattern of impacts is indicative of future patterns. A consequence of violating this assumption is that the model will erroneously identify areas as vulnerable. For example, much of the remaining forest in the Central Valley region is predicted to be vulnerable. Given that these forests have been highly accessible for much of the recent past, their persistence suggests that their vulnerability might have been overestimated. Conversely, the vulnerability of forest in the coastal range may have been underestimated, as recent conversions of native forest postdate the existing land cover map (on which the vulnerability assessment is based).

The priority areas for conservation should be the currently unprotected areas that are vulnerable to plantation conversion and that, if lost or degraded, will result in targets being compromised. Given our pre-specified targets for each forest type, we use the simulated annealing algorithm in the decision-support tool, Marxan, to identify areas that are irreplaceable and therefore in need of conservation action. Simulated annealing will not always find the best solution to a complex problem, but will usually find a solution that is near to optimal and do so quickly (Possingham *et al.*, 1993, 2000). It can also provide many good solutions to large and complex problems and therefore has the additional benefit of offering flexible solutions. Alternative approaches for conservation planning, such as scoring areas on the basis of their biodiversity values, have repeatedly been shown to provide inefficient assessments of conservation priorities as they do not account for the complementarity of areas in terms of their biological composition (Pressey and Nicholls, 1989). Furthermore, scoring approaches do not provide a transparent assessment of conservation priorities as a similar score for an area can be obtained by a variety of different means (Possingham *et al.*, 2006).

Eight of the 12 additional planning units identified to be required to meet the conservation targets are both irreplaceable and vulnerable to conversion (fall within the top right-hand corner of the priority plot). The remainder are irreplaceable but not vulnerable (fall within the top left-hand corner of the priority plot). Lower vulnerability replacements for the irreplaceable and vulnerable planning units were sought. However, in order to meet the targets for the Evergreen and Coigue–Raulí–Tepa/Roble–Raulí–Coigue forest types, some of the vulnerable areas will need to be included in the reserve network.

The Evergreen forest type is a priority for conservation owing to its lack of representation in the existing reserve network and the degree to which it has been cleared. The biggest impediment to obtaining a representative reserve network in the study region is being able to meet the conservation target for the Evergreen forest type. The most critical areas for meeting the conservation target for this forest type are located in the coastal range. The coastal evergreen forests are known as the Valdivian evergreen forests and they extend for 250 km from the Toltén River (39°S) to south of the Llico River (41.4°S). In the coastal range, a total of 621 native plant species have been recorded (61 pteridophytes, eight gymnosperms and 552 angiosperms) as have many rare species of reptiles and amphibians, together with bird species that are restricted to the coastal forests (Smith-Ramírez, 2004).

These results concur with Armesto *et al.* (1992, 1996a), who found that the most critical areas in the temperate rainforest region of southern South America, in terms of species richness, endemism and direct or indirect threat by humans, occur in the Chilean coastal range. Our results also confirm the conclusions of Smith-Ramírez (2004), who stated that 'the establishment of new conservation areas in the coastal evergreen forest region, where the largest areas of continuous old-growth forest still remain, is required for the long-term protection of biodiversity in the coastal range'.

It may be infeasible to add all the high-priority areas identified in the coastal range to the reserve network, owing to the social and financial difficulties of acquiring and managing these areas. Where flexibility is required, high-priority areas could be allocated to a variety of land tenure types, varying from strict protected areas to areas with off-reserve management arrangements with private land holders (Pauchard and Villarroel, 2002). A large majority of the remaining forested areas in Chile are on private property (Neira *et al.*, 2002), including approximately 50% of the vegetation associations that are under-represented in the existing reserve network (Calcagni *et al.*, 1999). Further, the government has limited resources to purchase land and landowners are placing ever higher monetary values on their properties

(Neira *et al.*, 2002). Therefore, the long-term conservation and protection of native forest in Chile will likely require inclusion of areas in both the public and private reserve network.

Focusing conservation efforts on areas with the greatest conservation value and the highest likelihood of losing significant portions of this value should achieve maximum impact for conservation investment, and maximize the extent to which conservation goals are achieved (Pressey, 1997). However, the conservation value and vulnerability of an area is only part of the information needed for prioritizing areas for conservation. Other factors, such as land cost and cultural and societal values, will also be important and private conservation areas are likely to play an important role in achieving conservation goals in Chile. These additional factors could also be incorporated into the framework for systematic conservation planning.

Acknowledgement

The assistance of colleagues in the Universidad Austral de Chile is gratefully acknowledged with respect to accessing relevant data for southern Chile.

Appendix

Table A14.1. Definition of forest types included in the analysis (following Neira *et al.*, 2002).

Forest type	Dominant and key associated tree species
Alerce	*Fitzroya cupressoides* (alerce), *Nothofagus betuloides* (Magellanic coigue), *Nothofagus nitida* (Chiloé coigue), *Podocarpus nubigena* (prickly leaved mañio), *Pilgerodendron uviferum* (ciprés de las Guaitecas)
Araucaria	*Araucaria araucana* (monkey puzzle, pehuen), *Nothofagus dombeyi* (coigue), *Nothofagus alpina* (roble), *Nothofagus antarctica* (ñirre), *Drimys winteri* (canelo), *Nothofagus pumilio* (lenga)
Cordilleran Cypress	*Austrocedrus chilensis* (ciprés de la cordillera), *Cryptocarya alba* (peumo), *Peumus boldo* (boldo), *Maytenus boaria* (maitén), *Quillaja saponaria* (quillay)
Magellanic Coigue	*Nothofagus betuloides* (Magellanic coigue), *Nothofagus pumilio*, *Weinmannia trichosperma* (tineo), *Podocarpus nubigena*, *Pilgerodendron uviferum*
Coigue–Raulí–Tepa/ Roble–Raulí–Coigue	*Nothofagus dombeyi*, *Nothofagus alpina* (raulí), *Laureliopsis philippiana* (tepa), *Aextoxicon punctatum* (olivillo)
Evergreen	*Laureliopsis philippiana*, *Amomyrtus luma* (luma), *Drimys winteri*, *Weinmannia trichosperma*
Lenga	*Nothofagus pumilio*, *Nothofagus dombeyi*, *Nothofagus alpina*, *Araucaria araucana*, *Nothofagus antarctica*, *Nothofagus betuloides*
Ñirre	*Nothofagus antarctica*

References

Andelman, S.J. and Willig, M.R. (2003) Present patterns and future prospects for biodiversity in the Western Hemisphere. *Ecology Letters* 6, 818–824.

Armesto, J.J., Smith-Ramírez, C., Leon, P. and Arroyo, M.T.K. (1992) Biodiversidad y conservación del bosque templado en Chile. *Ambiente y Desarrollo* 8, 19–24.

Armesto, J., Aravena, J.C., Villagrán, C., Pérez, C. and Parker, G.G. (1996a) Bosques templados de la cordillera de la costa. In: Armesto, J.J. Villagrán, C. and Arroyo, M.K. (eds) *Ecología de los Bosques Nativos de Chile*. Editorial Universitaria, Santiago, Chile, pp. 199–213.

Armesto, J., Rozzi, R. and Leon-Lobos, P. (1996b) Ecologia de los bosques chilenos, sintesis y proyecciones. In: Armesto, J.J., Villagrán, C. and Arroyo, M.K. (eds) *Ecología de los Bosques Nativos de Chile*. Editorial Universitaria, Santiago, Chile, pp. 405–421.

Armesto, J.J., Rozzi, R., Smith-Ramírez, C. and Arroyo, M.T.K. (1998) Conservation targets in South American temperate forests. *Science* 282, 1271–1275.

Ball, I.R. and Possingham, H.P. (2000) *Marxan (v 1.8.6): Marine Reserve Design Using Spatially Explicit Annealing. User Manual*. The University of Queensland, Brisbane, Australia.

Bedward, M., Pressey, R.L. and Keith, D.A. (1992) A new approach for selecting fully representative reserve networks: addressing efficiency, reserve design and land suitability with an iterative analysis. *Biological Conservation* 62, 115–125.

Calcagni, R., Yunis, C., García, D. and Villarroel, P. (1999) Lugares naturales y calidad de vida: una propuesta para integrar 'lo natural' y 'lo social'. *Ambiente y Desarrollo* 15, 93–103.

CODEFF (1992) *El futuro del Bosque Nativo Chileno: Un Desafio de Hoy*. CODEFF, Santiago, Chile.

CONAF-CONAMA-BIRF (1997) *Catastro y Evaluación de los Recursos Vegetacionales Nativos de Chile*. Corporación Nacional Forestal (CONAF), Chilean Forest Service, Santiago, Chile.

CONAF-CONAMA-BIRF (1999) *Catastro y Evaluación de los Recursos Vegetacionales Nativos de Chile: Monitoreo de Cambios*. Corporación Nacional Forestal (CONAF), Chilean Forest Service, Santiago, Chile.

Contreras, M., de la Maza, C., Merino, R., Morales, A., Barros, P. and Weintraub, A. (1979) *Evaluación Económica de Parques Nacionales: el Sistema de Parques Nacionales en Chile, Resumen de Metodologías*. Investigación y Desarrollo Forestal. CONAF/FAO/PNUD, Santiago de Chile, Chile.

Davis, S., Heywood, V.H. and Hamilton, A.C. (1994–1997) *Centres for Plant Diversity: A Guide and Strategy for their Conservation*. World Wide Fund for Nature and the International Union for Conservation of Nature and Natural Resources, Gland, Switzerland.

Faith, D.P. and Walker, P.A. (1994) *Diversity: A Software Package for Sampling Phylogenetic and Environmental Diversity*. Reference and User's Guide Vol. 2.1. Division of Wildlife and Ecology, CSIRO, Canberra, Australia.

Faith, D.P., Carter, G., Cassis, G., Ferrier, S. and Wilkie, L. (2003) Complementarity, biodiversity viability analysis, and policy-based algorithms for conservation. *Environmental Science and Policy* 6, 311–328.

Gajardo, R. (1983) *Sistema Básico de Clasificación de la Vegetación Nativa Chilena*. Universidad de Chile, CONAF, Santiago, Chile.

Groves, C. (2003) *Drafting a Conservation Blueprint: A Practitioner's Guide to Planning for Biodiversity*. Island Press, Washington, DC.

Kelley, C., Garson, J., Aggarwal, A. and Sarkar, S. (2002) Place prioritization for biodiversity reserve network design: a comparison of the SITES and ResNet software packages for coverage and efficiency. *Diversity and Distributions* 8, 297–306.

Lara, A., Solari, M.E., Rutherford, P., Thiers, O., Trecaman, R., Prieto, R. and Montory, C. (1999) *Cobertura de la Vegetación Original de la Ecoregión de los Bosques Valdivianos en Chile Hacia 1550*. Informe Técnico. Proyecto FB 49-WWF/Universidad Austral de Chile, Valdivia, Chile.

Margules, C.R. and Pressey, R.L. (2000) Systematic conservation planning. *Nature* 405, 243–253.

Margules, C.R. and Redhead, T.D. (1995) *BioRap: Guidelines for Using the BioRap Methodology and Tools*. CSIRO, Canberra, Australia.

Meneses, M. (2001) *Cambios en el Uso del Suelo y su Relación con la Expansión de Plantaciones en las Regiones VIII Y X*. Facultad Ciencias Forestales, Instituto de Manejo Forestal, Valdivia, Chile.

Neira, E., Verscheure, H. and Revenga, C. (2002) *Chile's Frontier Forests: Conserving a Global Treasure*. World Resources Institute, Comité Nacional Pro Defensa de la Fauna y Flora, University Austral of Chile, Chile.

Noss, R.F., Carroll, C., Vance-Borland, K. and Wuerthner, G. (2002) A multicriteria assessment of the irreplaceability and vulnerability of sites in the Greater Yellowstone Ecosystem. *Conservation Biology* 16, 895–908.

NSW NPWS (1999) *C-Plan Conservation Planning Software*. User Manual for C-Plan Version 3.2. New South Wales National Parks and Wildlife Service, Government of Australia, Canberra.

Olson, D.M. and Dinerstein, E. (1998) The Global 200: a representation approach to conserving the Earth's most biologically valuable ecoregions. *Conservation Biology* 12, 502–515.

Ormazábal, C. (1986a) El sistema nacional de áreas silvestres de Chile. *Flora, Fauna y Áreas Silvestres* 1, 10–15.

Ormazábal, C. (1986b) *Preservación de recursos fitogenéticos in situ a través de parques nacionales y otras áreas protegidas. Importancia, avances, limitaciones y proyección futura*. Corporación Nacional Forestal, Ministerio de Agricultura, Santiago, Chile.

Pauchard, A. and Villarroel, P. (2002) Protected areas in Chile: history, current status and challenges. *Natural Areas Journal* 22, 318–330.

Possingham, H., Day, J. Goldfinch, M. and Salzborn, F. (1993) The mathematics of designing a network of protected areas for conservation. In: Sutton, D., Cousins, E. and Pearce, C. (eds) *Decision Sciences: Tools for Today, Proceeding of the 12th Australian Operations Research Conference*. ASOR, University of Adelaide, Adelaide, Australia, pp. 536–545.

Possingham, H., Ball, I. and Andelman, S. (2000) Mathematical methods for identifying representative reserve networks. In: Ferson, S. and Burgman, M. (eds) *Quantitative Methods for Conservation Biology*. Springer, New York, pp. 291–305.

Possingham, H.P., Wilson, K.A., Andelman, S.J. and Vynne, C.H. (2006) Protected areas: goals, limitations, and design. In: Groom, M.J., Meffe, G.K. and Carroll, C.R. (eds) *Principles of Conservation Biology*. Sinauer Associates Inc., Sunderland, Massachusetts, pp. 509–533.

Pressey, R.L. (1997) Priority conservation areas: towards an operational definition for regional assessments. In: Pigram, J.J. and Sundell, R.C. (eds) *National Parks and Protected Areas: Selection, Delimitation, and Management*. Centre for Water Policy Research, Armidale, New South Wales, Australia, pp. 337–357.

Pressey, R.L. (1998) Algorithms, politics and timber: an example of the role of science in a public, political negotiation process over new conservation areas in production forests. In: Wills, R.T. and Hobbs, R.J. (eds) *Ecology for Everyone: Communicating Ecology to Scientists, the Public and the Politicians*. Chipping Norton, Surrey, UK and Beatty and Sons, New South Wales, Australia, pp. 73–87.

Pressey, R.L. and Nicholls, A.O. (1989) Efficiency in conservation planning: scoring versus iterative approaches. *Biological Conservation* 50, 199–218.

Pressey, R.L. and Taffs, K.H. (2001) Scheduling conservation action in production landscapes: priority areas in western New South Wales defined by irreplaceability and vulnerability to vegetation loss. *Biological Conservation* 100, 355–376.

Pressey, R.L., Humphries, C.J., Margules, C.R., Vanewright, R.I. and Williams, P.H. (1993) Beyond opportunism – key principles for systematic reserve selection. *Trends in Ecology and Evolution* 8, 124–128.

Pressey, R.L., Cowling, R.M. and Rouget, M. (2003) Formulating conservation targets for biodiversity pattern and process in the Cape Floristic Region, South Africa. *Biological Conservation* 112, 99–127.

RACAC (1996) *Draft Interim Forestry Assessment Report*. Resource and Conservation Assessment Council, Sydney, Australia.

Rodrigues, A.S.L., Tratt, R., Wheeler, B.D. and Gaston, K.J. (1999) The performance of existing networks of conservation areas in representing biodiversity. *Proceedings of the Royal Society of London, Series B. Biological Sciences* 266, 1453–1460.

Rozzi, R., Silander, J., Armesto, J.J., Feinsinger, P. and Massardo, F. (2000) Three levels of integrating ecology with the conservation of South American temperate forests: the initiative of the Institute of Ecological Research Chiloé, Chile. *Biodiversity and Conservation* 9, 1199–1217.

Sarakinos, H., Nicholls, A.O., Tubert, A., Aggarwal, A., Margules, C.R. and Sarkar, S. (2001) Area prioritisation for biodiversity conservation in Québec on the basis of species distributions: a preliminary analysis. *Biodiversity and Conservation* 10, 1419–1472.

Schlatter, J.E. and Gerding, V. (1995) Métodos de clasificación de sitios para la producción forestal, ejemplo en Chile. *Bosque* 16, 13–20.

Schlatter, J.E., Gerding, V. and Huber, H. (1995) *Sistema de Ordenamiento de la Tierra: Herramienta para la Planificación Forestal Aplicado a la X Región*. Universidad Austral de Chile, Facultad de Ciencias Forestales, Valdivia, Chile.

Smith-Ramírez, C. (2004) The Chilean coastal range: a vanishing center of biodiversity and endemism in southern temperate rain forests. *Biodiversity and Conservation* 13, 373–393.

Stattersfield, A.J., Crosby, M.J., Long, A.J. and Wege, D.C. (1998) *Endemic Bird Areas of the World: Priorities for Biodiversity Conservation*. Birdlife International, Cambridge, UK.

Stoms, D.M., Chomitz, K.M. and Davis, F.W. (2004) TAMARIN: a landscape framework for evaluating economic incentives for rain forest restoration. *Landscape and Urban Planning* 68, 95–108.

Veblen, T.T. (1983) Degradation of native forest resources in southern Chile. In: Steen, H.K. (ed.) *History of Sustained-Yield Forests: A Symposium*. Forest History Society, Durham, North Carolina.

Wilcox, K. (1995) *Chile's Native Forests: A Conservation Legacy*. Ancient Forest International, Redway, California.

Williams, P.H., Moore, J.L., Kamden Toham, A., Brooks, T.M., Strand, H., D'Amico, J., Wisz, M., Burgess, N.D., Balmford, A. and Rahbek, C. (2003) Integrating biodiversity priorities with conflicting socio-economic values in the Guinean–Congolian forest region. *Biodiversity and Conservation* 12, 1297–1320.

Wilson, K.A., Newton, A.C., Echeverría, C., Weston, C.J. and Burgman, M.A. (2005a) A vulnerability analysis of the temperate forests of south central Chile. *Biological Conservation* 122, 9–21.

Wilson, K.A., Pressey, R.L., Newton, A.C., Burgman, M.A., Possingham, H.P. and Weston, C.J. (2005b) Measuring and incorporating vulnerability into conservation planning. *Environmental Management* 35, 527–543.

15 Restoration of Forest Ecosystems in Fragmented Landscapes of Temperate and Montane Tropical Latin America

M. González-Espinosa, N. Ramírez-Marcial, A.C. Newton, J.M. Rey-Benayas, A. Camacho-Cruz, J.J. Armesto, A. Lara, A.C. Premoli, G. Williams-Linera, A. Altamirano, C. Alvarez-Aquino, M. Cortés, C. Echeverría, L. Galindo-Jaimes, M.A. Muñiz-Castro, M.C. Núñez-Ávila, R.A. Pedraza, A.E. Rovere, C. Smith-Ramírez, O. Thiers and C. Zamorano

Photographs of Mr Alfredo Núñez illustrating the vegetation recovery and growth of *Fitzroya cupressoides* seedlings between 1998 and 2004 as part of an ecological restoration programme conducted by UACH researchers in Nuñez's property. Photos: Cristian Echeverría

Summary

Temperate and tropical montane forests in Latin America represent a major natural resource at both regional and national levels for a number of reasons – biological, climatic, economic, cultural. Native tree species in these forests share conservation problems because of deforestation, habitat degradation, overall biodiversity loss and integrity of landscape structure. However, literature on forest restoration research and practices in these ecosystems is scanty and dispersed. We integrate forest restoration experiences aimed at a variety of purposes that allow us to gain insight over several years under contrasting ecological, social and economic conditions in six study regions: the Argentinian Andes, the IX and X Regions in Chile (including northern Chiloé Island), and central Veracruz and the central and northern Highlands of Chiapas (Mexico). By comparing analogous conditions and highlighting differences among the study sites, current pitfalls can be identified and used to define a minimum set of elements to be considered in a protocol for restoration practices. The restoration studies reviewed here include a wide variety of ecological and socio-economic circumstances that allow the identification of broad guidelines, criteria and indicators for planning, implementing and monitoring ecological restoration programmes. We conclude with statements that suggest approaches, strategies and concrete actions that might be considered as lessons learned and inputs for best practice in forest restoration research and programmes conducted in other developing regions.

Introduction

Temperate or tropical montane habitats occur in densely populated areas of most Latin American and Caribbean countries. These forests are not the most extensive types of forest ecosystems in Latin America, but their biodiversity, endemism and conservation threats are unusually high (Rzedowski, 1978, 1993; Donoso-Zegers, 1993; Hamilton *et al.*, 1995; Webster, 1995; Brown and Kappelle, 2001; Kappelle, 2004, 2006). The temperate and mountain forestlands represent a major natural resource at both regional and national levels for a number of reasons (biological, climatic, economic and cultural). In addition to their remarkable biological diversity, these forest communities are embedded within very different development contexts that must be considered in restoration programmes aimed at their sustainable use.

The temperate Andean forests of Chile and Argentina constitute a biogeographically isolated biome along both slopes of the Andes Cordillera, surrounded by the Pacific Ocean, the central Chilean Mediterranean scrub and the Atacama Desert farther north, the vast treeless semidesert and humid steppes and *pampas* east of the Cordillera, and subantarctic habitats in the southernmost lowlands of the continent (Cabrera and Willink, 1973; Armesto *et al.*, 1997). As observed in other temperate forest ecosystems of the world (broadly defined as those located at latitudes >30° either N or S of the equator), these forests have a relatively high productivity and show high regeneration dynamics (Donoso-Zegers, 1993; Donoso and Lara, 1998). However, these southern forests harbour more plant forms than their northern hemisphere counterparts, and a high level of endemism of vascular plants is one of their most striking attributes (e.g. 34% of the angiosperm genera; Armesto *et al.*, 1997).

The mountains of Veracruz and Chiapas (eastern and southern Mexico, respectively) include a number of highly diverse forest formations (Gómez-Pompa, 1973; Rzedowski, 1978; Breedlove, 1981; González-Espinosa *et al.*, 2004, 2005), from highly seasonal pine forest and pine–oak forest to formations such as montane rainforest (800–2500 m elevation) and evergreen cloud forest (>2500 m; Breedlove, 1981; Ramírez-Marcial, 2001; González-Espinosa *et al.*, 2006). The seasonal formations of Chiapas extend over a rather continuous distribution in the mountain systems of Guatemala, El Salvador, Honduras and northern Nicaragua (Kappelle, 2006). The optimal formations have a highly patchy distribution from subtropical areas in southern Tamaulipas through the Central American mountain ranges and the northern Andes, and are related in the south to the subtropical *Yungas* forests of southern Bolivia and north-east Argentina (Puig and Bracho, 1987; Brown and Grau, 1995; Hamilton *et al.*, 1995; Brown and Kappelle, 2001; Luna *et al.*, 2001). These forests harbour an outstandingly high biodiversity and contribute significant local inputs of water through fog condensation. Although it is recognized that they have a relatively poor primary productivity (Silver *et al.*, 2001), a considerable number of timber and non-timber products are obtained by local people, notably fuelwood (Brown and Kappelle, 2001).

Forest ecosystems represent a most valuable resource for people inhabiting the above-mentioned regions. Yet different social and economic contexts define distinct problems for conservation, sustainable use and restoration of their forest ecosystems. Rural communities in the mountains of Chiapas have some of the lowest well-being indices within Mexico, and their forest resources are currently used by a large part of the local population to provide them with non-commercial timber and firewood (Montoya-Gómez, 1998; Montoya-Gómez *et al.*, 2003). In contrast, in central Veracruz a mid-class group of landholders has become increasingly aware about the long-term benefits of conserving isolated remnant forest fragments for the provision of ecosystem services (Manson, 2004). In Chile forestlands are subjected to intensive management and provide forest products for global markets (Lara, 2004). Yet only 10% of the total rural communities in the country participate in this forestry industry, primarily involving those living where industrial plantations of exotic species have been promoted. Furthermore, many of these communities are among the most marginalized in Chile and have poverty indicators that have more than tripled in comparison to people living in urban regions (Sánchez *et al.*, 2002). In all countries here considered an overall legal framework is available to ensure the conservation and sustainable use of forests; yet they display considerable differences: law enforcement is still badly needed in southern Mexico, while in Chile a second-generation legislation process is currently under way in the Congress to protect native forests in particular (Lara, 2004).

Forests in these regions share a number of threats for the conservation of viable populations of native tree species and their sustainable use, including deforestation, habitat degradation, overall biodiversity loss and integrity of the landscape structure (Aldrich *et al.*, 1997; Ramírez-Marcial *et al.*, 2001, 2005; Galindo-Jaimes *et al.*, 2002; Williams-Linera, 2002; Newton *et al.*, 2004; Cayuela *et al.*, 2005, 2006a, b; and others in this volume). Native forest cover in the VII Region of central Chile has been reduced by 67% between 1975 and

2000, at an annual forest loss rate of 4.5%; corresponding figures for the more southern X Region during the same period are 24% of forest cover at an annual rate of 1.2% (see Chapter 2; Echeverría, 2005). In the VII Region, during the last three decades the native forest area has been mostly converted into forest plantations of exotic species, such as pines and eucalypts. In the X Region, loss of native forest has been associated with an expansion of agricultural land and forest logging for firewood and woodchips (Echeverría, 2005). In the central highlands of Chiapas deforestation has also been intense, but highly variable during the last three decades (de Jong *et al.*, 1999; Ochoa-Gaona and González-Espinosa, 2000); annual deforestation rates ranged from 0.46% up to 3.42%. However, estimates for the last decade, which includes the start of the Zapatista revolt in 1994, indicate considerably higher rates: up to 4.98% (Ochoa-Gaona and González-Espinosa, 2000), and even higher than 6% (Cayuela *et al.*, 2005). Nevertheless, loss of forest cover does not account for structural and floristic impoverishment in the remaining forest patches, which has also been considerable (González-Espinosa *et al.*, 1995, 2006; Ramírez-Marcial *et al.*, 2001; Galindo-Jaimes *et al.*, 2002; Williams-Linera, 2002; Ochoa-Gaona *et al.*, 2004; Chapter 3).

These considerations led us to conclude that forest restoration projects are badly needed in Latin America. Yet it should be recognized that a number of forest restoration initiatives have been undertaken. Furthermore, forest restoration projects in the region may represent some of the oldest (e.g. Janzen, 1987, 2002) or most ambitious in extent worldwide (e.g. Kageyama and Gandara, 2000; Wuethrich, 2007). Yet tropical lowland forests, mainly rainforests, have received most of the attention with respect to restoration projects in the region (Guariguata *et al.*, 1995; Kageyama and Gandara, 2000; Janzen, 2002; Meli, 2003). In most cases the focus has been on the recovery of degraded rainforest stands; in other cases the establishment of selected tree populations, or forest cover, in old pasture or agricultural lands (Guevara *et al.*, 1986, 1992; Aide *et al.*, 2000; Janzen, 2002; Florentine and Westbrooke, 2004). Much emphasis has been placed on the role of vertebrates (including domesticated animals; Posada *et al.*, 2000) in seed dispersal from naturally established standing remnant trees (e.g. Otero-Arnáiz *et al.*, 1999; Toh *et al.*, 1999; Cubiña and Aide, 2001). Less common have been efforts involving enrichment planting in stands with degraded floristic, structural and functional attributes (e.g. Ramos and del Amo, 1992; Montagnini *et al.*, 1995).

In this chapter, we draw upon forest restoration experiences aimed at a variety of purposes pursued for several years under contrasting ecological, social and economic conditions in six temperate or tropical mountain study regions located in Argentina, Chile and Mexico. By comparing analogous conditions or stressing differences among the study sites we suggest approaches, strategies and concrete actions that might be considered as lessons learned and best practice in forest restoration. Starting from the discussion of results obtained, we aim to identify general issues that might offer insights for planning, implementing and monitoring restoration programmes in other developing regions that share socio-economic and natural attributes with our study sites.

Definition and Description of Forest Restoration

Forest restoration in our study sites may potentially include a variety of practices and purposes, but two have been more frequently defined as a goal: (i) establishment of native tree species in open areas, frequently after agricultural use; and (ii) floristic enrichment of impoverished secondary stands, frequently after selective logging of timber trees and saplings for firewood. We adopt the concept that forest restoration should be defined broadly, with an aim towards the eventual attainment of environmental health as indicated by forest structure, floristic composition and ecosystem functioning, along with social and financial viability of forest utilization. In the long term, we propose for our study regions that restoration of forest habitats should aim to support the enhancement of a respectful attitude towards nature and culture, social welfare, political coexistence and tolerance, and aesthetic and historical values, among others (Higgs, 1997; Cairns, 2002; González-Espinosa *et al.*, 2007).

Forest restoration practices attempt to simulate ecological processes influential during secondary succession (Bradshaw, 1987, 2002). In each particular case study, the practices we have used follow different approaches to simulate mechanisms of succession. In the central Highlands of Chiapas and central Veracruz, a major concern has been the utilization of a large number of native tree species in order to restore the high local diversity. This approach has required the experimental study of germination requirements and response of seedlings and juveniles of key species to gradients of shade and temperature (Alvarez-Aquino *et al.*, 2004; Ramírez-Marcial *et al.*, 2005). Although restrictions on genetic variation imposed by secondary forest regeneration in highly diverse forests are recognized (Sezen *et al.*, 2005), this issue has not yet been a major concern in our Mexican studies (but see Rowden *et al.*, 2004). On the other hand, in the Chilean and Argentinian sites, interest has concentrated on threatened endemic conifer species. Species have been investigated singly, and emphasis has been given to conserving the genetic variation of highly threatened populations (Premoli *et al.*, 2001, 2003; Bekessy *et al.*, 2002; Allnutt *et al.*, 2003) and to identify particular environmental factors limiting recruitment and establishment (e.g. seed dispersal, water-table fluctuations).

Study Regions

The 33 sites within the six study regions encompass a considerable range of ecological conditions in areas close to the northern limits of the mountain cloud forests (Williams-Linera, 2002) down to the central distribution of the South American temperate forests in northern Chiloé Island (Armesto *et al.*, 1998). An envirogram plotting the values of mean annual rainfall and mean annual temperature for the 33 study sites in all regions indicates that – with the possible exception of very dry and cold sites – most of the combinations between 1000 and 2200 mm of annual rainfall and 8°C and 22°C of mean annual temperature have been included in our restoration essays (Fig. 15.1). The South American sites are

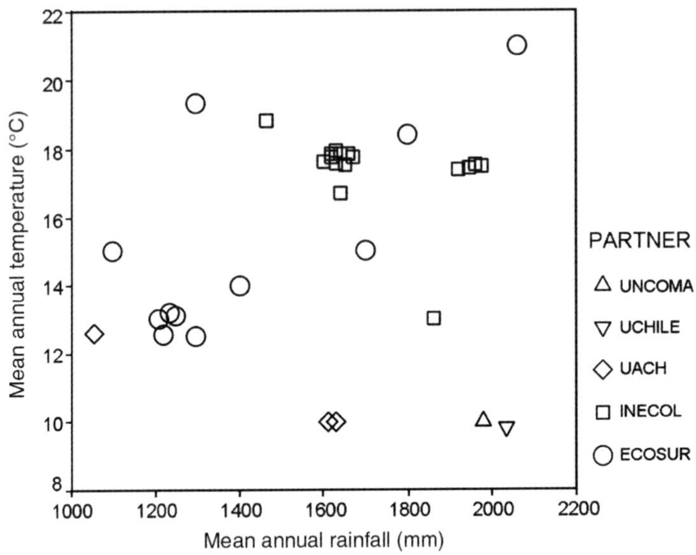

Fig. 15.1. Scatter plot of mean annual rainfall (MAR, mm year^{-1}) and mean annual temperature (MAT, °C) for the 32 field sites where the BIOCORES project partners have established forest restoration essays. UNCOMA, Universidad Nacional del Comahue (Bariloche, Argentina); UCHILE, Universidad de Chile (Chiloé Island, Chile); UACH, Universidad Austral de Chile (IX and X Regions, Chile); INECOL, Instituto de Ecología (central Veracruz, Mexico); ECOSUR, El Colegio de la Frontera Sur (central and northern Highlands of Chiapas, Mexico). See Tables 15.1 and 15.2 for additional details.

within a belt of cold temperatures (10–13°C) at relatively low elevations, and represent a set of low-energy sites (annual actual evapotranspiration, *AAET*, mostly lower than 600 mm year^{-1}; Table 15.1). Most of the sites in central Veracruz are located in habitats within a very narrow belt of mean annual temperatures and an annual rainfall range of *c.*1500 mm. Finally, the Chiapas sites include the widest range of probed environmental conditions, including the warmest and wettest sites among the whole set. Moreover, estimates of biologically useful energy (Rosenzweig, 1968) in Chiapas sites range from *c.*1000 up to >2100 mm year^{-1} and facilitate comparisons among all the study sites (Table 15.1).

Case Studies

Argentina (Site 1)

Restoration trials with Nothofagus pumilio *(Lenga)*

We established a long-term reciprocal transplant experiment to compare seedling growth (height, basal diameter and architecture) in 220 plants from two elevations: 1500 and 1000 m (Table 15.2). Seedlings were planted at both

Table 15.1. Location name, geographical coordinates, mean elevation (m), mean annual temperature (*MAT*, °C), mean annual rainfall (*MAR*, mm $year^{-1}$), annual actual evapotranspiration (*AAET*, mm $year^{-1}$; estimated with model by Turc (1954)), predominant landform and soil types/attributes of the study sites.

Partner	Site no.	Name of site	Latitude	Longitude	Elevation	*MAT*	*MAR*	*AAET*	Landform	Soil type/attributes
UNCOMA	1	Nahuel Huapi National Park	41° 08' S	71° 19' W	1000–1500	10.0	1980	577	Hillside	Andosols
UCHILE	2	Senda Darwin	45° 53' S	73° 40' W	40	9.8	2035	571	Flat area	Ñadi type (shallow flooded peat bog) on Fe/Si duripans
UACH	3	Villa Las Araucarias	38° 30' S	73° 16' W	630	12.6	1055	601	Flat area, hillside, steep slope	Shallow fertile derived from granite
UACH	4	Fundo Núñez	41° 26' S	73° 07' W	65	10.0	1610	566	Flat area	Ñadi type (shallow flooded peat bog) on Fe/Si duripans
UACH	5	Lahuen Ñadi	41° 26' S	73° 07' W	62	10.0	1630	566	Flat area	Same as in Site 4
INECOL	6	Rancho Viejo	19° 30' N	97° 00' W	1500	17.8	1650	884	Gentle and steep slopes	Acrisols, with high or very high organic matter content
INECOL	7	Xolostla 1	19° 32' N	96° 58' W	1450	17.8	1650	884	Same as above	Same as above
INECOL	8	Xolostla 2	19° 32' N	96° 58' W	1450	17.8	1650	884	Same as above	Same as above
INECOL	9	Xolostla 3	19° 32' N	96° 58' W	1450	17.8	1650	884	Same as above	Same as above
INECOL	10	Rancho Raúl	19° 31' N	96° 58' W		17.8	1650	884	Gentle slopes	Same as above
INECOL	11	Nicoletta	19° 32' N	96° 58' W		17.8	1650	884	Gentle slopes	Same as above
INECOL	12	Casazzas	19° 32' N	96° 58' W		17.8	1650	884	Gentle slopes	Same as above

Continued

Table 15.1. *Continued*

Partner	Site no.	Name of site	Latitude	Longitude	Elevation	*MAT*	*MAR*	*AAET*	Landform	Soil type/attributes
INECOL	13	El Cedro	19° 32’ N	96° 58’ W		17.8	1650	884	Gentle slopes	Same as above
INECOL	14	Capulines	19° 31’ N	96° 59’ W		17.8	1650	884	Gentle slopes	Same as above
INECOL	15	Rancho Olinca	19° 32’ N	96° 59’ W		17.8	1650	884	Gentle slopes	Same as above
INECOL	17	Las Cañadas 2	19° 11’ N	96° 59’ W	1340	17.4	1960	899	Same as above	Same as above
INECOL	18	Las Cañadas 3	19° 11’ N	96° 59’ W	1340	17.4	1960	899	Same as above	Same as above
INECOL	19	Las Cañadas 4	19° 11’ N	96° 59’ W	1340	17.4	1960	899	Same as above	Same as above
INECOL	20	La Martinica	19° 35’ N	96° 57’ W	1470	18.8	1460	896	Gentle slopes	Same as above
INECOL	21	Mesa de La Yerba	19° 33’ N	97° 01’ W	1875	13.0	1860	688	Gentle slopes	Same as above
ECOSUR	22	Rancho Merced-Bazom 1	16° 44’ N	92° 29’ W	2350	13.0	1250	642	Flat area, gentle slopes	Luvisol, rendzina
ECOSUR	23	Rancho Merced-Bazom 2	16° 44’ N	92° 29’ W	2300	13.0	1250	642	Flat area, steep slopes	Luvisol, rendzina
ECOSUR	24	Rancho Merced-Bazom 3	16° 44’ N	92° 29’ W	2350	13.0	1250	642	Flat areas	Luvisol, rendzina

ECOSUR	25	Rancho Merced-Bazom 4	16° 44' N	92° 29' W	2400	13.0	1280	642	Flat areas	Luvisol, rendzina
ECOSUR	26	Corazón de María	16° 41' N	92° 32' W	2380	15.0	1100	682	Flat areas	Vertic cambisol, gleysol
ECOSUR	27	Universidad Lindavista	17° 10' N	92° 54' W	1720	15.0	1700	763	Steep slope	Luvisol, lithosol, rendzina
ECOSUR	28	Estación Biológica Huitepec	16° 44' N	92° 41' W	2500	12.5	1300	631	Gentle and steep slopes	Cambisol, acrisol
ECOSUR	29	Moxviquil	16° 45' N	92° 38' W	2130	13.0	1200	635	Steep slopes	Lithosol
ECOSUR	30	Mitzitón	16° 40' N	92° 33' W	2400	14.0	1400	695	Flat area, gentle slopes	Luvisol, rendzina
ECOSUR	31	San Cayetano	16° 57' N	92° 46' W	1620	18.4	1800	933	Steep slopes	Phaeozem
ECOSUR	32	La Trinitaria	16° 08' N	92° 04' W	1590	19.3	1300	877	Flat	Vertisol
ECOSUR	33	Montebello	16° 04' N	91° 37' W	1520	21.0	2060	1,108	Flat and gentle slopes	Luvisol, rendzina

UNCOMA, Universidad Nacional del Comahue (Bariloche, Argentina); UCHILE, Universidad de Chile (Chiloé Island, Chile); UACH, Universidad Austral de Chile (IX and X Regions, Chile); INECOL, Instituto de Ecología (Xalapa, Mexico); ECOSUR, El Colegio de la Frontera Sur (central and northern Highlands of Chiapas, Mexico).

Table 15.2. Forest field restoration experiments conducted in 33 study sites. Months of duration up to May 2005 in each of the study sites in all regions, plot sizes, number of species included, initial condition being restored (AP, abandoned pasture; BF, bog field; DF, degraded forest; ESF, early secondary forest; FE, forest edges; FF, fallow field; FI, recent fire; MSF, mid-successional forest; OA, open area; OF, old-growth forest; SH, shrubland), plant performance variables measured (G, % germination; R, natural recruitment; S, % survival; H, stem height; B, basal stem diameter), conclusions on possible mechanisms or ecological interactions supposed to be implied, and identity of studied species. Numbers in parentheses within the same site (see Table 15.1) refer to particular studies in the site as indicated in the text.

Site	Months	No. plots	Plot size (m^2)	No. species	No. plants	Initial condition	Variables measured	Conclusions on mechanisms or interactions involved	Species included
1 (1)	12	1	120	1	220	DF	S, H, B, architecture	Adaptation to contrasting elevations	*Nothofagus pumilio*
1 (2)	12	1	50,000	1	3,000	DF, FI	R, S, H, B	Facilitation by shrubs and herbs after fire	*Austrocedrus chilensis*
2 (1)	33	4	< 2,500	1	392	BF, FI, OA, SH	S, H	Interference by *Sphagnum* sp. moss	*Pilgerodendron uviferum*
2 (2)	24	1	5,000	4	?	AP	R	Increased seed density by perching birds	*Amomyrtus luma*, *Berberis buxifolia*, *Berberis darwinii*, *Drimys winteri*
3 (1)	10	–	–	1	800 seeds	OA, SH	G	Seed stratification at 4°C	*Araucaria araucana*
3 (2)	24	2	10,000 & 5,000	1	200	OA, SH	R, S, H	Cyclical seed production	*Araucaria araucana*
3 (3)	8	4 (?)	?	1	?	BF, DF	S, H	Effects of root pruning and mycorrhization	*Araucaria araucana*
4 (1)	68	1	2,650	1	700	AP, BF	S, H, B	Negative effect of drainage on plant performance	*Fitzroya cupressoides*
5 (1)	32	1	?	1	1,076	AP, BF	S, H, B	Negative effect of drainage on plant performance	*Fitzroya cupressoides*

6–9, 16–19 (1)	70, 30	9	?	7	1,680	ESF, DF, FF	S, H, B	Identification of functional groups: light demanding, shade tolerant, intermediate	*Carpinus caroliniana*, *Fagus grandifolia* var. *mexicana*, *Juglans pyriformis*, *Liquidambar styraciflua*, *Podocarpus matudai*, *Quercus acutifolia*, *Symplocos coccinea*
10–15, 20, 21 (2)	18	6	?	6	?	AP, FE, FF	S, H, B	Competition with grasses, shading by established trees, underground herbivory by pocket gophers (*Thomomys* sp.)	*Fagus grandifolia* var. *mexicana*, *Quercus germana*, *Quercus xalapensis*, *Trema micrantha*, *Heliocarpus donnell-smithii*, *Rapanea myricoides*
22 (4)	54	6	100	9	486	DF, FE, FF	R, S, H, B	Facilitation by light-demanding species is not a requisite for enrichment of secondary stands	*Arbutus xalapensis*, *Clethra pachecoana*, *Cornus disciflora*, *Olmediella betschleriana*, *Prunus rhamnoides*, *Prunus serotina* ssp. *capuli*, *Quercus crassifolia*
23	60	2	2,500	11	?	Pine-dominated ESF	R, S, H, B	A pine-dominated canopy benefits broadleaved late successional species	*Acer negundo* ssp. *mexicana*, *Buddleja cordata*, *Liquidambar styraciflua*, *Magnolia sharpii*, *Photinia microcarpa*, *Prunus lundelliana*, *Quercus crispipilis*, *Quercus laurina*, *Quercus rugosa*, *Styrax magnus*, *Ternstroemia lineata* ssp. *chalicophyla*
24	48	6	2,100	5	1,470	DF	S, H, B	Different responses of oak species across the forest-edge–grassland gradient	*Quercus candicans*, *Quercus crassifolia*, *Quercus laurina*, *Quercus rugosa*, *Quercus segoviensis*

Continued

Table 15.2. *Continued*

Site	Months	No. plots	Plot size (m^2)	No. species	No. plants	Initial condition	Variables measured	Conclusions on mechanisms or interactions involved	Species included
25	72	8	Variable, mostly *c*.400	5	596	MSF, OF	S, H, G	Plant performance depends on relative conditions of the light environment in addition to point-level values	*Alnus acuminata* ssp. *arguta*, *Cornus excelsa*, *Liquidambar styraciflua*, *Persea americana*, *Quercus laurina*
26	120	8	400	4	1,656	OA, SH	R, S, H, B	Dominant shrub *Baccharis vaccinioides* functions as a nurse plant for trees	*Pinus ayacahuite*, *Pinus pseudostrobus* var. *apulcensis*, *Quercus crassifolia*, *Quercus rugosa*
27	48	4	1,800	16		OA	S, H, B	Facilitation is possible; a requisite for restoration of open areas	*Acer negundo* ssp. *mexicana*, *Clethra pachecoana*, *Cleyera theaeoides*, *Cornus disciflora*, *Liquidambar styraciflua*, *Magnolia sharpii*, *Oreopanax xalapensis*, *Quercus candicans*, *Pinus chiapensis*, *Podocarpus matudai*, *Prunus rhamnoides*, *Psychotria galeottiana*, *Zanthoxylum melanostictum*, *Styrax magnus*, *Symplocos limoncillo*
28	96(180)	6	1,000	7		AP/SH, MSF, OF	S, H, B	Conifers perform well in open areas; broad-leaved understorey tree species require facilitation by other species providing partial shade	*Abies guatemalensis*, *Oreopanax xalapensis*, *Pinus ayacahuite*, *Pinus pseudostrobus* var. *apulcensis*, *Rhamnus sharpii*, *Ternstroemia lineata* ssp. *chalicophyla*

29	60	4	1,500	25		OA	R, S, H, B	Facilitation is possible; required for restoration of open areas	*Acer negundo* ssp. *mexicana*, *Alnus acuminata* ssp. *arguta*, *Arbutus xalapensis*, *Buddleja cordata*, *Chiranthodendron pentadactylon*, *Clethra pachecoana*, *Cleyera theaeoides*, *Cornus disciflora*, *Cornus excelsa*, *Ehretia thinifolia*, *Ilex vomitoria*, *Liquidambar styraciflua*, *Olmediella betschleriana*, *Persea americana*, *Pinus pseudostrobus* var. *apulcensis*, *Psychotria galeottiana*, *Prunus brachybotria*, *Prunus rhamnoides*, *Prunus serotina* ssp. *capuli*, *Quercus crispipilis*, *Quercus rugosa*, *Rapanea juergensenii*, *Rhamnus sharpii*, *Styrax magnus*, *Zanthoxylum melanostictum*
30 (4)	54	4	100	9	324	FE, FF, DIF	R, S, H, B	Facilitation by light-demanding species is not a requisite for enrichment of secondary stands	*Arbutus xalapensis*, *Clethra pachecoana*, *Cornus disciflora*, *Olmediella betschleriana*, *Prunus rhamnoides*, *Prunus serotina* ssp. *capuli*, *Quercus crassifolia*, *Quercus laurina*, *Quercus rugosa*
30 (6)	22	21	100	10	1,656	AP, SH	S, H, B	Above- and below-ground competition with grasses; different between grasslands and shrublands. Facilitation of grass cover on seedlings observed in a few cases	*Alnus acuminata* ssp. *arguta*, *Garrya laurifolia*, *Nyssa sylvatica*, *Pinus ayacahuite*, *Pinus pseudostrobus* var. *apulcensis*, *Pinus serotina* ssp. *capuli*, *Quercus crispipilis*, *Rapanea juergensenii*, *Styrax magnus*, *Ternstroemia lineata* ssp. *chalicophyla*

Continued

Table 15.2. *Continued*

Site	Months	No. plots	Plot size (m^2)	No. species	No. plants	Initial condition	Variables measured	Conclusions on mechanisms or interactions involved	Species included
31	12	1	500	16	200	OA	S, H, B	Facilitation is possible; a requisite for restoration of open areas	*Cornus excelsa*, *Fraxinus uhdei*, *Juniperus gamboana*, *Olmediella betschleriana*, *Pinus ayacahuite*, *Pinus montezumae*, *Prunus serotina* ssp. *capuli*, *Quercus crassifolia*, *Quercus crispipilis*, *Quercus rugosa*, *Quercus sapotifolia*, *Quercus segoviensis*, *Quercus* sp., *Randia aculeata*, *Turpinia tricornuta*
32	9	1	400	16	1,032	OA	S, H, B	Facilitation is possible and a requisite for restoration of open areas	Same as above
33	21	8	2,500	16	3,200	FI, ESF, OA, SH	R, S, H, B	Facilitation is possible and a requisite for restoration of open areas	*Ilex vomitoria*, *Myrica cerifera*, *Olmediella betschleriana*, *Oreopanax xalapensis*, *Prunus brachybotria*, *Prunus lundelliana*, *Psychotria galeottiana*, *Quercus sapotifolia*, *Quercus* sp., *Randia aculeata*, *Rapanea myricoides*, *Rhamnus capraeifolia*, *Styrax magnus*, *Synardisia venosa*, *Turpinia tricornuta*

elevations in Chall-Huaco Valley, Nahuel Huapi National Park in May 2005. Previous studies indicate that individuals from subalpine contrasting elevations may be genetically different due to reproductive barriers to gene flow exerted by phenological differences (Premoli, 2003). Furthermore, greenhouse experiments have shown heritable variation in ecophysiological traits along with morphological and phenological differences associated with elevation (Premoli, 2004). Results of the reciprocal transplant experiments will allow the testing of adaptive differences between plants from different provenances that will guide restoration trials.

Restoration trial with Austrocedrus chilensis *(Ciprés de la Cordillera)*
A restoration essay was established on *c.*5 ha of hillside originally covered by monospecific *Austrocedrus chilensis* forest near the Nahuel Huapi National Park. The entire area was burnt four years before the start of the study and then illegally logged. *Austrocedrus* is affected by fire and herbivore browsing, and early regeneration stages are highly dependent on facilitating shrubs (Kitzberger *et al.*, 2000; Rovere *et al.*, 2005). Various interest groups are participating in the study including: (i) the private sector, represented by a company that provides the study site; (ii) the Provincial Government, represented by Servicio Forestal de la Provincia de Río Negro (Río Negro Province Forest Service), which supplies plants and provides logistic support; and (iii) Universidad Nacional del Comahue, responsible for designing and monitoring the study, as well as for organizing activities aimed at environmental education in the local community. Vegetation and forest floor cover, and natural regeneration of *A. chilensis* were initially assessed. We planted 3000 trees during winter 2004 (Table 15.2). Preliminary results indicate that shrub cover after fire is high (54%). Natural regeneration of *A. chilensis* has been very low (less than one sapling per ha), but preliminary results indicate that survival and establishment are facilitated by shrubs and herbs.

Chile: Northern Chilóe Island (Site 2)

Long-term restoration of Pilgerodendron uviferum *(Ciprés de las Guaitecas)*
The experiment was established in August 2002, in an open area that was subjected to a fire and became wet shrubland afterwards (Table 15.2). Little regeneration and slow succession are currently observed. Seasonal flooding caused by logging and burning of the forest favours invasion by *Sphagnum*. The study assesses the effects of the substrate of *Sphagnum* moss on growth and survival of *Pilgerodendron uviferum* in areas disturbed by human impact. The experiment includes two sites with four plots in each within a multifactorial design; plants were spaced at 1 m distance ($N = 49$ in each plot). The plants were obtained from cuttings and grown for two years in the nursery at Senda Darwin Biological Station. Plants of different origins and known gender were randomly allocated among plots. The sites were with and without *Sphagnum*. Growth of *P. uviferum* was similar during the first years of the study, yet plant responses were significantly different

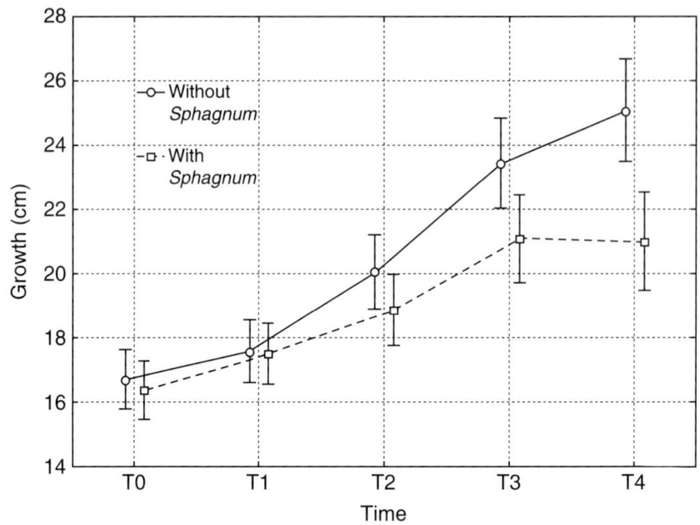

Fig. 15.2. Growth response of *Pilgerodendron uviferum* in plots with and without *Sphagnum* sp. moss at Senda Darwin Biological Station, northern Chiloé Island, Chile. T0 is August 2002; T4 is February 2005.

by early 2005: saplings in plots without *Sphagnum* grew more than those in plots with *Sphagnum* (Fig. 15.2). A repeated measures ANOVA on $\log_{10}$ growth showed significant interaction between substrate and time ($P < 0.001$). However, per cent survival was not significantly different in plots with *Sphagnum* treatments. The preliminary results suggest that *Sphagnum* cover seems to have a negative effect on growth of *P. uviferum*; so far survival seems to be unrelated to substrate.

Effects of coarse woody debris and bird perches on tree recruitment in artificial prairies

A number of studies in the temperate rainforest of Chiloé Island show that many trees, shrubs and vines display a bird-dispersal syndrome. It is also known that seed rain is much lower in shrublands and prairies than in forest fragments. This study aims to assess: (i) the effects of different substrates on the establishment of woody species in anthropogenic prairies; and (ii) the effect of artificial perches that could be used by birds in facilitating the establishment of bird-dispersed plants. Different substrates (logs, woody detritus of *Drimys winteri* (Canelo) and *Nothofagus dombeyi* (Coigüe común)) and prairie soil with or without perches were randomly distributed in artificial prairies at Senda Darwin Biological Station ($N = 180$). Seed deposition has only been observed on woody detritus and log substrates. To evaluate the function of perches, we sampled seed rain in traps with and without perches in the same artificial prairies. After four months, we found seeds in all traps with perches ($N = 15$) and only in eight traps without perches. The species found were *D. winteri*, *Amomyrtus luma* (Luma), *Berberis buxifolia* and *Berberis darwinii* (all dispersed by birds; Table 15.2). Number of seeds per trap was different between perch and non-perch treatments ($P < 0.0001$). These data indicate that the presence of perches may increment the seed rain of bird-dispersed woody species in prairies of Chiloé Island.

Chile: Region IX (Site 3)

Activities have been conducted in two sites of the Cordillera de la Costa (Coastal Range): Villa Las Araucarias and Nahuelbuta National Park (Table 15.2). Tree cores ($N = 200$) and chunks ($N = 15$) collected at Villa Las Araucarias are being cross-dated to date the occurrence of fires and to generate a fire chronology of *Araucaria araucana*. However, cross-dating has been troublesome because the trees are in flat areas and fires are highly frequent. Fire scars are produced on the trunk perimeter, and not in a particular area of the stem as in hilly areas. Only a few samples from Nahuelbuta National Park have been cross-dated due to the difficulty in differentiating the tree rings. Additional samples are currently being collected to obtain an improved fire chronology.

In March 2004 we collected seeds of *A. araucana* to produce plants for restoration and research activities. In October 2004 the seeds were sown using four different germination treatments (four replicates of 50 seeds each). High germination was observed in control seeds; the seeds were stored at 4°C from March through October, which could cause their stratification and therefore reduce the effect of the pre-germination treatments. Also, no control was implemented on treatment location inside the greenhouse; germination of the untreated seeds could be enhanced in the south side.

In 2003 two plantations of *A. araucana* from seeds collected at Villa Las Araucarias were established in two permanent plots of 1.0 and 0.5 ha (labelled as plots 1 and 2, $N = 100$ per plot). Survival and growth of seedlings and saplings were assessed in 30 and 20 subplots distributed within the two plots. Mortality of *A. araucana* plants in April 2005 was higher in plot 2 (25%) than in plot 1 (12%). These trends in mortality are similar to those recorded in March 2004 (17% and 20%, respectively). This variation in mortality rate between sites could be explained by differences in the site and canopy cover. Plot 1 is on a steep slope and has some canopy protection from remaining trees; plot 2 is a flat, open site. Scarce natural regeneration has been observed, most probably due to an extremely low production of seeds during the last two years in Nahuelbuta National Park and null in Villa Las Araucarias. Given the biannual seeding cycles of *A. araucana*, we anticipate higher seed production in 2006 and 2007.

In addition, in 2004 new plantations were established in sites with different levels of forest cover: Site A, a gap within a plantation of the exotic *Pinus radiata*; Site B, under the canopy of *P. radiata* trees; Site C, with side-protection by *N. dombeyi* and *A. araucana*; and Site D, a small depression covered by peat bog. The lowest and highest mortalities were obtained in sites D (4%) and A (8%).

These low mortality values are considered as quite favourable, given the extremely harsh climate of the study area. Currently, we are planning to improve the survival rates of *A. araucana* seedlings by applying cultural treatments such as root pruning, mycorrhization and *in situ* production of plants.

Chile: Region X (Sites 4 and 5)

Long-term restoration of Fitzroya cupressoides *(Alerce)*

Ecological restoration works have been conducted in two study areas in the X Region. The first plantation of *Fitzroya cupressoides* was at the property of Mr Alfredo Núñez (hereafter Fundo Núñez) in 1998 (Table 15.2). Plants were produced from seeds and cuttings collected in the local area. Monitoring activities such as assessments of survival and growth in height and diameter have been undertaken each year. In September 2002, another plantation was established at the Lahuen Ñadi Park with cuttings from a local population. Until April 2005, plant mortality at Fundo Núñez was 12%. Mean increase of stem height of *F. cupressoides* at Fundo Núñez has been 10.3 cm $year^{-1}$ between 1999 and 2005. Yet, in well-drained areas within the plot, mean growth rate has been 31.8 cm $year^{-1}$. At Lahuen Ñadi, mean growth has been 4.4 cm $year^{-1}$. These marked differences may be explained by the drainage conditions where the plants are established, as most microsites at Lahuen Ñadi are poorly drained. A total of 160 seed traps were installed in June 2003 at Fundo Núñez to collect seeds of *F. cupressoides* as a function of wind direction. Seed production is highly variable among years: a total of 29,477 seeds were collected in 2003, but only 217 and 257 seeds in 2004 and 2005, most of them moved by winds with N–S or S–N orientation. To analyse the effect of water-table fluctuations on the establishment and growth of *F. cupressoides* plants, several piezometers have been installed (22.6 devices ha^{-1}) at Fundo Nuñez. Results have revealed that variations in plant growth have been associated with fluctuations in the water-table level.

Mexico: Central Veracruz (Xalapa; Sites 6–21)

Restoration of tropical montane cloud forests

A major goal of restoration activities in central Veracruz has been the maintenance of regional diversity. Since 1998 a number of tree restoration plots have been established and monitored to determine the potential of ecological restoration with selected native tree species and to define criteria for matching these species with particular microhabitat conditions. The native tree species used were *Carpinus caroliniana, Fagus grandifolia* var. *mexicana, Juglans pyriformis, Liquidambar styraciflua, Podocarpus matudai, Quercus acutifolia* and *Symplocos coccinea* (Table 15.2). The restoration experiments were conducted in three forest fragment interiors, three post-agriculture fallow fields adjacent to the forest fragments, and three early secondary forest stands (*acahuales*). Results were compared with on-farm plantations established by private landowners. Plant performance was evaluated as survival, and increment in stem height and basal stem diameter. Other variables monitored were natural recruitment, soil pH, organic matter and compaction. Responses were integrated using functional groups (light-demanding, shade-tolerant and intermediate). Initial age and seedling height had a significant effect on survival, but not on height or diameter increment across all species and sites. Overall survival was highest in early secondary forests (70%), followed by forest interior (42%) and fallow field (36%). Maximum height was recorded outside the forest. Average stem height was greater in the adjacent agricultural fields (4.6 m) and in early secondary forests (3.6 m) than in the forest fragment interiors (0.62 m). Annual diameter increment rate was lower in forest interior (0.22 cm year^{-1} in 2000, and 0.04 cm year^{-1} in 2004) than in adjacent field (1.04 and 0.64 cm year^{-1}) and in old-field sites (0.66 and 0.50 cm year^{-1}). *Juglans*, *Podocarpus* and *Quercus* exhibited the greatest survival (62–80%), but intermediate relative growth rates in stem height (26–57 cm year^{-1}; Fig. 15.3); *Carpinus* and *Liquidambar* showed intermediate survival (50–54%), but high growth increments (45–96 cm year^{-1}); and *Fagus* and *Symplocos* displayed low survival (18–20%) and low height increments (13–29 cm year^{-1}). We conclude that performance of different tree species depends on specific level of disturbance exhibited at each site, suggesting the importance of accurate species–site matching to obtain optimum rates of survival and growth in particular scenarios. *Juglans* and *Quercus* have the potential to be used in the rehabilitation of degraded and disturbed areas, respectively; *Podocarpus* can be used in plantation enrichment; *Liquidambar* and *Carpinus* may be used to expand the extent of cloud forest; *Fagus* and *Symplocos* can survive and grow in forests other than those in which they are naturally present.

Forest restoration in abandoned pastures

Land clearing to establish pastures with non-native grasses and urban/suburban development has been a common practice in central Veracruz over the last 50 years. Yet opportunities to restore montane cloud forests from abandoned pastures exist as land use changes due to low productivity. We established six restoration plantations by planting seedlings of three primary

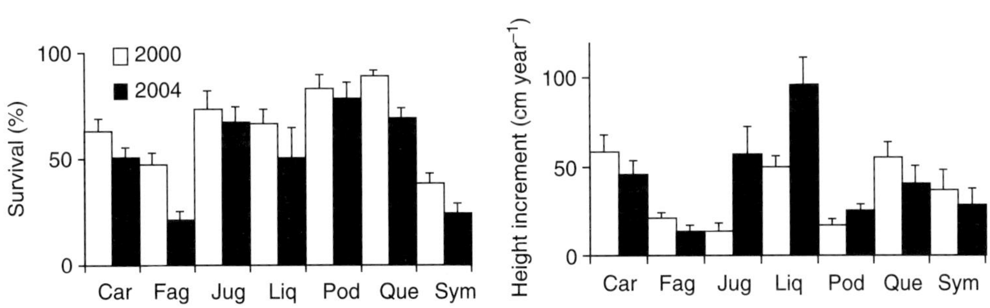

Fig. 15.3. Per cent survival between 2000 and 2004 and growth rates in stem height (cm year^{-1}) at 2000 and 2004 for native tree species used in restoration field experiments in central Veracruz, Mexico. Car, *Carpinus caroliniana*; Fag, *Fagus grandifolia* var. *mexicana*; Jug, *Juglans pyriformis*; Liq, *Liquidambar styraciflua*; Pod, *Podocarpus matudai*; Que, *Quercus acutifolia*; Sym, *Symplocos coccinea*.

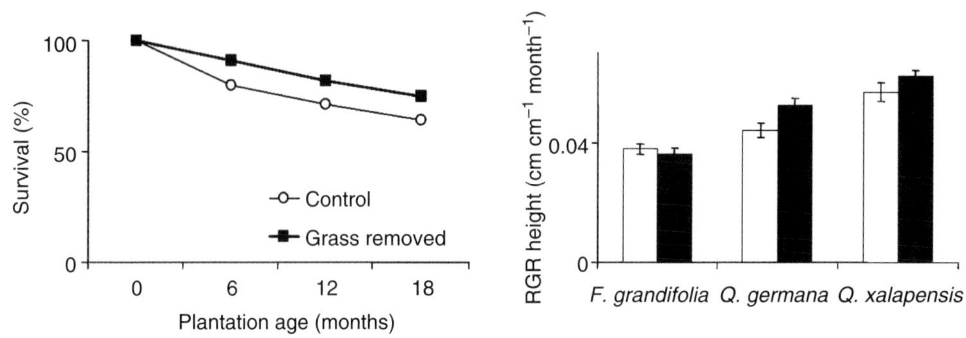

Fig. 15.4. Per cent survival of seedlings with and without surrounding grass cover and relative growth rates in stem height (cm cm^{-1} month^{-1}) of three primary tree species (*Fagus grandifolia* var. *mexicana*, *Quercus germana* and *Quercus xalapensis*) used in restoration experiments in abandoned pastures in central Veracruz, Mexico.

tree species (*Fagus grandifolia* var. *mexicana*, *Quercus germana* and *Q. xalapensis*) in three recently (<1 year) and three long-abandoned pastures (12–17 years); the seedlings were planted at 0, 10 and 40–50 m from the forest border. A treatment of removal of herbaceous vegetation was included. Sapling survival was higher when grasses (mostly the stoloniferous exotic *Pennisetum clandestinum*) were removed than in controls. All species attained larger diameter and height growth in plots with grass removed in comparison to controls (Fig. 15.4.). Survival of *F. grandifolia* and *Q. germana* was higher in older fields, while *Q. xalapensis* displayed a similar survival in recent and long-abandoned pastures, but higher mortality close to the forest border.

Mexico: Central and Northern Highlands of Chiapas (Sites 22–33)

Functional groups of native tree species

Matching the tolerance of native tree species with environmental gradients that operate at the microsite level is required for successful forest restoration (Ramírez-Marcial *et al.*, 2005). Conditions occurring in restoration sites represent environmental filters that define the assembly rules of a plant community (Temperton *et al.*, 2004). Forest restoration should be based on the grouping of sets of species into functional groups whose life history attributes and population dynamics are sufficiently consistent to guide restoration actions at the plot, landscape and regional spatial scales in high diversity areas. Therefore, we have studied the main germination requirements of a large number of species (140 taxa; Ramírez-Marcial *et al.*, 2003, 2005) while producing seedlings to be used in field experiments on plantation enrichment. Some of the species studied have been classified as endangered taxa in national or international lists (Oldfield *et al.*, 1998; SEMARNAT, 2002). The tolerance to partial shade (or intolerance to open conditions) of more than 40 species has been evaluated under common nursery conditions; boxes covered with black net mesh of different openings allowing variable light incidence on the experimental plants have been used (Fig. 15.5A).

Nursery experiments on seedling response to light and water gradients

Understanding the responses of key species to environmental gradients is a crucial piece of knowledge to model and guide practices aimed at restoration of forest communities. We conducted a nursery experiment to elucidate the specific responses of seedlings of three *Pinus* spp., three *Quercus* spp. and six other understorey broadleaved tree species in a common garden: *Alnus acuminata, Cornus disciflora, Garrya laurifolia, Olmediella betschleriana, Prunus lundelliana* and *Styrax magnus*. Selection of species was based on our advances in a classification scheme of tree seedling functional groups, which considers attributes pertaining to their regeneration niche as well as to availability of seeds. The experiment started in March 2003 and included three conditions (25, 75 and 100%) of photosynthetic active radiation (PAR) and three soil moisture levels: field capacity (24%), intermediate (18%) and permanent wilting point (13%). A number of 12 replicates (20 for conifers and oaks) for each treatment combination and species were established. A total of 2064 seedlings were planted in independent plots within a common garden of *c.*500 m^2 located at the ECOSUR facilities in San Cristóbal de Las Casas, Chiapas. The experiment ended at the start of the rainy season (end of May 2003), but some lower levels of direct sunlight (8, 15 and 25) were assessed with *Pinus* spp. and *Quercus* spp. in March–May 2004. We measured seedling survival every 2 weeks, and stem height, basal stem diameter, number of leaves, number of recently emerged leaves and leaf size of the three largest leaves. At the beginning of the rainy season, we harvested four out of ten seedlings to analyse patterns of resource allocation to different plant organs. We left six seedlings in the nursery to provide information on long-term responses to radiation (water cannot be controlled during the rainy season). Seedlings of five out of six species (but not *A. acuminata*) subjected to drier and more open conditions had higher mortality than those with heavier shade and wetter soil. Stem height, basal diameter and number of leaves were affected by shade intensity. Light conditions had the highest effect on the distribution of dry biomass in all tree species.

Underground herbivory and seedling establishment

Establishment of enrichment plantings may be affected by herbivores and root feeders. Root damage by larvae of *Phyllophaga* spp. (Coleptera: Melolonthidae) has been observed to affect seedling survival and establishment. We evaluated below-ground herbivory by two *Phyllophaga* species (*P. obsoleta* and *P. tumulosa*) on seedlings of ten native tree species (*Arbutus xalapensis, Litsea glaucescens, Myrica cerifera, Nyssa sylvatica, Persea americana, Quercus crassifolia, Quercus skutchii, S. magnus, Synardisia venosa* and *Ternstroemia lineata* ssp. *chalicophyla*). A total of 550 plants were included in the experiment and 300 seedlings were inoculated with one larva of each *Phyllophaga* species. Plants were maintained under nursery conditions for two months. Plants were harvested and oven-dried to obtain biomass of aerial and below-ground plant organs. The results indicate that herbivory of roots was significantly different for eight of the ten studied species (except *P. americana* and *S. venosa*) and damage intensity by *P. obsoleta* was higher in five tree species.

Restoration of forest edges (Sites 22 and 30)

Forest clearing in Chiapas is mostly related to establishment of slash-and-burn *milpa* agriculture (maize–beans–squash). The system may last for 2–4 years, but the use of fertilizer and herbicides may allow for permanent agriculture (González-Espinosa *et al.*, 1991, 2006; García-Barrios and González-Espinosa, 2004). Secondary forests usually develop with a variable dominance of *Pinus* spp. due to selective logging of *Quercus* spp. and other broadleaved species that are preferentially used for firewood; on the other hand, *Pinus* spp. are allowed to grow until they attain adequate sizes for timber extraction and can reproduce several times. Forest restoration opportunities arise when fallow fields, pastures and early secondary forests are left for succession to progress. In 1998 we started a study with experimental clearings (ten plots, 10 m × 10 m each; Table 15.2) at the border of forests with variable dominance by *Pinus* spp., subsequently followed by two agricultural cycles, fallow field and enrichment of shrublands. After 54 months of transplanting the saplings, the nine broadleaf tree species that were introduced (mostly old-growth and intermediate successional species; Table 15.2) show an average survival of 73% (590 alive plants out of 810). The greatest relative growth rate in height and diameter has been observed in *Arbutus*, *Clethra*, *Cornus* and *Quercus laurina*. These preliminary results indicate that enrichment of forest edges in a forested landscape does not seem to require a previous facilitation stage with light-demanding species.

Restoration essays in a variety of field conditions (Sites 23–29 & 31–33)

The central and northern Highlands of Chiapas include a wide variety of environmental conditions and the distribution of many native tree species samples these conditions extensively. To probe the involved gradients we have been keen to take advantage of offerings from interested groups to establish

Fig. 15.5. Relationship between relative growth rates in stem height (RGR_{height}) under partial shade (25% of direct light) and at full direct light in open areas for seedlings of 42 native tree species of the Highlands of Chiapas (Mexico) under nursery or common garden conditions (A), and for 24 tree species under field conditions (B). Abigua, *Abies guatemalensis*; Acapen, *Acacia pennatula*; Alnacu, *Alnus acuminata* ssp. *arguta*; Arbxal, *Arbutus xalapensis*; Budcor, *Buddleja cordata*; Chipen, *Chiranthodendron pentadactylon*; Clepac, *Clethra pachecoana*; Clethe, *Cleyera theaeoides*; Cordis, *Cornus disciflora*; Garlau, *Garrya laurifolia*; Ilevom, *Ilex vomitoria*; Liqsty, *Liquidambar styraciflua*; Magsha, *Magnolia sharpii*; Myrcer, *Myrica cerifera*; Olmbet, *Olmediella betschleriana*; Orexal, *Oreopanax xalapensis*; Perame, *Persea americana*; Pinaya, *Pinus ayacahuite*; Pinpse, *Pinus pseudostrobus* ssp. *apulcensis*; Pintec, *Pinus tecunumanii*; Plamex, *Platanus mexicana*; Prulun, *Prunus lundelliana*; Prurha, *Prunus rhamnoides*; Pruser, *Prunus serotina* ssp. *capuli*; Psygal, *Psychotria galeottiana*; Queaca, *Quercus acatenangensis*; Quecan, *Quercus candicans*; Quecra, *Quercus crassifolia*; Quecri, *Quercus crispipilis*; Quelau, *Quercus laurina*; Querug, *Quercus rugosa*; Quesap, *Quercus sapotifolia*; Queseg, *Quercus segoviensis*; Quesku, *Quercus skutchii*; Quesp., *Quercus* sp.1; Ranacu, *Randia aculeata*; Rapjue, *Rapanea juergensenii*; Rapmyr, *Rapanea myricoides*; Rhacap, *Rhamnus capraeifolia* var. *grandifolia*; Rhasha, *Rhamnus sharpii*; Simlim, *Symplocos limoncillo*; Stymag, *Styrax magnus*; Synven, *Synardisia venosa*; Terlin, *Ternstroemia lineata* ssp. *chalicophila*; Terooc, *Ternstroemia oocarpa*; and Zanmel, *Zanthoxylum melanostictum*.

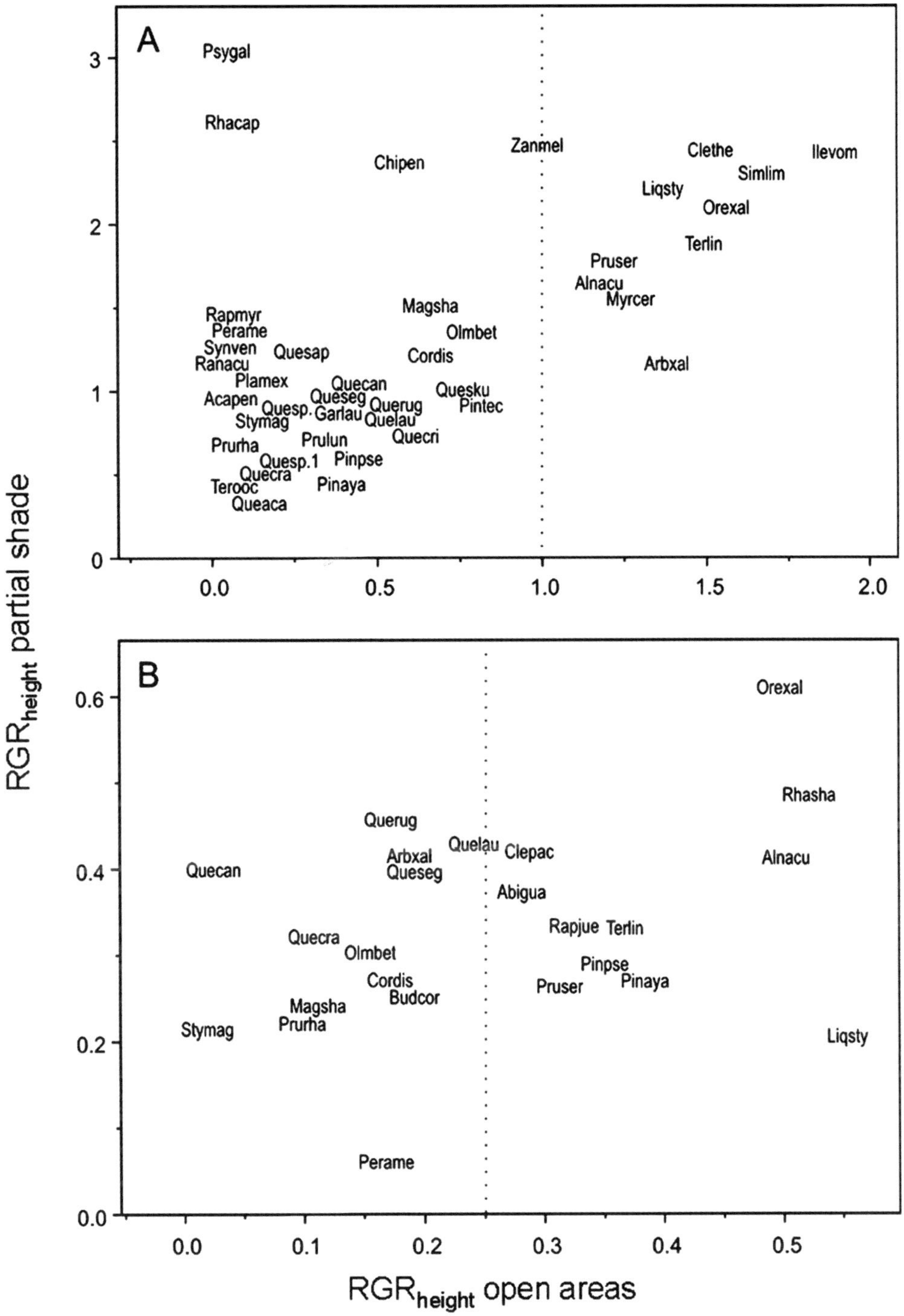
A
Psygal
Rhacap
Zanmel
Chipen
Clethe
Ilevom
Simlim
Liqsty
Orexal
Terlin
Pruser
Alnacu
Myrcer
Magsha
Rapmyr
Perame
Olmbet
Synven
Quesap
Cordis
Ranacu
Arbxal
Plamex
Quecan
Quesku
Acapen
Queseg
Querug
Pintec
Quesp.
Garlau
Quelau
Stymag
Prurha
Prulun
Quecri
Quesp.1
Pinpse
Quecra
Terooc
Pinaya
Queaca
B
Orexal
Rhasha
Querug
Quelau
Clepac
Arbxal
Alnacu
Quecan
Queseg
Abigua
Rapjue
Terlin
Quecra
Olmbet
Pinpse
Cordis
Pruser
Pinaya
Budcor
Magsha
Stymag
Prurha
Liqsty
Perame
RGR_{height} partial shade
RGR_{height} open areas
3
2
1
0
0.0
0.5
1.0
1.5
2.0
0.6
0.4
0.2
0.0
0.1
0.2
0.3
0.4
0.5

restoration essays in their lands. Therefore, a number of restoration plantations have been established and monitored (survival and growth of stem height and basal diameter) using a set of 60 species including conifers, *Quercus* spp. and other broadleaved species that can be considered as early, intermediate or late successional (Sites 26–33 in Table 15.2). The essays have been started at different times (mostly 1–3 years ago, and one study has been monitored for 15 years; Quintana-Ascencio *et al.*, 2004). Although the species were introduced in sites with different disturbance regimes, it is clear that survival after 3 years may be 30–40% in open areas, but >90% under induced pine-dominated canopies. Some species can be distinguished for their growth potential under a variety of environments (e.g. *A. acuminata, Buddleja cordata, Chiranthodendron pentadactylon, Pinus* spp., *L. styraciflua, O. betschleriana*), and it is possible to propose some species groups. For example, *Oreopanax xalapensis, Rhamnus sharpii* and *A. acuminata* are easy to propagate by seed and can establish well in open areas, in early successional forests and under *Baccharis vaccinioides* shrubs (a typical nurse plant; Ramírez-Marcial *et al.*, 1996). *Pinus* spp., *Buddleja* spp., *L. styraciflua* and *Prunus serotina* ssp. *capuli* are shade-intolerant species that can establish easily in open areas; their high growth rates induce facilitation processes for late successional species that require a previous canopy such as *Magnolia sharpii, P. americana, P. lundelliana, Prunus rhamnoides, S. magnus,* and others (Fig. 15.5). A first detailed account of the invertebrate soil fauna has been obtained in the eight restoration plots established in Site 33, which were subjected to severe fire disturbance in 1998. Abundance and diversity of the soil fauna showed marked seasonality and it includes 187 morphological species belonging to 58 families and 20 zoological orders within six classes and three phyla.

Interactions between tree seedlings and herbaceous cover (Site 30)

Forest restoration in abandoned pastures could be accelerated or arrested if tree seedling establishment is affected by competition from the surrounding herbaceous cover. Seedlings of ten native tree species (Site 30, Table 15.2) were used in experiments. In July 2003 we established 21 experimental plots (10 m × 10 m) in three grassland and four shrubland sites. Each plot included 6–10 seedlings of each species (a total of 1656 plants). In each grassland or shrubland, one plot served as control, a second one was subjected to a treatment of aerial herb removal (clipping herbs within a radius of 30 cm around each seedling), and a third plot was subjected to total herb removal (both above and underground tissues killed with herbicide application). After 22 months, the preliminary results suggest that grasses may have different competitive effects on seedlings, both above and below ground, in grasslands and shrublands.

Discussion

This integrated and synoptic report pinpoints some valuable experiences that can be considered as lessons learned, and can contribute to the development of best practice in forest restoration in our study sites and other similar areas. The large range of environmental conditions included in these studies is matched by a wide array of socio-economic factors. Their joint consideration may lead to broad guidelines, criteria and indicators for ecological restoration that may represent many of the conditions prevailing in other developing regions. Current pitfalls can be identified and used to define a minimum set of elements to be considered in a protocol for a more widespread assessment of restoration experiences, both scientific and practical.

Ecological issues and forest restoration

Forest restoration aims to reproduce and enhance ecological processes that drive community development through time. Ecological models incorporating general principles that drive the organization of ecosystem diversity during succession are particularly relevant in this context (Bradshaw, 1987; Ramírez-Marcial *et al.*, 2005; Ruiz-Jaén and Aide, 2005). So far most of our studies have concentrated on assessment of plant performance (mostly at the seedling stage, rarely with saplings) in response to either one or many variables. As an example of this latter case we can mention treatments with and without grasses (or moss), which most probably trigger a number of non-specified interacting variables such as: (i) competition for nutrients, water and light; (ii) modification of temperature and humidity gradients in the immediate neighbourhood of the target plants; (iii) differential effects of the biota below ground, and so on. In the end, we may still be presented with major problems in explaining the results obtained and therefore in defining the best restoration practice for a particular site, i.e. conducting actual restoration. These experiences highlight the need for more inclusive research models about the most crucial processes involved. There is a lack of models that can be used to explore the assembly rules involved in the stratification of forest communities and shade and (or) drought tolerance along environmental gradients at landscape and regional spatial scales (Hobbs, 2002). Some promising models may be those aiming to explain broad macroecological patterns of diversity based on life history and population attributes (e.g. Huston and Smith, 1987; Storch *et al.*, 2005).

Successful forest restoration depends on the appropriate matching of environment with species tolerance. It is not coincidental that all of our research teams began with trying to understand the germination or vegetative propagation requirements of individual species or groups of species. This has been pursued in the first place to secure provision of adequate experimental material, but also to define protocols for widespread application of propagation techniques. Yet, unless several environmental variables are studied in a factorial way (e.g. light and water availability), our experiences with common garden or nursery experiments indicate that only preliminary and relative conclusions can be reached in comparison to field experiments. For example, relative growth rates of a considerable number of tree species were 4–5 times higher in the nursery than under a variety of field conditions in the Highlands of Chiapas (Ramírez-Marcial *et al.*, 2005), suggesting also the need for better experimental control in experiments under actual canopies (Fig. 15.5).

A neglected issue that may have implications for forest restoration practices results from traditional forest use patterns: low intensity but long duration human disturbance associated with selective and scattered logging of small trees, or harvesting of branches and resprouting stems (for firewood or non-commercial timber use; Vetaas, 1997; Ramírez-Marcial *et al.*, 2001, 2005; Barrón-Sevilla, 2002; Martorell and Peters, 2005). This may create environmental gradients inside the forest that do not match either those associated with disturbance patterns in old-growth stands (either forest gaps or sunflecks) or those involving widespread forest clearing (Méndez-Dewar, 2000). This little-studied aspect of forest heterogeneity may influence individual plant responses in restoration practices aimed at species enrichment of degraded stands.

Socio-economic issues and forest restoration

Until recently the strategies followed for conservation and sustainable use of forests, and also the role of forest restoration, have differed among the study regions. The South American cases exemplify a conservation strategy largely dependent on the availability of national parks and/or biological reserves for the conservation of particular species (Table 15.2) vis-à-vis native forest destruction driven by logging companies, establishment of industrial plantations with exotic species and activities of small farmers (the frontier model *sensu* Rudel and Roper, 1997). It would seem that coexistence between biodiversity and increased demand for agricultural products is being solved mostly through adoption of the model that couples land-sparing with high-yield farming (Green *et al.*, 2005).

In contrast, conditions prevailing in Chiapas point towards different avenues for development and conservation. Forest loss can be mostly explained by the so-called immiserization model (Rudel and Roper, 1997), which involves increasing populations of poor peasants who have scarce economic opportunities besides clearing additional land for agriculture. Yet this does not mean that the frontier model did not play a major role in the region in the 1970s and early 1980s, particularly in lowland areas (Montoya-Gómez *et al.*, 2003). In addition, many indigenous Mayan communities or their organizations have a strong interest in increasing their political self-determination over their relatively densely populated territories (Burguete-Cal y Mayor, 1999; Cartagena-Licona *et al.*, 2005). Conservation is not seen by these communities as an alternative viable land use if no short-term economic benefits are envisaged to support local development initiatives. Under this predominant scenario, which may continue for some decades into the future, forest restoration could play a crucial role in forest conservation and sustainable use, as it could contribute to wildlife-friendly farming in high diversity and complex forest landscapes (Bray and Merino, 2004; Holder, 2004; Bray *et al.*, 2005; Green *et al.*, 2005). Sustainable use and conservation of forested landscapes will depend, therefore, on coalescing scattered forested areas through new social contracts among communities that frequently compete for economic opportunities, and finding new market values for traditional products provided by highly diverse mixed forests (including timber, non-timber and ecosystem services). It is in this context that calculating the current and future cost of restoration practices becomes an issue of utmost importance, and one for which unfortunately all contributing teams have so far only scanty information – if any.

Current trends suggest that, in the mid-term, forest restoration practice in the South American study regions and in southern Mexico may have a few more common elements than those they now share. On the one hand, original indigenous groups may be called upon along with other social actors to play an unprecedented role in forest planning in Chile (Lara, 2004), and local entrepreneurs may increasingly participate in financing a more intensive agriculture and social welfare in Chiapas that may allow setting aside larger forest areas for biodiversity conservation and ecosystem services (Cartagena-Licona *et al.*, 2005; Ixtacuy-López *et al.*, 2006). On the other hand, as has happened in Chile before (Armesto *et al.*, 1998), the participation of rather resourceful and well-educated social groups in the cities may become a key factor in local forest restoration efforts. In central Veracruz, a number of social groups based in the city of Xalapa have supported forest conservation actions and environmental education, including rehabilitation of evergreen cloud forest species and habitats (Pedraza and Williams-Linera, 2003; Williams-Linera *et al.*, 2003; Alvarez-Aquino *et al.*, 2004; Benítez-Badillo *et al.*, 2004; Suárez-Guerrero and Equihua-Zamora, 2005).

Academic institutions and forest restoration

Academic groups have to define their role as intermediate actors within the complex social scenario that forest restoration may imply (Lyall *et al.*, 2004; Castillo *et al.*, 2005). The wide spectrum of social conditions under which our restoration research has been conducted provides opportunities to focus on the activities of the research group once results have been validated and can be transferred to users and interest groups. In the IX and X Regions of Chile, the academic groups based at the Universidad Austral de Chile and Universidad Católica de Temuco have been able to organize an inclusive network of public and private stakeholders with an interest in *in situ* conservation. Their results in conservation biology research have provided the basis on which to conduct educational and outreach activities involving governmental and non-governmental organizations, university researchers and local people. A similar experience between academic groups and private landholders has occurred in central Veracruz. Progress in Chiapas is still some steps behind such outreach activities and widespread adoption of forest restoration practices. Yet, as in the Chilean case, in both regions of Mexico there is coincidence in the view that high-diversity native forest restoration and long-term and widespread conservation will only be attained if representatives of all involved social actors participate in what should be an ecologically defined common venture (González-Espinosa *et al.*, 2007). The academic group may currently have much of the technological know-how to promote and carry out widespread restoration actions. Mature research groups may play a crucial role in providing strategic links for other social actors involved because of informal networks maintained by their senior members (Guimerà *et al.*, 2005). Yet, unless forest restoration achieves the sustained support of the thousands of people that live in and own the forest-lands in question, its efforts will hardly surpass the stage of being a mere academic exercise and will fall short of impacting on public policy and decision-making circles.

Conclusions: Some Lessons Learned

We suggest that the following biological and socio-economic criteria could usefully be included among elements of best practice when starting a forest restoration programme, for either experimental or other purposes:

1. *To ensure that the biological material being used includes as much genetic variation as possible.* Recent studies provide evidence of the long-term reduced genetic variation that a founder population can impose on a regenerating secondary forest (Sezen *et al.*, 2005). Efforts should be made to ensure that any planting material used is well adapted to the sites where restoration is to take place.

2. *To obtain a reliable baseline estimate of the carbon content in the soil.* The global soil C pool is estimated to be 3.3 times the size of the atmospheric C pool and 4.5 times the size of the biotic sink (Lal, 2004). However, forest stands restored with different dominant species may differ in their potential root production and inputs to the soil C pool (e.g. pines lower than broadleaved native trees; Schlesinger and Lichter, 2001; Matamala *et al.*, 2003). As forest restoration is widely accepted as a viable alternative to increase C pools, its financial and social support can only benefit from being able to clearly show its potential advantage after some years.

3. *To approach the assessment of species with a gradient framework.* Species are usually distributed over a larger area than those used for restoration trials. Trees are long-lived species that may experience changing environments throughout their lifespan. Restoration predictions generated by models dealing with large spatial and temporal scales would benefit from a gradient approach to assess species responses.

4. *To consider major ecological principles and concepts;* in particular, assays designed to define the assembly rules of natural communities (e.g. plant succession, inter- and intraspecific competition, gene flow and inbreeding depression, nutrient cycling).

5. *To allow the potential users to define and take the first steps in the process of adopting results towards their application.* Forest restoration may be expensive, and potential users or landholders should be aware and ready to accept that application of their results may imply financial risks. Monitoring the effectiveness of restoration over large areas may only be possible if individuals or community landholders participate in the process after receiving adequate training and capacity building.

6. *To be aware of novel or non-conventional statistical approaches for analysis that can help to make sense out of data obtained under very different conditions.* Not all restoration experiences will contribute to developing scientific understanding, but long-term data under a variety of conditions may support meta-analysis approaches. In many cases establishing forest restoration trials and experiments has depended on opportunities offered by potential users or groups of interest that set challenges beyond conventional experimental layouts. All contributing research teams have been keen to identify interest groups that are willing to support restoration activities; in fact, access to several of the study sites listed in Table 15.1 was negotiated with private or community landholders.

7. *To adopt an adaptive management approach that can take advantage of changing values of the land and the tree species being used.* The academic groups should take the responsibility of identifying and promoting new technologies that could be used to improve the resource base of their partners.

8. *To assess the current and future finances of alternative restoration programmes.* In order to be adopted, ecological restoration must be environmentally and economically sound.

9. *To use native tree species in forest restoration programmes, preferably in mixed plantations.* The original and traditionally managed forest ecosystems of southern and eastern Mexico include a very high diversity of tree species. On the other hand, the temperate forests of Chile and Argentina include a large number of endemics. Yet this guideline may enter into conflict with the increasing interest or need to establish plantations with exotic species in highly productive sites; this should be resolved stressing regional and long-term sustainability criteria, and not predominantly with local and short-term cost–benefit planning.

10. *To use low cost alternatives in the first place.* There are many situations where it may be preferable to allow forests to recover naturally through secondary succession. Yet this may be a slower process and may not include the complete regional pool of species if dispersal limitations prevail in some taxa. Restoration for stand enrichment may be complemented with the provision and valuation of ecosystem services, including non-conventional timber and non-timber products in order to provide a pay-off for the long-term process.

Acknowledgements

Research supported by the Comisión Nacional para el Conocimiento y Uso de la Biodiversidad (CONABIO, L-031), the Fondo Mexicano para la Conservación de la Naturaleza (A2-99-006), the Consejo de Ciencia y Tecnología de Estado de Chiapas (FOMIX-CHIS-2002-C01-4640 and FOMIX-CHIS-2005-C03-010), the Secretaría del Medio Ambiente, Recursos Naturales and the Consejo Nacional de Ciencia y Tecnología (SEMARNAT-CONACYT C01-2002-048) and the Commission of the European Communities through the BIOCORES project (INCO Programme Framework 5, Contract No. ICA4-CT-2001-10095).

We appreciate the help over a number of years of many students and colleagues, in particular Juan Antonio Barrón-Sevilla, Martín Carmona, Luis Cayuela, Cristian Echeverría, Víctor Gerding, Pedro Girón Hernández, Duncan Golicher, Silvia Holz, Elke Huss, Fabiola López-Barrera, Alfonso Luna-Gómez, Paula Mathiasen, Guadalupe Méndez-Dewar, Lera Miles, Manuel R. Parra-Vázquez and Leonora Rojas.

References

Aide, T.M., Zimmerman, J.K., Pascarella, J.B., Rivera, L. and Marcano-Vega, H. (2000) Forest regeneration in a chronosequence of tropical abandoned pastures: implication for restoration ecology. *Restoration Ecology* 8, 328–338.

Aldrich, M., Billington, C., Edwards, M. and Laidlaw, R. (1997) *Tropical Montane Cloud Forests: An Urgent Priority for Conservation*. WCMC Biodiversity Bulletin No. 2. World Conservation Monitoring Centre, Cambridge, UK.

Allnutt, T.R., Newton, A.C., Premoli, A. and Lara, A. (2003) Genetic variation in the threatened South American conifer *Pilgerodendron uviferum* (Cupressaceae), detected using RAPD markers. *Biological Conservation* 114, 245–253.

Alvarez-Aquino, C., Williams-Linera, G. and Newton, A.C. (2004) Experimental native tree seedling establishment for the restoration of a Mexican cloud forest. *Restoration Ecology* 12, 412–418.

Armesto, J.J., León-Lobos, P. and Kalin-Arroyo, M. (1997) Los bosques templados del sur de Chile y Argentina: una isla biogeográfica. In: Armesto, J.J., Villagrán, C. and Kalin-Arroyo, M. (eds) *Ecología de los Bosques Nativos de Chile*. Editorial Universitaria, Santiago, Chile, pp. 23–28.

Armesto, J.J., Roíz, R., Smith-Ramírez, C. and Arroyo, M.T.K. (1998) Conservation targets in South American temperate forests. *Science* 282, 1271–1272.

Barrón-Sevilla, J.A. (2002) Efecto del disturbio antropogénico sobre la estructura y riqueza arbórea en bosques de pino encino de Los Altos de Chiapas, México. MSc thesis. El Colegio de la Frontera Sur, San Cristóbal de las Casas, Chiapas, Mexico.

Bekessy, S.A., Allnutt, T.R., Premoli, A.C., Lara, A., Ennos, R.A., Burgman, M.A., Cortés, M. and Newton, A.C. (2002) Genetic variation in the vulnerable and endemic monkey puzzle tree, detected using RAPDs. *Heredity* 88, 243–249.

Benítez-Badillo, G., Pulido-Salas, M.T.P. and Equihua-Zamora, M. (2004) *Árboles Multiusos Nativos de Veracruz para Reforestación, Restauración y Plantaciones*. Instituto de Ecología, A.C., Xalapa, Veracruz, Mexico.

Bradshaw, A.D. (1987) Restoration: an acid test for ecology. In: Jordan, W.R., Gilpin, M.E. and Aber, J.D. (eds) *Restoration Ecology: A Synthetic Approach to Ecological Research*. Cambridge University Press, Cambridge, UK, pp. 23–29.

Bradshaw, A.D. (2002) Introduction and philosophy. In: Perrow, M.R. and Davy, A.J. (eds) *Handbook of Ecological Restoration, Volume 1: Principles of Restoration*. Cambridge University Press, Cambridge, UK, pp. 3–9.

Bray, D.B. and Merino, L. (2004) *La Experiencia de las Comunidades Forestales en México: Veinticinco Años de Silvicultura y Construcción de Empresas Forestales Comunitarias*. Instituto Nacional de Ecología and Consejo Civil Mexicano para la Silvicultura Sostenible, Mexico City, Mexico.

Bray, D.B., Merino-Pérez, L. and Barry, D. (2005) *The Community Forests of Mexico: Managing for Sustainable Landscapes*. University of Texas Press, Austin, Texas.

Breedlove, D. (1981) *Flora of Chiapas. Part I: Introduction to the Flora of Chiapas*. California Academy of Sciences, San Francisco, California.

Brown, A.D. and Grau, H.R. (1995) *Investigación, Conservación y Desarrollo en Selvas Subtropicales de Montaña*. Laboratorio de Investigaciones Ecológicas de Las Yungas, Universidad Nacional de Tucumán, Tucumán, Argentina.

Brown, A.D. and Kappelle, M. (2001) Introducción a los bosques nublados del neotrópico: una síntesis regional. In: Kappelle, M. and Brown, A.D. (eds) *Bosques Nublados del Neotrópico*. Editorial INBio, Santo Domingo de Heredia, Costa Rica, pp. 25–40.

Burguete-Cal y Mayor, A. (1999) *México: Experiencias de Autonomía Indígena*. Grupo Internacional de Trabajo sobre Asuntos Indígenas (IWGIA), Copenhagen, Denmark.

Cabrera, A.L. and Willink, A. (1973) *Biogeografía de América Latina, Monografía No. 13, Serie Biología*. Secretaría General de la Organización de Estados Americanos, Washington, DC.

Cairns, J. Jr. (2002) Rationale for restoration. In: Perrow, M.R. and Davy, A.J. (eds) *Handbook of Ecological Restoration*. Cambridge University Press, Cambridge, UK, pp. 10–23.

Cartagena-Licona, R.P., Parra-Vázquez, M.R., Burguete-Cal y Mayor, A. and López-Meza, A. (2005) Participación social y toma de decisiones en los Consejos Municipales de Desarrollo Rural Sustentable de los Altos de Chiapas. *Gestión y Política Pública* 14, 341–398.

Castillo, A., Torres, A., Velázquez, A. and Bocco, G. (2005) The use of ecological science by rural producers: a case study in Mexico. *Ecological Applications* 15, 745–756.

Cayuela, L., González, M., Rey-Benayas, J.M., Ramírez, N. and Martínez, M. (2005) Imágenes de satélite revelan cómo desaparece el bosque en Chiapas. *Quercus* 232, 60–61.

Cayuela, L., Rey-Benayas, J.M. and Echeverría, C. (2006a) Clearance and fragmentation of tropical montane forests in the highlands of Chiapas, Mexico (1975–2000). *Forest Ecology and Management* 226, 208–218.

Cayuela, L., Golicher, D. and Rey-Benayas, J.M. (2006b) The extent, distribution, and fragmentation of vanishing montane cloud forest in the highlands of Chiapas, Mexico. *Biotropica* 38, 544–554.

Cubiña, A. and Aide, T.M. (2001) The effects of distance from forest edge on seed rain and soil seed bank in a tropical pasture. *Biotropica* 33, 260–267.

de Jong, B.H.J., Cairns, M.A., Haggerty, P.K., Ramírez-Marcial, N., Ochoa-Gaona, S., Mendoza-Vega, J., González-Espinosa, M. and March-Mifsut, I. (1999) Land-use change and carbon flux between 1970s and 1990s in the central highlands of Chiapas, Mexico. *Environmental Management* 23, 373–385.

Donoso, C. and Lara, A. (1998) *Silvicultura de los Bosques Nativos de Chile*. Editorial Universitaria, Santiago, Chile.

Donoso-Zegers, C. (1993) *Bosques Templados de Chile y Argentina: Variación, Estructura y Dinámica,* 4th edn. Editorial Universitaria, Santiago, Chile.

Echeverría, C.M. (2005) Fragmentation of temperate rain forests in Chile: patterns, causes, and impacts. PhD thesis, University of Cambridge, Cambridge, UK.

Florentine, S.K. and Westbrooke, M.E. (2004) Restoration on abandoned tropical pasture lands (do we know enough?). *Journal of Nature Conservation* 12, 85–94.

Galindo-Jaimes, L., González-Espinosa, M., Quintana-Ascencio, P. and García-Barrios, L. (2002) Tree composition and structure in disturbed stands with varying dominance by *Pinus* spp. in the highlands of Chiapas, Mexico. *Plant Ecology* 162, 259–272.

García-Barrios, L.E. and González-Espinosa, M. (2004) Change in oak to pine dominance in secondary forests may reduce shifting agriculture yields: experimental evidence from Chiapas, Mexico. *Agriculture, Ecosystems and Environment* 102, 389–401.

Gómez-Pompa, A. (1973) Ecology of the vegetation of Veracruz. In: Graham, A. (ed.) *Vegetation and Vegetational History of Northern Latin America*. Elsevier, Amsterdam, The Netherlands, pp. 73–148.

González-Espinosa, M., Quintana-Ascencio, P.F., Ramírez-Marcial, N. and Gaytán-Guzmán, P. (1991) Secondary succession in disturbed *Pinus-Quercus* forests of the highlands of Chiapas, Mexico. *Journal of Vegetation Science* 2, 351–360.

González-Espinosa, M., Ochoa-Gaona, S., Ramírez-Marcial, N. and Quintana-Ascencio, P.F. (1995) Current land-use trends and conservation of old-growth forest habitats in the highlands of Chiapas, Mexico. In: Wilson, M.H. and Sader, S.A. (eds) *Conservation of Neotropical Migratory Birds in Mexico*. Miscellaneous publication 727. Maine Agriculture and Forest Experiment Station, Orono, Maine, pp. 190–198.

González-Espinosa, M., Rey-Benayas, J.M., Ramírez-Marcial, N., Huston, M.A. and Golicher, D. (2004) Tree diversity in the northern Neotropics: regional patterns in highly diverse Chiapas, Mexico. *Ecography* 27, 741–756.

González-Espinosa, M., Ramírez-Marcial, N., Méndez-Dewar, G., Galindo-Jaimes, L. and Golicher, D. (2005) Riqueza de especies de árboles en Chiapas: variación espacial y dimensiones ambientales asociadas al nivel regional. In: González-Espinosa, M., Ramírez-Marcial, N. and Ruiz-Montoya, L. (eds) *Diversidad Biológica en Chiapas*. Plaza y Valdés, Mexico City, Mexico, pp. 81–125.

González-Espinosa, M., Ramírez-Marcial, N. and Galindo-Jaimes, L. (2006) Secondary succession in montane pine–oak forests of Chiapas, Mexico. In: Kappelle, M. (ed.) *Ecology and Conservation of Neotropical Montane Oak Forests. Ecological Studies 185*. Springer, Berlin, Germany, pp. 209–221.

González-Espinosa, M., Ramírez-Marcial, N., Camacho-Cruz, A., Holz, S.C., Rey-Benayas, J.M. and Parra-Vázquez, M.R. (2007) Restauración de bosques en territorios indígenas de Chiapas: modelos ecológicos y estrategias de acción. *Boletín de la Sociedad Botánica de México* 80 (Suplemento), 11–23.

Green, R.E., Cornell, S.J., Scharlemann, J.P.W. and Balmford, A. (2005) Farming and the fate of wild nature. *Science* 307, 550–555.

Guariguata, M.R., Rheingans, R. and Montagnini, F. (1995) Early wood invasion under tree plantations in Costa Rica: implications for forest restoration. *Restoration Ecology* 3, 252–260.

Guevara, S., Purata, S.E. and Vander Maarel, E. (1986) The role of remnant forest trees in tropical secondary succession. *Vegetatio* 66, 77–84.

Guevara, S., Meave, J., Moreno-Casasola, P. and Laborde, J. (1992) Floristic composition and structure of vegetation under isolated standing trees in Neotropical pastures. *Journal of Vegetation Science* 3, 655–664.

Guimerà, R., Uzzi, B., Spiro, J. and Nunes-Amaral, L.A. (2005) Team assembly mechanisms determine collaboration network structure and team performance. *Science* 308, 697–702.

Hamilton, L.S., Juvik, J.O. and Scatena, F.N. (1995) *Tropical Montane Cloud Forests. Ecological Studies 110*. Springer, New York.

Higgs, E.S. (1997) What is good ecological restoration? *Conservation Biology* 11, 338–348.

Hobbs, R.J. (2002) The ecological context: a landscape perspective. In: Perrow, M.R. and Davy, A.J. (eds) *Handbook of Ecological Restoration, Volume 1: Principles of Restoration*. Cambridge University Press, Cambridge, UK, pp. 24–45.

Holder, C.D. (2004) Changes in structure and cover of a common property pine forest in Guatemala, 1954–1996. *Environmental Conservation* 31, 22–29.

Huston, M.A. and Smith, T. (1987) Plant succession: life history and competition. *American Naturalist* 130, 168–198.

Ixtacuy-López, O., Estrada-Lugo, E. and Parra-Vázquez, M.R. (2006) Organización social en la apropiación del territorio: Santa Martha Chenalhó Chiapas. *Relaciones* 106, 183–219.

Janzen, D.H. (1987) How to grow a tropical national park: basic philosophy for Guanacaste National Park, northwestern Costa Rica. *Experientia* 43, 1033–1038.

Janzen, D.H. (2002) Tropical dry forest: Área de Conservación Guanacaste, northwestern Costa Rica. In: Perrow, M.R. and Davy, A.J. (eds) *Handbook of Ecological Restoration,* Volume 2: *Restoration in Practice*. Cambridge University Press, Cambridge, UK, pp. 559–583.

Kageyama, P. and Gandara, F.V. (2000) Recuperaçao de areas ciliares. In: Ribeiro-Rodriguez, R. and de Freitas-Leitao, H. (eds) *Matas Ciliares: Conservaçao e Recuperaçao*. Editora da Universidade de São Paulo, São Paulo, Brazil, pp. 249–269.
Kappelle, M. (2004) Tropical montane forests. In: Burley, J., Evans, J. and Youngquist, J.A. (eds) *Encyclopaedia of Forest Sciences, Volume 4*. Elsevier, Oxford, UK, pp. 1782–1793.
Kappelle, M. (2006) *Ecology and Conservation of Neotropical Montane Oak Forests. Ecological Studies 185*. Springer, Berlin, Germany.
Kitzberger, T., Steinaker, D.F. and Veblen, T.T. (2000) Effects of climatic variability on facilitation of tree establishment in northern Patagonia. *Ecology* 81, 1914–1924.
Lal, R. (2004) Soil carbon sequestration impacts on global climate change and food security. *Science* 304, 1623–1627.
Lara, A. (2004) Conservación de los sistemas boscosos: algunas lecciones de los últimos 20 años. *Ambiente y Desarrollo* 20, 111–115.
Luna, I., Velázquez, A. and Velásquez, E. (2001) México. In: Kappelle, M. and Brown, A.D. (eds) *Bosques Nublados del Neotrópico*. Editorial INBio, Santo Domingo de Heredia, Costa Rica, pp. 183–229.
Lyall, C., Bruce, A., Firn, J., Firn, M. and Tait, J. (2004) Assessing end-use relevance of public sector research organisations. *Research Policy* 33, 73–87.
Manson, R.H. (2004) Los servicios hidrológicos y la conservación de los bosques de México. *Madera y Bosques* 10, 3–20.
Martorell, C. and Peters, E.M. (2005) The measurement of chronic disturbance and its effects on the threatened cactus *Mammillaria pectinifera. Biological Conservation* 124, 199–207.
Matamala, R., González-Meler, M.A., Jastrow, J.D., Norby, R.J. and Schlesinger, W.H. (2003) Impacts of fine root turnover on forest NPP and soil C sequestration potential. *Science* 302, 1385–1387.
Meli, P. (2003) Restauración ecológica de bosques tropicales, veinte años de investigación académica. *Interciencia* 28, 2–24.
Méndez-Dewar, G. (2000) Contrastes espaciales de luz en claros, bordes y hábitats perturbados en Los Altos de Chiapas, México. MSc thesis. El Colegio de la Frontera Sur, San Cristóbal de Las Casas, Chiapas, Mexico.
Montagnini, F., Fanzeres, A. and Guimaraes da Viña, S. (1995) The potentials of 20 indigenous tree species for soil rehabilitation in the Atlantic Forest region of Bahia, Brazil. *Journal of Applied Ecology* 19, 386–390.
Montoya-Gómez, G. (1998) *El Subsector Forestal en México y Chiapas: Breve Análisis Económico de Largo Plazo*. Universidad Autónoma de Chiapas, Tuxtla Gutiérrez, Chiapas, Mexico.
Montoya-Gómez, G., Hernández-Ruiz, F. and Mandujano-Granados, M. (2003) Frontera Sur: de la riqueza de sus recursos naturales a la pobreza de sus habitantes. In: Montoya, G., Bello, E., Parra, M. and Mariaca, R. (eds) *La Frontera Olvidada entre Chiapas y Quintana Roo*. Consejo Estatal para la Cultura y las Artes de Chiapas and El Colegio de la Frontera Sur, Tuxtla Gutiérrez, Chiapas, Mexico, pp. 33–68.
Newton, A.C., Wilson, K. and Echeverría, C.M. (2004) Assessing the vulnerability of forests to environmental change. In: Smithers, R. (ed.) *Landscape Ecology of Trees and Forests, Proceedings of the 12th Annual International Association for Landscape Ecology (IALE UK) Conference*. IALE UK, Nottingham, UK, pp. 176–186.
Ochoa-Gaona, S. and González-Espinosa, M. (2000) Land-use and deforestation in the highlands of Chiapas, Mexico. *Applied Geography* 20, 17–42.
Ochoa-Gaona, S., González-Espinosa, M., Meave, J.A. and Sorani dal Bon, V. (2004) Effect of forest fragmentation on the woody flora of the highlands of Chiapas, Mexico. *Biodiversity and Conservation* 13, 867–884.

Oldfield, S., Lusty, C. and MacKinven, A. (1998) *The World List of Threatened Trees*. World Conservation Press, Cambridge, UK.

Otero-Arnáiz, A., Castillo, S., Meave, J. and Ibarra-Manríquez, G. (1999) Isolated pasture trees and the vegetation under their canopies in the Chiapas coastal plain, Mexico. *Biotropica* 31, 243–254.

Pedraza, R.A. and Williams-Linera, G. (2003) Evaluation of native tree species for the rehabilitation of deforested areas in a Mexican cloud forest. *New Forests* 26, 83–99.

Posada, J.M., Aide, T.M. and Cavelier, J. (2000) Cattle and weedy shrubs as restoration tools of tropical montane rainforest. *Restoration Ecology* 8, 370–379.

Premoli, A.C. (2003) Isozyme polymorphisms provide evidence of clinal variation with elevation in *Nothofagus pumilio. Journal of Heredity* 94, 218–226.

Premoli, A.C. (2004) Variación en *Nothofagus pumilio* (Poepp. et Endl.) Krasser. In: Donoso, C., Premoli, A.C., Gallo, L. and Ipinza, R. (eds) *Variación Intraespecífica en las Especies Arbóreas de los Bosques Templados de Chile y Argentina*. Editorial Universitaria, Santiago, Chile, pp. 145–172.

Premoli, A.C., Souto, C.P., Allnutt, T.R. and Newton, A.C. (2001) Effects of population disjunction on isozyme variation in the widespread *Pilgerodendron uviferum. Heredity* 87, 337–343.

Premoli, A.C., Vergara, R., Souto, C.P., Lara, A. and Newton, A.C. (2003) Lowland valleys shelter the ancient conifer *Fitzroya cupressoides* in the Central Depression of southern Chile. *Journal of the Royal Society of New Zealand* 33, 623–631.

Puig, H. and Bracho, R. (1987) *El Bosque Mesófilo de Montaña de Tamaulipas*. Instituto de Ecología, Mexico City, Mexico.

Quintana-Ascencio, P.F., Ramírez-Marcial, N., González-Espinosa, M. and Martínez-Icó, M. (2004) Sapling survival and growth of conifer and broad-leaved trees in successional habitats in the highlands of Chiapas, Mexico. *Applied Vegetation Science* 7, 81–88.

Ramírez-Marcial, N., González-Espinosa, M. and García-Moya, E. (1996) Establecimiento de *Pinus* spp. y *Quercus* spp. en matorrales y pastizales de los altos de Chiapas. *Agrociencia* 30, 249–257.

Ramírez-Marcial, N., González-Espinosa, M. and Williams-Linera, G. (2001) Anthropogenic disturbance and tree diversity in montane rain forests in Chiapas, Mexico. *Forest Ecology and Management* 154, 311–326.

Ramírez-Marcial, N., Camacho-Cruz, A. and González-Espinosa, M. (2003) *Guía para la Propagación de Especies Leñosas Nativas de los Altos y Montañas del Norte de Chiapas*. El Colegio de la Frontera Sur, San Cristóbal de las Casas, Chiapas, Mexico.

Ramírez-Marcial, N., Camacho-Cruz, A. and González-Espinosa, M. (2005) Potencial florístico para la restauración de bosques en Los Altos y las Montañas del Norte de Chiapas. In: González-Espinosa, M., Ramírez-Marcial, N. and Ruiz-Montoya, L. (eds) *Diversidad Biológica en Chiapas*. Plaza y Valdés, Mexico City, Mexico, pp. 325–363.

Ramos, J. and del Amo, S. (1992) Enrichment planting in a secondary forest in Veracruz, Mexico. *Forest Ecology and Management* 54, 289–304.

Rosenzweig, M.L. (1968) Net primary productivity of terrestrial environments: predictions from climatological data. *American Naturalist* 102, 67–84.

Rovere, A., Gobbi, M. and Relva, A. (2005) Regeneración de *Austrocedrus chilensis*. In: Arturi, M.F., Frangi, J.L. and Goya, J.F. (eds) *Ecología y Manejo de Bosques de la Argentina*. Editorial de la Universidad Nacional de La Plata, La Plata, Argentina, pp. 1–16.

Rowden, A., Robertson, A., Allnutt, T., Heredia, S., Williams-Linera, G. and Newton, A.C. (2004) Conservation genetics of Mexican beech, *Fagus grandifolia* var. *mexicana. Conservation Genetics* 5, 475–484.

Rudel, T. and Roper, J. (1997) The paths to rain forest destruction: crossnational patterns of tropical deforestation, 1975–1990. *World Development* 25, 53–65.

Ruiz-Jaén, M.C. and Aide, T.M. (2005) Vegetation structure, species diversity, and ecosystem processes as measures of restoration success. *Forest Ecology and Management* 218, 159–173.
Rzedowski, J. (1978) *Vegetación de México*. Limusa, Mexico City, Mexico.
Rzedowski, J. (1993) Diversity and origins of the phanerogamic flora of Mexico. In: Ramamoorthy, T.P., Bye, R., Lot, A. and Fa, J. (eds) *Biological Diversity of Mexico: Origins and Distribution*. Oxford University Press, New York, pp. 129–144.
Sánchez, X., González, C. and Amtmann, C. (2002) *Escenarios de la Nueva Ruralidad en Chile*. Universidad de Valparaíso, Valparaíso, Chile.
Schlesinger, W.H. and Lichter, J. (2001) Limited carbon storage in soil and litter experimental forest plots under increased atmospheric CO_2. *Nature* 411, 466–468.
SEMARNAT (2002) Norma Oficial Mexicana NOM-059-ECOL-2001: Protección ambiental; Especies nativas de México de flora y fauna silvestres; Categorías de riesgo y especificaciones para su inclusión, exclusión o cambio; Lista de especies en riesgo. Diario Oficial de la Federación, miércoles 6 de marzo de 2002. Secretaría de Medio Ambiente y Recursos Naturales, Estados Unidos Mexicanos, Mexico City, Mexico.
Sezen, U.U., Chazdon, R.L. and Holsinger, K.E. (2005) Genetic consequences of tropical second-growth forest regeneration. *Science* 307, 891.
Silver, W.L., Marín-Spiotta, E. and Lugo, A.E. (2001) El Caribe. In: Kappelle, M. and Brown, A.D. (eds) *Bosques Nublados del Neotrópico*. Editorial INBio, Santo Domingo de Heredia, Costa Rica, pp. 155–181.
Storch, D., Marquet, P.A. and Gaston, K.J. (2005) Untangling an entangled bank. *Science* 307, 684–686.
Suárez-Guerrero, A.I. and Equihua-Zamora, M.E. (2005) Experimental tree assemblages on the ecological rehabilitation of a cloud forest in Veracruz, Mexico. *Forest Ecology and Management* 218, 329–341.
Temperton, V.M., Hobbs, R.J., Nuttle, T. and Halle, S. (2004) *Assembly Rules and Restoration Ecology: Bridging the Gap Between Theory and Practice*. Island Press, Washington, DC.
Toh, I., Gillespie, M. and Lamb, D. (1999) The role of isolated trees in facilitating tree seedling recruitment at a degraded sub-tropical rainforest site. *Restoration Ecology* 7, 288–297.
Turc, L. (1954) Le bilan d'eau des sols: relations entre les précipitation, l'évaporation et l'écoulement. *Annales Agronomiques* 5, 491–596.
Vetaas, O.R. (1997) The effect of canopy disturbance on species richness in a central Himalayan oak forest. *Plant Ecology* 132, 29–38.
Webster, G.L. (1995) The panorama of neotropical cloud forest. In: Churchill, S.P., Balslev, H., Forero, E. and Luteyn, J.L. (eds) *Biodiversity and Conservation of Neotropical Montane Forests*. The New York Botanical Garden Press, New York, pp. 53–77.
Williams-Linera, G. (2002) Tree species richness complementarity, disturbance and fragmentation in a Mexican tropical montane cloud forest. *Biodiversity and Conservation* 11, 1825–1843.
Williams-Linera, G., Rowden, A. and Newton, A.C. (2003) Distribution and characteristics of relict populations of Mexican beech (*Fagus grandifolia* var. *mexicana*). *Biological Conservation* 109, 27–36.
Wuethrich, B. (2007) Reconstructing Brazil's Atlantic Forest. *Science* 315, 1070–1072.

16 Future Scenarios for Tropical Montane and South Temperate Forest Biodiversity in Latin America

L. Miles, A.C. Newton, C. Alvarez-Aquino, J.J. Armesto, R.F. del Castillo, L. Cayuela, C. Echeverría, M. González-Espinosa, A. Lara, F. López-Barrera, R.H. Manson, G. Montoya-Gómez, M.A. Muñiz-Castro, M.C. Núñez-Ávila, R.A. Pedraza, J.M. Rey-Benayas, A.E. Rovere, N. Rüger, C. Smith-Ramírez, C. Souto and G. Williams-Linera

Aerial photograph illustrating riparian native forest surrounded by a matrix of agricultural land in Region X, Chile. Photo: Cristian Echeverría

Summary

This chapter presents results of a scenario-building exercise, designed to explore future trends in forest biodiversity in four forest areas, and the potential implications for policy development and implementation. An expert consultation conducted in a workshop environment identified 11 principal pressures responsible for biodiversity loss in Latin America, namely land-cover change, fire, invasive species, browsing animals, pollution, mining, development of infrastructure (roads, pipelines, dams), logging/fuelwood extraction, habitat fragmentation, climate change and loss of keystone species and ecological structures. The relative importance of these different pressures was assessed in each of four study areas, namely Central Veracruz (Mexico), the Highlands of Chiapas (Mexico), Rio Maule-Cobquecura (Region VII, Chile) and Los Muermos-Ancud (Region X, Chile). Scores were generated for each area describing both variation in intensity of the pressures over time and their potential impacts on different components of biodiversity. The scoring process was used to support development of three scenario narratives for each area, namely *business as usual, deepening conservation crisis* and *effective conservation*. Recommendations for policy development and implementation are presented for each study area, based on these scenarios. The results indicate that action on global commitments to reduce biodiversity loss must take account of the geographical variation in the relative importance of different pressures and their varying impacts on different biodiversity components. Policy developments and practical conservation action will need to be tailored for individual areas, defined at the sub-national level.

Introduction

The development of effective conservation strategies and plans requires information not only on current status and trends in biodiversity, but on how biodiversity might change in the future. For example, if a species is declining in abundance, then a conservation intervention might be planned based on an assumption that this decline is likely to continue unless some form of action is taken. Models of ecological dynamics provide a set of tools for exploring potential future trends in the structure and composition of ecological communities (see Chapters 9–11). However, such models are based on a range of assumptions and uncertainties, which make it difficult to predict the future with any degree of accuracy. Conservation plans tend to ignore such uncertainties, and fail to consider the possibility of novel situations or surprises occurring, despite their potential importance (Scott, 1998). As a result, conservation planning may often risk costly failure (Holling and Meffe, 1996).

Scenario planning offers a tool for supporting conservation decision making under such uncertain conditions. A scenario can be defined in this context as an account of a plausible future (Peterson *et al.*, 2003). The development of scenarios is a recognized tool in business planning and economic forecasting (Wack, 1985a, b; Schwartz, 1991; van der Heijden, 1996), but only recently has it begun to be applied to biodiversity conservation. Peterson *et al.* (2003) provide a valuable introduction to the use of scenarios in this context. A first attempt to develop global biodiversity scenarios was presented by Sala *et al.* (2000), and elaborated further by Chapin *et al.* (2001). Scenarios are increasingly being used in environmental assessments at global and regional scales, such as the Millennium Ecosystem Assessment (http://www.maweb.org/)

(Carpenter *et al.*, 2005), the Global Environment Outlook (http://www.unep.org/geo/) (UNEP, 2003) and the International Assessment on Agricultural Science, Technology and Knowledge Development (http://www.agassessment.org). This reflects their value in communicating complex scientific information to policy makers.

Scenarios can be used to explore the uncertainty surrounding the future consequences of a decision, by developing a small number of contrasting scenarios. Generally, scenarios are developed by a group of people in a workshop who engage in a systemic process of collecting, discussing and analysing information. The scenarios may draw upon a variety of quantitative and qualitative data, such as the results of ecological surveys and outputs from modelling exercises. Peterson *et al.* (2003) suggest that the major benefits of scenario planning for conservation are: (i) increased understanding of key uncertainties; (ii) incorporation of alternative perspectives into conservation planning; and (iii) greater resilience of decisions to surprise events.

This chapter presents the results of a scenario-building exercise undertaken for four forested regions of Latin America. The aim was to identify and start to quantify present and potential future human impacts on genetic, species and habitat components of biodiversity. The goal was to identify priorities for conservation action and produce recommendations for policy makers.

Development of Scenarios

Scenarios were developed in a workshop activity involving a team of researchers drawn from a variety of different ecological fields, who were invited to provide an expert assessment of current and potential future pressures and their potential impacts on biodiversity. The exercise covered the forested parts of four study areas: Central Veracruz (Mexico), the Highlands of Chiapas (Mexico), Rio Maule-Cobquecura (Region VII, Chile) and Los Muermos-Ancud (Region X, Chile). The four regions were assessed separately by different groups of experts, both because the intensity of the pressures experienced differs between regions, and because the impact of a similar intensity of pressure varies between forest ecosystems. Two separate workshop activities were undertaken: (i) a scoring exercise, considering the processes influencing biodiversity and their potential impacts; and (ii) the development of scenario narratives.

Scoring exercise

Working in four regional groups, researchers were invited to identify a list of pressures (or processes influencing biodiversity) considered to be important in the forests with which they were familiar. They were then asked to perform a numerical scoring exercise, to indicate the likely intensity of each pressure at three dates: 2005 (the present time), 2010 and 2050. Intensity was scored as mean intensity over the forested parts of the study area, using a five-point scale: *Zero* (0), *Relatively Low* (1), *Moderate* (2), *Relatively High* (3) and *Very High* (4). For land-cover change, this would equate to: Zero, 0%; Low, 1–25%; Moderate, 26–50%; High, 51–75%; and Very High, 76–100% of the forest area converted to another land-cover type. Whilst land-cover change does not vary in intensity at a given location, other pressures may vary in intensity at a given location (i.e. fires may be either low or high temperature; logging can be selective or total). For these, the following guidance for scoring was developed:

Zero: no area of the region affected by the pressure under consideration.
Low: either a small area (<25%) at high intensity, or larger area at low intensity affected.
Moderate: either a moderate area (26–50%) at high intensity, or larger area at low intensity affected.
High: either a large area (51–75%) at high intensity or larger area at low intensity affected.
Very High: a very large area (>75%) at high intensity affected.

Participants were asked to refer to research results or supporting data if available, but otherwise to provide an estimate based on their expert judgement. Participants were then asked to estimate the potential impact of each intensity value of the pressure on each of the three components of biodiversity (genetic, species and habitat diversity). The scoring system for these impacts was as follows: *Zero* (0), *Low* (1), *Moderate* (2), *Relatively High* (3), *Very High* (4), *Complete Loss of Biodiversity* (5). This part of the exercise recognizes that different pressures have different relative impacts on biodiversity, and this may also vary between study areas and forest types. For example, fire of a moderate intensity might be expected to have a far greater impact on a montane cloud forest than on a lowland tropical dry forest, as in the latter case many species may have evolved in the presence of fire and display adaptations to it (Chapter 13). This is further illustrated by the results obtained. In 2005 in the Highlands of Chiapas, there was considered to be a high level of land-cover change. This was estimated to have a very high impact on habitat diversity, a moderate impact on species diversity and a moderate impact on genetic diversity. In comparison, in the Rio Maule-Cobquecura region, there was also considered to be a high level of land-cover change in 2005. This was estimated to have a very high impact on genetic diversity, and to be leading to a complete loss of both species and habitat diversity.

Development of narratives

The scoring exercise was used to prompt a discussion, within each group of experts in the workshop, leading to the development of scenario narratives for each of the study areas. To support this process, researchers were asked to consider the following questions:

- How might different pressures interact?
- What are the underlying factors responsible for these pressures, and how might they be addressed?

- Which specific species and habitats/ecosystems are at particular risk, and from which pressures?
- What specific recommendations can be made for policy makers, including those in national and local government, conservation organizations and the private sector?

For each region, three contrasting and yet plausible scenario narratives were then developed: *business as usual*, *deepening conservation crisis* and *effective conservation*. These were defined as follows:

1. *Business as usual.* What might happen to biodiversity if things continue as they are at present?
2. *Deepening extinction crisis.* What might happen to biodiversity if the current situation deteriorates? Why might this occur?
3. *Effective conservation.* What might happen to biodiversity if effective conservation action were to be implemented? How might this be brought about?

Under each scenario, the experts were invited to consider the following questions:

- What will happen to the pressures (and underlying drivers) responsible for biodiversity change?
- What will the impacts be on different components of biodiversity?
- What are the implications for human responses, including policy development and implementation, and practical conservation action?

Each regional group of experts was also invited to suggest one or more possible surprise events that would modify the course of the narrative scenarios, and to note the critical uncertainties for the future of that forest area. Interactions between pressures were considered in the narratives, but not in the preceding numerical exercise. The narratives and results of the scoring exercise are presented below, considering each of the four study areas individually.

Scenarios for Central Veracruz (Mexico)

Present trends and pressures

The major pressures and drivers (or ultimate causes) of biodiversity loss in the tropical montane cloud forest region between 1000 and 2000 m altitude in Central Veracruz were identified as operating at local, national and global scales. Land-cover change is currently the most significant direct cause of biodiversity loss in the area, followed by infrastructural development, logging/fuelwood extraction and habitat fragmentation (Table 16.1). At a global scale, underlying drivers include fluctuations in international markets, including those for coffee, sugar cane and beef, and the impacts on markets of international treaties such as the North American Free Trade Agreement (NAFTA).

Table 16.1. Results of a numerical scoring exercise to indicate the likely intensity of different pressures at three dates: 2005 (the present time), 2010 and 2050, in four study areas: Central Veracruz (Mexico), Highlands of Chiapas (Mexico), Rio Maule-Cobquecura (Region VII, Chile) and Los Lagos (Region X, Chile). Intensity was scored as mean intensity over the forested parts of the study area, using a five-point scale: *Zero* (0), *Relatively Low* (1), *Moderate* (2), *Relatively High* (3), *Very High* (4) (see text for details). The values presented here relate to the 'business as usual' scenario; in other words, projected future values are those based on current trends.

	Study area and year											
	Veracruz			Chiapas			Maule (VII)			Lagos (X)		
Pressure	2005	2010	2050	2005	2010	2050	2005	2010	2050	2005	2010	2050
Land-cover change	3	2	2	4	3	1	4	3	1	3	3	2
Fire	0	0	0	1	1	1	3	3	2	3	3	3
Invasive species	0	0	0	0	0	0	3	3	2	2	3	3
Browsing animals	1	2	0	1	0	0	2	2	2	4	4	3
Pollution	0	0	0	0	0	0	4	4	4	1	1	2
Mining	1	1	0	0	0	0	0	0	0	0	0	0
Infrastructure (roads, pipelines, dams)	2	2	2	1	1	1	1	1	1	1	2	3
Logging/fuelwood extraction	2	3	2	2	3	2	2	1	1	4	4	3
Fragmentation	2	2	3	4	3	2	4	4	3	4	4	3
Climate change	1	2	3	0	0	1	2	4	4	1	2	2
Loss of keystone species and structures	1	1	2	0	0	0	4	4	4	2	3	4

Mega-projects such as the Plan Puebla Panama pose a specific threat to regional biodiversity through plans for an expanded infrastructure of ports, roads, airports and railways, with associated land-use changes. National-scale drivers include subsidies for forest clearance for economically productive purposes, and a lack of enforcement of environmental legislation, including that on the implementation of environmental impact assessment studies.

Local-scale direct drivers include unsustainable harvests of forest products, through both selective logging and the collection of orchids, bromeliads and birds, infrastructure development and strip mining for gravel and sand. As forest cover disappears, biodiversity conservation depends strongly on species survival in other land uses. However, traditional agroecosystems (such as shade coffee, home gardens and traditional intercropping systems (*milpas*)) that harbour wild forest species are being transformed to low- or eroded-diversity systems (such as sugar cane, non-shade coffee, pastures or urban areas). Indirect drivers include population growth (especially in large cities), a lack of security in land tenure, a culture of land clearing and the social and economic marginalization of local people. This final factor is also leading to migration from rural areas and land abandonment, which can have a positive effect on biodiversity.

Scenarios

From 2005 to 2050, under a *business as usual* scenario, there are few changes in environmental legislation and subsidies. However, community groups continue to exert pressure on the government to implement the existing laws. New protected areas are created, but receive little planning or financial support, and so their impact on biodiversity conservation is limited.

Most pressures remain unchanged (Table 16.1), with an initial increase in the rate of logging, and a long-term increase in the rate of fragmentation. Initially, the rapid loss of undisturbed forest cover continues. Many forest fragments may not disappear in the short term, but will experience accelerated changes in structure and composition as a result of the selective removal of certain tree sizes and non-timber products, and of cattle grazing, especially close to forest edges. In 2005, there was already evidence that the hydrological system (rivers and springs) is affected by forest loss and degradation, and that small wetland areas are disappearing (Bruijnzeel, 2001). There is some evidence that deforestation will slow as the most accessible forest cover is removed, with the remaining forests being restricted to steeper slopes (Manson *et al.*, 2007). As a result of economic pressures, some large areas of agricultural land were being abandoned in 2005, resulting in an increase in old-field systems and secondary forests.

Within the reduced area of forest, patterns of fragmentation remain relatively unchanged under a *business as usual* scenario. Most forest fragments are currently small. Given the intensity of the land use under this scenario, the level of isolation of forest patches remains relatively high, thus reducing the level of gene flow between populations. Thus, endogamy and gene drift reduce genetic diversity, especially for species that require large habitats. To some degree, the impact of forest conversion on biodiversity depends on the dominant land use in the region. Some agroecosystems (e.g. shade coffee) harbour more forest biodiversity and conserve more ecosystem services than others (e.g. sugar cane or pasture); these systems also offer better connectivity between forest patches.

Species with very specific habitat requirements, especially including shade-tolerant primary forest species preferring high humidity conditions, continue to be threatened by these pressures. Whilst a large number of endemic forest species are present, local extinctions are difficult to document and had not been observed in the region by 2005. The long lifespan of many tree species means that an extinction debt can build up through limited regeneration opportunities, even though viable adult individuals are still present (Hanski and Ovaskainen, 2002; Helm *et al.*, 2006).

In a *deepening extinction crisis* scenario, changes in international markets lead to a dramatic increase in intensive land use such as sugar cane plantations, cattle pasture and urban areas. The associated increase in use of agrochemicals leads to high levels of pollution in water, soils and air, and resulting species loss in non-forest ecosystems. Forest fragments become smaller, more isolated and more disturbed, with edge effect penetration increasing. Eventually, only remnant fragments on very steep slopes and in protected reserves remain. In response to the decreased area of exploitable forest, groups illegally extracting timber, firewood and non-timber products from the remaining forests become better organized, more powerful and more difficult to control. The more frequent drier and warmer weather resulting from global warming and the increased concentration of human activity lead to fire becoming an important pressure within these patches.

Light-demanding and invasive species colonize the small, degraded forest patches, and most plant species dependent upon forest interior conditions become locally extinct. Animal species share a similar fate as a result of reduced opportunities for dispersal and reproduction. The remaining populations of native species suffer from founder effects and inbreeding.

As soil fertility, water quality and quantity decline, policy makers are eventually forced to act. The resulting reforestation programme is too late to save much of the region's biodiversity, with the new forests being little more than plantations.

An *effective conservation* scenario arises when markets for ecosystem services become increasingly important. The main drivers of biodiversity loss are controlled, as economic opportunities relating to water capture, carbon sequestration and ecotourism become more attractive than agricultural activities. Secondary succession becomes possible within degraded land areas, and much lost forest is recovered. Protected areas finally receive adequate government support in the form of staff, budgets and management plans, and are linked together with the remaining forest fragments through biological corridors.

Agricultural subsidies are redirected to sustainable organic agriculture. There is an increase in certified forestry (including plantations), which involves the implementation of management plans that pay attention to biodiversity conservation. The use of native species and application of ecological knowledge within the forestry sector reduces the costs of restoration efforts and ensures that most, if not all, forest structure and forest types are conserved. Traditional ecological knowledge is incorporated directly into management plans for forests, plantations and diversified farming. Finally, the development of other industries in the region reduces pressure on forest resources.

As a result of these timely developments, the populations of many species, including those thought to have become locally extinct, begin to recover. Gene flow between fragments increases, and the risk of random extinctions reduces.

Xalapa is declared a model sustainable city according to the UNESCO criteria. Politicians learn that conservation pays, and that balancing biodiversity conservation and productivity is both possible and very popular, as societal benefits rather than private interests are maximized.

Surprise events and critical uncertainties

Possible surprises

- Volcanic activity increases in unexpected locations, affecting land uses/land cover.
- A strong earthquake results in the destruction of human infrastructure; reconstruction costs are too high and population densities are drastically reduced.
- Water sources available to Xalapa are reduced and the city is forced to adopt drastic measures to take advantage of local water sources including local rivers, springs and rainwater (cisterns). At present, 60% of Xalapa's water supply comes from the state of Puebla.
- A major highway is built in the area of the remnant cloud forest fragments, thus affecting the biodiversity and hydrology of the region.
- A fall in the price of coffee drastically changes regional land uses; for example, an increase in the production of sugar cane would reduce biodiversity and increase forest fragmentation.
- New legislation in relation to payment for environmental services focuses on the sustainable use of forest natural resources.
- Large areas of forest are protected and restored by private interests, without government support, either with conservation in mind or in the expectation of ecosystem services payments.

Critical uncertainties

Global warming may lead to regional climate change such that large areas of cloud forest are succeeded by a different type of forest or become more attractive for different land uses. Even with a smaller magnitude of change, cloud forest species may be outcompeted by species originally belonging to other ecosystems.

Recommendations

- A network of connected protected cloud forest areas could be created and administered and managed by local owners.
- Considerably more financial, professional and governmental support could be provided for the current protected areas.
- An environmental zoning assessment could support land-use planning.
- Intensification of cattle ranching could reduce the overall land area required.
- Sustainable land use could be encouraged through:
  - Investment in organic coffee production and marketing.
  - Promotion of sustainable use of non-timber forest products.
  - Promotion of sustainable ecotourism.

Scenarios for the Highlands of Chiapas (Mexico)

Present trends and pressures

Land-cover change and habitat fragmentation are currently the most significant causes of biodiversity loss in the area, with logging/fuelwood extraction ranking as the next most important pressure (Table 16.1). The intensities of these pressures are determined by a suite of indirect drivers related to population density and growth rates, markets, culture, land tenure, weak and disorganized environmental governance, and a lack of institutional reliability and trust. Poor, marginalized people in this region have little incentive to conserve forests. There is a lack of markets for forest products, and a culture favouring agricultural activity over forestry, so that forest is perceived as potential agricultural land rather than a resource in itself. Indigenous groups place a particular cultural value on maizefields.

Scenarios

In a *business as usual* scenario, deforestation, fragmentation and loss of biodiversity continue at present rates until forest area becomes a limiting factor (Table 16.1). As the area of agriculture land increases, the agriculture frontier expands, and there is a concomitant decrease in biodiversity. The structure and composition of remaining forests is simplified as a result of fragmentation, edge effects and increased accessibility. The most dominant habitats are then pastures, maizefields and agroecosystems; there are fewer areas recovering from agriculture, and a reduction in overall habitat diversity.

Species of Andean and Neotropical affinity are often replaced in the landscape by species of Holarctic origin, although some native opportunistic and pioneer species will continue to be widespread. Species of limited range are particularly vulnerable to losses in genetic diversity as populations are lost. The effect on tree species is delayed in comparison to species with shorter life cycles (Helm *et al.*, 2006).

By 2050, as the same trends continue, there is an economic and social collapse, with accompanying emigration, and abandonment of rural areas. In particular, men emigrate in search of viable employment, breaking up family units. Policy changes occur in response, with the NGO sector playing a key role, but it is too late to alter the land-use changes and their impacts.

Rural conditions have been poor since before the social conflict in 1994, which led to the Zapatista uprising. Indigenous peoples continue to be marginalized, and suffer high rates of poverty. These conditions have driven the accelerated process of forest area loss (González-Espinosa, 2005; Cayuela *et al.*, 2006). A *deepening extinction crisis* could be produced if levels of conflict were to rise again. The processes described in the *business as usual* scenario would be accelerated, and the breakdown of the regional economy occur more rapidly.

For an *effective conservation* scenario to come into being, novel social and resource management policies, with national investment in ecosystem restoration are required. People are enabled to make use of the forest as an important resource rather than converting it to agriculture. Payments for environmental services such as water and carbon storage, and taxation of polluting industries, help to bring this about. Interventions to create these economic opportunities are needs-based, apply locally appropriate approaches developed through participation and foster equal opportunities, including between genders. As the economic situation improves, population growth rates decline, decreasing the level of pressure on natural resources.

Under this scenario, all components of biodiversity are better conserved. The productive role of biodiversity is better recognized as new products are discovered and exploited, and ecological and conservation values are integrated into international ecosystem service markets. Species-oriented restoration is rapid, and we expect the gradual recovery of ecosystems over time.

Surprise events and critical uncertainties

Possible surprises

- One or more earthquakes could create such destruction in the cities that urban populations migrate to rural areas, increasing the rate of forest loss and pressure on biodiversity.
- Erosion and mud slides produced by increased precipitation and changes in the rain regimes could lead to direct loss of forest area, and loss of crops resulting in increased poverty and land-cover change.
- Forest health could suffer as a result of desiccation and loss of biodiversity, allowing insects (e.g. bark drillers) or new plant diseases to have serious effects on tree canopy dominants.
- A new indigenous uprising could result in land invasions, resulting land-cover changes, new black markets for rural products and migration from conflict areas to forests.

- An increase in the level of drug production and traffic could lead to deforestation for marijuana cultivation.

Critical uncertainties

The long-term impacts of fire on vegetation cover and the forest's ability to recover following anthropogenic disturbance are uncertain.

Recommendations

- Invest in local development, employment and welfare; simultaneously enforce relevant legislation on forests and narcotic cultivation.
- Promote a new rurality, with new relationships between natural areas and development, between cities and rural areas.
- Redevelop political institutions, seeking a regulated autonomy within a national legal framework, to help satisfy indigenous people's demands for self-governance.
- Promote the peaceful coexistence of cultures, with education towards political, cultural and religious tolerance.
- Establish predictive models that relate climate variables, fire occurrence and vulnerability, in order to inform the regulation of fire setting in agricultural areas adjacent to wild lands.
- Identify areas vulnerable to mud slides, and initiate preventive forest restoration and hydrological management.

Scenarios for Rio Maule-Cobquecura (Region VII, Chile)

Present trends and pressures

Land-cover change, pollution, loss of keystone species and habitat fragmentation are currently the most significant causes of biodiversity loss in the area, with fire and invasive species ranking as the next most important pressures (Table 16.1).

Scenarios

Under a *business as usual* scenario, each of the major pressures continues to be important, with intensity only reducing by 2050 because the area of forest available to be affected has been so substantially reduced (Table 16.1). The area and spatial continuity of exotic pine plantations expands. The scarcity of land available for commercial afforestation in the western portion of the Rio Maule-Cobquecura region means that the process of conversion of native forests to plantations shifts to the eastern slope of the Coastal range. Here, there are reductions in the area of mixed sclerophyllous Mediterranean-type forests and shrublands, which harbour several endemic tree species. Firewood and timber for charcoal continue to be harvested from the remaining forest.

The increasing plantation area and connectivity of plantations leads to an increase in the area and severity of anthropogenic fires. Reduced precipitation as a result of climate change, and an increased frequency and intensity of ENSO-driven droughts (La Niña events) also leads to increased fire frequency. As the area of native forest decreases, the area affected by fire also decreases, but the proportional area affected remains high.

Other effects of the massive expansion of forest plantations include reductions in river flow and water availability, and an increase in soil erosion associated with the 12–20-year clear-cut cycle (Varas and Riquelme, 2002; CIREN, 2004; CONAMA, 2004). Together, these pressures lead to biodiversity losses in riparian habitats, wetlands, rivers and streams.

Rates of forest fragmentation and its impacts on genetic and species diversity initially increase, and then decrease as the area of native forest available for conversion is reduced. Invasive animals and plants increase in number as native forest fragments become smaller and are surrounded by a matrix of non-native plantations. These invaders include *Canis domesticus* (dog), *Pinus radiata* (Monterey pine), *Teline monspessulana* (broom) and *Acacia dealbata* (silver wattle).

The combination of land-use change, fire, logging and fragmentation leads to local extinctions and regional reductions in genetic biodiversity, especially for threatened species. The most threatened groups include amphibians, freshwater fish, crustaceans, aquatic insects, some mammal species such as *Pudu pudu* (pudu deer) and birds.

A *deepening extinction crisis* scenario results from a rise in the international price of wood pulp, bringing about a faster expansion of forest plantations, and related pressures on wetlands, rivers and streams. Large timber companies gain control of ever-increasing areas of land. In this market-oriented scenario, there is a strong pressure to weaken the legislation protecting the threatened species and habitats of Chile. An increase in the level of various threats that are present at low levels, such as infrastructure development, mining and industrial pollution, could also lead to a deepening extinction crisis.

In an *effective conservation* scenario, subsidies for exotic plantations are replaced with financial incentives for the sustainable management of forest ecosystems, and the restoration of native forests in priority areas. This policy change results from new legislation intended to implement biodiversity conservation and sustainable development targets. One major aim is to restore the area of native forest to 1975 levels, with managed forest plantations also established to provide access to firewood and non-timber forest products. Technical assistance is provided to landowners, forest certification employed more effectively and applied ecological research promoted. By 2050, there is an improvement in the conservation status of all biodiversity elements, with the exception of some irrecoverable losses of genetic diversity in threatened species.

Surprise events and critical uncertainties

Possible surprises

- A huge wildfire is looking increasingly possible, as a result of: (i) the massive expansion of contiguous fire-prone exotic pine and eucalyptus plantations; and (ii) reductions in precipitation and intensification of droughts in the region.
- The introduction of new invasive insect species and fungal diseases could affect the native plant and animal species. On the other hand, if these new pests or fungi preferentially infest pine or eucalyptus plantations, this could facilitate the recovery of native tree species.
- An earthquake could trigger landslides and debris flows, devastating some of the current forest stands that would be replaced by pioneer species. This happened after the Valdivia Earthquake of 1960.
- A tsunami would not only greatly disturb coastal marine and estuarine ecosystems, but in combination with subsidence from earthquakes could increase the area covered by wetlands, as was also seen after the Valdivia Earthquake of 1960.
- A large volcanic eruption could destroy thousands of hectares of forests and Andean grasslands and shrublands, covering them in tephra and pumice. This happened in 1957, after the eruption of Quizapu Volcano in the Andes of the Rio Maule-Cobquecura region. Snow persistence would be reduced, and the albedo and nutrient loss would significantly increase. Under the new Mediterranean-type climate, stream-flow variability between summer and winter would increase.

Critical uncertainties

The intensification of global climatic change might lead not only to temperature increases but to an increased inter-annual and intra-annual variability in precipitation, with a trend towards reduced annual totals. This Mediterranean-type climate would cause a decrease in water availability, with water restrictions imposed in summer.

Recommendations

- Forest conservation and research goals could include:
  - A base inventory of biodiversity for the region, with ongoing monitoring planned and implemented.
  - Better dialogue, collaboration and negotiation for conflict resolution between the various stakeholders dealing with the management and conservation of native forests including: the government, forest companies, rural communities, researchers and NGOs at national, regional, municipal and local levels.
  - Landscape planning, including the establishment of a network of new protected areas to improve the connectivity at a landscape scale,

  - would help to conserve all the remaining forest and shrub fragments, wetlands, Andean grasslands and shrublands.
  - The targeted restoration of those species, habitats and ecosystems that are assigned a high conservation priority would increase the chances of their long-term persistence in the region.
- For success in this region, the main change needed is the approval of a Law on Native Forests, which would bring economic incentives to the sustainable management and conservation of native forests. This law has been discussed since 1992, and its approval is considered the single most important policy measure towards forest conservation in Chile.
- A complementary goal is the elimination of the subsidies to exotic plantations in this region, since there is a need to reduce the planted area, which is already excessive and incompatible with maintaining a desirable level of biodiversity and ecosystem services as a basis for economic development and population welfare.
- Over 90% of the plantations are owned by two major private holdings, and virtually all of them are certified through existing certification systems (i.e. FSC, ISO 14001, Certfor). Therefore, the standards and procedures of certifiers need to be revisited, with the full participation of forest conservationists and researchers. Both accountability and compliance with certification, including adequate criteria to ensure the conservation of biodiversity, are crucial.
- Adequate planning of the roads, dams, irrigation channels and other infrastructure is also needed, in order to reverse unnecessary fragmentation caused by poorly planned works. A better coordination of the government services in charge of forests and biodiversity (CONAF, CONAMA) with the Public Works Ministry and the private sector is necessary.
- The use of pesticides, herbicides and other agrochemicals should be reduced in the forest plantations, and the most toxic agrochemicals should be eliminated.
- Opportunities for alternative projects involving the local and rural communities need to be promoted, in order to reverse deforestation trends and to promote socio-economic development in rural communities. These projects could include agroforestry, ecotourism, conservation and restoration programmes, and the sustainable use of native forests and other ecosystems, for example through the harvesting of non-timber forest products (NTFPs).
- The budget and resources devoted to the prevention and extinction of wildfires need to be strengthened, emphasizing the protection of native forests, and allocating a high priority to areas with threatened flora and fauna, or threatened ecosystems.
- The local education system, from kindergarten to high school, should promote awareness of the importance and uniqueness of the biodiversity of the Rio Maule-Cobquecura region, and how to effectively contribute to its conservation. Training at university level, and of professionals, workers, rural communities and other target groups, should also be considered, as well as public campaigns focused on critical issues such as fire prevention. The aim is to improve the attitudes and behaviour

of the population towards the conservation and sustainable use of the resources in the region.

- Measures to improve regional natural hazard response could include:
  - Working to better prevent and suppress wildfires in areas of high conservation value.
  - Improvement of the current germplasm banks and *ex situ* conservation efforts for the most threatened native species.
  - Further studies on long-term climatic variability using data derived from tree-rings, pollen, charcoal and lake sediments.
  - The development of better geological hazard maps for earthquakes, debris flows and volcanism.
  - Design action plans and expert systems to respond to natural hazards and to reduce the vulnerability of biodiversity, human populations and economic activities to such hazards, in coordination with the government institution that deals with hazards (Oficina Nacional de Emergencia – ONEMI).

Scenarios for Los Lagos (Region X, Chile)

Present trends and pressures

Browsing by livestock, logging/fuelwood extraction and habitat fragmentation are currently the most significant causes of biodiversity loss in the area, with land-cover change and fire ranking as the next most important pressures (Table 16.1).

Scenarios

Under a *business as usual* scenario each of the major drivers continues to be important in 2010 (Table 16.1). Fire frequency in natural forests remains high, as the expansion of fire-prone eucalyptus plantations continues. The fires that are used to convert native forests to pasture land also have a tendency to spread beyond their intended area.

Within natural forest areas, selective felling for timber using mobile sawmills continues, leaving only the less commercial trees. Emergent trees are lost, and structural diversity decreases. These levels of extraction will increase the probabilities of fires, presence of invasive species and cattle grazing in the understorey. As logging levels approach clearcutting, forest areas are converted to arborescent shrubland.

The extraction of peat and *Sphagnum* mosses from bogs and moorland also increases, being enabled by new mining laws, and leads to considerable biodiversity loss and erosion.

The commercial success of the Puelo and Huilo-Huilo hydroelectric dams results in plans for many more being drawn up. Visitor numbers decline at the Huilo-Huilo nature reserve, but energy companies are unconvinced by tourism arguments. Riverine fragmentation increases, overall water velocity decreases, and local extinctions of aquatic species result.

Invasive species spread through the landscape, as forest fragments become smaller and surrounded by invaded spaces. Non-native plant species already gaining a hold in the region in 2005 include *Ulex europaeus* (gorse), *Sarotamnus scoparius* (broom), *Rubus constrictus* (blackberry) and *Acer platanoides* (Norwegian maple). Invasive animal species include *Canis domesticus* (dog), *Mustela visori* (American mink) and *Salmo trutta* (brown trout).

By 2050, the levels of pressure from land-cover change, fire, logging and fragmentation all decrease, primarily because habitat loss leaves little native forest to be affected. The overall impacts on biodiversity of these events are negative. From a genetics perspective, there is a moderately severe impact on threatened species of flora and fauna. At a species level, the rates of loss are moderate to high. The only species to remain widespread in the region are those that are most resilient to change. By 2025, 1% of species are lost from the region. Particularly threatened tree species include *Pilgerodendron uvifera* (ciprés de las Guaytecas), *Persea lingue* (lingue), *Laurelia philippiana* (tepa) and *Eucryphia cordifolia* (ulmo). Vertebrate groups that suffer major losses include nutrias (otters), *Pudu pudu* (Pudú deer), birds, reptiles, fish and amphibians, especially including *Rhinoderma darwinii* (Darwin's frog) and *Bufo rubropunctatus* (red-spotted toad).

Aquatic ecosystem diversity is also affected by the increased area of plantations. Reductions in the availability and quality of water in streams and rivers lead to desiccation of riparian habitats and peat bogs.

A *deepening extinction crisis* scenario is brought about through increases in the market value of woody fibre, *Sphagnum* and other forest products. As fossil fuel prices increase without alternative energy sources coming online, the demand for fuelwood increases pressure on the forests. As there are no incentives to conserve the forest, these extractive pressures speed the rate of loss. If the regional impacts of climate change, water and air pollution are more rapid than anticipated, this will also hasten the rate of loss.

A number of factors combine to produce an *effective conservation* scenario. The promotion of ecotourism and regulation of existing tourism bring sustainable income that is dependent upon nature conservation to the region. Certification schemes for timber extraction, hand-in-hand with education at different levels, technical assistance to landowners and the elimination of subsidies for plantations of non-native species, combine to create an enabling environment for sustainable forestry. With subsidies being redirected to support sustainable management, and legislation being enacted to encourage habitat restoration for the conservation of target threatened species, pine forests are restored to native forests in selected areas. Other plantation areas are managed for long-term fuelwood supply.

Research funding under this scenario is directed towards biodiversity inventory, monitoring, restoration techniques and land-use planning with different future scenarios in mind. The aim might be to restore the area of forest to that existing in 1975, within the framework of a network of protected areas to improve connectivity on a landscape scale. Most elements of biodiversity and associated ecosystem services would be expected to recover under these circumstances, though some losses of genetic diversity are irrecoverable.

Surprise events and critical uncertainties

Possible surprises

- A huge wildfire could result from the combination of the spread of the invasive *Ulex europaeus* and the occurrence of extreme climate events such as prolonged or repeated droughts.
- The approval of legislation favourable to the environment would be surprising, but welcome.
- The approval of mega-projects in the region, bringing infrastructural development and greater market connectivity, would be likely to have additional negative effects on biodiversity.
- The introduction of different non-native plants could bring with it a new disease to which either native species or non-native species are vulnerable, resulting in widespread die-offs in native forests or plantations. A comparable example is the *Phytophthora ramorum* fungus, which is causing 'sudden oak death' in Europe and the USA (Henricot and Prior, 2004).
- An earthquake, perhaps accompanied by a tsunami, or a large volcanic eruption, would have unpredictable but calamitous effects (as described for the Rio Maule-Cobquecura region).
- Low-probability, high-impact events of global resonance could include impact of a large meteorite or spread of a novel disease.

Critical uncertainties

- The rate of future land-use change is extremely uncertain, as a result of the lack of clear governing legislation, and the overwhelming influence of future markets.
- Climate change patterns bring a great source of uncertainty, with models simulating greater climate variability into the future. If climate change results in decreased levels of precipitation and greater variability in precipitation, fires within plantations and natural forests can be expected to become more frequent and stronger in times of drought.
- The effectiveness of policy response measures in preventing biodiversity loss is highly dependent upon the resources invested and the strength of the prevailing pressures. Example measures include *ex situ* conservation including germplasm banks, an improved fire-fighting programme, and *in situ* conservation initiatives including restoration and harvest management.

Recommendations

- Forest conservation and research goals could include:
  - A focus on policy-relevant research, and improved dialogue between researchers and decision makers.

  - Simulation and risk mapping of the potential impacts of extreme events such as large fires, tsunamis or earthquakes.
  - Development of expert systems to respond to wildfire emergencies, in coordination with the government institution that deals with hazards (Oficina Nacional de Emergencia – ONEMI).
- As for the Rio Maule-Cobquecura region, the approval of a Law on Native Forests would provide economic incentives for the sustainable management and conservation of native forests.
- Redirecting subsidies from support to non-native plantations and from agriculture (including the drainage of swamp forests) towards sustainable management of native forest would provide great assistance for forest conservation. Reduction in pesticide use might accompany these changes.
- The expansion of plantation area would then cease. In particular, it would be useful to prevent the establishment of further *Eucalyptus* plantations in the region's seasonally flooded forest areas (*ñadis*).
- The promotion and regulation of forest certification schemes would support sustainable forest management and open different markets to local timber products.
- Investment in new pulp plants, which stimulates demand for plantations, could be ceased, and existing plants could be converted to conform to environmental standards.
- Improved land-use planning, taking multiple benefits of forest into account, would: (i) designate areas for forest restoration with connectivity, watershed management, biodiversity value and a forest coverage target in mind; and (ii) reduce the potential for further unnecessary fragmentation caused by the construction of roads and canals.
- Greater investment in and planning of fire control.
- Alongside these practical measures, education is key to gaining community cooperation with conservation. The local education system could help by disseminating knowledge about the region's unique biodiversity, and its contribution to ecosystem services.
- Opportunities for the rural population to avoid destructive logging and other impacts could be provided through creation of alternative employment, especially in the growing ecotourism industry, and within forest restoration initiatives.

Combining Intensity and Impacts of Pressures: A Modelling Approach

Results from the scoring exercise indicated that, as the intensity of pressures increases, the impacts on biodiversity are generally likely to increase (Table 16.2). However, the impacts of different pressures differed between biodiversity components and between study areas, depending on the pressure concerned. For example, a very high intensity of land-cover change was considered to have at least a very high impact on all three components of biodiversity in Rio Maule-Cobquecura, but only moderate impacts on genetic and species diversity in the Highlands of Chiapas.

Table 16.2. Impacts of different anthropogenic pressures on biodiversity, considering each of the three components of biodiversity (genetic, species and habitat diversity) individually. The scoring system for impacts was as follows: *Zero* (0), *Low* (1), *Moderate* (2), *Relatively High* (3), *Very High* (4), *Complete Loss of Biodiversity* (5). For details of scoring system for intensity of pressures, see text and Table 16.1.

		Study area											
		Veracruz			Chiapas			Maule (VII)			Lagos (X)		
Pressure	Intensity of pressure	Genetic diversity	Species diversity	Habitat diversity	Genetic diversity	Species diversity	Habitat diversity	Genetic diversity	Species diversity	Habitat diversity	Genetic diversity	Species diversity	Habitat diversity
Land-cover change	0	0	0	0	0	0	0	0	0	0	0	0	0
	1	2	1	1	1	1	1	3	3	4	2	2	2
	2	2	1	2	1	1	2	3	4	4	2	2	3
	3	3	2	3	2	2	3	4	4	5	3	3	4
	4	4	3	4	2	2	4	4	5	5	3	3	4
Fire	0	0	0	0	0	0	0	0	0	0	0	0	0
	1	0	1	1	1	1	2	2	3	3	1	2	3
	2	3	3	1	1	1	3	2	3	3	2	3	4
	3	4	3	4	2	2	4	3	3	4	2	3	4
	4	5	5	4	2	2	4	4	4	4	3	4	4
Invasive species	0	0	0	0	0	0	0	0	0	0	0	0	0
	1	0	1	1	1	1	2	2	3	2	1	2	2
	2	1	1	1	1	1	3	3	3	3	1	2	2
	3	1	1	2	2	2	3	4	4	4	2	3	3
	4	2	4	2	2	2	4	4	5	5	3	3	3
Browsing animals	0	0	0	0	0	0	0	0	0	0	0	0	0
	1	1	1	1	1	1	1	1	2	1	1	2	2
	2	3	2	1	1	1	1	2	2	2	2	3	2
	3	4	3	1	1	1	2	2	3	3	3	4	3
	4	5	4	2	2	2	3	3	4	3	3	4	3
Pollution	0	0	0	0	0	0	0	0	0	0	0	0	0
	1	1	1	0	1	1	1	1	1	1	1	1	1
	2	3	2	1	1	1	1	2	2	2	1	2	2
	3	4	4	3	1	1	2	2	3	3	2	3	3
	4	5	5	4	2	2	2	3	4	4	2	3	4

Continued

Table 16.2. *Continued*

Pressure	Intensity of pressure	Study area: Veracruz Genetic diversity	Veracruz Species diversity	Veracruz Habitat diversity	Chiapas Genetic diversity	Chiapas Species diversity	Chiapas Habitat diversity	Maule (VII) Genetic diversity	Maule (VII) Species diversity	Maule (VII) Habitat diversity	Lagos (X) Genetic diversity	Lagos (X) Species diversity	Lagos (X) Habitat diversity
Mining	0	0	0	0	0	0	0	0	0	0	0	0	0
	1	0	1	1	1	1	1	1	1	1	1	1	1
	2	1	1	1	1	1	1	2	2	2	1	2	2
	3	3	1	1	1	1	2	2	3	4	2	3	4
	4	5	5	5	2	2	2	3	4	5	2	4	4
Infrastructure	0	0	0	0	0	0	0	0	0	0	0	0	0
	1	0	1	1	1	1	1	1	2	2	1	1	2
	2	1	1	1	1	1	1	2	3	2	1	2	3
	3	1	3	2	1	1	2	3	4	4	2	2	3
	4	3	4	3	2	2	2	4	5	5	2	3	4
Logging	0	0	0	0	0	0	0	0	0	0	0	0	0
	1	0	1	1	1	1	1	1	1	1	2	3	3
	2	2	1	1	2	2	2	2	3	3	3	3	3
	3	3	2	2	2	3	3	3	4	4	3	4	4
	4	5	3	3	3	4	4	3	4	4	3	4	4
Fragmentation	0	0	0	0	0	0	0	0	0	0	0	0	0
	1	0	1	1	1	1	2	2	2	3	1	3	3
	2	0	1	1	1	2	3	3	3	3	2	3	3
	3	3	2	2	2	3	4	3	4	4	3	4	4
	4	4	3	2	2	4	4	4	4	5	3	4	4
Climate change	0	0	0	0	0	0	0	0	0	0	0	0	0
	1	0	1	1	1	1	1	2	1	2	2	2	2
	2	1	2	2	1	1	2	2	2	3	2	3	3
	3	2	3	2	2	2	3	3	3	4	3	3	4
	4	4	5	3	2	3	3	4	4	4	3	4	4
Keystone loss	0	0	0	0	0	0	0	0	0	0	0	0	0
	1	1	1	1	1	1	1	2	2	2	1	2	2
	2	2	2	1	2	2	2	2	2	3	2	2	2
	3	3	3	2	2	3	3	3	3	3	3	3	3
	4	4	4	3	3	4	4	4	4	4	3	4	3

Impacts of invasive species were also higher in Rio Maule-Cobquecura than in the other study areas, whereas impacts of mining were considered to be higher in Central Veracruz than in the other areas. In contrast, other pressures such as climate change and loss of keystone species were considered to have similar impacts for a given intensity in all of the study areas. Impacts also differed between biodiversity components; in general, impacts on habitat diversity were considered to be higher than on species or genetic diversity, for a given intensity of a particular pressure. However, these relationships again differed between study areas; for example impacts on genetic diversity were generally considered to be higher in Central Veracruz than in the other study areas.

Such scores of the potential impacts of different pressures can be combined with the scores describing the intensity of pressures. This could potentially provide a tool for projecting trends in biodiversity, based on an assessment of pressures. Such a tool could then be used interactively to examine the potential impact of different policy or management interventions. As a first attempt towards developing such an approach, we constructed a Bayesian Belief Network (BBN) incorporating the results of the scoring exercises. A BBN can be considered as a tool for exploring the probabilistic relationships between (usually categorical) variables. A BBN is constructed first by developing a graphical model illustrating the relationships between the variables of interest. These relationships are then defined in terms of the probabilities associated with the states of the variables concerned. Further information about the method is provided by Castillo *et al.* (1997) and Jensen (2001). An example of the application of BBNs to exploring the sustainable management of Latin American forests is provided by Newton *et al.* (2006).

Four major causes of biodiversity change, as identified during the workshop, were incorporated in the BBN: habitat fragmentation, logging, fire and land-cover change. These four pressures ranked among the most important when scores obtained for the four study areas were pooled together. The BBN was constructed using Hugin Developer 6.3 (Hugin Expert A/S, Aalborg, Denmark). In this model, the impact of the four pressures is considered separately on the three components of biodiversity: genetic, species and habitat diversity. A separate BBN was constructed for each of the four study areas, enabling scores to be integrated in the model for the relative impact of each level of each pressure on each of the three components of biodiversity. The impacts of different pressures were treated additively. The model was then explored using the scores for the estimated intensities of each pressure within each study area for three dates: 2005 (present day), 2010 and 2050. The model was run by selecting the appropriate state of each factor on a scale of 0–4, using the scores provided in the workshop.

The BBN infers outcomes based on the probabilistic relationships represented in the conditional probability tables (CPTs) associated with the individual variables. In this model, the probabilities incorporated in the CPTs were based on the scores provided in the workshop. Using the BBN, it is possible to alter the intensity value of any of the pressures included, to assess the impacts on different components of biodiversity. In this way, it is possible to explore how the impacts vary according to the level of different pressures, and to combine pressures together.

Initial results obtained using the model are presented in Fig. 16.1. These outputs highlight a number of interesting features:

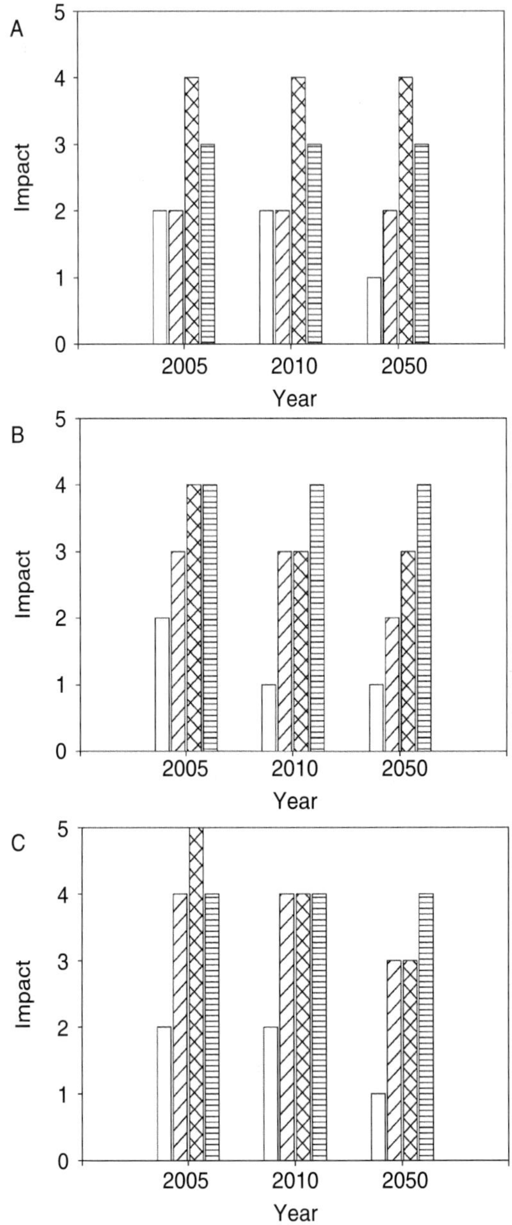

Fig. 16.1. Projected impacts of four combined pressures on different components of biodiversity under the 'business as usual' scenario, based on continuing current trends, using a Bayesian Belief Network (BBN) incorporating workshop scores (see text). Impact scale (loss of biodiversity): *Zero* (0); *Low* (1); *Moderate* (2); *Relatively High* (3); *Very High* (4); *Complete Loss* (5). Impact values presented are those inferred as most likely by the BBN, from combined scores of four pressures: fragmentation, logging, fire and land-cover change. Open bars, Veracruz; hatched bars, Highlands of Chiapas; cross-hatched bars, Rio Maule-Cobquecura; horizontally hatched bars, Los Lagos. Impacts presented relate to (A) genetic diversity, (B) species diversity and (C) habitat diversity.

- Projections differ between the three components of biodiversity. In general, more severe impacts are anticipated on habitat diversity than on genetic diversity, for example.
- Current rates of biodiversity loss appear to be higher, according to the results of this exercise, in the South American study areas (Los Lagos, Rio Maule-Cobquecura) than in the Mexican study areas (the Highlands of Chiapas, Central Veracruz).
- Across three of the four study areas, the impacts of these pressures on biodiversity are projected to diminish with time. This counter-intuitive result is explicable in terms of the current high rates of biodiversity loss. Such high rates of loss cannot be sustained indefinitely into the future: if biodiversity continues to decline at present rates, then by 2050 there will be relatively little left to lose. Therefore rates of loss will decline.

Conclusions

These examples highlight the value of scenarios as a tool for conservation planning, enabling the implications of research results to be communicated in a way that can readily be understood by decision makers. The contrasting narratives produced for each of the four study areas illustrate how the particular circumstances differ between areas, highlighting the need for specific conservation actions to be developed at local or sub-regional scales. While the 'extinction crisis' can be considered as a global phenomenon (Ceballos and Ehrlich, 2002; Thomas *et al.*, 2004), the precise causes of biodiversity loss and the severity of their potential impacts vary substantially from place to place. This variation is perhaps unsurprising given the ecological, political and socio-economic differences between the study areas, but the finding has important implications for policy initiatives developed at the international scale. For example, parties to the Convention on Biological Diversity (CBD) have endorsed a far-reaching Programme of Work relating specifically to conservation of forest biodiversity, which defines broad activities for the assessment and reduction of threats to forest biodiversity and suggests that guidance should be developed and implemented 'to help the selection of suitable forest management practices for specific forest ecosystems' (CBD, 2002). The implication of the current research is that, although the general causes of biodiversity loss may be common to many areas, each forest area will differ in terms of: (i) the precise combination of different pressures; (ii) their varying intensities over space and time; and (iii) their contrasting potential impacts on different components of biodiversity. Distinctive approaches to addressing these problems will therefore need to be developed for each area individually, as illustrated by the recommendations for policy and action presented here.

Peterson *et al.* (2003) suggest the particular circumstances under which scenario-based conservation planning might be preferred, relating to the degree of uncertainty and the degree to which a system can be controlled.

When control of a situation is difficult and uncertainty is high, these authors suggest that scenario planning is an effective approach. The study areas considered here could certainly be considered as meeting these criteria; much of the environmental change that is occurring is uncontrollable and uncertain in terms of outcome. However, in other situations, alternative approaches such as adaptive management planning (Margoluis and Salafsky, 1998; Salfasky *et al.*, 2001, 2002) might be more appropriate. At present, there is little evidence of adaptive management in any of the study areas. This partly reflects a lack of resources and capacity among the institutions that might implement it. As illustrated by the recommendations presented here, the policy environment relating to natural forests is poorly developed, non-existent or even actively antagonistic to conservation. The problem therefore lies deeper than a consideration of the most appropriate approach to conservation planning and management. Rather, the priority is to strengthen the political will, institutional capacity and financial support for forest conservation – whether the institutions be government agencies, conservation NGOs, community-based organizations or private sector enterprises.

Peterson *et al.* (2003) also highlight some of the problems of the scenario-planning approach, such as the reliance on expert opinion. It is conceivable that the predictions of experts may be no better than those of non-experts. Certainly it could be argued that the reliance on one type of expert, namely research scientists with expertise in forest ecology, limits the value of the scenarios presented here. This could be addressed by involving a broader range of stakeholders in scenario-building exercises. Such an approach, implemented at the level of the individual study areas, could provide a useful means of strengthening dialogue between research scientists and other stakeholders, including local communities, government representatives and non-government organizations. Methods could also be used to elicit information from a broader range of experts, for example specialists in mammals, insects or other species groups (see Burgman, 2005). However, it is salutary to consider the extent of uncertainty surrounding biodiversity in these study areas. The precise patterns of distribution, abundance and population trends of the vast majority of species remain poorly defined, and, as a result, it is unclear precisely how many species are currently threatened with extinction. Given this lack of baseline information, estimates of potential future change in biodiversity can remain little more than informed guesswork.

The development of effective conservation strategies depends on a comprehensive assessment of different pressures or threats (Salafsky *et al.*, 2002). It is notable that little progress has been made in developing appropriate methods for assessing such pressures. Wilson *et al.* (2005) provide a recent review of relevant approaches by considering the concept of vulnerability, which may be defined as the likelihood or imminence of biodiversity loss to current or impending threatening processes (Pressey *et al.*, 1996). Wilson *et al.* (2005) differentiate different elements of vulnerability, including the intensity of a threatening process in an area, and the effects of a threatening process on particular features of biodiversity. As described by Wilson *et al.* (2005) and in Chapter 14 of the present volume, spatially explicit statistical or process-based modelling approaches offer methods for assessing the exposure of areas to threatening processes. However, few examples of this approach are available that consider multiple threats (e.g. Miles *et al.*, 2006; ten Brink *et al.*, 2006), and still fewer consider impacts on multiple components of biodiversity. Bayesian Belief Networks, as described here, could potentially provide a tool for such analyses. BBNs possess the advantage of modelling probabilistic relationships, enabling the uncertainty surrounding pressures and their impacts to be explicitly incorporated and explored. Ideally, future developments of this approach might be informed by further quantitative analysis of threatening processes (see Chapter 14), and by information regarding the interactive effects of different pressures on biodiversity. The lack of information about such interactions is one of the most serious areas of uncertainty. It is conceivable that future losses of biodiversity could be substantially more rapid than envisaged here, because of novel interactions that might occur in future (e.g. climate change influencing spread of pests and diseases and interacting with the fire regime).

Use of such analytical approaches could provide a means of further developing biodiversity scenario approaches. For example, Sala *et al.* (2000) adopted a simple multiplication procedure for combining scores describing the magnitude of expected changes in drivers of biodiversity change and their potential impacts. While conceptually simple, this method offers limited scope for analysing or exploring the uncertainty surrounding the scores, which, in common with the current investigation, were based entirely on expert knowledge. Another interesting contrast between the global analysis presented by Sala *et al.* (2000) and the current investigation relates to the pressures (or drivers, *sensu* Sala *et al.*, 2000) that were identified. These authors considered five pressures: land-use change, climate, nitrogen deposition, biotic exchange and atmospheric CO_2. In the current analysis, 11 pressures were identified through discussion as significant current causes of biodiversity loss within the study areas. Atmospheric CO_2 was not included (except with respect to its role in climate change). Given their importance within the areas assessed here, it is surprising that pressures such as habitat fragmentation, overharvesting (i.e. logging), infrastructural development and fire were not considered by Sala *et al.* (2000).

One of the outcomes of the modelling analysis presented here is that the rate of biodiversity loss might actually decline in some areas in coming decades, even if current trends in pressures continue (Fig. 16.1). This reflects the current high rates of biodiversity loss, and the fact that these rates cannot be maintained indefinitely. In other words, the rate of biodiversity loss will decline when there is little biodiversity left to be lost. This has significant implications for the current international policy goal of reduced rate of loss of biodiversity ('the 2010 biodiversity target'), as endorsed by the CBD (Balmford *et al.*, 2005). It is ironic that this policy objective might be met only when its ultimate aim, to conserve biodiversity, has failed.

References

Balmford, A., Bennun, L., ten Brink, B., Cooper, D., Côté, I.M., Crane, P., Dobson, A., Dudley, N., Dutton, I., Green, R.E., Gregory, R.D., Harrison, J., Kennedy, E.T., Kremen, C., Leader-Williams, N., Lovejoy, T.E., Mace, G., May, R., Mayaux, P., Morling, P., Phillips, J., Redford, K., Ricketts, T.H., Rodríguez, J.P., Sanjayan, M., Schei, P.J., van Jaarsveld, A.S. and Walther, B.A. (2005) The Convention on Biological Diversity's 2010 target. *Science* 307(5707), 212–213.

Bruijnzeel, L.A. (2001) Hydrology of tropical montane cloud forests: a reassessment. *Land Use and Water Resources Research* 1, 1–18.

Burgman, M. (2005) *Risks and Decisions for Conservation and Environmental Management*. Cambridge University Press, Cambridge, UK.

Carpenter, S.R., Pingali, P.L., Bennett, E.M. and Zurek, M.B. (2005) *Millennium Ecosystem Assessment. Ecosystems and Human Well-Being: Scenarios. Findings of the Scenarios Working Group*. Millennium Ecosystem Assessment Series, Island Press, Washington, DC. Available at: http://www.maweb.org/

Castillo, E., Gutierrez, J.M. and Hadi, A.S. (1997) *Expert Systems and Probabilistic Network Models*. Springer, New York.

Cayuela, L., Rey-Benayas, J.M. and Echeverría, C. (2006) Clearance and fragmentation of tropical montane forests in the Highlands of Chiapas, Mexico (1975–2000). *Forest Ecology and Management* 226, 208–218.

CBD (2002) Decision VI/22: forest biological diversity. In: *Decisions Adopted by the Conference of the Parties to the Convention on Biological Diversity at its Eighth Meeting*. Curitiba, 20–31 March 2006. UNEP/CBD/COP/8/31. Available at: http://www.biodiv.org/decisions/default.aspx?dec=VI/22 (accessed 8 November 2006).

Ceballos, G. and Ehrlich, P.R. (2002) Mammal population losses and the extinction crisis. *Science* 296(5569), 904–907.

Chapin, F.S., Sala, O.E. and Huber-Sannwald, E. (2001) *Global Biodiversity in a Changing Environment. Scenarios for the 21st Century*. Ecological Studies 152. Springer, New York.

CIREN (2004) *Recuperación secano costero de VI y VII regiones*. Available at: http://www.agricultura.gob.cl/noticias/detallenoticia_print.php?cod_not_p=1226 (accessed 8 November 2006).

CONAMA (2004) *Política Ambiental de la Región del Maule*. Available at: http://www.conama.cl/portal/1255/article-26195.html (accessed 8 November 2006).

González-Espinosa, M. (2005) Forest use and conservation implications of the Zapatista rebellion in Chiapas, Mexico. In: Kaimowitz, D. (ed.) *Forests and Conflicts*. ETFRN News No. 43–44 (European Tropical Forest Research Network), Wageningen, The Netherlands, pp. 74–76.

Hanski, I. and Ovaskainen, O. (2002) Extinction debt at extinction threshold. *Conservation Biology* 16, 666–673.

Helm, A., Hanski, I. and Pärtel, M. (2006) Slow response of plant species richness to habitat loss and fragmentation. *Ecology Letters* 9, 72–77.

Henricot, B. and Prior, C. (2004) *Phytophthora ramorum*, the cause of sudden oak death or ramorum leaf blight and dieback. *Mycologist* 18, 151–156.

Holling, C.S. and Meffe, G.K. (1996) Command and control and the pathology of natural resource management. *Conservation Biology* 10, 328–337.

Jensen, F.V. (2001) *Bayesian Networks and Decision Graphs*. Springer, Berlin, Germany.

Manson, R.H., Williams-Linera, G. and Carter, J. (2007) Current and future land-use change in the central mountains of Veracruz, Mexico: implications for the conservation of neotropical cloud forest. (In preparation.)

Margoluis, R. and Salafsky, N. (1998) *Measures of Success: Designing, Managing, and Monitoring Conservation and Development Projects*. Island Press, Washington, DC.

Miles, L., Newton, A.C., DeFries, R., Ravilious, C., May, I., Blyth, S., Kapos, V. and Gordon, J. (2006) A global overview of the conservation status of tropical dry forests. *Journal of Biogeography* 33, 491–505.

Newton, A.C., Marshall, E., Schreckenberg, K., Golicher, D., te Velde, D.W., Eduoard, F. and Arancibia, E. (2006) Use of a Bayesian Belief Network to predict the impacts of commercializing non-timber forest products on livelihoods. *Ecology and Society* 11, 24.

Peterson, G.D., Cumming, G.S. and Carpenter, S.R. (2003) Scenario planning: a tool for conservation in an uncertain world. *Conservation Biology* 17, 358–366.

Pressey, R., Ferrier, S., Hager, T., Woods, C., Tully, S. and Weinman, K. (1996) How well protected are the forests of north eastern New South Wales? Analyses of forest environments in relation to formal protection measures, land tenure and vulnerability to clearing. *Forest Ecology and Management* 85, 311–333.

Sala, O.E., Chapin, F.S., III, Armesto, J.J., Berlow, R., Bloomfield, J., Dirzo, R., Huber-Sanwald, E., Huenneke, L.F., Jackson, R.B., Kinzig, A., Leemans, R., Lodge, D., Mooney, H.A., Oesterheld, M., Poff, N.L., Sykes, M.T., Walker, B.H., Walker, M. and Wall, D.H. (2000) Global biodiversity scenarios for the year 2100. *Science* 287, 1770–1774.

Salafsky, N., Margoluis, R. and Redford, K. (2001) *Adaptive Management: A Tool for Conservation Practitioners*. Biodiversity Support Program, Washington, DC.

Salafsky, N., Margoluis, R., Redford, K. and Robinson, J. (2002) Improving the practice of conservation: a conceptual framework and agenda for conservation science. *Conservation Biology* 16, 1469–1479.

Schwartz, P. (1991) *The Art of the Long View: Paths to Strategic Insight for Yourself and Your Company*. Doubleday, New York.

Scott, J.C. (1998) *Seeing Like a State: How Certain Schemes to Improve the Human Condition have Failed*. Yale University Press, New Haven, Connecticut.

ten Brink, B., Alkemade, R., Bakkenes, M., Eickhout, B., de Heer, M., Kram, T., Manders, T., van Oorschot, M., Smout, F., Clement, J., van Vuuren, D., Westhoek, H., Miles, L., Lysenko, I., Fish, L., Nellemann, C., van Meijl, H. and Tabeau, A. (2006) *Cross-roads of Planet Earth's Life. Exploring Means to Meet the 2010-biodiversity Target*. Netherlands Environmental Assessment Agency, Bilthoven.

Thomas, J.A., Telfer, M.G., Roy, D.B., Preston, C.D., Greenwood, J.J.D., Asher, J., Fox, R., Clarke, R.T. and Lawton, J.H. (2004) Comparative losses of British butterflies, birds, and plants and the global extinction crisis. *Science* 303, 1879–1881.

UNEP (2003) *Global Environment Outlook 3*. United Nations Environmental Programme, Nairobi, Kenya.

van der Heijden, K. (1996) *Scenarios: The Art of Strategic Conversation*. Wiley, New York.

Varas, E. and Riquelme, J. (eds) (2002) *Tecnologías apropiadas para el manejo sustentable de los suelos de la Región del Maule*. Serie Actas Instituto de Investigaciones Agropecuarias, Santiago, Chile No. 17, 148 pp.

Wack, P. (1985a) Scenarios: uncharted waters ahead. *Harvard Business Review* 63, 72–89.

Wack, P. (1985b) Scenarios: shooting the rapids. *Harvard Business Review* 63, 139–150.

Wilson, K., Pressey, B., Newton, A., Burgman, M., Possingham, H. and Weston, C. (2005) Measuring and incorporating vulnerability into conservation planning. *Environmental Management* 35, 527–543.

17 Synthesis

A.C. Newton

Typical fragmented forest landscape in the central Highlands of Chiapas, Mexico. Photo: Mario Gonzalez-Espinosa

Introduction

This chapter provides a brief summary of key research findings, presented as a series of propositions based on the evidence provided in the preceding chapters. The objective of this summary is to identify findings that have relevance beyond the study areas that were the immediate focus of the research, and that may be applicable to other forest areas subjected to intense human pressure. In this way, it is hoped that the research will contribute to a general understanding of the impacts of anthropogenic disturbance on forest biodiversity.

The text is organized according to the four overarching research questions that were posed at the outset:

1. To what extent have forest loss and fragmentation occurred in the study areas during recent decades?
2. What other forms of anthropogenic disturbance have these forests been subjected to?
3. If forest loss, fragmentation and degradation have occurred, how have they affected different components of biodiversity?
4. Given current trends, how can biodiversity be conserved effectively in forest landscapes subjected to human use?

Forest Loss and Fragmentation

Forest losses have been substantial. Analysis of satellite remote sensing imagery indicated that substantial forest loss has occurred in each of the areas assessed over the past three decades. Across the study period, there was a reduction in natural forest area of 67% in Rio Maule-Cobquecura (Chile), 57% in the Highlands of Chiapas (Mexico), 26% in Central Veracruz (Mexico) and 23% in Los Muermos-Ancud (Chile). These losses are equivalent to annual forest loss rates of 4.5%, 3.4%, 2.0% and 1.1% per year, respectively. The rate of forest loss in Chiapas during the period 1990–2000 was even higher, at 6.2% per year. This value and that recorded for Rio Maule-Cobquecura represent two of the highest deforestation rates ever recorded.

Most forest has been lost through conversion to agriculture. Although immediate causes of deforestation differed between the study areas, analyses indicate that two factors were significant in all four locations: slope and distance to forest patch edge. Results also revealed that the clearance of forests was concentrated around edges of forest fragments in all of the study areas. This highlights the importance of accessibility in determining the pattern and rate of deforestation, and reflects the importance of conversion to agriculture as the main factor responsible for forest loss. In general, deforestation was associated with conversion to either crop or pasture lands. However, in the Rio Maule-Cobquecura region of Chile, the major factor responsible for deforestation was conversion to plantations of exotic tree species, principally *Pinus radiata* and *Eucalyptus* spp.

Forest loss has been accompanied by substantial forest fragmentation. Deforestation in the study areas was accompanied by a decrease in forest patch size, a rapid increase in the density and isolation of forest patches and a decline in area of patch interiors and large patches. In all of the study areas except Central Veracruz, the mean size of forest fragments declined consistently over time. Interior forest habitat decreased progressively over time in all of the study areas. Also, forest fragments become more isolated as other land-cover types occupied the deforested areas in the landscapes. Rio Maule-Cobquecura and the Highlands of Chiapas were characterized by substantial reductions in the total core area (96% and 90%, respectively) over the past three decades, while Los Muermos-Ancud and Central Veracruz presented lower reductions (51% and 26%, respectively). Reductions in the mean proximity of forest fragments over the study periods were also higher in Rio Maule-Cobquecura and the Highlands of Chiapas, with values of 98.7% and 98.6%, respectively. On the basis of current trends, we predict further spatial changes in forest cover will occur in coming decades within each of the study areas.

The underlying drivers of deforestation are political and economic in nature. Chile is characterized by a strong free market economy, dominated by economically powerful private domestic and international pulp and paper companies. This has led to a market-friendly forest policy, leading to violation of many principles of sustainable development and negative environmental impacts. This emphasis on economic development also accounts for the political paralysis in developing a Law on Native Forests, which could provide economic incentives for the sustainable management and conservation of native forests. This law has been discussed since 1992, but is still not approved. Compared to Chile, Mexican industrial timber interests are relatively weak, but rural communities living in forest areas are much better organized. In Mexico, the *ejido* land tenure system provides a platform for organizing political and economic activity that is not available in Chile. However, in both countries, deforestation reflects social and economic policies, such as provision of subsidies for native forest clearance and alternative land uses. Another problem is lack of enforcement of the environmental legislation that does exist. The contrasting cases of Chile and Mexico therefore provide significant insight into the conditions needed for an improvement in national forest policies. Although the socio-economic and political circumstances are very different in the two countries, the end result – high rates of deforestation and forest fragmentation – have been the same.

Other Anthropogenic Disturbance

Anthropogenic disturbance is chronic, widespread and increasing in intensity. In each of the study areas, the native forests that remain are being subjected to intense human pressures. Most significant of these are harvesting of trees for timber and fuelwood, browsing by livestock and the use of fire, all of which are widespread. In Oacaxa and Chiapas, Mexico, forests have traditionally been subjected to slash-and-burn agriculture (*milpa*), involving localized forest clearance (by cutting and burning) to enable crop cultivation for a limited period, which is subsequently followed by abandonment. Such activities have produced pronounced spatial heterogeneity in forest structure and composition. However, the tradition is now breaking down in many areas; agriculture is now being intensified in fixed locations, and some areas abandoned entirely. In some areas, such as southern Mexico, increasing anthropogenic disturbance is associated with an increasing human population. In others, such as southern Chile, infrastructural development (such as road construction) is increasing accessibility to native forest areas that were formerly remote.

Different forms of anthropogenic disturbance interact. Deforestation can increase access to forest areas that were previously difficult to reach, increasing opportunities for tree harvesting and livestock browsing. Similarly, effects of logging and browsing animals can be more intense in smaller forest fragments. Fire regime can be influenced by other forms of disturbance, such as the impact of logging on fuel availability. Fire is frequently used to increase resources for browsing animals. Disturbance may also interact with other environmental pressures. For example, the distribution and intensity of fire events is influenced by climate, and forest disturbance may also influence the spread of invasive species.

Impacts on Biodiversity

Losses to date have occurred at the population rather than at the species level. Research to date has provided no evidence that any species has gone extinct during the past three decades. Those species most at risk of extinction are likely to be narrow endemics associated with old-growth forest. The declines in area of relatively undisturbed, natural forest that have occurred in all study areas must have been accompanied by significant declines in abundance of such species. However, these declines are difficult to document, because information is lacking about the current status of many endemic species, and very little information is available regarding population trends over time. Evidence from genetic research indicates that any decline in abundance or geographical range is likely to have been accompanied by a loss of genetic variation, but, again, it is difficult to estimate the extent of such losses solely from current assessments. However, the relatively high degree of population differentiation recorded in the tree species investigated indicates that loss of any population may result in the loss of distinctive variation. The fact that centres of genetic diversity, such as the Chilean coastal range, are currently experiencing high rates of forest loss and degradation is of particular concern.

Future losses at the species level may be substantial. Statistical models were used to estimate potential future losses of floristic diversity, based on field survey and estimates of deforestation rates derived from satellite imagery, and assuming that the drivers of deforestation will not change in the future.

Results indicated pronounced differences between study regions, with a value close to zero recorded for Oaxaca and estimates of more than 40% species likely to be lost (or at least to become seriously threatened with extinction) by 2025 in Chiapas. This reflects the extraordinarily high recent deforestation rates and the high species diversity in this region. Late-successional, relatively slow-growing species may be at particularly high risk of extinction.

Forest fragmentation can lead to losses of tree diversity. Significant differences were recorded between the study areas. Whereas in the Highlands of Chiapas and Central Veracruz the effects of fragmentation on tree diversity were not directly observable, in Los Muermos-Ancud correlations between mean species richness and fragment metrics were all significant. Because of the slow response of tree populations to recent disturbances, it is likely that the full impact of human activities on tree diversity will not become apparent for some time. It is notable that significant impacts of fragmentation were recorded only in the investigation in which only fragments created at least 23 years ago were assessed. Consequently, it may take two or more decades for the impacts of fragmentation on tree diversity to become apparent.

Anthropogenic disturbance changes forest structure and composition. Disturbance, particularly logging, triggers secondary succession and hence a change in community composition. Consistent patterns were recorded in all of the study areas. Intense human disturbance is causing widespread conversion of old-growth forests to stands dominated by early-successional tree species. This conversion is associated with a simplification of forest structure, as large-diameter trees are lost from the landscape, leading to a loss of diversity at the community scale and a decline in habitat quality for many organisms. However, the process of succession differed between study areas. For example, highest species richness was recorded in early- or mid-successional forests in the Highlands of Chiapas and Central Veracruz, but in late-successional forests in Oaxaca. Soil characteristics also vary along successional gradients, although responses differed among study areas. Such variation limits scope for developing generally applicable indicators for monitoring forest biodiversity.

Rates of forest recovery from disturbance are low. Modelling analyses were supported by observations of successional chronosequences, suggesting that the recovery of late-successional forest following anthropogenic disturbance may require a timescale of centuries, even when a source of colonists is available nearby. Evidence suggests that recovery of soil macrofauna communities may be even slower than that of vegetation. Such results highlight the importance of conserving those relatively undisturbed forest stands that remain.

Edge effects influence forest dynamics and biodiversity. The abundance of both plant and animal species was influenced by the characteristics of forest edges, leading to changes in plant–animal interactions across edges. Edge characteristics can thereby influence ecological processes such as pollination, gene flow, seed germination, seed removal and/or predation by birds and rodents, and regeneration of tree species. For example, pollinator assemblages may differ between the edge and interior of forest patches, as floral displays are more attractive in edges. Similarly frugivorous birds were found to deposit a greater number and diversity of seeds in edges than in the forest interior. In both temperate and tropical montane forest types, when tree seedlings were established experimentally there was an overall positive edge effect on seedling survival and growth. However, field observations suggest that, for some species, seedling densities are lower in edges than in forest interior habitat, perhaps reflecting an interaction between edge effects and other processes such as disturbance or herbivory. Despite this, in some forest ecosystems edge habitats provide opportunities for forest regeneration and forest fragment expansion. Understanding forest dynamics therefore depends on an understanding of the influence of edge effects on ecological processes.

Many processes are responsible for biodiversity loss. An expert consultation conducted in a workshop environment identified 11 principal processes or pressures responsible for biodiversity loss, namely land-cover change, fire, invasive species, browsing animals, pollution, mining, development of infrastructure (roads, pipelines, dams), logging/fuelwood extraction, habitat fragmentation, climate change and loss of keystone species and ecological structures. The relative importance of these different pressures differed among study areas. However, loss of habitat resulting from land-cover change was consistently the most important cause of biodiversity loss, followed by forest fragmentation. Many threatening processes are considered likely to intensify in coming decades.

Responses

Protected area networks need to be strengthened. Many areas of high biodiversity value, at both species and genetic levels, lie outside existing protected area networks. There is therefore an urgent need to establish new protected areas within each of the study areas, to safeguard forest biodiversity. Systematic approaches to conservation planning, as illustrated here in southern Chile, could be of value in this context. Management of existing protected areas needs to be strengthened to ensure that they are effective, for example by developing plans to detect and counter threatening processes such as fire. As many remaining forest areas lie on communally or privately owned land, there is a need to support the development of forest reserves in areas outside state ownership.

There is potential for sustainable timber production from native forests, but this potential may be difficult to realize in practice. Currently, most native forests are being exploited for timber rather than managed sustainably. Simulation results showed that native forest types have potential for sustainable wood production. However, every anthropogenic intervention in the form of wood extraction, even at very low levels, has an ecological impact on the forests. Comparing all logging scenarios, the overall ecological impact increased linearly with the amount of extracted wood. The most notable effect of wood extraction on forest structure was the loss of large old trees, resulting in a simplification of forest structure and a decline in the quality of habitat for forest-dwelling organisms. Successful management of native species-rich forests will require an adaptation of currently available management approaches to the ecological properties of the target species. Particular attention should be given to management of secondary forests, as their area is increasing, their growth rates are relatively high, and their structure and species composition are less vulnerable to tree harvesting than old-growth forests.

Forest restoration is technically feasible, but expensive and slow. Research has demonstrated that native tree species can be successfully propagated and established in field conditions, offering the possibility of restoring degraded forest areas. Most research investigations performed to date have concentrated on assessing plant performance (mostly at the seedling stage) in response to environmental variables. However, successful forest restoration depends on restoring functional ecological communities, not individual trees. For large-scale forest restoration to be achieved, partnerships need to be developed between public and private stakeholders, supported by research, education and outreach activities. Experience to date illustrates how researchers can support the development of such partnerships, yet political and economic support is required if large-scale forest restoration is to be achieved, involving the thousands of people that live and own the forestlands in question. In this context, the economics of forest restoration are of paramount importance; provision of financial incentives is likely to be essential if restoration initiatives are to become widespread. Although relatively low-cost approaches exist, such as enabling forests to recover naturally through successional processes, many decades or even centuries may be required to re-establish late-successional forest communities.

Political commitments to forest conservation need to be strengthened. Rapid progress could be made in conserving forest biodiversity within the study areas, if appropriate political support were provided. This could include a commitment to strengthen and further develop protected area networks, removal of incentives for alternative land uses (such as plantation forestry in Chile), and promotion of alternative livelihood strategies such as organic coffee production, sustainable use of non-timber forest products and sustainable ecotourism. There is a widespread need to improve existing legislation relating to conservation of native forests, and to ensure its effective implementation. The financial and technical resources devoted to the prevention and control of wildfires also need to be strengthened, with the aim of protecting native forests. Improved land-use planning would also be highly beneficial, linking forest conservation and restoration with watershed management, and reducing the potentially negative effects of infrastructural development. Ultimately, if forest conservation is to be successful, economic incentives for conservation will need to be provided.

Whatever action is taken, further losses of forest biodiversity are very likely. Researchers identified a wide range of different uncertainties relating to the process of biodiversity change, the most important of which relates to the potential impact of climate change. Current climate scenarios suggest that substantial climatic shifts could occur throughout much of Latin America. The impacts of this climatic change on biodiversity are difficult to predict, but are likely to be profound. In particular, there is the prospect of surprising or novel events, perhaps involving interactions between different threatening processes. Rainforests are dependent on maintenance of high rainfall and high humidity, and may therefore be particularly vulnerable to climate change. Evidence suggests that fire regime is intimately associated with climate. If precipitation is reduced, increasing frequency and intensity of fires are highly likely, and large-scale catastrophic fires become a possibility, which may accelerate the conversion of forest to other land uses. A shift in weather patterns could also cause spread of pests, diseases or invasive exotic species. Whatever climate change occurs, the effects on forest biodiversity are likely to be strongly negative. Rates of biodiversity loss could therefore increase substantially in the near future, despite international policy commitments to the contrary. This possibility emphasizes how urgently practical conservation action is needed, if major losses of forest biodiversity are to be averted.

Index